AF885879

Infosys Science Foundation Series

Infosys Science Foundation Series in Mathematical Sciences

Series Editors

Gopal Prasad, University of Michigan, Ann Arbor, USA

Irene Fonseca, Carnegie Mellon University, Pittsburgh, PA, USA

Editorial Board

Chandrashekhar Khare, University of California, Los Angeles, USA

Mahan Mj, Tata Institute of Fundamental Research, Mumbai, India

Manindra Agrawal, Indian Institute of Technology Kanpur, Kanpur, India

Ritabrata Munshi, Tata Institute of Fundamental Research, Mumbai, India

S. R. S. Varadhan, New York University, New York, USA

Weinan E, Princeton University, Princeton, USA

The **Infosys Science Foundation Series in Mathematical Sciences,** a Scopus-indexed book series, focuses on high-quality content in the domain of pure and applied mathematics, biomathematics, financial mathematics, operations research and theoretical computer science. With this series, Springer hopes to provide readers with monographs, textbooks, edited volume, handbooks and professional books of the highest academic quality on current topics in relevant disciplines. Literature in this series will appeal to a wide audience of researchers, students, educators and professionals across the world.

Durdimurod K. Durdiev · Zhanna D. Totieva

Kernel Determination Problems in Hyperbolic Integro-Differential Equations

Springer

Durdimurod K. Durdiev
The Institute of Mathematics
Academy of Sciences of the Republic
of Uzbekistan
Tashkent, Uzbekistan

Zhanna D. Totieva
Southern Mathematical Institute
Vladikavkaz Scientific Center
Russian Academy of Sciences
Vladikavkaz, Russia

ISSN 2363-6149 ISSN 2363-6157 (electronic)
Infosys Science Foundation Series
ISSN 2364-4036 ISSN 2364-4044 (electronic)
Infosys Science Foundation Series in Mathematical Sciences
ISBN 978-981-99-2259-8 ISBN 978-981-99-2260-4 (eBook)
https://doi.org/10.1007/978-981-99-2260-4

Mathematics Subject Classification: 35L20, 35R30, 35Q99

© The Editor(s) (if applicable) and The Author(s), under exclusive license to Springer Nature Singapore Pte Ltd. 2023
This work is subject to copyright. All rights are solely and exclusively licensed by the Publisher, whether the whole or part of the material is concerned, specifically the rights of translation, reprinting, reuse of illustrations, recitation, broadcasting, reproduction on microfilms or in any other physical way, and transmission or information storage and retrieval, electronic adaptation, computer software, or by similar or dissimilar methodology now known or hereafter developed.
The use of general descriptive names, registered names, trademarks, service marks, etc. in this publication does not imply, even in the absence of a specific statement, that such names are exempt from the relevant protective laws and regulations and therefore free for general use.
The publisher, the authors, and the editors are safe to assume that the advice and information in this book are believed to be true and accurate at the date of publication. Neither the publisher nor the authors or the editors give a warranty, expressed or implied, with respect to the material contained herein or for any errors or omissions that may have been made. The publisher remains neutral with regard to jurisdictional claims in published maps and institutional affiliations.

This Springer imprint is published by the registered company Springer Nature Singapore Pte Ltd.
The registered company address is: 152 Beach Road, #21-01/04 Gateway East, Singapore 189721, Singapore

To the memory of our parents

Preface

Mathematical physics usually studies well-posed problems that have solutions, which are unique and stable to small changes in the data in suitable functional spaces. Such problems, as a rule, are called direct problems. In this case, it is assumed that a differential equation is specified, as well as certain initial and boundary conditions. However, in applications there are interesting problems, where the differential (or integro-differential) equation is only partially specified, namely some functions that are part of the differential equation (either in the right-hand side, integrant in integro-differential equations or the initial and boundary conditions) remain unknown. The problems, in which these unknowns are to be determined on some information about the solutions of direct problems for differential equations, are called inverse problems.

According to the current accepted terminology, the inverse problem is called one-dimensional if unknown functions depend only on one variable, and multidimensional if they are functions of several variables.

Inverse problems of mathematical physics can be divided into groups depending on the additional information that is given; which functions are unknown; what differential equations are considered. The most frequently used is the classification of additional information. According to the type of additional information asked about the solution of the direct problem, the inverse problems for hyperbolic equations can be divided into three main groups: kinematic, spectral, and dynamic [1, 2].

In this book, the questions of constructing methods for solving one-dimensional and multidimensional inverse dynamical problems for hyperbolic equations with memory are studied, and furthermore, the theorems of uniqueness, stability, and existence of solutions of these inverse problems are obtained. The problems of memory restoration in hyperbolic equations are a relatively new direction in the theory of inverse problems, which arose at the end of the 80s of the last century. It is mainly represented by the works of Italian mathematicians A. Lorenzi, M. Grasselli, E. Paparoni, and others [3–7]. After 2000, this trend has been developed in the works of V. G. Romanov, A. L. Bukhgeym, S. I. Kabanikhin, V. G. Yakhno, J. Janno, L. von Wolfersdorf, and others [8–18, 6, 19–31]. Under certain conditions, convolution-type operators that describe "prehistory" or "memory phenomena" are added to the equations for the spread of elastic and electromagnetic waves.

Let us explain the example of the spread of electromagnetic waves. In rapidly varying electromagnetic fields, whose frequencies are not limited by the smallness condition as compared with the frequencies characteristic of establishing the electric and magnetic polarization of a substance, the unique dependence of D and B (induction of electric and magnetic fields, respectively) on the meanings of E and H (intensity of the corresponding fields) at the same time. It turns out that the meanings of D and B at a given time depend not only on E and H, but also on the whole background of the action of these fields (such a medium is called an environment with aftereffect) [32]. This circumstance is an expression of the fact that the establishment of the electric and magnetic polarization of a substance does not manage to follow the change in the electromagnetic field. The most general view of the linear relationship between $D(x, t)$, $B(x, t)$ and the corresponding meanings of the functions $E(x, t)$, $H(x, t)$ at all previous points in time can be written as integral relations [32]:

$$
\begin{aligned}
&D(x, t) = \varepsilon E + \int_0^t \varphi(t-\tau)E(x, \tau)d\tau \\
&B(x, t) = \mu H + \int_0^t \psi(t-\tau)H(x, \tau)d\tau, \\
&E = (E_1, E_2, E_3), H = (H_1, H_2, H_3), D = (D_1, D_2, D_3), \\
&B = (B_1, B_1, B_3), x = (x_1, x_2, x_3)
\end{aligned}
\tag{1}
$$

$\varphi(t) = diag\,(\varphi_1, \varphi_2, \varphi_3)$, $\psi(t) = diag\,(\psi_1, \psi_2, \psi_3)$ are diagonal matrix functions representing memory. These functions are finite for all values of their argument and tend to zero at $t \to \infty$. The latter circumstance is an expression of the fact that the values of $D(x, t)$, $B(x, t)$ at a given time cannot be noticeably influenced by the values of $E(x, t)$, $H(x, t)$ at very old moments. The physical mechanism underlying the integral dependencies of the form (1) is in the process of establishing electromagnetic polarization. Therefore, the interval of values in which the functions $\varphi(t)$, $\psi(t)$ differ noticeably from zero, the magnitude of the relaxation time characterizing the rate of these processes.

Let, in accordance with equalities (1), the set of vectors E, H be the solution of the Cauchy problem for the system of Maxwell equations with zero initial data:

$$
\nabla \times H = (\partial/\partial t)D(x, t) + \sigma E + j, \nabla \times E = -(\partial/\partial t)B(x, t), (x, t) \in \mathbb{R}^4,
$$

$$
(E, H)|_{t<0} \equiv 0. \tag{2}
$$

Here $j = j(x, t)$ is a given function characterizing the external current density. Further, we assume that the electrodynamic parameters ε, μ, σ are given positive constants and $\psi = 0$ (the absence of a prehistory of the magnetic field).

Under the above assumptions, we obtain from the system of Maxwell equations (2) a second-order equation for the vector H. For this, we apply the operation $\nabla\times$ to the first and $\partial/\partial t$ to the second equations of the system (2). In view of the equality $\nabla \times \nabla \times H = \nabla(div\, H) - \Delta H$, we find

$$\mu(\varepsilon H_{tt} + \sigma H_t) - \Delta H + \int_0^t \mu\varphi(\tau) H_{tt}(x, t-\tau)d\tau = \nabla \times j. \qquad (3)$$

The equality $divH = 0$ is used here, which comes from relations (2). Indeed, applying the operation *div* to the second equations (2), taking into account the equality $div(\nabla \times E) = 0$, we have $\partial/\partial t(divH) = 0$. Hence, integrating the last equality under the initial conditions from (2), we come to equality $divH = 0$.

In what follows, we will assume that $\varepsilon = \mu = 1$. Twice we integrate the integral in (3) by parts and introduce a new vector function $u(x, t) = (u_1, u_2, u_3), u_i = H_i \exp[-0.5(\sigma + \varphi_i(0))t]$, $i = 1, 2, 3$. Then we write the equation (3) and the initial condition in the form

$$u_{tt} - \Delta u + cu + \int_0^t k(t-\tau)u(x, \tau)d\tau = F(x, t), \qquad (4)$$

$$u|_{t<0} = 0. \qquad (5)$$

Here

$$c = diag(c_1, c_2, c_3), c_i := \varphi_{it}(0) - 0.25(\sigma + \varphi_i(0))^2,$$

$$k(t) = diag(k_1, k_2, k_3), k_i := \varphi'' \exp[-0.5(\sigma + \varphi_i(0)t)],$$
$$F(x, t) = (F_1, F_2, F_3), F_i := \exp[-0.5(\sigma + \varphi_i(0)t)](\nabla \times j)_i, \ i = 1, 2, 3,$$

where $(\nabla \times j)_i$, is i—component of the vector $(\nabla \times j)$.

An equation of the type (4) with an integral term of convolution type will be further called the equation with memory. This equation with condition (5) is called the generalized Cauchy problem of determining $u(x, t)$ for known functions $k(t)$, $F(x, t)$.

Another application of equations with memory occurs in geophysics, when the properties of the medium are investigated using seismic waves [33]. In fact, according to smoothness assumptions, the system of equations for the nonelastic Boltzmann model (one of the most common for a linear nonelastic medium) reduces to the equation [33]:

$$u_{tt}(x, t) = u_{xx}(x, t) + \sigma^{-1}(x)\sigma'(x)u_x(x, t) +$$
$$\omega(x)\int_0^t k((t-\tau)\big(u_{xx}(x, \tau) + \sigma^{-1}(x)\sigma'(x)u_x(x, \tau)\big)d\tau,$$
$$(x, t) \in \mathbb{R} \times [0, T]. \qquad (6)$$

The functions σ and ω, in this equation, are related to the Lame parameters and the density of the medium under consideration. Assuming that σ and ω are constant($\omega = 1$), the equation (6) can be reduced to

$$u_{tt}(x, t) = u_{xx}(x, t) + \int_0^t k(t-\tau)u_{xx}(x, \tau)d\tau, \; (x, t) \in R \times [0, T]. \quad (7)$$

Introducing the function $v(x, t)$ by the formula

$$v(x, t) = u(x, t) + \int_0^t k(t-s)u(x, s)ds,$$

it is easy to verify that $u(x, t)$ can be expressed via $v(x, t)$ as follows:

$$u(x, t) = v(x, t) + \int_0^t h(t-s)v(x, s)ds,$$

where h is the solution of the Volterra equation: $h(t) = -k(t) - \int_0^t k(t-s)h(s)ds, \quad t \in [0, T]$. Then, after simple calculations, from equation (7) we obtain one equation of the form (4) with respect to the functions $v(x, t)$, $h(t)$. So, we come to the equations with memory.

The considered inverse dynamic problems for differential equations consist in determining the function of memory of the medium according to some information about the generalized solutions of these equations. Here, by means of generalized solutions, the processes of the spread of elastic or electromagnetic waves arising from sources of the type of pulsed directional "impacts" or "explosions" are described. The considered generalized solutions of differential equations, as a rule, are fundamental solutions. Note that the fundamental solutions for a one second-order hyperbolic equation are the subject of the well-known theory of G. Hadamard [34]. In the works of authors of this book and their collaborators [35–73], inverse problems for equations with memory were investigated. The results of these papers compiled theorems on unique solvability, stability estimates for a number of statements of inverse problems for hyperbolic-type integro-differential equations. The reader can find numerical solutions to such problems in the works [74–76].

Here we also note that integro-differential equations of parabolic type describe the diffusion and heat propagation processes in media with memory. The problem of determining the memory kernel in these equations requires a different technique than in the hyperbolic case. The reader can familiarize himself with such problems in the works [77, 78, 17, 51, 52, 54, 58, 79–82] (see also the bibliography in them). Methods for solving coefficient inverse problems for various types of differential equations using Carleman estimates were developed in [83–86].

This book presents some methods for studying inverse problems associated with hyperbolic convolution-type integro-differential equations. The results presented here do not pretend to be exhaustive; they constitute only a small part of the rapidly developing theory of inverse problems for such equations. We recommend a series of books [87–91, 1, 92–95, 2, 96], in which the reader can find the most suitable method for studying a newly arisen problem.

Let us give a summary of the book. Consider the integro-differential wave equation

$$u_{tt} - \Delta u = \int_0^t k(x, t-\tau)u(x, \tau)d\tau + f(x, t), (x, t) \in \mathbb{R}^4, \tag{8}$$

$u = u(x, t)$ is a potential of velocities of medium points, $x = (x_1, x_2, x_3)$. Suppose that for $t < 0$ the space $\mathbb{R}^3$ was at rest, that is, the Cauchy data are given

$$u(x, t)|_{t<0} = 0. \tag{9}$$

This condition is the generalized Cauchy data [97]. If the medium fills the half-space $x_3 > 0$, then along with equation (8) and condition (9) the condition is given

$$u_{x_3}\big|_{x_3=0} = g(x_1, x_2, t). \tag{10}$$

In our studies, for the functions f and g often taken the functions

$$g = 0, f = \delta(x, t) = \delta(x_1, x_2, x_3, t)$$

or

$$f = 0, \ g = \delta(x_1, x_2, t),$$

where $\delta(\cdot)$ is the Dirac delta function. Functions of this type simulate an instantaneous force concentrated at the point (0, 0, 0) of the space $\mathbb{R}^3$.

The inverse problem is to determine the kernel of the integral operator $k(x, t),\ \ t \in [0, T]$, if we know the mode of oscillation of the point (0, 0, 0) for problems in the space

$$u(0, t) = F(t), \ \ t \in [0, T]$$

or the oscillation mode of the surface $x_3 = 0$ is known for problems in the half-space

$$u(x_1, x_2, 0, t) = F(x_1, x_2, t), (x_1, x_2, t) \in \mathbb{R}^2 \times [0, T].$$

Let us stop on the brief content of Chap. 1. Consider a second-order linear integro-differential equation of hyperbolic type and the generalized Cauchy problem for it with a concentrated effect

$$u_{tt} - Lu = \int_0^t k(\tau)u(x, t-\tau)d\tau + \delta(x, t), \tag{11}$$

$$u|_{t<0} = 0. \tag{12}$$

Here $u(x, t)$ is a required function, $x = (x_1, x_2, x_3) \in \mathbb{R}^3$, L is a differential operator

$$Lu = \Delta u + \sum_{i=1}^{3} b_i(x)u_{x_i} + c(x)u, \ \ \Delta = \sum_{i=1}^{3} \frac{\partial^2}{\partial x_i^2}.$$

The problem of finding the function $u(x, t)$, satisfying equalities (11), (12) for known functions b_i, c, k, presents so-called direct problem for the integro-differential equation. Problem (11), (12) is studied at $b_i = 0,\ c = 0,\ i = 1, 2, 3$ in the half-space $\mathbb{R}^3_+ = \{(x_1, x_2, x_3) \in \mathbb{R}^3 | x_3 > 0\}$ with zero condition on the boundary. In the inverse problem, the kernel $k(t),\ \ t > 0$ of the integral operator in (11) is determined for the known oscillation mode of the point $x = 0$ of the medium. In the study of problem (11), (12) for the nonzero functions b_i, c, $i = 1, 2, 3$ the method of S. L. Sobolev is used. In [98, 99], he developed a method of solving the Cauchy problem in three-dimensional space for a wave equation with variable coefficients. Further, in [100], this method was extended to general second-order linear equations of hyperbolic type. Following [101], we present some results from the mentioned works. Consider the Cauchy problem

$$u_{tt} = \sum_{i,j=1}^{3} a_{ij}(x)u_{x_i x_j} + \sum_{i=1}^{3} b_i(x)u_{x_i} + c(x)u + f(x, t). \tag{13}$$

$$u|_{t=0} = \varphi(x), \quad u_t|_{t=0} = \psi(x). \tag{14}$$

Here $a_{ij}(x)$ satisfy inequalities

$$\sum_{i, j=1}^{n} a_{ij}(x)\eta_i\eta_j \geq \mu_0 \sum_{i=0}^{n} \eta_i^2, \ \mu_0 > 0. \tag{15}$$

Additionally, it is assumed that the coefficients a_{ij} are sufficiently smooth and are such that for each pair of points $x, \ \xi; \ \xi = (\xi_1, \ \xi_2, \ \xi_3)$ a unique geodesic $\Gamma(x, \xi)$, is corresponded, connecting them in a metric in which an element of length ds is found by the formula

$$ds = \sqrt{\sum_{i, j=1} b_{ij}(x)dx_i dx_j}, \tag{16}$$

where b_{ij} are the matrix elements, which are the inverse of the matrix $A = \left\| a_{ij} \right\|$. We denote by $\tau(x, \ \xi)$ the distance in the metric (16) between the points $x, \ \xi$

$$\tau(x, \ \xi) = \int_{\Gamma(x, \xi)} \sqrt{\sum_{i,j=1} b_{ij}(x)dx_i dx_j}.$$

The physical meaning of the function $\tau(x, \ \xi)$is the travel time of the perturbation produced at the point ξ to the point x. Obviously $\tau(x, \ \xi) = \tau(\xi, \ x)$. S. L. Sobolev established the existence of the function $\sigma(x, \ \xi)$, which satisfies the equation in the first argument

$$2\sum_{i, j=1}^{3} a_{ij}\tau_{x_i}\sigma_{x_j} + \sigma\left[\sum_{i, j=1}^{3} a_{ij}\tau_{x_i x_j} + \sum_{i=1}^{3}\left(2\sum_{j=1}^{3}\frac{\partial a_{ij}}{\partial x_j} - b_j\right)\tau_{x_i}\right] = 0 \tag{17}$$

possessing a number of remarkable properties [101] (p. 63), due to which the problem (13), (15) is reduced to an integral equation

$$\begin{aligned}u(x,\,t) = &\frac{1}{4\pi}\iint_{S(x,\,t)}\{\sigma(\xi,\,x)N_\xi\varphi(\xi) - \varphi(\xi)N_\xi\sigma(\xi,\,x) + \sigma(\xi,\,x)\\ &\times\left[\psi(\xi)N_\xi\tau(\xi,\,x) + \varphi(\xi)R_\xi\right]\}dS_\xi\\ &+\frac{1}{4\pi}\iiint_{D(x,\,t)}\sigma(\xi,\,x)f(\xi,\,t-\tau(\xi,\,x))dv_\xi\\ &+\frac{1}{4\pi}\iiint_{D(x,\,t)}\tilde{L}_\xi^*\sigma(\xi,\,x)u(\xi,\,t-\tau(\xi,\,x))dv_\xi.\end{aligned}$$

Here ξ is the variable point of integration, $S(x, t)$ is the quasisphere $\tau(x,\xi) = t$, $D(x,t))$ is the volume bounded by this quasisphere, dS, dv are the elements of surface area and the volume respectively, $\tilde{L}^*$ is an adjoint operator, according to Lagrange, with an operator

$$\tilde{L}u = \sum_{i,j=1}^{3} a_{ij}(x)u_{x_ix_j} + \sum_{i=1}^{3} b_i(x)u_{x_i} + c(x)u, \tag{18}$$

N is the linear differential operator

$$N\varphi(x) = \sum_{i,\,j=1}^{3} a_{ij}\sum_{i,\,j=1}^{3} a_{ij}\varphi_{x_i}\cos(n,\,x_i),$$

The index ξ of the operators $\tilde{L}^*$, N shows that they are applied with respect to the variable ξ,

$$R = \sum_{i=1}^{3}\left(b_i - \sum_{j=1}^{3}\frac{\partial a_{ij}}{\partial x_i}\right)\cos(n,\,x_i),$$

n is the unit vector of the outer normal to the surface $S(x, t)$.

In the case when $\tilde{L} = L$ that is $a_{ij} = \delta_{ij}$ (δ_{ij} is the Kronecker symbol), to all the conditions listed in [101] (p. 63) and the equation (19) satisfies the function

$$\sigma(x,\,\xi) = \frac{1}{|x-\xi|}\exp\left\{\frac{1}{2}\int_0^1\sum_{i=1}^{3} b_i(\xi+\gamma(x-\xi))(x_i-\xi_i)d\gamma\right\}.$$

Based on the method of S. L. Sobolev in Sect. 1.2, it has been constructed Volterra-type integral equation which is an equivalent to problem (11), (12). In this case, it is shown that the constructed equation has a unique solution, and it can be obtained using the method of successive approximations. In the inverse problem, the kernel $k(t)$, $t > 0$ is found for a given mode of oscillation of the point $x = 0$.

Here, there will also be investigated three one-dimensional inverse problems for a system of thermoelasticity equations in a vertically inhomogeneous disconnected medium, an integro-differential equation of acoustics and a two-dimensional system of Maxwell equations with memory. The memory of the medium is determined by the concentrated source of disturbances for the first problem and with the distribution for the second. These problems are of both practical and theoretical interest, and the results of their research constitute local unique solvability theorems and estimates of the continuous dependence of a change in a solution on a change in information.

Chapter 2 is devoted to the study of multidimensional inverse problems. We consider the problem of finding the function $k(x, t)$, $t \geq 0$, $x \in \mathbb{R}^n$ from the equations

$$u_{tt} - u_{zz} - \Delta u = \int_0^t k(x, \tau)u(x, z, t - \tau)d\tau, \quad (x, t) \in \mathbb{R}^{n+1}, \quad z > 0, \tag{19}$$

$$u|_{t<0} \equiv 0, \quad u_z|_{z=0} = \delta'(t) + h(x, t)\theta(t), \quad (x, t) \in \mathbb{R}^{n+1}, \tag{20}$$

on the known information

$$u|_{z=0} = -\delta(t) + f(x, t)\theta(t), \quad (x, t) \in \mathbb{R}^{n+1}, \tag{21}$$

where $\theta(t) = 1,\ t \geq 0;\ \theta(t) = 0,\ t < 0;\ \delta'(t) = (d^2/dt^2)\theta(t),\ h(x, t),\ f(x, t)$ are given smooth functions.

The study of problem (19)–(21) was carried out on the basis of the scale method of Banach spaces of analytical functions. This method was developed in the works of L. V. Ovsyannikov [102, 103] and L. Nirenberg [104]. The possibilities of its application to questions of solvability of multidimensional inverse problems were studied in the book of V. G. Romanov [96] (Sect. 3). It is shown that problem (19)–(21) is uniquely solvable in the class of functions analytical with respect to the variable x, a stability estimate is obtained for the solution, and a uniqueness theorem is formulated. The above problem with the similar equations was also investigated, when the memory is multiplied by the second time derivative and the first-order differential operator with respect to the variable x, whose coefficients are real analytical functions. The differential properties of solving direct and inverse problems are studied. It also considered the inverse problem for a two-dimensional system of Maxwell equations with memory. It is shown that this problem, reformulated in several other terms, can be reduced to the investigated problem (19)–(21).

Chapter 3 discusses the issues of global solvability of memory determination problems. For many one-dimensional inverse problems, local solvability is a well-known fact. Even with arbitrarily smooth data, the solution to the inverse problem may not exist for any predetermined segment. The local nature of existence theorems is due to the fact that the Volterra integral equations obtained in the process of solving the inverse problem are nonlinear. In [11], it was noted that since the nonlinearity is convolutional, it is possible to solve the resulting Volterra equations in space with an exponential weight. This allowed to prove the global solvability of such problems.

In Chap. 3, the problems of determining memory from initial–boundary value problems for the wave equation with a given mode of oscillation of the medium boundary are investigated. The method used here is a development of the method used in [11, 14, 16]. Here, the memory is determined from the system of equations of electrodynamics under the known mode of oscillation of the point $x = 0$. In this section, the problem of simultaneously determining two unknowns is also considered. These functions are the regular part of the instantaneous source that physically simulates a point explosion at the boundary of the region and the memory of the medium. For their definition, two observations are used on how the boundary of the region oscillates. At the end of this chapter, it is considered the multidimensional inverse problem of determining the convolutional kernel of the integral term in an integro-differential wave equation. It is proven the global unique solvability of this problem in the class of continuous functions in the time variable and analytic in a part of the space variables.

Chapter 4 discusses two tasks of simultaneously determining two unknowns. In one of them, for the equation describing the process of wave spread in a half-space, with a filled medium, the problem of determining the velocity of wave propagation, which depends only on one variable y, and the function of medium memory is considered. In this case, a pulsed point source acts on the border of the half-space. In the other, the problem of finding the memory and the form of the signal from a point source of disturbances, causing oscillations of the medium, was investigated. In both cases, it is shown that both unknown functions of one variable are uniquely determined by specifying the Fourier transform of the direct problem on the boundary of the half-space with respect to the variable x for two different transformation parameters. The stability estimate of the solution to the problem is being set.

Chapter 5 deals with the inverse problems of recovering a memory kernel in hyperbolic integro-differential equations, having a special dependence with respect to one of the spatial variables. In the first section, it is supposed that the unknown kernel is a trigonometric polynomial in the spatial variables with continuous coefficients with respect to the time variable. In the second section, it is determined the two-dimensional spatial part of the three-dimensional kernel and wherein it is supposed that the unknown function is a trigonometric polynomial in the spatial variable y with continuous coefficients with respect to the variable x. The results of this chapter are published in [105, 106].

Chapter 6 devoted to various coefficients inverse problems for viscoelasticity system. In the first section of this chapter, it is considered a multidimensional inverse problem related to the determination of density function $\rho(x_2, x_3)$, which is the coefficient of the system of integro-differential equations of viscoelasticity (the system of differential Lame equations with memory) in the linear approximation. Some properties regarding the solution of the direct linearized problem are given.

The second section deals with the problem of determining the thermal expansion coefficient $\alpha(z),\ z \in [0,\ Z],\ Z > 0$ is a fixed number, occurring in the system of integro-differential thermoviscoelasticity equations. The medium density and the Lame parameters are assumed to be function of one variable. The integrand $h(t),\ t \in [0;\ T]$ is known. The inverse problem is replaced by the equivalent integral equation for unknown functions. The local theorem of unique solvability is proved, and the stability estimate of solving the inverse problem is obtained.

In the third section, it is investigated the problem of determining the moduli of elasticity $c_{11}(x_3),\ c_{12}(x_3),\ c_{44}(x_3), x_3 > 0$, occurring in the system of integro-differential viscoelasticity equations for homogenous anisotropic medium. It is assumed that density of medium is constant and matrix kernel $k(t) = diag(k_1,\ k_2,\ k_3)(t),\ t \in [0,\ T]$ is known. As additional information is the Fourier transform of the first and third component of the displacements vector for $x_3 = 0$. The results are the local theorems on the existence of a unique solution of the inverse problems and the theorems of stability.

Chapter 7 studies one and two-dimensional problems of determining the memory kernel from a system of viscoelasticity and thermoviscoelasticity equations. The first two chapters define one- and two-dimensional kernels of the system of viscoelastic equations. Wherein, under the assumption that the coefficients of the system depend on one spatial variable and the applied concentrated force is directed along one of the coordinate axes, the system is reduced to one second-order equation. The research of problems is based on the methods used in Chaps. 2 and 3.

In the third section, it is studied the problem of determining the one-dimensional matrix kernel of the system of viscoelasticity. This problem reduces to the sequential determination of the diagonal elements of the matrix kernel from second-order hyperbolic equations.

In the fourth section, the problem of recovering the memory kernel of thermoviscoelasticity equation is considered. Based on the methods of Chap. 1, a local existence and uniqueness theorem is proved and an estimate of the stability of the solution is obtained.

Chapter 8 devoted to the problems of reconstructing the coefficients of differential equations and the convolution kernel of hyperbolic integro-differential equations by the method of Dirichlet-to-Neumann operator. The rise of interest to this kind of problems is generally related to Calderdon who posed [107] the question whether the conductivity $\gamma(x)$ in the elliptic equation

$$\nabla(\gamma(x)\nabla u) = 0 \tag{22}$$

over a bounded domain $\Omega \subset \mathbb{R}^n$ is uniquely determined by the (stationary) Dirichlet-to-Neumann operator. The positive answer to this question for functions $\gamma(x) \in$

$C^\infty(\Omega)$ was given in the classical article [108] by Sylvester and Uhlmann who proved existence of special solutions to (22). Using the results of [108], Alessandrini [109] further proved conditional stability of a solution to the problem. Much information on this topic is presented in the survey [110] (see also the bibliography therein). Uniqueness and conditional stability were proven in [111, 110] respectively for one nonstationary problem, namely the problem of reconstructing the coefficient $q(x)$ in the hyperbolic equation

$$u_{tt} - \Delta u + q(x)u = 0$$

over the domain $\Omega \times (0,\ T)$, with Ω a bounded domain in $\mathbb{R}^n$, $n \geq 2$, and $T > 0$. This chapter deals with one problem of determining the potential and two problems of restoring memory from the integro-differential hyperbolic equation using the method of Dirichlet-to-Neumann operator.

Chapter 9 contains some opening problems.

Tashkent, Uzbekistan
Vladikavkaz, Russia
October 2022

Durdimurod K. Durdiev
Zhanna D. Totieva

References

1. Kabanikhin, S.I. 2009. Inverse and Ill–posed problems. Novosibirsk, Sib. Nauch. Izd. (Russian).
2. Romanov, V.G. 1987. *Inverse problems of mathematical physics*. Utrecht: VNU Science Press.
3. Grasselli, M., S.I. Kabanikhin, and A. Lorenzi. 1990. An inverse hyperbolic integro–differential problem arising in Geophysics II. *Nonlinear Analysis, Theory, Methods and Applications* 15: 283–298.
4. Lorenzi, A. 1994. An identification problem related to a nonlinear hyperbolic integro-differential equation. *Nonlinear Analysis: Theory, Methods and Applications* 22: 297–321.
5. Lorenzi, A., and E. Paparoni. 1992. Direct and inverse problems in the theory of materials with memory. Rendiconti del Seminario Matematico della Università di Padova 87: 105–138.
6. Lorenzi, A., J.Sh. Ulekova, and V.G. Yakhno. 1994. An inverse problem in viscoelasticity. *Journal of Inverse and Ill-Posed Problems* 2: 131–165.
7. Lorenzi, A., and V.I. Priimenko. 1996. Identification problems related to electro-magneto-elastic interactions. *Journal of Inverse and Ill-Posed Problems* 4 (2): 115–143.
8. Ayapbergenova, A.T., S.I. Kabanikhin, and A. Lorenzi. 2004. Justification of optimization methods for inverse integ–differential hyperbolic problems. *Journal of Inverse and Ill-Posed Problems* 12 (5): 457–479.
9. Blagoveshchenskii, A.S., and D.A. Fedorenko. 2008. The inverse problem for the acoustic equation in a weakly horizontally inhomogeneous medium. *Journal of Mathematical Sciences (N. Y.)* 155 (3): 379–389.
10. Blagoveshchenskii, A.S., Y.V. Kurylev, and V.V. Zalipaev. 2005. Inverse problem in a weakly horizontally inhomogeneous layered medium. Loughborough University preprint.
11. Bukhgeim, A.L. 1993. Inverse problems of memory reconstruction. *Journal of Inverse and Ill-Posed Problems* 1 (3): 193–206.

12. Bukhgeim, A.L, and G.V. Dyatlov. 2000. Inverse problems for equations with memory. In: *Inverse problems and related topics*, 19–35. CHAPMAN HALL/CRC Research notes in mathematics 419, Boca Raton, etc.
13. Bukhgeim, A.L, and G.V. Dyatlov. 1996. Uniqueness in one inverse problem of memory determination. *Siberian Mathematical Journal* 37 (3): 526–533.
14. Bukhgeim, A.L., and N.I. Kalinina. 1997. Global convergence of the Newton method in inverse problems of memory reconstruction. *Siberian Mathematical Journal* 38 (5): 881–895.
15. Bukhgeim, A.L., G.V. Dyatlov, and V.M. Isakov. 1997. Stability of memory reconstruction from the Dirichlet–Neumann operator. *Siberian Mathematical Journal* 38 (4): 636–646.
16. Bukhgeim, A.L, N.I. Kalinina, and V.B. Kardakov. 2000. Two methods for the inverse problem of memory reconstruction. *Siberian Mathematical Journal* 41 (4): 634–642.
17. Colombo, F., and D. Guidetti. 2011. Some results on the identication of memory kernels. *Operator Theory: Advances and Applications* 216: 121–138.
18. Janno, J., and L. von Wolfersdorf. 2001. An inverse problem for identification of a time- and space-dependent memory kernel in viscoelasticity. *Inverse Problems* 17 (1): 13–24.
19. Lorenzi, A., and V.G. Romanov. 2006. Stability estimates for an inverse problem related to viscoelastic media. *Journal of Inverse and Ill-Posed Problems* 14 (1): 57–82.
20. Lorenzi, A., F. Messina, and V.G. Romanov. 2007. Recovering a Lame kernel in a viscoelastic system. *Applicable Analysis* 86 (11): 1375–1395.
21. Lorenzi, A., and V.G. Romanov. 2011. Recovering two Lame kernels in a viscoelastic system. *Inverse Problems and Imaging* 5 (2): 431–464.
22. Lorenzi, A., and E. Sinestrari. 1988. An inverse problem in theory of materials with memory. *Nonlinear Analysis, Theory, Methods and Applications* 12: 411–423.
23. Lorenzi, A., and V.G. Yakhno. 1997. An identification problem related to an isotropic stratified viscoelastic cylindrical body. *Journal of Inverse and Ill-Posed Problems* 5 (1): 29–53.
24. Romanov, V.G. 2016. A problem of recovering a special two-dimention potential in a hyperbolic equation. *Eurasian Journal of Mathematical and Computer Applications* 7 (6): 573–588.
25. Romanov, V.G. 2014. On the determination of the coefficients in the viscoelasticity equations. *Siberian Mathematical Journal* 55 (3): 503–510.
26. Romanov, V.G. 2014. Inverse problems for differential equations with memory. *Eurasian Journal of Mathematical and Computer Applications* 2 (4): 51–80.
27. Romanov, V.G. 2013. An asymptotic expansion for a solution to viscoelasticy equations. *Eurasian Journal of Mathematical and Computer Applications* 1 (1): 42–62.
28. Romanov, V.G. 2012. Stability estimates for the solution in the problem of determining the kernel of the viscoelasticity equation. *Journal of Applied and Industrial Mathematics* 6 (3): 360–370.
29. Romanov, V.G. 2012. A two-dimensional inverse problem for the viscoelasticity equation. *Siberian Mathematical Journal* 53 (6): 1128–1138.
30. Romanov, V.G. 2011. The problem of determinig the kernel of electrodynamics equations for dispersion media. *Doklady Mathematics* 84 (2): 613–616.
31. Romanov, V.G., and M. Yamamoto. 2010. Recovering a Lame kernel in a viscoelastic equation by a single boundary measurement. *Applicable Analysis* 89 (3): 377–390.
32. Landau, L.D., and E.M. Lifshits. 1959. *Electrodynamics of continuous media.* M., Nauka (Russian).
33. Alekseev, A.S., and V.I. Dobrinsky. 1975. Some aspects of practical application of inverse dynamical problems of seismology. *Mathematical Problems of Geophysics* 6 (2): 7–53 (Russian). V. Ts. SO SN SSSR, Novosibirsk.
34. Hadamard, G. 1978. *The Cauchy problem for linear equations with partial derivatives of a hyperbolic type.* M.: Nauka (Russian).
35. Durdiev, D.K. 1989. Inverse problem for a three dimensional wave equation in a medium with memory, 19–27. In: Mat. An. Disk. Mat., Novosibirsk State University Press (Russian).

36. Durdiev, D.K. 1992. To the question of correctness of one inverse problem for hyperbolic integro–differential equation. *Sibirskii Matematicheskii Zhurnal* 33 (3): 69–77.
37. Durdiev, D.K. 1994. A multidimensional inverse problem for an equation with memory. *Sibirskii Matematicheskii Zhurnal* 35 (3): 574–582.
38. Durdiev, D.K. 2004. Local solvability of one inverse problem for an integro–differential equation with memory. *Uzbek Mathematical Journal* 1: 77–86 (Russian).
39. Durdiev, D.K. 2006. A memory determination problem in the two dimensional Maxwell system of equations. *Bulletin of NUUz* 2: 13–17 (Russian).
40. Durdiev, D.K. 2006. Differential properties of an inverse problem solution for equation with memory. *Uzbek Mathematical Journal* 3: 26–36 (Russian).
41. Durdiev, D.K. 2007. Some multidimensional inverse problems of memory determination in hyperbolic equations. *Journal of Mathematical Physics, Analysis, Geometry* 2: 441–423.
42. Durdiev, D.K. 2008. Global solvability of an inverse problem for an integro–differetial equation of electrodynamics. *Differential* Equations 44 (7): 867–873.
43. Durdiev, D.K. 2008. Problem of determining the nonstationary potential in a hyperbolic–type equation. *TMF* 156 (2): 220–225.
44. Durdiev, D.K. 2009. The questions of global solvability for two coefficients determining problem in the wave equation with memory. *Scientific Bulletin of Bukhara University* 1: 89–96 (Russian).
45. Durdiev, D.K. 2009. Global solvability of one inverse problem for integro–differential equation of electrodynamics. *Uzbek Mathematical Journal 3*: 36–45 (Russian).
46. Durdiev, D.K. 2009. The problem of determining a function of memory of a medium and the regular part of a pulsed source. *Matematicheskie Zametki* 86 (2): 202–212.
47. Durdiev, D.K. 2009. An inverse problem for determining two coefficients in an integro–differential wave equation. *Sibirskii Zhurnal Industrial'noi Matematiki* 12 (3) (2009), 28–40.
48. Durdiev, D.K. 2009. Inverse problem for the system of termoelasticity equations for a vertically inhomogeneous cohesionless medium with memory. *Differential* Equations 45 (9): 1229–1236.
49. Durdiev, D.K. 2009. Global solvability of two unknowns identification problem in one inverse problem for the integro–differential wave equation. Vest. Samar. Gos Tekhn. Univ., Ser. Fiz.–Mat. Nauki [J. Samara State Tech. Univ. Ser. Phys. Math. Sci.], 2 (19) (2009), 17–28 (Russian).
50. Durdiev, D.K. 2009. An identification problem of memory function of a medium and the form of an impulse source. *Journal of Siberian Federal University. Mathematics and Physics* 2 (2): 127–136.
51. Durdiev, D.K., and Zh.Zh. Zhumaev. 2019. Problem of determining a multidimensional thermal memory in a heat conductivity equation. Methods of Functional Analysis and Topology 25 (3): 219–226.
52. Durdiev, D.K., and Zh.Zh. Zhumaev. 2020. Problem of determining the thermal memory of a conducting medium. Differential Equations 56 (6): 785–796.
53. Durdiev, D.K., and A.Sh. Rashidov. 2014. Inverse problem of determining the kernel in an integrodifferential equation of parabolic type. Differential Equations 50 (1): 110–116.
54. Durdiev, D.K., and Zh.Z. Nuriddinov. 2021. Determination of a multidimensional kernel in some parabolic integro–differential equation. *Journal of Siberian Federal University. Mathematics and Physics* 14 (1): 117–127.
55. Durdiev, D.K., and Z.R. Bozorov. 2013. A problem of determining the kernel of integro–differential wave equation with weak horizontal properties. *Dal'nevostochnyi Matematicheskii Zhurnal* 13 (2): 209–221 (Russian).
56. Durdiev, D.K., and U.D. Durdiev. 2014. Stability of the inverse problem solution for Maxwell integro–differential in the homogenuous anisotropic medium. *Uzbek Mathematical Journal* 2: 25–34 (Russian).
57. Durdiev, D.K., and U.D. Durdiev. 2016. The problem of kernel determination from viscoelasticity system integro-differential equations for homogeneous anisotropic media. *Nanosystems: Physics, Chemistry, Mathematics* 7 (3): 405–409.

58. Durdiev, D.K., and A.A. Rahmonov. 2018. Inverse problem for the system integrodifferential equation Sh waves in a visco-elastic porous medium: Global solubility. *TMF* 195 (3): 925–940.
59. Durdiev, D.K., and Zh.Sh. Safarov. 2012. The local solvability of a problem of determining the spatial part of a multidimensional kernel in the integro–differential equation of hyperbolic type. Vest. Samar. Gos Tekhn. Univ., Ser. Fiz.–Mat. Nauki [*Journal of Samara State Technical University, Ser. Physical and Mathematical Sciences*] 4 (29): 37–47 (Russian).
60. Durdiev, D.K., and Zh.Sh. Safarov. 2015. Inverse problem of determining the one dimentional kernel of the viscoelasticity equation in a bounded domain. *Matematicheskie Zametki* 97 (6): 855–867.
61. Durdiev, D.K., and Z.D. Totieva. 2013. The problem of determining the onedimensional kernel of the viscoelasticity equation. *Sibirskii Zhurnal Industrial'noi Matematiki* 16 (2): 72–82 (Russian).
62. Durdiev, D.K., and Z.D. Totieva. 2015. The problem of determining the multidimensional kernel of the viscoelasticity equation. *Vladikavkaz Mathematical Journal* 17 (4): 18–43 (Russian).
63. Durdiev, D.K., and Z.D. Totieva. 2017. The problem of determining the onedimensional kernel of the electroviscoelasticity equation. *Siberian Mathematical Journal* 58: 427–444.
64. Durdiev, D.K., and Z.D. Totieva. 2018. The problem of determining the onedimensional matrix kernel of the system of viscoelasticity equations. *Mathematical Methods in the Applied Sciences* 41 (17): 8019–8032.
65. Durdiev, D.K., and Z.D. Totieva. 2020. The problem of determining the onedimensional kernel of viscoelasticity equation with a source of explosive type. *Journal of Inverse and Ill-Posed Problems* 28 (1): 43–52.
66. Totieva, Z.D. 2012. On the fundamental solution of the Cauchy problem for a hyperbolic operator. *Vladikavkazskii Matematicheskii Zhurnal* 14 (2): 45–49 (Russian).
67. Totieva, Z.D. 2016. The multidimensional problem of determining the density function for the system of viscoelasticity. *Siberian Electronic Mathematical Reports* 13: 635–644 (Russian).
68. Totieva, Z.D. 2017. The problem of determining the coefficient of thermal expansion of the equation of thermoviscoelasticity. *Siberian Electronic Mathematical Reports* 14: 1108–1119 (Russian).
69. Totieva, Z.D. 2018. The problem of determining the piezoelectric module of electroviscoelasticity equation. *Mathematical Methods in the Applied Sciences* 41 (16): 6409–6421.
70. Totieva, Z.D. 2019. The problem of determining the matrix kernel of the anisotropic viscoelasticity equations system. *Vladikavkazskii Matematicheskii Zhurnal* 21 (2): 58–66 (Russian).
71. Totieva, Z.D. 2019. One-dimensional inverse coefficient problems of anisotropic viscoelasticity. *Siberian Electronic Mathematical Reports* 16: 786–811 (Russian).
72. Totieva, Z.D., and D.K. Durdiev. 2018. The problem of finding the one-dimensional kernel of the thermoviscoelasticity equation. *Mathematical Notes* 103: 118–132.
73. Zh.Sh. Safarov, and D.K. Durdiev. 2018. Inverse problem for an integro–differential equation of acoustics. *Differential* Equations 44 (1): 136–144.
74. Bozorov, Z.R. 2020. Numerical determining a memory function of a horizontallystratified elastic medium with aftereffect. Eurasian Journal of Mathematical and Computer Applications 8 (2): 28–40.
75. Durdiev, U.D. 2020. Numerical method for determining the dependence of the dielectric permittivity on the frequency in the equation of electrodynamics with memory. *Sibirskie Èlektronnye Matematicheskie Izvestiya* 17: 179–189.
76. Kolmogorov, A.N., and S.V. Fomin. 1999. Elements of the theory of functions and functional analysis. Fizmatlit, Moscow (2004) (Russian); English transl. prev. ed., Dover, New York (1999).
77. Colombo, F. 2001. Direct and inverse problems for a phase-field model with memory. *Journal of Mathematical Analysis and Applications* 260: 517–54.

78. Colombo, F. 2007. A inverse problem for a parabolic integro-differential model in the theory of combustion. *Physica D: Nonlinear Phenomena* 236: 81–89.
79. Favaron, A. 2006. An identification problem arising in the theory of heat conduction for materials with memory. *Journal of Inverse and Ill-Posed Problems* 14 (3): 235–249.
80. Grasselli, M. 1992. An identification problem for a linear integro-differential equation occurring in heat flow. *Mathematical Methods in the Applied Sciences* 15: 167–186.
81. Karuppiah, K., J.K. Kim, and K. Balachandran. 2015. Parameter identification of an integro-differential equation. *Nonlinear Functional Analysis and Applications* 20 (2): 169–185.
82. Janno, J. 2004. Determination of a time- and space-dependent heat flux relaxation function by means of a restricted Dirichlet-to-Neumann operator. *Mathematical Methods in the Applied Sciences* 27 (11): 1241–1260.
83. Beilina, L., and M.V. Klibanov. 2012. *Approximate global convergence and adaptivity for coefficient inverse problems*. Springer.
84. Bellassoued, M., and M. Yamamoto. 2017. *Carleman estimates and applications to inverse problems for hyperbolic systems*. Springer.
85. Bukhgeim, A.L., and M.V. Klibanov. 1981. Uniqueness in the whole of one class of multidimensional inverse problems. *Doklady Akademii Nauk SSSR* 260 (2): 269–272.
86. Klibanov, M.V., and A.A. Timanov. 2004. *Carleman estimates for coefficient inverse problems and numerical applicationz*. Utrecht-Boston.
87. Anikonov, Yu.E. 1995. *Multidimensional inverse and Ill-posed problems for differential equations*. Utrecht: VSP.
88. Anikonov, Yu.E. 1997. *Formulas in inverse and Ill-posed problems*. Utrecht: VSP.
89. Bukhgeim, A.L. 2000. *Introduction to the theory of inverse problems*. Walter de Gruyter.
90. Cannon, J.R., and U. Hornung. 1986. *Inverse problems*. Basel-Boston-Stuttgart: Birkhauser.
91. Kabanikhin, S.I. 1988. *Projection—Difference methods for determining the coefficients of hyperbolic equations*. Novosibirsk: Nauka (Russian).
92. Kabanikhin, S.I., and A. Lorenzi. 1999. *Identification problems of wave phenomena: Theory and numerics*. Utrecht: VSP.
93. Newton, R.G. 1989. *Inverse Schrodinger scattering in Tree dimensions, texts and monographs in physics*. Berlin: Springer.
94. Prilepko, A.I., D.G. Orlovsky, and I.A. Vasin. 1999. *Methods for solving inverse problems in mathematical physics*. New York: Marcel Dekker, Inc.
95. Ramm, A.G. 1994. *Multidimensional inverse scattering problems*. Moswcow: Mir (Russian).
96. Romanov, V.G. 2005. *Stability in inverse problems*. M.: Nauchniy Mir (Russian).
97. Vladimirov, V.S. 1971. *Equations of mathematical physics*. New York: Marcel Dekker, INC.
98. Sobolev, S.L. 1930. Wave equation for an inhomogenuous medium. *Trudy Seismologicheskogo Instituta* 6: 1–57 (Russian).
99. Sobolev, S.L. 1934. To the question on integration of wave equation in an inhomogenuous medium. *Trudy Seismologicheskogo Instituta* 1 (42) (Russian).
100. Sobolev, S.L. 1936. New method of solving Cauchy's problem for partial differential equations of normal hyperbolic form. *Matematicheskii Sbornik* 1 (43) (Russian).
101. Romanov, V.G. 1972. *Some inverse problems for hyperbolic–type equations*. Novosibirsk: Nauka (Russian).
102. Ovsyannikov, L.V. 1965. Singular operator in the scale of the Banach spaces. *Doklady Akademii Nauk SSSR* 163 (4): 819–822 (Russian).
103. Ovsyannikov, L.V. 1971. Nonlinear Cauchy problem in the scales of the Banach spaces. *Doklady Akademii Nauk SSSR* 200 (4): 789–792 (Russian).
104. Nirenberg, L. 1974. *Topics in nonlinear functional analysis*. New York: Courant Institute Mathematical Sciences, New York University.
105. Durdiev, U.D. 2019. A problem of identification of a special 2D memory kernel in an integro-differential hyperbolic equation. Eurasian Journal of Mathematical and Computer Applications 7 (2): 4–19.
106. Durdiev, U.D., and Z.D. Totieva. 2019. A problem of determining a special spatial part of 3D memory kernel in an integro-differential hyperbolic equation. *Mathematical Methods in Applied Sciences* 42 (18): 7440–7451.

107. Calderon, A.P. 1980. On an inverse boundary value problem. Seminar on Numerical Analysis and Its Applications to Continuum Physics (Rio de Janeiro, 1980), Soc. Brasil. Mat., Rio de Janeiro, (1980), 65–73.
108. Sylvester, J., and G. Uhlmann. 1987. A global uniqueness theorem for an inverse boundary value problem. *Annals of Mathematics* (2) 251 (1): 153–169.
109. Alessandrini, G. 1988. Stable determination of conductivity by boundary measurements. *Applicable Analysis* 1–3 (27): 153–172.
110. Nachman, A. 1988. Reconstructions from boundary measurements. *Annals of Mathematics* (2) 128 (3): 531–576.
111. Rakesh, and W.W. Symes. 1988. Uniqueness for an inverse problem for the wave equation. *Communications in Partial Differential* Equations 13 (1): 87–96.

Contents

Chapter 1
Local Solvability of One-Dimensional Kernel Determination Problems

This chapter discusses one-dimensional inverse problems of determining the memory of a medium entering into an integral term of convolution type in hyperbolic equations. In Sect. 1.1, the problem of finding a kernel from an integro-differential wave equation in a half-space $\mathbb{R}^3_+ = \{(x_1, x_2, x_3) \in \mathbb{R}^3 \big| \, x_3 > 0\}$ is investigated. Further, in Sect. 1.2, the results obtained in the previous section are generalized to the case of a second-order hyperbolic-type equation with variable coefficients at lower derivatives with respect to $x = (x_1, x_2, x_3)$, the main part of which coincides with the wave operator. Vibration of the medium comes from a point source located in the point $x = 0$ at the moment $t = 0$. In both sections, the property of direct problems solution, which are well-posed boundary–initial value problems, is investigated. The mode of oscillation of point $x = 0$ is set to solve the inverse problem. In Sect. 1.3, the one-dimensional kernel memory is determined from the system of thermoelasticity equations for a vertically inhomogeneous cohesionless medium. In Sect. 1.4, it is investigated the local solvability of inverse problem for an integro-differential equation of acoustics. Section 1.5 studies a problem of determining of two functions representing the electrical and magnetic prehistory of the medium, respectively. These unknown functions are found from the two-dimensional Maxwell system of integro-differential equations with a distributed source of perturbation.

The unique solvability theorems are proved, and the conditional stability estimates for the solutions of the direct and the inverse problems are obtained.

© The Author(s), under exclusive license to Springer Nature Singapore Pte Ltd. 2023

D. K. Durdiev and Z. D. Totieva, *Kernel Determination Problems in Hyperbolic Integro-Differential Equations*, Infosys Science Foundation Series in Mathematical Sciences, https://doi.org/10.1007/978-981-99-2260-4_1

1.1 The Inverse Problem for the Wave Equation in the Medium with Memory

For $x_3 > 0$ we consider the equation

$$u_{tt} = \Delta u + \int_0^t k(\tau)u(x, t - \tau)d\tau + \delta(x, t) \tag{1.1}$$

with zero initial and boundary conditions

$$u\big|_{t<0} = 0, \quad u_{x_3}\big|_{x_3=0} = 0, \tag{1.2}$$

where $x = (x_1, x_2, x_3)$ is a point of three-dimensional space $\mathbb{R}^3$; Δ is the Laplace operator with variables $x = (x_1, x_2, x_3)$, $\delta(x, t)$ is the four-dimensional Dirac delta function, $\delta(x, t) = \delta(x_1)\delta(x_2)\delta(x_3)\delta(t)$.

We pose the inverse problem as follows: define the function $k(t), t > 0$, that is in the integral term of Eq. (1.1) given at the point $x = 0$, a generalized solution of problem (1.1) and (1.2):

$$u\big|_{x=0} = f(t), \quad t > 0. \tag{1.3}$$

Definition 1.1 The solution of the inverse problem will be called function $k(t)$, such that the corresponding solution to problem (1.1) and (1.2) satisfies condition (1.3).

We continue the function $u(x, t)$ to the half-space $x_3 < 0$ as an even function with respect x_3. The solution of the problem (1.1) and (1.2) we present in the following form:

$$u(x, t) = -2\varepsilon(x, t) + v(x, t),$$

where

$$\varepsilon(x, t) = \frac{\theta(t)}{2\pi}\delta\left(t^2 - |x|^2\right)$$

is the fundamental solution of the wave equation. Then, inverse problem (1.1)–(1.3) for a scattered wave $v(x, t)$ is reduced to the form

$$v_{tt} - \Delta v = \frac{\theta(t - |x|)}{2\pi|x|}k(t - |x|) + \int_0^t k(\tau)v(x, t - \tau)d\tau, \quad (x, t) \in \mathbb{R}^4, \tag{1.4}$$

$$v|_{t<0} = 0, \tag{1.5}$$

$$v|_{x=0} = f(t), \quad t > 0. \tag{1.6}$$

Consider the Cauchy problem (1.4) and (1.5). Obviously, $v|_{t<|x|} = 0$.

Using the Kirchhoff formula, we obtain the integral equation in the domain $t \geq |x|$

$$v(x,t) = \frac{1}{8\pi^2} \iiint\limits_{|\xi|+|\xi-x|\leq t} \frac{k(t-|\xi|-|\xi-x|)}{|\xi||\xi-x|} d\xi + \frac{1}{4\pi} \iiint\limits_{|\xi|+|\xi-x|\leq t} \int\limits_0^{t-|\xi|-|\xi-x|} \frac{k(\tau)v(\xi, t-|\xi-x|-\tau)}{|\xi-x|} d\tau d\xi. \quad (1.7)$$

Let us transform the integrals in Eq. (1.7). With this for a fixed x, we consider the family of confocal ellipsoids

$$G(x,\alpha) = \{\xi : |\xi| + |\xi - x| = \alpha\}, \quad |x| \leq \alpha \leq t$$

and the spherical coordinate system r, θ, φ, whose pole is at the point 0, and the polar axis passes through points $0, x$. Point $\xi \in G(x,\alpha)$ will be characterized by curvilinear coordinates: $(\theta, \varphi, \alpha)$. Use the equation of ellipsoid $G(x,\alpha)$ of the form

$$r(\theta,\alpha) = \frac{\alpha^2 - |x|^2}{2(\alpha - |x|\cos\theta)}. \quad (1.8)$$

For points $\xi \in G(x,t)$, by virtue of the relations

$$|\xi| = r,\ |x-\xi| = \alpha - r = \frac{\alpha^2 + |x|^2 - 2\alpha\cos\theta}{2(\alpha - |x|\cos\theta)},$$

$$d\xi = r^2(\theta,\alpha)\sin\theta\, d\theta\, d\varphi\, dr = \frac{\alpha^2 + |x|^2 - 2\alpha\cos\theta}{2(\alpha - |x|\cos\theta)^2} r^2(\theta,\alpha)\sin\theta\, d\theta\, d\varphi\, d\alpha,$$

we obtain the following equalities [1]:

$$\frac{d\xi}{|\xi - x|} = \frac{2r^3(\theta,\alpha)}{\alpha^2 - |x|^2}\sin\theta\, d\theta\, d\varphi\, d\alpha, \quad (1.9)$$

$$\frac{d\xi}{|\xi||\xi - x|} = \frac{\alpha^2 - |x|^2}{2(\alpha - |x|\cos\theta)^2}\sin\theta\, d\theta\, d\varphi\, d\alpha. \quad (1.10)$$

In view of this, we write Eq. (1.7) in the form

$$v(x,t) = \frac{1}{4\pi}\int_0^{t-|x|} k(\alpha)\mathrm{d}\alpha$$
$$+\frac{1}{2\pi}\int_{|x|}^{t}\frac{\mathrm{d}\alpha}{\alpha^2-|x|^2}\int\int_{G(x,\alpha)} r^3(\theta,\alpha)\int_0^{t-\alpha} k(\tau)v(\xi,t-\alpha+r-\tau)\mathrm{d}\tau\,\mathrm{d}\omega,$$
$$t \geq |x|, \tag{1.11}$$

where $\mathrm{d}\omega = \sin\theta\,\mathrm{d}\theta\,\mathrm{d}\varphi$ is the element of the solid angle with the center at the point 0.

For brevity, we introduce a linear integral operator A_k by the formula

$$A_k[v(x,t)]$$
$$= \frac{1}{2\pi}\int_{|x|}^{t}\frac{\mathrm{d}\alpha}{\alpha^2-|x|^2}\iint_{G(x,\alpha)} r^3(\theta,\alpha)\int_0^{t-\alpha} k(\tau)v(\xi,t-\alpha+r-\tau)\mathrm{d}\tau\,\mathrm{d}\omega.$$

The index k of the operator A_k indicates the function $k(t)$ that enters in its definition. Then, Eq. (1.11) can be written as

$$v(x,t) = \frac{1}{4\pi}\int_0^{t-|x|} k(\alpha)\mathrm{d}\alpha + A_k[v(x,t)],\, t \geq |x|. \tag{1.12}$$

Differentiating this equation with respect to the variable t, we get

$$w(x,t) = \frac{1}{4\pi}k(t-|x|) + A_k[w(x,t)], \quad t \geq |x|, \tag{1.13}$$

where $w(x,t) = v_t(x,t)$.

We first focus on the properties of the functions $v(x,t)$, $w(x,t)$, defined by Eqs. (1.12) and (1.13). Suppose that the function $k(t)$ is given and belongs to the class of continuous functions. In this case, we show that Eqs. (1.12) and (1.13) have the unique solution in the class of continuous functions, and it can be obtained by the method of successive approximations.

Consider the sphere $S(T) = \xi : |\xi| = T$, where T is some positive number, and choose pairs of points (x,t), such that the corresponding to them spheres $|\xi - x| = t$ do not extend beyond the surface of the sphere $S(T)$. For fulfillment of this requirement it is sufficient that the aggregate of (x,t) belonged to a four-dimensional domain

$$D(T) = \{(x,t) : |x| \leq t \leq T - |x|\}.$$

Let

$$\|v\| = \max_{(x,t)\in D(T)} |v(x,t)|, \quad \|w\| = \max_{(x,t)\in D(T)} |w(x,t)|,$$
$$\|k\| = \max_{t\in[0,T]} |k(t)|.$$

Let us choose for Eq. (1.12) a process of the successive approximations according to the scheme

$$v_0(x,t) = \frac{1}{4\pi} \int_0^{t-|x|} k(\alpha)\mathrm{d}\alpha,$$

$$v_n(x,t) = v_0(x,t) + A_k\left[v_{n-1}(x,t)\right], \quad t \geq |x|, \ n = 1, 2, 3, \ldots. \tag{1.14}$$

Obviously, each of the functions $v_n(x,t)$ is continuous in the domain $D(T)$. At the same time, from the linearity of the operator A_k, it follows the estimates

$$|v_0(x,t)| \leq \frac{1}{4\pi}\|k\|(t-|x|),$$

$$|v_n(x,t)| \leq \frac{\|k\|}{4\pi} \sum_{i=0}^{n} \left(\frac{\|k\|T}{4}\right)^i \frac{(t-|x|)^{2i+1}}{(2i+1)!},$$
$$n = 1, 2, 3, \ldots, \ (x,t) \in D(T). \tag{1.15}$$

We write for Eq. (1.12), according to the corresponding scheme (1.15), Neumann series

$$v_0(x,t) + \sum_{n=1}^{\infty} \left(v_n(x,t) - v_{n-1}(x,t)\right). \tag{1.16}$$

From the above estimates, it follows that the members of this series are majorized by the members of series

$$\frac{\|k\|}{4\pi} \sum_{n=0}^{\infty} \left(\frac{\|k\|T}{4}\right)^n \frac{(t-|x|)^{2n+1}}{(2n+1)!},$$

which, in turn, for all $(x,t) \in D(T)$ is majorized by the following convergent numerical series:

$$\frac{\|k\|}{4\pi} \sum_{n=0}^{\infty} \left(\frac{\|k\|T}{4}\right)^n \frac{T^{2n+1}}{(2n+1)!}.$$

Therefore, according to the Weierstrass criterion the Neumann series (1.16), whose nth partial sum coincides with $v_n(x,t)$, converges uniformly: $v_n(x,t) \Rightarrow v_n(x,t)$, where $v_n(x,t)$ is a continuous function in the domain $D(T)$. Passing in the last of inequalities (1.15) to the limit as $n \to \infty$, we obtain the estimate for function $v(x,t)$:

$$\|v\| \leq \frac{\|k\|}{4\pi} \sum_{n=0}^{\infty} \left(\frac{\|k\|T}{4}\right)^n \frac{T^{2n+1}}{(2n+1)!}. \tag{1.17}$$

In view of the uniform convergence $u_n(x,t)$, we can pass in (1.14) to the limit as $n \to \infty$. Then, we obtain relation (1.12). In this way, the function $v(x,t)$ as the limit of a uniformly convergent sequence $v_n(x,t)$ is a solution of Eq. (1.12). Let's prove the uniqueness of the solution of Eq. (1.12). Let $v_1(x,t)$, $v_2(x,t)$ be two continuous solutions of Eq. (1.12) and $\bar{v}(x,t)$ is their difference. Then, $\bar{v}(x,t)$ satisfies the homogeneous equation

$$\bar{v}(x,t) = A_k[\bar{v}(x,t)], \quad (x,t) \in D(T). \tag{1.18}$$

Denote $\max_{(x,t)\in D(T)} |\bar{v}(x,t)| =: C$. Estimating by the majorant method right-hand side of (1.18), we get $|\bar{v}(x,t)| \leq (\|k\|T)/4 \cdot (t-|x|)/3!$. Substituting this estimate to the right side of (1.18), we obtain a new estimate

$$|\bar{v}(x,t)| \leq C\left(\frac{\|k\|T}{4}\right)^2 \frac{(t-|x|)^5}{5!}$$

and so on. After the n-step, we have the estimate

$$|\bar{v}(x,t)| \leq C\left(\frac{\|k\|T}{4}\right)^n \frac{(t-|x|)^{2n+1}}{(2n+1)!}. \tag{1.19}$$

If $\bar{v}(x,t) \neq 0$, then, there is $(x_0, t_0) \in D(T)$ for which $\bar{v}(x,t)$ is different from zero. On the other hand, for any $\epsilon > 0$ and for sufficiently large n the inequality

$$C\left(\frac{\|k\|T}{4}\right)^n \frac{(t-|x|)^{2n+1}}{(2n+1)!} < \epsilon$$

is valid. From (1.19) we get that $\bar{v}(x_0, t_0) < \epsilon$, where ϵ is an arbitrarily small quantity. For instance, choosing $0 < \epsilon < |\bar{v}(x_0, t_0)|/2$, we get a contradiction. Therefore, $\bar{v}(x,t) = 0$ and the solution of Eq. (1.12) is unique.

We now turn to Eq. (1.13). For this equation we use the process of successive approximations according to the following scheme:

$$\begin{aligned} w_0(x,t) &= \frac{k(t-|x|)}{4\pi}, \\ w_n(x,t) &= A_k\left[w_{n-1}(x,t)\right], \quad t \geq |x|,\ n = 1,2,3,\ldots \end{aligned} \tag{1.20}$$

We show that the series

$$\sum_{n=0}^{\infty} w_n(x,t)$$

converges uniformly in the domain $D(T)$. In this domain the following estimates take place

$$\|w_n\| \le \frac{\|k\|}{4\pi} \frac{\left[\sqrt{\|k\|T}(t-|x|)\right]^{2n}}{(2n)!2^{2n}}, \qquad n = 1, 2, 3, \ldots.$$

Hence the series $\sum_{n=0}^{\infty} w_n(x,t)$ is majorized by series

$$\frac{\|k\|}{4\pi} \sum_{n=0}^{\infty} \frac{\left[\sqrt{\|k\|T}(t-|x|)\right]^{2n}}{(2n)!2^{2n}},$$

which, in turn, for all $(x,t) \in D(T)$ is majorized by the convergent numerical series

$$\frac{\|k\|}{4\pi} \sum_{n=0}^{\infty} \frac{\left[\sqrt{\|k\|T^3}\right]^{2n}}{(2n)!2^{2n}} = \frac{\|k\|}{4\pi} \mathrm{ch}\left(\frac{1}{2}\sqrt{\|k\|T^3}\right).$$

The series $\sum_{n=0}^{\infty} w_n(x,t)$ converges to a unique continuous solution of Eq. (1.13). The following estimate is valid:

$$\|w\| \le \frac{\|k\|}{4\pi} \mathrm{ch}\left(\frac{1}{2}\sqrt{\|k\|T^3}\right). \tag{1.21}$$

Now we pass to the consideration of the inverse problem and suppose, as before, that $k(t)$ is a continuous function. It is clear from this that the function $f(t)$ for $t \to +0$ must satisfy the additional condition $f(+0) = 0$, and there must be a continuous differentiable at $t > 0$. Putting in (1.12) $x = 0$ and using condition (1.6), we obtain

$$f(t) = \frac{1}{4\pi} \int_0^t k(\alpha)d\alpha + A_k[v(x,t)]|_{x=0}, \quad t > 0. \tag{1.22}$$

Differentiating equality (1.22) with respect to the variable t and introducing the notation

$$F(t) = 4\pi f'(t),$$

we find

$$k(t) = F(t) - 4\pi A_k\,[w(x,t)]\,|_{x=0}. \tag{1.23}$$

Let $\|F\| = \max_{t\in[0,T]} |F(t)|$. Fix $\epsilon > 0$ and consider the ball

$$\Phi^T(F,\epsilon)=\{r(t):r\in C[0,T],\ \|F-r\|\le\epsilon]\}.$$

We introduce the operator by formula

$$U[k(t)]=F(t)-4\pi A_k[w(x,t)]|_{x=0}. \tag{1.24}$$

Then, Eq. (1.23) is written in the more convenient form

$$k(t)=U[k(t)]. \tag{1.25}$$

Let $T^*=\min(T_1,T_2)$, where

$$T_1=\sqrt[3]{\frac{4\gamma_1^2}{\|F\|+\epsilon}},\quad T_2=\sqrt[3]{\frac{4\gamma_2^2}{\|F\|+\epsilon}}, \tag{1.26}$$

and γ_1 is the positive root of the equation

$$\gamma^2\mathrm{ch}\gamma=\frac{2\epsilon}{\|F\|+\epsilon}, \tag{1.27}$$

γ_2 is the positive root of the equation

$$\gamma^2\mathrm{ch}\gamma\left(1+\frac{\gamma^2}{4}\mathrm{ch}\gamma\right)=1. \tag{1.28}$$

Theorem 1.1 *Let $f(t)\in C^1[0,T]$ and $f(+0)=0$. Then, for $t\in[0,T]$, $T\in(0,T^*)$, there exists a unique continuous solution of Eq. (1.25) belonging to a ball $\Phi^T(F,\epsilon)$.*

We **prove** this theorem using contraction mapping principle. The operator U is a contraction operator if the following two conditions are fulfilled:

(1) If $\varphi\in\Phi^T(F,\epsilon)$, then $U\varphi\in\Phi^T(F,\epsilon)$,
(2) If φ^1 and φ^2 are arbitrary elements in $\Phi^T(F,\epsilon)$, then the following inequality holds true:

$$||U\varphi^1-U\varphi^2||\le\rho||\varphi^1-\varphi^2||,$$

where $0<\rho<1$.

First, we show that for any $T\in(0,T^*)$ the operator U maps the ball $\Phi^T(F,\epsilon)$ into itself, i.e., from the condition $k(t)\in\Phi^T(F,\epsilon)$ it follows $U[k(t)]\in\Phi^T(F,\epsilon)$. Indeed, for any continuous function $k(t)$ the function $U[k(t)]$, calculated by the formula (1.24), will be continuous function. In addition, estimating the norm of differences, we find

$$\|U\,[k(t)] - F(t)\| \leq 4\pi \|k\| \|w\| A_1[1]|_{x=0} \leq \frac{T^3}{8} \|k\|^2 \mathrm{ch}\left(\frac{1}{2}\sqrt{\|k\|T^3}\right).$$

Here we used the estimate (1.21) and easily verifiable equality $|A_1[1]|_{x=0} = t^3/8$. We note that the function standing in the right side of this inequality, monotonically increases with increasing T, and if the function $k(t)$ belongs to the ball $\Phi^T(F, \epsilon)$, then

$$\|k\| \leq \epsilon + \|F\|. \tag{1.29}$$

Therefore, for all $T \in (0, T^*)$ this inequality will only intensify if instead of T we substitute T_1, and instead of $\|k\|$ we substitute the expression $\epsilon + \|F\|$. Carrying out this and using formulas (1.26) and (1.27), we get

$$\|U\,[k(t)] - F(t)\| \leq \frac{1}{2}(\epsilon + \|F\|)\left[\gamma^2 \mathrm{ch}\gamma\right]_{\gamma=\gamma_1} = \epsilon.$$

Thus, we have proved that $U\,[k(t)] \in \Phi^T(F, \epsilon)$.

Now show the fulfillment of the second condition of the contraction mapping principle. Consider two functions $\tilde{k}(t), k(t)$, belonging to the ball $\Phi^T(F, \epsilon)$, and estimate the difference of their images $U[\tilde{k}(t)]$, $U[k(t)]$. The solution of Eq. (1.13), corresponding to function $\tilde{k}(t)$, we denoted by $\tilde{w}(x,t)$. Thus, $\tilde{w}(x,t)$ satisfies the equation

$$\tilde{w}(x,t) = \frac{1}{4\pi}\tilde{k}(t - |x|) + A_{\tilde{k}}[\tilde{w}(x,t)], \quad t \geq |x|. \tag{1.30}$$

By using this equation and (1.13), we estimate the norm of difference

$$\begin{aligned} \left\|U\,[k(t)] - U\left[\tilde{k}(t)\right]\right\| &\leq 4\pi\left(\|k - \tilde{k}\|\|w\| + \|\tilde{k}\|\|w - \tilde{w})\|\right) A_1[1]|_{x=0} \\ &\leq \frac{T^3\pi}{2}\left(\|k - \tilde{k}\|\|w\| + \|\tilde{k}\|\|w - \tilde{w}\|\right). \end{aligned} \tag{1.31}$$

Using Eqs. (1.13) and (1.30), we derive an estimate for $\|w - \tilde{w}\|$ through $\|k - \tilde{k}\|$. From equalities (1.13) and (1.30) we make an integral equation for $w - \tilde{w}$:

$$\begin{aligned} w(x,t) - \tilde{w}(x,t) = &\frac{1}{4\pi}\left[k(t - |x|) - \tilde{k}(t - |x|)\right] + A_{k-\tilde{k}}\,[w(x,t)] \\ &+ A_{\tilde{k}}\left[w(x,t) - \tilde{w}(x,t)\right]. \end{aligned} \tag{1.32}$$

Considering $|w(x,t) - \tilde{w}(x,t)|$, from Eq. (1.32) we get the following linear inequality with respect to $|w(x,t) - \tilde{w}(x,t)|$:

$$|w(x,t)-\tilde{w}(x,t)| \le \|k-\tilde{k}\| \left[\frac{1}{4\pi}+\frac{T^3\|k\|}{32\pi}\text{ch}\left(\frac{1}{2}\sqrt{\|k\|T^3}\right)\right]$$
$$+\|\tilde{k}\|A_1\left[|w-\tilde{w}|\right].$$

Applying to this inequality the method of successive approximations, we find

$$\|w(x,t)-\tilde{w}(x,t)\|$$
$$\le \left\|k-\tilde{k}\right\| \left[\frac{1}{4\pi}+\frac{T^3\|k\|}{32\pi}\text{ch}\left(\frac{1}{2}\sqrt{\|k\|T^3}\right)\right]\text{ch}\left(\frac{1}{2}\sqrt{\|k\|T^3}\right). \tag{1.33}$$

Substituting the estimates (1.21) and (1.33) into inequality (1.31), we get

$$\left\|U\left[k(t)\right]-U\left[\tilde{k}(t)\right]\right\| \le \|k-\tilde{k}\|\frac{T^3\|k\|}{8}\text{ch}\left(\frac{1}{2}\sqrt{\|k\|T^3}\right)$$
$$+\|k-\tilde{k}\|\frac{T^3\|k\|}{8}\text{ch}\left(\frac{1}{2}\sqrt{\|\tilde{k}\|T^3}\right)\left[1+\frac{T^3\|\tilde{k}\|}{8}\text{ch}\left(\frac{1}{2}\sqrt{\|k\|T^3}\right)\right]. \tag{1.34}$$

The functions $k(t)$, $\tilde{k}(t)$ belong to the ball $\Phi^T(F,\epsilon)$, and, consequently, for each of them the inequality (1.29) holds. We notice that the function on the right-hand side of inequality (1.34) increases monotonically with increasing $\|k\|$, $\|\tilde{k}\|$ and T. Therefore, if $\|k\|$, $\|\tilde{k}\|$, in inequality (1.34), we replace by $\epsilon+\|F\|$, and instead of T we set T_2, the inequality will only intensify. Having done this, we find

$$\left\|U\left[k(t)\right]-U\left[\tilde{k}(t)\right]\right\| \le \rho(T)\|k-\tilde{k}\|,$$
$$\rho(T)=\gamma^2\text{ch}\gamma\left[1+\frac{1}{4}\gamma^2\text{ch}\gamma\right]_{\gamma=\frac{1}{2}\sqrt{(\epsilon+\|F\|)T^3}}$$
$$<\gamma^2\text{ch}\gamma\left[1+\frac{1}{4}\gamma^2\text{ch}\gamma\right]_{\gamma=\gamma_2}=1.$$

From this inequality it follows that the distance in the functional space C between functions $U[k(t)]$, $U[\tilde{k}(t)]$ is less than the distance between the functions $k(t)$, $\tilde{k}(t)$ in the same space. Therefore, the operator U in the ball $\Phi^T(F,\epsilon)$ is a contraction operator. Then, according to Banach theorem, the operator U in the ball $\Phi^T(F,\epsilon)$ has a unique fixed point, that is, there is only one solution of Eq. (1.25). Theorem 1.1 is proven.

Now we give a stability estimate of the solution of inverse problem and prove the uniqueness theorem for an arbitrary fixed $T>0$. Denote by $K(k_0)$ the set of functions $k(t)\in C[0,T]$, satisfying the inequality $|k(t)|\le k_0$ $t\in[0,T]$ with fixed positive constant k_0.

Theorem 1.2 *Let $k(t) \in K(k_0)$, $\tilde{k}(t) \in K(k_0)$ be solutions of the inverse problem (1.4)–(1.6) with the data $f(t)$, $\tilde{f}(t)$, respectively. Then, there exists a positive number $C = C(k_0, T)$, such that the following stability estimate is valid:*

$$\|k(t) - \tilde{k}(t)\|_{C[0,T]} \leq C\|f(t) - \tilde{f}(t)\|_{C^1[0,T]}. \tag{1.35}$$

Proof Using equality (1.23), we compose an equation for difference $k(t) - \tilde{k}(t)$

$$k(t) - \tilde{k}(t) = 4\pi\left[f'(t) - \tilde{f}'(t)\right]$$
$$-4\pi\left\{A_{k-\tilde{k}}\left[w(x,t)\right]_{x=0} + A_{\tilde{k}}\left[w(x,t) - \tilde{w}(x,t)\right]_{x=0}\right\}, \quad (x,t) \in D_T. \tag{1.36}$$

Here the function $\tilde{w}$ satisfies equality (1.30).

We note that the domain D_T allows an equivalent description:

$$D_T = \left\{(x,t) : 0 \leq |x| \leq \frac{T}{2} - \left|t - \frac{T}{2}\right|, 0 \leq t \leq T\right\}.$$

Let

$$\psi(t) = \max\left[\max\left|k(t) - \tilde{k}(t)\right|, \max_{0 \leq |x| \leq T/2 - |t - T/2|} |w(x,t) - \tilde{w}(x,t)|\right],$$

where $t \in [0, T]$. Using Eqs. (1.32) and (1.36), we find that for $0 \leq t \leq |x| \leq T/2 - |t - T/2|$ inequalities

$$|w(x,t) - \tilde{w}(x,t)| \leq \frac{1}{4\pi}\left|k(t) - \tilde{k}(t)\right| + \frac{T^2}{4}(\mu_0 + k_0)\int_0^t \psi(\alpha)d\alpha, \tag{1.37}$$

$$\left|k(t) - \tilde{k}(t)\right| \leq 4\pi\left\|f(t) - \tilde{f}(t)\right\|_{C^1[0,T]} + \pi T^2(\mu_0 + k_0)\int_0^t \psi(\alpha)d\alpha \tag{1.38}$$

are valid, where $\mu_0 := (k_0/4\pi)\text{ch}\left(\frac{1}{2}\sqrt{\|k\|T^3}\right)$. From (1.37), using inequality (1.38), we get

$$|w(x,t) - \tilde{w}(x,t)| \leq \left\|f(t) - \tilde{f}(t)\right\|_{C^1[0,T]} + \frac{T^2}{2}(\mu_0 + k_0)\int_0^t \psi(\alpha)d\alpha. \tag{1.39}$$

It follows from the relations (1.38) and (1.39) that $\psi(t)$ satisfies the following integral inequality

$$\psi(t) \leq 4\pi \left\| f(t) - \tilde{f}(t) \right\|_{C^1[0,T]} + \pi T^2(\mu_0 + k_0) \int_0^t \psi(\alpha)\mathrm{d}\alpha, \quad t \in [0, T].$$

Using the *Gronwall inequality,* we obtain the estimate

$$|\psi(t)| \leq 4\pi \exp\left(\pi T^2(\mu_0 + k_0)\right) \left\| f(t) - \tilde{f}(t) \right\|_{C^1[0,T]}, \quad t \in [0, T]. \tag{1.40}$$

Inequality (1.35) is equivalent to (1.40) if

$$C = 4\pi \exp\left(\pi T^2(\mu_0 + k_0)\right).$$

The uniqueness theorem for any fixed $T > 0$ follows from the stability Theorem 1.2.

Theorem 1.3 *Suppose that the functions* $\{k(t), \tilde{k}(t)\} \in C[0, T])$ *and* $\{f(t), \tilde{f}(t)\} \in C^1[0, T]$ *have the same meaning as in Theorem 1.2. If moreover* $f(t) = \tilde{f}(t)$, $t \in [0, T]$, *then* $k(t) = \tilde{k}(t)$ *for* $t \in [0, T]$.

1.2 The Inverse Problem for Hyperbolic Second-Order Integro-differential Equation with Variable Coefficients for Lower Derivatives

The present section deals with a memory determination problem of a hyperbolic second-order integro-differential equation with constant principal part and variable coefficients at lower derivatives. The research method is based on reducing the problem to nonlinear system of integral equations of the second kind and uses the fundamental solution for a hyperbolic equation with variable coefficients, constructed by Sobolev [1–3]. The theorem of uniqueness and stability in global and the existence theorem in local are proved.

1.2.1 Formulation of the Problem. The Derivation of Integral Equations

Consider the equation

$$u_{tt} - Lu = \int_0^t k(\tau)u(x, t - \tau)\mathrm{d}\tau + \delta(x, t) \tag{1.41}$$

at the condition

$$u|_{t<0} = 0, \tag{1.42}$$

where $t \in \mathbb{R}^1$, $x \in \mathbb{R}^3$, L is differential operator, which has the form

$$Lu = \Delta u + \sum_{i=1}^{3} b_i(x) u_{x_i} + c(x)u, \tag{1.43}$$

$$\Delta u = \left(\frac{\partial^2}{\partial x_1^2} + \frac{\partial^2}{\partial x_2^2} + \frac{\partial^2}{\partial x_3^2} \right) u;$$

$b_i(x)$ $(i = 1, 2, 3)$, $c(x)$ are given smooth functions.

The inverse problem is as follows: define the functions $k(t)$ by the generalized solution of Cauchy problem (1.41) and (1.42) known at $x = 0$:

$$u(0, t) = f(t), \quad t > 0. \tag{1.44}$$

We investigate the problem (1.41), (1.42), and (1.44) by reducing it to a system of nonlinear integral equations with respect to unknown functions. Using the results of [1] (p. 64), we reverse the operator on the left side of (1.41) on the Cauchy data (1.42):

$$\begin{aligned} u(x, t) &= \frac{1}{4\pi} \sigma(x) \delta(t - |x|) + \frac{1}{4\pi} \int\limits_{|\xi - x| \le t} \Big[L_\xi^* \sigma(x, \xi) u(\xi, t - |\xi - x|) \\ &\quad + \sigma(x, \xi) \int\limits_0^{t - |\xi - x|} k(\nu) u(\xi, t - |\xi - x| - \nu) d\nu \Big] d\xi, \end{aligned}$$

where

$$\sigma(x, \xi) = \frac{1}{|\xi - x|} \exp \left\{ \frac{1}{2} \int\limits_0^1 \sum_{i=1}^{3} b_i[\xi + \alpha(x - \xi)](x_i - \xi_i) d\alpha \right\}, \tag{1.45}$$

$\sigma(x) = \sigma(x, 0)$.

L_ξ^* denotes an operator that is adjoint in the sense of Lagrange with the operator L. Equality holds true for it

$$L_\xi^* [|\xi - x| \sigma(x, t)] = |\xi - x| L_\xi^* \sigma(x, \xi), \tag{1.46}$$

which is easily verified directly.

Introducing a new function $v(x, t)$ by formula

$$u(x,t) = v(x,t) + \frac{1}{4\pi}\sigma(x)\delta\left(t - |x|\right),$$

we get equation for it

$$\begin{aligned}
v(x,t) &= \frac{1}{16\pi^2}\int\limits_{D(x,t)} \sigma(\xi)L_\xi^*\sigma(x,\xi)\delta(t-|\xi|-|\xi-x|)\mathrm{d}\xi \\
&+\frac{1}{16\pi^2}\int\limits_{D(x,t)} \sigma(\xi)\sigma(x,\xi)\int\limits_0^{t-|\xi-x|} k(\nu)\delta(t-|\xi|-|\xi-x|-\nu)\mathrm{d}\nu\,\mathrm{d}\xi \\
&+\frac{1}{4\pi}\int\limits_{D(x,t)} \Big[L_\xi^*\sigma(x,\xi)v(\xi,t-|\xi-x|) \\
&+\sigma(x,t)\int\limits_0^{t-|\xi-x|} k(\nu)v(\xi,(t-|\xi-x|-\nu)\mathrm{d}\nu\Big]\mathrm{d}\xi,
\end{aligned}$$

where $D(x,\ t)$ is the domain in the space of variables ξ bounded by surface $S(x,\ t) = \left\{x \middle| \ |\xi - x| + |\xi = t|\right\}$. Using property of δ-function, we rewrite the last equality in the equivalent form

$$\begin{aligned}
v(x,t) &= \int\limits_{S(x,t)} \frac{\Sigma_0(x,\xi)}{|\xi||\xi-x|}\frac{\mathrm{d}s}{\left|\nabla_\xi(|\xi|+|\xi-x|)\right|} \\
&+\int\limits_{|x|}^{t} k(t-\tau)\int\limits_{S(x,t)} \frac{\Sigma_1(x,\xi)}{|\xi||\xi-x|}\frac{\mathrm{d}s\,\mathrm{d}\tau}{\left|\nabla_\xi(|\xi|+|\xi-x|)\right|} \\
&+\int\limits_{D(x,t)} \Big[\Sigma_2(x,\xi)\frac{v(\xi,t-|\xi-x|)}{|\xi-x|} \\
&+\frac{\Sigma_3(x,\xi)}{|\xi-x|}\int\limits_0^{t-|\xi-x|} k(\nu)v(t-|\xi-x|-\nu)\mathrm{d}\nu\Big]\mathrm{d}\xi.
\end{aligned} \tag{1.47}$$

Here $\mathrm{d}s$ is the element of the surface area $S(x,\ t)$;

$$\begin{aligned}
\Sigma_0(x,\xi) &= \frac{1}{16\pi^2}|\xi|\sigma(\xi)|\xi-x|L_\xi^*\sigma(x,\xi), \\
\Sigma_1(x,\xi) &= \frac{1}{16\pi^2}|\xi|\sigma(\xi)|\xi-x|\sigma(x,\xi), \\
\Sigma_2(x,\xi) &= \frac{1}{4\pi}|\xi-x|L_\xi^*\sigma(x,\xi), \quad \Sigma_3(x,\xi) = \frac{1}{4\pi}|\xi-x|\sigma(x,\xi).
\end{aligned} \tag{1.48}$$

Let us transform the integrals entering into the formula (1.47). We introduce, as in the previous section, at the same time with the coordinate system ξ, the Cartesian system η, for which the center is placed at the point 0, and the axis η_3 is directed along the straight line connecting the points 0 and x toward the point x. We use a system in which ξ and η are related by the formula conversions [4]:

$$\xi = Q\eta, \tag{1.49}$$

$$Q = \begin{pmatrix} \cos\theta_0 \cos\varphi_0 & -\sin\theta_0 \sin\varphi_0 & \sin\theta_0 \cos\varphi_0 \\ \cos\theta_0 \sin\varphi_0 & \sin\theta_0 \cos\varphi_0 & \sin\theta_0 \sin\varphi_0 \\ -\sin\theta_0 & 0 & \cos\theta_0 \end{pmatrix}.$$

Here θ_0, φ_0 are the angular coordinates of the unit vector $x/|x|$ at spherical coordinate system connected with the Cartesian system coordinates in the usual way. In addition, with this coordinate system, we associate in the usual way a spherical coordinate system r, θ, φ. Then, denoting the first term in (1.47) by J, we find

$$J(x,t) = \frac{1}{|x|} \int\limits_0^{2\pi} \mathrm{d}\varphi \int\limits_{\frac{t-|x|}{2}}^{\frac{t+|x|}{2}} \Sigma_0(x,\xi)\mathrm{d}r.$$

In this formula the variable point ξ is connected with η by formulas (1.49). Further,

$$\eta = r(\sin\theta\cos\varphi, \sin\theta\sin\varphi, \cos\theta),$$

where $r = (t^2 - |x^2|)/2(t - |x|\cos\theta)$. In the inner integral of J, we make the change of the integration variable r with the variable z as follows:

$$r = \frac{1}{2}(t + |x|z).$$

Then

$$J(x,t) = \frac{1}{2} \int\limits_0^{2\pi} \mathrm{d}\varphi \int\limits_{-1}^{1} \Sigma_0(x,\xi)\mathrm{d}z,$$

and η is expressed in terms of the integration variables φ, z by the formulas

$$\eta_1 = \frac{1}{2}\sqrt{t^2 - |x|^2}\sqrt{1 - z^2}\cos\varphi, \quad \eta_2 = \frac{1}{2}\sqrt{t^2 - |x|^2}\sqrt{1 - z^2}\sin\varphi,$$

$$\eta_3 = \frac{1}{2}(tz + |x|).$$

It is easy to see that the integral over $S(x,\tau)$ in the second term of (1.47) differs from J only by the integrand. Therefore, the transformation of this integral is quite analogous to the transformation of $J(x,t)$. In order to transform the remaining integral terms of (1.47), we use the equation of the ellipsoid $S(x,\tau)$ in the form

$$r(\theta,\tau)=\frac{\tau^2-|x|^2}{2(\tau-|x|\cos\theta)}.$$

As a result, it is easy to get equality [4]:

$$\frac{\mathrm{d}\xi}{|\xi-x|}=\frac{2r^3(\theta,\tau)}{\tau^2-|x|^2}\mathrm{d}\omega\,\mathrm{d}\tau,$$

where $\mathrm{d}\omega=\sin\theta\,\mathrm{d}\theta\,\mathrm{d}\varphi$ is the element of the solid angle with the center at the point 0.

In view of the above, Eq. (1.47) takes the form

$$\begin{aligned} v(x,t) &= \frac{1}{2}\int\limits_0^{2\pi}\mathrm{d}\varphi\int\limits_{-1}^{1}\Sigma_0(x,\xi)\mathrm{d}z+\frac{1}{2}\int\limits_{|x|}^{t}k(t-\tau)\int\limits_0^{2\pi}\mathrm{d}\varphi\int\limits_{-1}^{1}\Sigma_1(x,\xi)\mathrm{d}z\,\mathrm{d}\tau \\ &+2\int\limits_{|x|}^{t}\frac{\mathrm{d}\tau}{\tau^2-|x|^2}\int\limits_{S(x,\tau)}r^3(\theta,\tau)\Big[v(\xi,t-\tau+r)\Sigma_2(x,\xi) \\ &+\Sigma_3(x,\xi)\int\limits_0^{t-\tau}k(\nu)v(\xi,t-\tau+r-\nu)\mathrm{d}\nu\Big]\mathrm{d}\omega. \end{aligned} \tag{1.50}$$

Thus, the regular part of the solution of the problem (1.41) and (1.42) satisfies the integral equation (1.50). We denote $G_T=\{x:|x|\le T\}$, T is an arbitrary positive number. Let

$$b_i(x)\in C^2(G_T),\ i=1,2,3,\ \ c(x)\in C(G_T),\ \ k(t)\in C(0,T],$$

where $C^2(G_T)$, $C(G_T)$ are classes of twice continuously differentiable and continuous functions in the domain G_T, respectively. Then, we show that the function $v(x,t)$ is continuous with respect to the variables x,t in the domain

$$D_T=:\{(x,t):|x|\le t\le T-|x|\}$$

and it can be found from Eq. (1.50) by the method of successive approximation.

In [1] (p. 63) it was received the estimate

$$\left|L_\xi^*\sigma(x,\xi)\right|\le\frac{M_0}{|x-\xi|},$$

where M_0 is a positive constant depending only on the global properties of the coefficients of the operator L^*. Using the definitions of the functions $\Sigma_i, i = 0, 1, 2, 3,$ through formulas (1.48) and the last inequality, we find

$$|\Sigma_i(x, \xi)| \leq M$$

for all $i = 0, 1, 2, 3$, where a constant M depends only on T and the norms of the functions $b_i, i = 1, 2, 3, c$ in the spaces $C^2(G_T), C(G_T)$, respectively.

We apply to Eq. (1.50) the method of successive approximation according to the following scheme:

$$\begin{aligned}
v^0(x,t) &= \frac{1}{2}\int_0^{2\pi} d\varphi \int_{-1}^{1} \Sigma_0(x,\xi)dz + \frac{1}{2}\int_{|x|}^{t} k(t-\tau)\int_0^{2\pi} d\varphi \int_{-1}^{1} \Sigma_1(x,\xi)dz\, d\tau,\\
v^n(x,t) &= v^0(x,t) + 2\int_{|x|}^{t} \frac{d\tau}{\tau^2 - |x|^2} \int_{S(x,\tau)} r^3(\theta,\tau)\Big[v^{n-1}(\xi, t-\tau+r)\Sigma_2(x,\xi)\\
&\quad + \Sigma_3(x,\xi)\int_0^{t-\tau} k(\nu)v^{n-1}(\xi, t-\tau+r-\nu)d\nu\Big]d\omega,\\
&\quad n = 1, 2, \ldots.
\end{aligned} \tag{1.51}$$

It is obvious that each of the functions $v^n(x,t)$ is continuous in the domain D_T. Denoting $k_0 = \max_{t\in[0,T]} |k(t)|$, we estimate functions $v^n(x,t)$ for $(x,t) \in D_T$:

$$\begin{aligned}
\left|v^0(x,t)\right| &\leq 2\pi M\left[1 + k_0(t - |x|)\right],\\
\left|v^1(x,t)\right| &\leq 2\pi M\left[1 + k_0(t - |x|)\right]\\
&\quad + 4\pi^2 M^2 T\left[t - |x| + k_0(t-|x|)^2 + k_0^2\frac{(t-|x|)^3}{2!}\right].
\end{aligned}$$

Thus

$$\left|v^n(x,t)\right| \leq 2\pi M \sum_{m=0}^{n}\sum_{j=0}^{m+1} (2\pi MT)^m k_0^j \frac{(t-|x|)^{m+j}}{m!j!}.$$

For Eq. (1.50), we write the Neumann series corresponding to the scheme (1.51)

$$v^0(x,t) + \sum_{n=1}^{\infty}\left(v^n(x,t) - v^{n-1}(x,t)\right).$$

Its partial sum coincides with the function $v^n(x,t)$, and therefore, this series can be majorized by

$$2\pi M \sum_{n=0}^{\infty}\sum_{j=0}^{n+1} (2\pi MT)^n\, k_0^j \frac{(t-|x|)^{n+j}}{n!j!},$$

which, in turn, for all $(x,t)\in D_T$ is majorized by convergent numerical series

$$\sum_{n=0}^{\infty}\sum_{j=0}^{n+1} (2\pi M)^{n+1}\, k_0^j \frac{T^{2n+j}}{n!j!}.$$

Thus, the Neumann series converges uniformly, and therefore, its sum is a continuous function in the domain D_T. As usual, it is not difficult to prove that the sum of a series is a solution of Eq. (1.50) and it is unique.

Differentiating equality (1.50) with respect to the variable t, we find

$$\begin{aligned}
&v_t(x,t)\\
&= \frac{t}{4\sqrt{t^2-|x|^2}}\int_0^{2\pi} d\varphi \int_{-1}^{1} \big\langle \nabla_\xi \Sigma_0, Q\eta'(t,x,\varphi,z)\big\rangle\, dz\\
&+ \pi k(t-|x|)\int_{-1}^{1} \Sigma_1\left(x,0,0,\frac{z+1}{2}|x|\right) dz\\
&+ \frac{1}{4}\int_{|x|}^{t} \frac{k(t-\tau)\tau\, d\tau}{\sqrt{\tau^2-|x|^2}}\int_0^{2\pi} d\varphi\int_{-1}^{1}\big\langle \nabla_\xi \Sigma_1, Q\eta'(\tau,x,\varphi,z)\big\rangle\, dz\\
&+ \frac{2}{t^2-|x|^2}\int_{S(x,t)} r^3(\theta,t)v(\xi,r)\Sigma_2(x,\xi)d\omega\\
&+ 2\int_{|x|}^{t}\frac{d\tau}{\tau^2-|x|^2}\int_{S(x,\tau)} r^3(\theta,\tau)\Big\{v_t(\xi,t-\tau+r)\Sigma_2(x,\xi)\\
&+ \Sigma_3(x,\xi)\left[k(t-\tau)v(\xi,r)+\int_0^{t-\tau} k(\nu)v_t(\xi,t-\tau+r-\nu)d\nu\right]\Big\}\, d\omega. \qquad (1.52)
\end{aligned}$$

Here

$$\nabla_\xi \Sigma_j(x,\xi) = \left(\Sigma_{j\xi_1}, \Sigma_{j\xi_2}, \Sigma_{j\xi_3}\right)(x,\xi), \quad j=0,1,$$

$\langle \cdot, \cdot \rangle$ is a scalar product in $\mathbb{R}^3$;

$$\eta'(t, x, z, \varphi) = (\eta'_1, \eta'_2, \eta'_3) = \left(\sqrt{1-z^2}\cos\varphi, \sqrt{1-z^2}\sin\varphi, \frac{z}{t}\sqrt{t^2-|x|^2}\right).$$

We introduce linear integral operators by the formulas

$$A[k(t), v(x,t)] = 2\int_{|x|}^{t} \frac{d\tau}{\tau^2 - |x|^2} \int_{S(x,\tau)} r^3(\theta, \tau)\Big[v(\xi, t-\tau+r)\Sigma_2(x,\xi)$$
$$+ \Sigma_3(x,\xi)\int_0^{t-\tau} k(\nu)v(\xi, t-\tau+r-\nu)d\nu\Big]d\omega,$$

$$B[k(t), v(x,t)]$$
$$= 8\sqrt{t^2-|x|^2}\int_{|x|}^{t} \frac{k(t-\tau)d\tau}{\tau^2-|x|^2}\int_{S(x,\tau)} r^3(\theta,\tau)v(\xi,r)\Sigma_3(x,\xi)d\omega, \quad (1.53)$$

$$E[v(x,t)] = \frac{8}{\sqrt{t^2-|x|^2}}\int_{S(x,\tau)} r^3(\theta,t)v(\xi,r)\Sigma_2(x,\xi)d\omega.$$

Denote

$$v_0(x,t) = \frac{1}{2}\int_0^{2\pi} d\varphi \int_{-1}^{1} \Sigma_0(x,\xi)dz,$$
$$w_0(x,t) = t\int_0^{2\pi} d\varphi \int_{-1}^{1} \langle \nabla_\xi \Sigma_0, \; Q\eta' \rangle dz. \quad (1.54)$$

We also introduce a new function $w(x,t)$ by the formula

$$w(x,t) = 4\sqrt{t^2-|x|^2}\, v_t(x,t).$$

Then, Eqs. (1.50) and (1.52) we write in the form

$$v(x,t) = v_0(x,t) + \frac{1}{2}\int_{|x|}^{t} k(t-\tau)\int_0^{2\pi} d\varphi \int_{-1}^{1} \Sigma_1(x,\xi)dz\, d\tau$$
$$+ A[k(t), v(x,t)], \quad (1.55)$$

$$
\begin{aligned}
w(x,t) &= w_0(x,t) + 4\pi\sqrt{t^2-|x|^2}k(t-|x|)\int_{-1}^{1}\Sigma_1\left(x,0,0,\frac{z+1}{2}|x|\right)\mathrm{d}z \\
&+\sqrt{t^2-|x|^2}\int_0^{\sqrt{t^2-|x|^2}} k\left(t-\sqrt{\beta^2-|x|^2}\right)\mathrm{d}\beta\int_0^{2\pi}\mathrm{d}\varphi \\
&\times\int_{-1}^{1}\langle\nabla_\xi\Sigma_1, Q\eta'\rangle|_{\tau=\sqrt{\beta^2-|x|^2}}\mathrm{d}z + B[k(t), v(x,t)] \\
&+ E[v(x,t)] + \sqrt{t^2-|x|^2}A\left[k(t), \frac{w(x,t)}{\sqrt{t^2-|x|^2}}\right].
\end{aligned}
\tag{1.56}
$$

Passing in the last equality to the limit at $x \to 0$, and taking into account the conditions (1.44) we obtain the equation for the unknown function $k(t)$

$$
\begin{aligned}
k(t) &= F(t) + \bar{w}_0(t) - 2\pi\int_0^t k(t-\beta)\mathrm{d}\beta\int_0^{2\pi}\mathrm{d}\varphi\int_{-1}^{1}\langle\nabla_\xi\Sigma_1, Q\eta'\rangle|_{\tau=\beta,x=0}\mathrm{d}z \\
&-\frac{2\pi}{t}\Bigg\{B[k(t), v(x,t)] + E[v(x,t)] \\
&+\sqrt{t^2-|x|^2}A\left[k(t), \frac{w(x,t)}{\sqrt{t^2-|x|^2}}\right]\Bigg\}_{x=0},
\end{aligned}
\tag{1.57}
$$

where

$$
F(t) = 8\pi f'(t), \quad \bar{w}_0(t) = -\frac{2\pi}{t}w_0(0,t).
$$

The order of quantity, included in braces when $x = 0$, is equal to t^2. Because of this, in particular, it follows from (1.57) the following equality

$$
k(+0) = F(+0). \tag{1.58}
$$

1.2.2 The Existence and Uniqueness Theorem

The system of Eqs. (1.55)–(1.57) is a closed system of integral equations with respect to functions $v(x,t)$, $w(x,t)$, $k(t)$. We write it in the form of an operator equation

$$
\psi = U\psi, \tag{1.59}
$$

where ψ is vector function of variables x, t with components $\psi_i, i = 1, 2, 3$ at that $\psi_1(x,t) = v(x,t)$, $\psi_2(x,t) = w(x,t)$, $\psi_3(x,t) \equiv \psi_3(t) = k(t)$. The operator U is defined on a set of functions $\psi \in C[D_T]$, and in accordance with equalities (1.55)–(1.57) has the form

$$U = (U_1, U_2, U_3),$$

$$U_1\psi = v_0(x,t) + \frac{1}{2}\int_{|x|}^{t} \psi_3(t-\tau)\int_0^{2\pi} d\varphi \int_{-1}^{1} \Sigma_1(x,\xi)dz\, d\tau + A\left[\psi_3(t), \psi_1(x,t)\right],$$

$$\begin{aligned} U_2\psi &= w_0(x,t) + 4\pi\sqrt{t^2-|x|^2}\psi_3(t-|x|)\int_{-1}^{1} \Sigma_1\left(x, 0, 0, \frac{z+1}{2}|x|\right) dz \\ &+\sqrt{t^2-|x|^2}\int_0^{\sqrt{t^2-|x|^2}} \psi_3\left(t-\sqrt{\beta^2-|x|^2}\right) d\beta \int_0^{2\pi} d\varphi \\ &\times \int_{-1}^{1} \left\langle \nabla_\xi \Sigma_1, Q\eta'\right\rangle|_{\tau=\sqrt{\beta^2-|x|^2}} dz + B[\psi_3(t), \psi_1(x,t)] \\ &+E[\psi_1(x,t)] + \sqrt{t^2-|x|^2}A\left[\psi_1(t), \frac{\psi_2(x,t)}{\sqrt{t^2-|x|^2}}\right], \end{aligned} \tag{1.60}$$

$$\begin{aligned} U_3\psi &= F(t) + \bar{w}_0(t) - 2\pi\int_0^t \psi_3(t-\beta)d\beta \int_0^{2\pi} d\varphi \int_{-1}^{1} \left\langle \nabla_\xi \Sigma_1, Q\eta'\right\rangle|_{\tau=\beta, x=0} dz \\ &-\frac{2\pi}{t}\left\{B[\psi_3(t), \psi_1(x,t)] + E[\psi(x,t)]\right. \\ &\left.+\sqrt{t^2-|x|^2}A\left[\psi_3(t), \frac{\psi_2(x,t)}{\sqrt{t^2-|x|^2}}\right]\right\}_{x=0}. \end{aligned}$$

We make the following assumptions regarding the functions $b_i(x), i = 1, 2, 3$, $c(x)$:

$$b_i(x) \in C^3[G_T], \quad c(x) \in C[G_T].$$

Introduce notations

$$\|\psi\| = \max_{1\le i\le 3}\left[\max_{(x,t)\in D_T} |\psi_i(x,t)|\right],$$

$$\psi_0(x,t) = (\psi_{01}, \psi_{02}, \psi_{03}) = (v_0(x,t),\ w_0(x,t),\ \bar{w}_0(t)),$$

$$b_0 = \max_{1 \le i \le 3} \left\{ \|b_i\|_{C^3[G_T]} \right\}, \quad c_0 = \|c\|_{C^1[G_T]}.$$

In the future, we need estimates for the functions

$$\Sigma_i(x,\xi),\ i = 0,1,2,3,\ \left\langle \nabla_\xi \Sigma_j(x,\xi), Q\eta'(t,x,\varphi,z) \right\rangle,\ j = 0,1$$

for

$$(x,\xi) \in G_T \times G_T,\quad t \in [0,T],\quad \varphi \in [0,2\pi],\quad z \in [-1,1].$$

Therefore, we will carry out some calculations related to the definition of the dependence of the functions $\Sigma_i(x,\xi), i = 0,1,2,3$ on $b_i, i = 0,1,2, c$. Using (1.45), (1.46), and (1.48), we find

$$|\xi - x| L_\xi^* \sigma(x,\xi) = L_\xi^* \exp\left\{ \frac{1}{2} \int_0^1 \sum_{i=1}^3 b_i\left[\xi + \alpha(x-\xi)\right](x_i - \xi_i)\mathrm{d}\alpha \right\}$$

$$= \Delta_\xi \exp\left\{ \frac{1}{2} \int_0^1 \sum_{i=1}^3 b_i[\xi + \alpha(x-\xi)](x_i - \xi_i)\mathrm{d}\alpha \right\}$$

$$- \sum_{i=1}^3 \frac{\partial}{\partial \xi_i} \left\{ b_i(\xi) \exp\left\{ \frac{1}{2} \int_0^1 \sum_{i=1}^3 b_i[\xi + \alpha(x-\xi)](x_i - \xi_i)\mathrm{d}\alpha \right\}\right\}$$

$$+ c(\xi) \exp\left\{ \frac{1}{2} \int_0^1 \sum_{i=1}^3 b_i[\xi + \alpha(x-\xi)](x_i - \xi_i)\mathrm{d}\alpha \right\}$$

$$:= \exp\left\{ \frac{1}{2} \int_0^1 \sum_{i=1}^3 b_i\left[\xi + \alpha(x-\xi)\right](x_i - \xi_i)\mathrm{d}\alpha \right\} \Psi(x,\xi),$$

where

$$\Psi(x,\xi)$$

$$= \frac{1}{4}\left\{ \sum_{i=1}^3 \int_0^1 \left[\sum_{j=1}^3 b_{j\xi_i}\left[x + \alpha(\xi - x) \right](x_j - \xi_j)\alpha \right.\right.$$

$$\left.\left. - b_i\left[x + \alpha(\xi - x)\right] \right]\mathrm{d}\alpha \right\}^2$$

$$+ \frac{1}{2} \int_0^1 \left\{ \sum_{i=1}^3 \Delta_\xi b_i\left[x + \alpha(\xi - x)\right](x_i - \xi_i)\alpha^2 - 2\mathrm{div}_\xi b\left[x + \alpha(\xi - x)\right] \right\} \mathrm{d}\alpha$$

$$-\frac{1}{2}\sum_{i=1}^{3} b_i(\xi) \int_0^1 \left[\sum_{j=1}^{3} b_{j\xi_i}\left[x+\alpha(\xi-x)\right](x_j-\xi_j)\alpha - b_i\left[x+\alpha(\xi-x)\right] \right] d\alpha$$

$$-\sum_{i=1}^{3} b_{j\xi_i}(\xi) + c(\xi). \tag{1.61}$$

Here the low index ξ in the operators Δ, div indicates that they are applied by variable ξ. Consequently,

$$\Sigma_0(x,\xi) = \Sigma_1(x,\xi)\Psi(x,\theta), \quad \Sigma_2(x,\xi) = \Sigma_3(x,\xi)\Psi(x,\xi). \tag{1.62}$$

Further,

$$\frac{\partial}{\partial \xi_\nu}\Sigma(x,\xi)$$

$$= \frac{1}{2}\Sigma_0(x,\xi) \int_0^1 \left\{ \sum_{i=1}^{3} \left[b_{i\xi_\nu}(\alpha\xi)\alpha\xi_i + b_{i\xi_\nu}\left[x+\alpha(\xi-x)\right]\alpha(x_i-\xi_i) \right] \right.$$

$$\left. + b_\nu(\alpha\xi) - b_\nu\left[x+\alpha(\xi-x)\right] \right\} d\alpha + \Sigma_1(x,\xi)\frac{\partial}{\partial \xi_\nu}\Psi(x,\xi), \quad \nu = 1,2,3,$$

$$\frac{\partial}{\partial \xi_\nu}\Psi(x,\xi)$$

$$= \frac{1}{4}\sum_{i=1}^{3} \int_0^1 \left\{ \sum_{j=1}^{3} b_{j\xi_i}\left[x+\alpha(\xi-x)\right](x_j-\xi_j) - b_i\left[x+\alpha(\xi-x)\right] \right\} d\alpha$$

$$\times \sum_{i=1}^{3} \int_0^1 \left\{ \sum_{j=1}^{3} b_{j\xi_i\xi_\nu}\left[x+\alpha(\xi-x)\right](x_j-\xi_j) - b_{\nu\xi_i}\left[x+\alpha(\xi-x)\right]\alpha \right.$$

$$\left. - b_{i\xi_\nu}\left[x+\alpha(\xi-x)\right]\alpha \right\} d\alpha$$

$$+ \frac{1}{2} \int \left\{ \Delta_\xi b_{i\xi_\nu}\left[x+\alpha(\xi-x)\right]\alpha^3 - \Delta_\xi b_\nu\left[x+\alpha(\xi-x)\right]\alpha^2 \right.$$

$$\left. - 2\operatorname{div}_\xi b_{\xi_\nu}\left[x+\alpha(\xi-x)\right] \right\} d\alpha$$

$$-\frac{1}{2}\sum_{i=1}^{3} b_{i\xi_\nu}(\xi) \int_0^1 \left\{ \sum_{j=1}^{3} b_{j\xi_i\xi_\nu}\left[x+\alpha(\xi-x)\right](x_j-\xi_j)\alpha^2 \right.$$

$$\left. - b_{\nu\xi_i}\left[x+\alpha(\xi-x)\right]\alpha - b_{i\xi_\nu}\left[x+\alpha(\xi-x)\right]\alpha \right\} d\alpha$$

$$-\sum_{i=1}^{3} b_{j\xi_i\xi_\nu}(\xi) + c_{\xi_\nu}(\xi). \tag{1.63}$$

$$\Sigma_{1\xi_\nu}(x,\xi) = \Sigma_1(x,\xi)\int\limits_0^1 \left\{ \sum_{i=1}^{3} \left[b_{i\xi_\nu}(\alpha\xi)\alpha\xi_i + b_{i\xi_\nu}\left[x+\alpha(\xi-x)\right]\alpha(x_i-\xi_i) \right] \right.$$

$$\left. + b_\nu(\alpha\xi) + b_\nu\left[x+\alpha(\xi-x)\right] \right\} d\alpha.$$

From formulae (1.61) and (1.63) we obtain

$$|\Psi(x,\xi)| \le \frac{3b_0^2}{4}\left(\frac{3}{2}T+1\right)\left(\frac{3}{2}T+3\right)$$
$$+\frac{3b_0}{2}(T+4)+c_0 := \chi_0,\ (x,\xi)\in G_T\times G_T,$$
$$\left|\frac{\partial}{\partial\xi_\nu}\Psi(x,\xi)\right| \le \frac{3b_0^2}{4}(3T+2)(9T+4)+\frac{3b_0^2}{2}(T+2)+\frac{13b_0}{3}+c_0 := \chi_1,$$
$$\nu=1,2,3,\ \ (x,\xi)\in G_T\times G_T.$$

In view of these inequalities, we find the following estimates:

$$|\Sigma_1(x,\xi)| \le \frac{1}{16\pi^2}\exp(3b_0T) := \alpha_1,$$

$$|\Sigma_0(x,\xi)| \le |\Sigma_1(x,\xi)|\,|\Psi(x,\ t)| \le \alpha_1\chi_0 := \alpha_0,$$

$$|\Sigma_3(x,\xi)| \le \frac{1}{4\pi}\exp\left(\frac{3}{2}b_0T\right) := \alpha_3,$$

$$|\Sigma_2(x,\xi)| \le |\Sigma_3(x,\xi)|\,|\Psi(x,t)| \le \alpha_3\chi_0 := \alpha_2. \tag{1.64}$$

From relations (1.63) we get

$$\left|\left\langle\nabla_\xi\Sigma_j(x,\xi),\,Q\eta'\right\rangle\right| \le \max_{1\le i\le 3}\left\{\|\Sigma_{j\xi_i}\|\right\}\max_{1\le i\le 3}\left\{1,\ \|\eta_i'\|\right\}$$
$$\le \frac{3}{2}\alpha_0 b_0 T + \alpha_1\chi_1 := \beta_0,$$

$$j=0,1,\ (x,\xi)\in G_T\times G_T,\quad t\in[0,T],\ \varphi\in[0,2\pi],\ z\in[-1,1].$$

Fix the number $M>0$ and consider the ball

$$\Omega(\psi_0,M) = \left\{g(x,t)\in C\left[D_T\right] \big| \|\psi_0-g\| \le M\right\}.$$

It is easy to see that for $\psi\in\Omega(\psi_0,M)$ it takes place the estimate

$$\|\psi\| \le \|\psi_0\| + M \equiv N. \tag{1.65}$$

Let

$$T^* = \min_{1 \le i \le 6} \{\gamma_i\},$$

where $\gamma_i, i = 1, 2, 3, 4, 5, 6$ are positive roots of the following equations, respectively:

$$\gamma \left(\gamma^2 + \frac{2\alpha_2}{\alpha_3 N}\gamma + \frac{4\alpha_1}{\alpha_3 N}\right) = \frac{M}{2\pi\alpha_3 N^2},$$

$$\gamma \left[\left(\frac{4}{3} + \frac{\ln\gamma}{6\alpha_3}\right)\gamma^2 + \frac{8\alpha_2 + 6\beta_0}{3\alpha_3 N}\gamma + \frac{4\alpha_1}{\alpha_3 N}\right] = \frac{M}{2\pi\alpha_3 N^2}, \tag{1.66}$$

$$\gamma \left(\frac{4}{3}\gamma + \frac{8\alpha_2}{3\alpha_3 N} + \frac{\beta_0}{\pi\alpha_3 N^2} + \frac{2\beta_0}{\alpha_3 N}\right) = \frac{M}{4\pi^2\alpha_3 N^2},$$

$$\gamma \left(\gamma^2 + \frac{3\alpha_2}{4\alpha_3 N}\gamma + \frac{3\alpha_1}{\alpha_3 N}\right) = \frac{3}{2\pi\alpha_3 N},$$

$$\gamma \left[\left(1 + \frac{N}{3} + \frac{N}{6\alpha_3}\ln\gamma\right)\gamma^2 + \frac{\alpha_2 + \beta_0}{\alpha_3 N}\gamma + \frac{\alpha_1}{\alpha_3 N}\right] = \frac{1}{4\pi\alpha_3 N}, \tag{1.67}$$

$$\gamma \left(4\gamma + \frac{4\alpha_2 + 3\beta_0}{\alpha_3 N}\right) = \frac{3}{8\pi^2\alpha_2 N}.$$

Theorem 1.4 *Let all above-mentioned conditions about functions $b_i(x)$, $i = 1, 2, 3$, $c(x)$ be fulfilled. Besides, $f(t) \in C^1[0, T]$, and the following matching condition is satisfied:*

$$c(0) + \sum_{i=1}^{3} \left[\frac{3}{4} b_i^2(0) - 2 b_{i x_i}(0)\right] = 8\pi f(0). \tag{1.68}$$

Then in the domain $D_T, T \in (0, T^*)$, there exists a unique continuous solution of Eq. (1.59) belonging to a ball $\Omega(\psi_0, M)$.

For *proof* of Theorem 1.4, we use the following inequalities:

$$|A[1, 1]| \le \frac{\pi}{6}(t - |x|)\,[3\alpha_2(t + |x|) + \alpha_3(t - |x|)(t + 2|x|)]$$

$$|B[1, 1]| \le 2\pi\alpha_3\sqrt{\left(t^2 - |x|^2\right)^3},$$

$$|E[1]| \le 4\pi\alpha_2 t\sqrt{t^2 - |x|^2}, \tag{1.69}$$

$$\left| A\left[\psi_3(t), \frac{1}{\sqrt{t^2-|x|^2}}\right]\right| \leq \frac{3\pi}{4}\left[\sqrt{t^2-|x|^2}\left(\alpha_2 + \frac{\alpha_3\,\|\psi_3\|\,t}{2}\right)\right.$$
$$\left. + \frac{\|\psi_3\|\,x^2|}{2}\ln\frac{|x|}{t+\sqrt{t^2-|x|^2}}\right],$$

which easily follow from formulas (1.53) for all $(x,t) \in D_T$.

Let us prove, for example, the validity of the first of inequalities (1.69). By definition

$$A[1,1] = 2\int\limits_{|x|}^{t} \frac{d\tau}{\tau^2-|x|^2} \int\limits_{S(x,\tau)} r^3(\theta,\tau)\Big[\Sigma_2(x,\xi) + \Sigma_3(x,\xi)(t-\tau)\Big]d\omega,$$

we use the formula

$$r^2(\theta,\tau)\sin\theta\,d\theta = -\frac{\tau^2-|x|^2}{2|x|}dr,$$

which follows from the equation of the family of confocal ellipsoids $S(x,\tau)$ in the polar coordinate system:

$$r(\theta,\tau) = \frac{\tau^2-|x|^2}{2(\tau-|x|\cos\theta)}.$$

We proceed to the integration variable r and, taking into account the estimates from (1.64) for Σ_2, Σ_3, obtain

$$|A[1,1]| \leq \frac{2\pi}{|x|}\int\limits_{|x|}^{t}(\alpha_2+\alpha_3(t-\tau))\,d\tau \int\limits_{\frac{\tau-|x|}{2}}^{\frac{\tau+|x|}{2}} r\,dr$$
$$= \frac{\pi}{6}(t-|x|)\left[3\alpha_2(t+|x|)+\alpha_3(t-|x|)(t+2|x|)\right].$$

We show that for any $T \in (0,T^*)$ the operator U maps the ball $\Omega(\psi_0, M)$ into itself. Obviously, the operator U transforms $\psi(x,t) \in C[D_T]$ into functions also belonging to space $C[D_T]$. In addition, composing the norm of differences and using formulas (1.60), inequalities (1.64), (1.65), and (1.69), we find

$$\|U_1\psi - \psi_{01}\| \leq 2\pi\alpha_1\,\|\psi_3\|\,(t-|x|) + \|\psi_1\|\,|A\left[\|\psi_3\|,1\right]|$$
$$\leq \frac{\pi\alpha_3}{2}N^2T^3 + \pi\alpha_2 NT^2 + 2\pi\alpha_1 NT,$$

$$\|U_2\psi - \psi_{02}\| \leq 8\pi\alpha_1\,\|\psi_3\|\,\frac{t^2-|x|^2}{+}4\pi\beta_0\,\|\psi_3\|\,(t^2+|x|^2)$$
$$+\|\psi_1\|\,\|\psi_3\|\,B[1,1]| + \|\psi_1\|\,|E[1]|$$

$$+\|\psi_2\|\left|A\left[\|\psi_3\|,\frac{1}{\sqrt{t^2-|x|^2}}\right]\right|\sqrt{t^2-|x|^2}$$
$$\leq 2\pi\alpha_3 N^2T^3+\frac{4\pi NT}{3}\left[T\left(\alpha_2+\frac{\alpha_3 NT}{2}\right)+\frac{NT^2}{4}\ln T\right]$$
$$+4\pi\alpha_2 NT^2+4\pi\beta_0 NT^2+8\pi\alpha_1 NT,$$
$$\|U_3\psi-\psi_{03}\|\leq 4\pi\beta_0 t+8\pi^2\beta_0\|\psi_3\|\,t+\frac{2\pi}{t}\Big\{\|\psi_1\|\,\|\psi_3\|\,B[1,\ 1]|$$
$$+\|\psi_1\|\,|E[1]|+\|\psi_2\|\left|A\left[\|\psi_3\|,\frac{1}{\sqrt{t^2-|x|^2}}\right]\right|\sqrt{t^2-|x|^2}\Big\}_{x=0}$$
$$\leq 4\pi^2\alpha_3 N^2T^2+\frac{8\pi^2 NT}{3}\left(\alpha_2+\frac{\alpha_3 N}{2}\right)+8\pi^2\alpha_2 NT$$
$$+8\pi^2\beta_0 NT+4\pi\beta_0 T.$$

For all $T\in(0,T^*)$ from these inequalities in view of (1.66) it follows $\|U\psi-\psi_0\|\leq M$, i.e., $U\psi\in\Omega(\psi_0,M)$.

Now suppose that $\psi^{(1)}$, $\psi^{(2)}$ are any two elements of $\Omega(\psi_0,M)$, $T<T^*$. Then, using auxiliary inequalities of the following type

$$\left|\psi_i^{(1)}\psi_j^{(1)}-\psi_i^{(2)}\psi_j^{(2)}\right|\leq\left|\psi_i^{(1)}\right|\left|\psi_j^{(1)}-\psi_j^{(2)}\right|$$
$$+\left|\psi_j^{(2)}\right|\left|\psi_i^{(1)}-\psi_i^{(2)}\right|\leq 2N\left\|\psi^{(1)}-\psi^{(2)}\right\|,$$

we obtain

$$\left\|U_1\psi^{(1)}-U_1\psi^{(2)}\right\|\leq 2\pi\alpha_1(t-|x|)\left\|\psi_3^{(1)}-\psi_3^{(2)}\right\|$$
$$+\left|A\left[\psi_3^{(1)},\psi_1^{(1)}\right]-A\left[\psi_3^{(2)},\ \psi_1^{(2)}\right]\right|$$
$$\leq\left(2\pi\alpha_1 T+\frac{\pi}{2}\alpha_2T^2+\frac{2\pi}{3}\alpha_3NT^3\right)\left\|\psi^{(1)}-\psi^{(2)}\right\|,$$

$$\left\|U_2\psi^{(1)}-U_2\psi^{(2)}\right\|\leq 8\pi\alpha_1\sqrt{t^2-|x|^2}\left\|\psi_3^{(1)}-\psi_3^{(2)}\right\|$$
$$+4\pi\beta_0(t^2-|x|^2)\left\|\psi_3^{(1)}-\psi_3^{(2)}\right\|+\left|\psi_1^{(1)}\psi_3^{(1)}-\psi_1^{(2)}\psi_3^{(2)}\right|\,|B[1,1]|$$
$$+\left\|\psi_1^{(1)}-\psi_1^{(2)}\right\|\,|E[1]|+\sqrt{t^2-|x|^2}\,||\,A\left[\psi_3^{(1)},\frac{\psi_2^{(1)}}{\sqrt{t^2-|x|^2}}\right]$$
$$-A\left[\psi_3^{(2)},\frac{\psi_2^{(2)}}{\sqrt{t^2-|x|^2}}\right]\Bigg|\leq\Bigg[8\pi\alpha_1T+4\pi\beta_0T^2+4\pi\alpha_2T^2$$
$$+4\pi\alpha_3NT^3+\frac{8\pi N^2T^2}{3}\left(\alpha_2+\frac{\alpha_3T}{2}+\frac{T}{4}\ln T\right)\Bigg]\left\|\psi^{(1)}-\psi^{(2)}\right\|,$$

$$\left\|U_3\psi^{(1)} - U_3\psi^{(2)}\right\| \le 8\pi^2\beta_0 t \left\|\psi_3^{(1)} - \psi_3^{(2)}\right\|$$
$$+\frac{2\pi}{t}\Bigg\{\left|\psi_1^{(1)}\psi_3^{(1)} - \psi_1^{(2)}\psi_3^{(2)}\right| |B[1,1]| + \left\|\psi_1^{(1)} - \psi_1^{(2)}\right\| |E[1]|$$
$$+\sqrt{t^2-|x|^2}\,\left|\left|A\left[\psi_3^{(1)}, \frac{\psi_2^{(1)}}{\sqrt{t^2-|x|^2}}\right] - A\left[\psi_3^{(2)}, \frac{\psi_2^{(2)}}{\sqrt{t^2-|x|^2}}\right]\right|\right|\Bigg\}_{x=0}$$
$$\le \left[8\pi^2\alpha_2 T + 8\pi^2\beta_0 T + 8\pi^2\alpha_3 N T^2 + \frac{8\pi^2 T}{3}(\alpha_2 + \alpha_3 N T)\right.$$
$$\times \left\|\psi^{(1)} - \psi^{(2)}\right\|.$$

If we take into account (1.67), then from the last inequalities one gets

$$\left\|U\psi^{(1)} - U\psi^{(2)}\right\| \le \frac{T}{T^*}\left\|\psi^{(1)} - \psi^{(2)}\right\|.$$

This implies that for any $T \in (0, T^*)$ the operator U implements a contraction mapping of the ball $\Omega(\psi_0, M)$ onto itself. Then, by the Banach theorem in the ball $\Omega(\psi_0, M)$, there exists a unique continuous solution of Eq. (1.59). Therefore, solving the system of Eqs. (1.55)–(1.57), for example, by the method of successive approximations (from the Banach theorem it follows its convergence to the solution), we uniquely construct in the domain D_T for $T \in (0, T^*)$ the functions $v(x,t)$, $w(x,t)$, $k(t)$. By that it is determined the solution of the inverse problem, that is, the function $k(t)$ on the interval $[0, T]$.

Theorem 1.5 *Under the conditions of Theorem 1.4, the function $k(t) \in C[0, T]$ is uniquely determined by the information (1.44).*

The *proof* of this theorem is carried out by contradiction. Then, by virtue of Theorem 1.4, there exists $T \in (0, T^*)$, such that

$$k_1(t) = k_2(t), \quad t \in [0, T]. \tag{1.70}$$

We denote by T_0 the exact upper bound of all possible numbers $T \in (0, T^*)$, for which equality (1.67) holds. Now we consider a similar with (1.41), (1.42), and (1.44) problem, in which the Cauchy data are given on the hyperplane $t = T_0$. Since

$$u_1\big|_{t<T_0} \equiv u_2\big|_{t<T_0} \equiv 0,\ u_1\big|_{x=0} = u_2\big|_{x=0} = f(t),$$

then, by virtue of Theorem 1.4, there is a number $T^{**} \in (T_0, T)$, such that $k_1(t) = k_2(t)$. And this contradicts the definition of T_0. The resulting contradiction proves the validity of Theorem 1.5.

1.2.3 The Stability Estimate

Let T^* be from the statement of Theorem 1.4. Consider the set $\Phi(T)$, $T \in (0, T^*)$ of functions (b, c, f) for which $b(x) = (b_1, b_2, b_3) \in C^3[G(T)]$, $c(x) \in C^1[G(T)]$, $f(t) \in C^1[0, T]$ and the relation (1.68) is fulfilled.

Theorem 1.6 *Let* $(b, c, f) \in \Phi(T)$, $(\bar{b}, \bar{c}, \bar{f}) \in \Phi(T)$. *Then for the corresponding solutions* $k(t), \bar{k}(t)$ *of the inverse problem (1.41), (1.42), and (1.44), the following estimate is valid:*

$$\|k(t) - \bar{k}(t)\|_{C[0,T]} \leq d\rho, \tag{1.71}$$

where

$$\rho = \|b - \bar{b}\|_{C^3[G(T)]} + \|c - \bar{c}\|_{C^1[G(T)]} + \|f - \bar{f}\|_{C^1[0,T]};$$

the constant d *depends only on* T, b_0, c_0, k_0

$$\bar{b}_0 = \max_{1 \leq i \leq 3} \left\{ \|\bar{b}_i\|_{C^3[G(T)]} \right\}, \quad \bar{c}_0 = \|\bar{c}\|_{C^1[G(T)]}, \quad k_0 = \|k\|_{C[0,T]}.$$

Proof Let v, $\bar{v}$ denote the solutions of the integral equation (1.50) corresponding to the functions (b, c, k), $(\bar{b}, \bar{c}, \bar{k})$, respectively, and we will denote the difference between the two functions, by the same denotation only with the upper sign $\sim$. For example, $\tilde{v} = v - \bar{v}$, $\tilde{k} = k - \bar{k}$ etc. Then, from Eqs. (1.55)–(1.57) it is easy to get the following system of equalities:

$$\tilde{v}(x,t) = \tilde{v}_0(x,t) + \frac{1}{2} \int_{|x|}^{t} \bar{k}(t-\tau) \int_0^{2\pi} d\varphi \int_{-1}^{1} \tilde{\Sigma}_1(x,\xi) dz \, d\tau$$

$$+ \frac{1}{2} \int_{|x|}^{t} \tilde{k}(t-\tau) \int_0^{2\pi} d\varphi \int_{-1}^{1} \Sigma_1(x,\xi) dz \, d\tau$$

$$+ 2 \int_{|x|}^{t} \frac{d\tau}{\tau^2 - |x|^2} \int_{S(x,\tau)} r^3(\theta,\tau) \Sigma_3(x,\xi) \int_0^{t-\tau} \tilde{k}(\nu) v(\xi, t-\tau+r-\nu) d\nu \, d\omega$$

$$+ A[\bar{k}(t), \tilde{v}(x,t)] + A^{[\tilde{\Sigma}_2, \tilde{\Sigma}_3]}[\bar{k}(t), \bar{v}(x,t)], \tag{1.72}$$

$$\tilde{w}(x,t) = \tilde{w}_0(x,t) + 4\pi \sqrt{t^2 - |x|^2} \tilde{k}(t - |x|) \int_{-1}^{1} \Sigma_1 \left(x, 0, 0, \frac{z+1}{2} |x| \right) dz$$

$$+ 4\pi \sqrt{t^2 - |x|^2} \bar{k}(t - |x|) \int_{-1}^{1} \tilde{\Sigma}_1 \left(x, 0, 0, \frac{z+1}{2} |x| \right) dz$$

$$+\sqrt{t^2-|x|^2}\int\limits_0^{\sqrt{t^2-|x|^2}} \tilde{k}\left(t-\sqrt{\beta^2+|x|^2}\right)\mathrm{d}\beta$$

$$\times\int\limits_0^{2\pi}\mathrm{d}\varphi\int\limits_{-1}^{1}\left\langle\nabla_\xi\Sigma_1,\,Q\eta'\right\rangle\Big|_{\tau=\sqrt{\beta^2-|x|^2}}\mathrm{d}z$$

$$+\sqrt{t^2-|x|^2}\int\limits_0^{\sqrt{t^2-|x|^2}} \bar{k}\left(t-\sqrt{\beta^2-|x|^2}\right)\mathrm{d}\beta$$

$$\times\int\limits_0^{2\pi}\mathrm{d}\varphi\int\limits_{-1}^{1}\left\langle\nabla_\xi\tilde{\Sigma}_1,\,Q\eta'\right\rangle\Big|_{\tau=\sqrt{\beta^2-|x|^2}}\mathrm{d}z$$

$$+2\int\limits_{|x|}^{t}\frac{\mathrm{d}\tau}{\tau^2-|x|^2}\int\limits_{S(x,\tau)} r^3(\theta,\tau)\Sigma_3(x,\xi)$$

$$\times\int\limits_0^{t-\tau}\tilde{k}(\nu)\frac{w(\xi,t-\tau+r-\nu)\mathrm{d}\nu\,dw}{\sqrt{(t-\tau+r-\nu)^2-r^2}}+B\left[\tilde{k}(t),v(x,t)\right]$$

$$+B\left[\bar{k}(t),\tilde{v}(x,t)\right]+B^{\tilde{\Sigma}_3}\left[\bar{k}(t),\bar{v}(x,t)\right]+E\left[\tilde{v}(x,t)\right]$$

$$+E^{\tilde{\Sigma}_2}\left[\bar{v}(x,t)\right]+\sqrt{t^2-|x|^2}A\left[\tilde{k}(t),\frac{\bar{w}(x,t)}{\sqrt{t^2-|x|^2}}\right]$$

$$+\sqrt{t^2-|x|^2}A^{(\tilde{\Sigma}_2,\tilde{\Sigma}_3)}\left[\bar{k}(t),\frac{\bar{w}(x,t)}{\sqrt{t^2-|x|^2}}\right]. \tag{1.73}$$

$$\tilde{k}(t)=\tilde{F}(t)+\tilde{\bar{w}}_0(t)$$

$$-2\pi\int\limits_0^{t}\tilde{k}(t-\beta)\mathrm{d}\beta\int\limits_0^{2\pi}\mathrm{d}\varphi\int\limits_{-1}^{1}\left\langle\nabla_\xi\Sigma_1,\,Q\eta'\right\rangle|_{\tau=\beta,x=0}\mathrm{d}z$$

$$-2\pi\int\limits_0^{t}\bar{k}(t-\beta)\mathrm{d}\beta\int\limits_0^{2\pi}\mathrm{d}\varphi\int\limits_{-1}^{1}\left\langle\nabla_\xi\tilde{\Sigma}_1,\,Q\eta'\right\rangle|_{\tau=\beta,x=0}\mathrm{d}z$$

$$-\int\limits_0^{t}\tau\mathrm{d}\tau\int\limits_0^{2\pi}\int\limits_0^{\pi}\int\limits_0^{t-\tau}\tilde{k}(\nu)\left[\frac{\Sigma_3(x,\xi)w(\xi,t-\tau+r-\nu)\mathrm{d}\nu\,dw}{\sqrt{(t-\tau+r-\nu)^2-r^2}}\right]_{x=0}$$

$$-\frac{2\pi}{t}\Big\{B\left[\bar{k}(t), v(x,t)\right] + B\left[\bar{k}(t), \tilde{v}(x,t)\right] + B^{\tilde{\Sigma}_3}\left[\bar{k}(t), \bar{v}(x,t)\right]$$
$$+ E\left[\tilde{v}(x,t)\right] + E^{\tilde{\Sigma}_2}\left[\bar{v}(x,t)\right] + \sqrt{t^2-|x|^2}$$
$$\times\left(A\left[\tilde{k}(t), \frac{\tilde{w}(x,t)}{\sqrt{t^2-|x|^2}}\right] + A^{(\tilde{\Sigma}_2,\tilde{\Sigma}_3)}\left[\bar{k}(t), \frac{\bar{w}(x,t)}{\sqrt{t^2-|x|^2}}\right]\right)\Big\}_{x=0}. \tag{1.74}$$

Here the upper index in the operators A, B, E indicates that functions $\Sigma_i, i = 1, 2, 3$, included in the definition of these operators replaced by those functions that are listed in the upper index. For example, $A^{(\tilde{\Sigma}_2,\tilde{\Sigma}_3)}[\cdot,\cdot]$ is the operator defined first of relations (1.53), in which the functions Σ_2 and Σ_3 are replaced with $\tilde{\Sigma}_2$ and $\tilde{\Sigma}_3$, respectively.

In the study of the direct problem (1.41) and (1.42) was obtained the estimate

$$|v(x,t)| \le \sum_{n=0}^{\infty}\sum_{j=0}^{n+1}(2\pi M)^{n+1} k_0^j \frac{T^{2n+j}}{n!j!} := \mu_0, \tag{1.75}$$

where $M = \max\limits_{0\le i\le 3}\alpha_i$. Such estimate, in view of $\left(\bar{b}, \bar{c}, \bar{f}\right) \in \Phi(T)$, is fulfilled as well for $|\bar{v}(x,t)|$:

$$|\bar{v}(x,t)| \le \mu_0,\ (x,t) \in D(T).$$

Acting in the same way, as in obtaining the estimate (1.75), from Eq. (1.56) it is not difficult to derive inequalities

$$|w(x,t)| \le \mu_1,\ |\bar{w}(x,t)| \le \mu_1,\ (x,t) \in D_T,$$

where the constant μ_1 depends only on $T, b_0, c_0, k_0, \bar{b}_0, \bar{c}_0, \bar{k}_0$. We represent the domain D_T in the form

$$D_T = \left\{(x,t)\middle|0 \le |x| \le \frac{T}{2} - \left|t - \frac{T}{2}\right|,\ 0 \le t \le T\right\}.$$

Let

$$\omega(t) = \max\Bigg[\max_{0\le|x|\le T/2-|t-T/2|}|v(x,t) - \bar{v}(x,t)|,$$

$$\max_{0\le|x|\le T/2-|t-T/2|}|w(x,t) - \bar{w}(x,t)|, \max\left|k(t) - \bar{k}(t)\right|\Bigg],\quad t \in [0,T].$$

Let us estimate the functions in the system (1.72)–(1.74) for $(x,t) \in D_T$ through the value ρ defined in Theorem 1.6. Consider Eq. (1.72). We will estimate the first term in this equation. By virtue of the first equality of (1.54), we have

$$|\tilde{v}_0(x,t)| \le \frac{1}{2} \max_{(x,t)\in D_T} \left| \int_0^{2\pi} \int_{-1}^{1} \tilde{\Sigma}_0(x,\xi) \mathrm{d}z \right| \le 2\pi \max_{(x,\xi)\in[G_T\times G_T]} \left|\tilde{\Sigma}_0(x,\xi)\right|, \tag{1.76}$$

where

$$\tilde{\Sigma}_0(x,\xi) = \Sigma_0(x,\xi) - \bar{\Sigma}_0(x,\xi);$$

$\Sigma_0(x,\xi)$ is defined by the first equality of (1.48), and

$$\bar{\Sigma}_0(x,\xi) = \frac{1}{16\pi^2}|\xi|\bar{\sigma}(\xi)|\xi - x|\bar{L}^*_\xi \bar{\sigma}(x,\xi),$$

the function $\bar{\sigma}(x,\xi)$ has form (1.45) with $\bar{b}_i, \bar{\sigma}(x) = \bar{\sigma}(x,0)$, $\bar{L}^*_\xi$ is the operator L^*_ξ, with coefficients $\bar{b}_i$, $i = 1, 2, 3$, $\bar{c}$. By virtue of equalities (1.45) and (1.62), we have

$$\begin{aligned} \tilde{\Sigma}_0(x,\xi) &= \Sigma_0(x,\xi) - \bar{\Sigma}_0(x,\xi) = \Sigma_0(x,\xi) \\ &= \Sigma_1(x,\xi)\Psi(x,\xi) - \bar{\Sigma}_1(x,\xi)\bar{\Psi}(x,\xi) \\ &= \tilde{\Sigma}_1(x,\xi)\Psi(x,\xi) + \bar{\Sigma}_1(x,\xi)\tilde{\Psi}(x,\xi), \end{aligned} \tag{1.77}$$

where $\bar{\Sigma}_1(x,\xi)$, $\bar{\Psi}(x,\xi)$ are defined by (1.48) and (1.61), respectively, in which functions b_i, $i = 1, 2, 3$, c were replaced by $\bar{b}_i$, $i = 1, 2, 3$, $\bar{c}$. Estimates of the functions $\tilde{\Sigma}_1(x,\xi)$, $\tilde{\Psi}(x,\xi)$ are carried out as follows:

$$\begin{aligned} \left|\tilde{\Sigma}_1(x,\xi)\right| &= \left| \frac{1}{16\pi^2} \exp\left\{ \frac{1}{2} \int_0^1 \sum_{i=1}^{3} [b_i(\alpha\xi)\xi_i + b_i(\xi + \alpha(x-\xi))(x_i - \xi_i)]\,\mathrm{d}\alpha \right\} \right. \\ &\quad \left. - \exp\left\{ \frac{1}{2} \int_0^1 \sum_{i=1}^{3} \left[\bar{b}_i(\alpha\xi)\xi_i + \bar{b}_i(\xi + \alpha(x-\xi))(x_i - \xi_i)\right] \mathrm{d}\alpha \right\} \right| \\ &\le \frac{3T}{32\pi^2} \exp\left[3T \max(b_0, \bar{b}_0)\right] \left\|b - \bar{b}\right\|, \quad (x,\xi) \in G_T \times G_T. \end{aligned}$$

Using formula (1.61), similarly to the previous one, we find

$$\left|\tilde{\Sigma}(x,\xi)\right| = \left|\Psi(x,\xi) - \bar{\Psi}(x,\xi)\right| \le \psi_0 \left(\|b - \bar{b}\| + \|c - \bar{c}\|\right), \quad (x,\xi) \in G_T \times G_T,$$

where the constant ψ_0 depends only on T, b_0, c_0, $\bar{b}$, $\bar{c}$. Thus, from (1.77) and (1.76) follows the estimate of the first term on the right side of Eq. (1.72)

$$|\tilde{v}_0(x,t)| \le \psi_1 \rho, \quad (x,t) \in D_T, \tag{1.78}$$

where the constant ψ_1 also depends only on $T, b_0, c_0, \bar{b}, \bar{c}$. It is quite easy to obtain the estimates of the second and last terms in (1.72) in terms of ρ

$$\left| \frac{1}{2} \int\limits_{|x|}^{t} \tilde{k}(t-\tau) \int\limits_{0}^{2\pi} d\varphi \int\limits_{-1}^{1} \Sigma_1(x,\xi)dz\, d\tau \right| \leq \psi_1 \rho, \quad (x,t) \in D_T, \tag{1.79}$$

$$\left| A^{[\tilde{\Sigma}_2, \tilde{\Sigma}_3]} \left[\tilde{k}(t), \tilde{v}(x,t) \right] \right| \leq \psi_2 \rho, \ (x,t) \in D_T. \tag{1.80}$$

Here the constants ψ_1, ψ_2 depend on a set of numbers $T, b_0, c_0, \bar{b}_0, \bar{c}_0, \bar{k}_0$. The other part of terms in (1.72) are estimated by the following integral expression:

$$\psi_3 \int\limits_0^t \omega(\tau)d\tau, \quad t \in [0,T],$$

where the constant ψ_3 depends on the same numbers as ψ_1, ψ_2. Thus, from the above reasoning it follows that the function $v(x,t)$ admits the estimate

$$|\tilde{v}(x,t)| \leq \psi_4 \rho + \psi_3 \int\limits_0^t \omega(\tau)d\tau, \quad t \in [0,T], \tag{1.81}$$

where the constant ψ_4 depends on the same numbers as ψ_3.

Similar arguments for Eqs. (1.73) and (1.74) lead to the following estimates for the functions $\tilde{w}, \tilde{k}$:

$$|\tilde{w}(x,t)| \leq \psi_5 \rho + \psi_6 \int\limits_0^t \omega(\tau)d\tau, \quad t \in [0,T], \tag{1.82}$$

$$\left| \tilde{k}(x,t) \right| \leq \psi_7 \rho + \psi_8 \int\limits_0^t \omega(\tau)d\tau, \quad t \in [0,T], \tag{1.83}$$

where constants $\psi_i,\ i = 5, 6, 7, 8$, depend on a set of numbers $T, b_0, c_0, \bar{b}_0, \bar{c}_0, k_0, \bar{k}_0$. From relations (1.81)–(1.83) it follows that $\omega(t)$ satisfies the integral inequality

$$\omega(t) \leq \psi_{00} \rho + \psi_{11} \int\limits_0^t \omega(\tau)d\tau, \quad t \in [0,T],$$

in which ψ_{00} max(ψ_4, ψ_5, ψ_7), ψ_{11} max(ψ_3, ψ_6, ψ_8). Using Gronwall inequality, we get the estimate

$$|\omega(t)| \le \psi_{00} \exp(\psi_{11} t)\rho, \quad t \in [0, T]. \tag{1.84}$$

From inequality (1.84) follows the estimate (1.71), if we set $d = \psi_{00} \exp(\psi_{11} T)$.

1.3 Inverse Problem for the System of Thermoelasticity Equations for a Vertically Inhomogeneous Cohesionless Medium with Memory

For $x = (x_1, x_2, x_3) \in \mathbb{R}^3$, $t \in \mathbb{R}$, and $x_3 > 0$, consider the following system of cohesionless dynamic thermoelasticity differential equations with memory:

$$\rho \frac{\partial^2 u_i}{\partial t^2} = \sum_{j=1}^{3} \frac{\partial \sigma_{ij}}{\partial x_j} + \int_0^t h(\tau) u_i(x, t-\tau) d\tau, \quad i = 1, 2, 3, \tag{1.85}$$

$$\frac{\partial H}{\partial t} = k \Delta H, \tag{1.86}$$

where $H(x, t)$ is the temperature increment, $u(x, t) = (u_1(x, t), u_2(x, t), u_3(x, t))$ is the displacement vector, $h(t)$ is a scalar function characterizing the medium memory, and σ_{ij} are the stresses, which admit the representation [5] (p. 20) by Hooke's law:

$$\sigma_{ij} = \mu \left(\frac{\partial u_i}{\partial x_j} + \frac{\partial u_j}{\partial x_i} \right) + \delta_{ij} \left[\lambda \operatorname{div} u - (3\lambda + 2\mu) \int_0^{H(x,t)} \alpha(y) dy \right]. \tag{1.87}$$

Here δ_{ij} is the Kronecker delta, and $\operatorname{div} u = \sum_{i=1}^{3} \partial u_i / \partial x_i$. The system of differential equations (1.85) and (1.86) contains the temperature conduction coefficient k, the medium density ρ, the Lam'e parameters λ, and μ, and the medium thermal expansion coefficient α. In the sequel, we assume that $k > 0$ is constant and that $\rho = \rho(x_3)$, $\mu = \mu(x_3)$, and $\lambda = \lambda(x_3)$ are functions of one variable x_3 and satisfy the conditions $\rho(x_3) > 0$, $\mu(x_3) > 0$, and $\lambda(x_3) + 2\mu(x_3) > 0$.

Consider system (1.85) and (1.86) with the following initial and boundary conditions:

$$u_i \mid_{t<0} \equiv 0, i = 1, 2, 3, \tag{1.88}$$

$$H \mid_{t<0} \equiv 0, \tag{1.89}$$

$$\sigma_{3j} \mid_{x_3=+0} = \nu \delta'(t), \nu = \text{colon}(0, 0, 1), \quad j = 1, 2, 3, \tag{1.90}$$

$$\left(\frac{\partial H}{\partial x_3} - \gamma H\right)\Big|_{x_3=+0} = -\gamma(T_1 - T_0)\theta_1(t), \tag{1.91}$$

where T_1, T_0, and γ are some constants, $T_1 > T_0, \gamma > 0$; $\theta_1(t) = t\theta(t)$, $\theta(t) = 1$ for $t \geq 0$, $\theta(t) = 0$ for $t < 0$, and $\delta'(t)$ is the derivative of the Dirac delta function. The boundary condition (1.91) simulates a heat shock on the surface of the half-space $x_3 > 0$. Whereby the temperature on the boundary increases from the initial temperature T_0 to the value T_1. After that, there is convective heat exchange between the surface $x_3 = 0$ and the medium filling the half-space $x_3 > 0$.

The problem of finding the displacement vector $u(x, t)$ and the temperature increment $H(x, t)$, satisfying relations (1.85)–(1.91) (in the generalized sense) for given functions $\alpha(y)$, $h(t)$, $\rho(x_3)$, $\lambda(x_3)$, and $\mu(x_3)$ and given constants k, γ, T_0, and T_1 will be called the *direct problem*.

Let us state the *inverse problem*: find the function $h(t)$, $t \in [0, T]$, $T > 0$, occurring in Eq. (1.85) from the known additional condition

$$u_3(0, t) = g(t), \quad t \in [0, T] \tag{1.92}$$

of the solution of the direct problem. Thus, the functions $\alpha(y)$, $\rho(x_3)$, $\lambda(x_3)$, and $\mu(x_3)$ and the constants k, γ, T_0, and T_1 are assumed to be given. A study of various statements of inverse problems of determining the functions $\rho(x_3)$, $\lambda(x_3)$, $\mu(x_3)$, and $\alpha(y)$ from the system of elasticity equations can be found in the monograph [6]. In the following, we assume that the functions α, ρ, λ, and μ are sufficiently smooth. The order of smoothness will be specified below.

Problem (1.86), (1.89), and (1.91) can be viewed as a separate initial–boundary value problem for the heat equation. One can readily verify that it has the solution

$$\begin{aligned} H(z, t) &= \theta(t)(T_1 - T_0) \\ &\times \int_0^t \left[\operatorname{erfc}\left(\frac{z}{2\sqrt{k\tau}}\right) - \exp(\gamma z + \gamma^2 k\tau)\operatorname{erfc}\left(\frac{z}{2\sqrt{k\tau}} + \gamma\sqrt{k\tau}\right)\right] d\tau, \end{aligned}$$

where

$$\operatorname{erfc}(z) = 1 - \operatorname{erf}(z), \quad \operatorname{erf}(z) = (2/\sqrt{\pi}) \int_0^z \exp(-\xi^2) d\xi, \quad z = x_3.$$

Therefore, Eqs. (1.85), (1.88), and (1.90) for the function $H(z, t)$ are equivalent to the relations

$$\begin{aligned} \rho \frac{\partial^2 u_3}{\partial t^2} &= \frac{\partial}{\partial z}\left[(\lambda(z) + 2\mu(z))\frac{\partial u_3}{\partial z} - (3\lambda(z) + 2\mu(z))R(H(z, t))\right] \\ &+ \int_0^t h(\tau)u_3(z, t - \tau)d\tau, \end{aligned} \tag{1.93}$$

$$u_3\Big|_{t<0} \equiv 0, \tag{1.94}$$

$$\frac{\partial u_3}{\partial z}\,|_{z=+0}= \frac{\delta'(t)}{\lambda(0)+2\mu(0)} + \frac{3\lambda(0)+2\mu(0)}{\lambda(0)+2\mu(0)}R(H(0,t)), \tag{1.95}$$

$$u_1 = u_2 = 0, \quad u_3 = u_3(z,t), \quad z = x_3,$$

where $R(s) = \int_0^s \alpha(y)\mathrm{d}y$.

By replacing the variable x_3 by y and introducing a new function $w(y,t)$ instead of $u_3(z,t)$ by the formulas

$$y = \tau(z) \equiv \int_0^z \frac{\mathrm{d}\xi}{v(\xi)}, \quad v(\xi) := \sqrt{\frac{\lambda(z)+2\mu(z)}{\rho(z)}},$$

$$w(y,t) := u_3(\tau^{-1}(y),t)\frac{1}{S(y)}, \quad S(y) := \sqrt{\frac{a(y)}{a(0)}}, \tag{1.96}$$

$$a(y) := \left\{\left[\lambda(\tau^{-1}(y)) + 2\mu(\tau^{-1}(y))\right]\rho(\tau^{-1}(y))\right\}^{-\frac{1}{2}},$$

where $z = \tau^{-1}(y)$ is the inverse function of $y = \tau(z)$, we rewrite relations (1.93)–(1.95) and (1.92) in the form

$$\frac{\partial^2 w}{\partial t^2} = \frac{\partial^2 w}{\partial y^2} + q(y)w - \sqrt{a(y)a(0)}\frac{\partial}{\partial y}\left[(3\tilde{\lambda}(y)+2\tilde{\mu}(y))R(\tilde{H}(y,t))\right]$$
$$+\frac{1}{\tilde{\rho}(y)}\int_0^t h(\tau)w(y,t-\tau)\mathrm{d}\tau, \tag{1.97}$$

$$w\,|_{t<0}\equiv 0, \tag{1.98}$$

$$\frac{\partial w}{\partial y}\,|_{y=0}= a(0)\delta'(t) + a(0)\left[3\lambda(0)+2\mu(0)\right]R(H(0,t)) - \frac{a'(0)g(t)}{2a(0)}, \tag{1.99}$$

$$w\,|_{y=0}= g(t), \tag{1.100}$$

where

$$q(y) := \frac{S'(y)}{S(y)} - 2\left[\frac{S'(y)}{S(y)}\right]^2, \quad \tilde{\rho}(y) := \rho(\tau^{-1}(y)),$$

$$\tilde{\lambda}(y) := \lambda(\tau^{-1}(y)), \quad \tilde{\mu}(y) := \mu(\tau^{-1}(y)), \quad \tilde{H}(y,t) := H(\tau^{-1}(y),t).$$

Suppose that the function $g(t)$ has the structure:

$$g(t) = -a(0)\delta(t) + g_0(t)\theta(t), \tag{1.101}$$

where $g_0(t)$ is a given sufficiently smooth function. This is a consequence of the fact that the solution of the direct problem (1.97)–(1.99) admits the representation

$$w(y,t) = -a(0)\delta(t-y) + \tilde{w}(y,t)\theta(t-y).$$

Obviously, $\tilde{w} \equiv w$ for $t > y$. Problem (1.97)–(1.100) for $\tilde{w}$ in the domain $t \geq y \geq 0$ is equivalent to the problem

$$\begin{aligned}\frac{\partial^2 w}{\partial t^2} &= \frac{\partial^2 w}{\partial y^2} + q(y)w - \sqrt{a(y)a(0)}\frac{\partial}{\partial y}\left[(3\tilde{\lambda}(y) + 2\tilde{\mu}(y))R(\tilde{H}(y,t))\right] \\ &- \frac{a(0)}{\tilde{\rho}(y)}h(t-y) + \frac{1}{\tilde{\rho}(y)}\int\limits_0^{t-y} h(\tau)w(y,t-\tau)d\tau,\end{aligned} \tag{1.102}$$

$$w\mid_{y=0} = g_0(t), \tag{1.103}$$

$$\frac{\partial w}{\partial y}\mid_{y=0} = a(0)\left[3\lambda(0) + 2\mu(0)\right]R(H(0,t)) - \frac{a'(0)g_0(t)}{2a(0)}, \tag{1.104}$$

$$w\mid_{t=y+0} = \frac{a'(0)}{2} - \frac{a(0)}{2}\int\limits_0^y q(s)ds. \tag{1.105}$$

The inverse problem has thereby been reduced to the problem of finding the function $h(t)$ from relations (1.102)–(1.105). Starting from these relations, we construct a system of integral equations for the functions w and h. By using the d'Alembert formula, which provides the solution of the Cauchy problem for equation with the initial data (1.103) and (1.104) on the line $y = 0$ from (1.102)–(1.104), we obtain the equation

$$\begin{aligned}w(y,t) &= w_0(y,t) + \frac{1}{2}\int\int\limits_{\Delta(y,t)}\left[\frac{a(0)}{\tilde{\rho}(\xi)}h(\tau-\xi) - q(\xi)w(\xi,\tau)\right. \\ &\left. - \frac{1}{\tilde{\rho}(\xi)}\int\limits_0^{\tau-\xi} h(\beta)w(\xi,\tau-\beta)d\beta\right]d\tau\, d\xi,\ t > y > 0.\end{aligned} \tag{1.106}$$

Here

$$\Delta(y,t) := \{(\tau,\xi) | 0 \leq \xi \leq y, t-y+\xi \leq \tau \leq t+y-\xi\},$$

$$
\begin{aligned}
&w_0(y,t)\\
&=\frac{1}{2}\left[g_0(t+y)+g_0(t-y)\right]\\
&+\frac{1}{2}a(0)(3\lambda(0)+2\mu(0))\int\limits_{t-y}^{t+y}R\left(H(0,\beta)\right)\mathrm{d}\beta-\frac{a'(0)}{4a(0)}\int\limits_{t-y}^{t+y}g_0(\beta)\mathrm{d}\beta\\
&+\frac{1}{2}\int\int\limits_{\Delta(y,t)}\sqrt{a(\xi)a(0)}\frac{\partial}{\partial\xi}\left[(3\tilde{\lambda}(\xi)+2\tilde{\mu}(\xi))R(\tilde{H}(\xi,\tau))\right]\mathrm{d}\tau\,\mathrm{d}\xi. \quad (1.107)
\end{aligned}
$$

In relation (1.106), we pass to the limit as $t\to y+0$ and use relation (1.105):

$$
\begin{aligned}
&\int\limits_0^y\int\limits_{\xi}^{2y-\xi}\left[\frac{a(0)}{\tilde{\rho}(\xi)}h(\tau-\xi)-q(\xi)w(\xi,\tau)-\frac{1}{\tilde{\rho}(\xi)}\int\limits_0^{\tau-\xi}h(\beta)w(\xi,\tau-\beta)\mathrm{d}\beta\right]\mathrm{d}\tau\,\mathrm{d}\xi\\
&=a'(0)-a(0)\int\limits_0^y q(s)\mathrm{d}s-2w_0(y,y).
\end{aligned}
$$

In particular, this implies one of the matching conditions between problem data:

$$
2g_0(0)=a'(0). \quad (1.108)
$$

By differentiating the previous equation with respect to y, we obtain

$$
\begin{aligned}
&\int\limits_0^y\left[\frac{a(0)}{\tilde{\rho}(\xi)}h(2(y-\xi))-q(\xi)w(\xi,2y-\xi)-\frac{1}{\tilde{\rho}(\xi)}\int\limits_0^{2(y-\xi)}h(\beta)w(\xi,2y-\xi-\beta)\mathrm{d}\beta\right]\mathrm{d}\xi\\
&=-\frac{a(0)q(y)}{2}-\frac{\mathrm{d}}{\mathrm{d}y}w_0(y,y). \quad (1.109)
\end{aligned}
$$

Since

$$
\begin{aligned}
&\frac{\mathrm{d}}{\mathrm{d}y}w_0(y,y)=g_0'(2y)+a(0)\left[3\lambda(0)+2\mu(0)\right]R\left[H(0,2y)\right]-\frac{a'(0)}{2a(0)}g_0(2y)\\
&+\int\limits_0^y\sqrt{a(\xi)a(0)}\frac{\partial}{\partial\xi}\left[(3\tilde{\lambda}(\xi)+2\tilde{\mu}(\xi))R(\tilde{H}(\xi,2y-\xi))\right]\mathrm{d}\xi, \quad (1.110)
\end{aligned}
$$

from (1.109) we readily obtain the second matching condition

$$2g_0'(0)a(0) - a'g_0(0) = -a^2(0)\left[S'(0) - 2(S'(0))^2\right]. \tag{1.111}$$

Recall that the functions $a(y)$ and $S(y)$ occurring in (1.108) and (1.111) are defined in (1.96). To obtain an integral equation for h, we differentiate relation (1.109) with respect to y after the replacement of the variable $2(y-\xi)$ in the second integral by ξ' we have:

$$\begin{aligned}
&-\frac{a(0)}{\rho(0)}h(2y) - q(y)w(y,y)\\
&+\int_0^y \Bigg[\frac{a(0)\tilde{\rho}'(\xi)}{2\tilde{\rho}^2(\xi)}h(2(y-\xi)) - \frac{2}{\tilde{\rho}(\xi)}h(2(y-\xi))w(\xi,\xi)\\
&-2q(\xi)w_t(\xi,2y-\xi) - \frac{1}{\tilde{\rho}(\xi)}\int_0^{2(y-\xi)} h(\beta)w_t(\xi,2y-\xi-\beta)\mathrm{d}\beta\Bigg]\mathrm{d}\xi\\
&= -\frac{a(0)q'(y)}{2} - \frac{d^2}{\mathrm{d}y^2}w_0(y,y).
\end{aligned}$$

By using relation (1.105) for the computation of $w(y,y)$ and by solving the last relation for h, we obtain the equation

$$\begin{aligned}
h(y) = h_0(y) + \int_0^{y/2} \Bigg[&\frac{2\rho(0)}{a(0)\tilde{\rho}(\xi)}h(y-2\xi)w(\xi,\xi)\\
&-\frac{\rho(0)\tilde{\rho}'(\xi)}{2\tilde{\rho}^2(\xi)}h(y-2\xi) + \frac{2\rho(0)}{a(0)}q(\xi)w_t(\xi,y-\xi)\\
&+\frac{\rho(0)}{a(0)\tilde{\rho}(\xi)}\int_0^{y-2\xi} h(\beta)w_t(\xi,y-\xi-\beta)\mathrm{d}\beta\Bigg]\mathrm{d}\xi,
\end{aligned} \tag{1.112}$$

where

$$\begin{aligned}
h_0(y) &= q\left(\frac{y}{2}\right)\left[\frac{\rho(0)a'(0)}{2a(0)} - \frac{\rho(0)}{2}\int_0^{y/2} q(s)\mathrm{d}s\right]\\
&\quad -\left[\frac{q'(t)}{2\rho(0)} + \frac{\rho(0)}{a(0)}\frac{\mathrm{d}^2}{\mathrm{d}t^2}w_0(t,t)\right]_{t=y/2}.
\end{aligned} \tag{1.113}$$

To find the function $(\mathrm{d}^2/\mathrm{d}t^2)w_0(t,t)|_{t=y/2}$ from the definition of the function $h_0(y)$, one should integrate by parts in (1.110), differentiate both sides of the resulting relation with respect to y, and replace y by $y/2$. Then we obtain

$$\frac{\mathrm{d}^2}{\mathrm{d}t^2}w_0(t,t)|_{t=y/2} = 2g_0'(y) - \frac{a'(0)}{a(0)}g_0'(y)$$
$$+\sqrt{a\left(\frac{y}{2}\right)a(0)}\left[(3\tilde{\lambda}'(t)+2\tilde{\mu}'(t))\right]_{t=y/2} R\left(\tilde{H}\left(\frac{y}{2},\frac{y}{2}\right)\right)$$
$$+\sqrt{a(\frac{y}{2})a(0)}\left[\left(3\tilde{\lambda}\left(\frac{y}{2}\right)+2\tilde{\mu}\left(\frac{y}{2}\right)\right)\right]\alpha\left(\tilde{H}\left(\frac{y}{2},\frac{y}{2}\right)\right)\psi(y)$$
$$-\int_0^{y/2}\sqrt{\frac{a(0)}{a(\xi)}}a'(\xi)\left[(3\tilde{\lambda}(\xi)+2\tilde{\mu}(\xi))\right]\alpha(\tilde{H}(\xi,y-\xi))\tilde{H}_y(\xi,y-\xi)\mathrm{d}\xi, \quad (1.114)$$

where

$$\psi(y) := \frac{\mathrm{d}}{\mathrm{d}y}\tilde{H}\left(\frac{y}{2},\frac{y}{2}\right) = H_z(z,t)|_{z=\tau^{-1}(y/2),t=y/2}\frac{\mathrm{d}}{\mathrm{d}y}\tau^{-1}(\frac{y}{2})$$
$$+\frac{1}{2}H_t(z,t)|_{z=\tau^{-1}(y/2),t=y/2} = -\frac{1}{2}\gamma(T_1-T_0)$$
$$\times\int_0^{y/2}\left[\exp(\gamma z+\gamma^2 k\tau)\mathrm{erfc}\left(\frac{z}{2\sqrt{k\tau}}+\gamma\sqrt{k\tau}\right)v(z)\right]_{z=\tau^{-1}(y/2)}\mathrm{d}\tau$$
$$+\frac{1}{2}(T_1-T_0)\Big[\mathrm{erfc}\left(\frac{z}{2\sqrt{kt}}\right)$$
$$-\exp(\gamma z+\gamma^2 kt)\mathrm{erfc}\left(\frac{z}{2\sqrt{kt}}+\gamma\sqrt{kt}\right)\Big]_{z=\tau^{-1}(y/2),t=y/2},$$
$$\tilde{H}_y(\xi,y-\xi) = (T_1-T_0)$$
$$\times\Bigg[\mathrm{erfc}\left(\frac{z}{2\sqrt{k(y-\xi)}}\right)-\exp(\gamma z+\gamma^2 k(y-\xi))$$
$$\times\mathrm{erfc}\left(\frac{z}{2\sqrt{k(y-\xi)}}+\gamma\sqrt{k(y-\xi)}\right)\Bigg]_{z=\tau^{-1}(\xi)}.$$

To close the system (1.106) and (1.112) with respect to the unknown functions, one needs an equation for $w_t(y,t)$. We use relation (1.106) for the computation of $w_t(y,t)$. By using the equivalent description of the domain $\Delta(y,t)$ in the form

$$\Delta(y,t) = \{(\xi,\tau)|t-y\le\tau\le t+y, 0\le\xi\le y-|t-\tau|\},$$

we differentiate equation (1.106) with respect to t:

$$w_t(y,t) = w_{0t}(y,t) + \frac{1}{2}\int\limits_{t-y}^{t+y}\Bigg[\frac{a(0)}{\tilde{\rho}(y-|t-\tau|)}h(\tau+|t-\tau|-y)$$

$$- q(y-|t-\tau w(y-|t-\tau|,\tau) - \frac{1}{\tilde{\rho}(y-|t-\tau|)}$$

$$\int\limits_0^{\tau+|t-\tau|-y} h(\beta)w(y-|t-\tau|,\tau-\beta)\mathrm{d}\beta\Bigg]\mathrm{sgn}(t-\tau)\mathrm{d}\tau, \qquad (1.115)$$

where, according to (1.107), w_{0t} has the form

$$w_{0t}(y,t)$$

$$= \frac{1}{2}\left[g_0'(t+y) + g_0'(t-y)\right] + \frac{1}{2}a(0)(3\lambda(0)+2\mu(0))$$

$$\times\left[R\left(H(0,t+y)\right) - R\left(H(0,t-y)\right)\right] - \frac{a'(0)}{4a(0)}\left[g_0(t+y)-g_0(t-y)\right]$$

$$+\frac{1}{2}\int\limits_0^y \sqrt{a(\xi)a(0)}\frac{\partial}{\partial\xi}\Bigg\{(3\tilde{\lambda}(\xi)+2\tilde{\mu}(\xi))\Big[R(\tilde{H}(\xi,t+y-\xi))$$

$$-R(\tilde{H}(\xi,t-y+\xi))\Big]\Bigg\}\mathrm{d}\xi$$

$$= \frac{1}{2}\left[g_0'(t+y) + g_0'(t-y)\right] - \frac{a'(0)}{4a(0)}\left[g_0(t+y)-g_0(t-y)\right]$$

$$-\frac{1}{4}\int\limits_0^y \sqrt{\frac{a(0)}{a(\xi)}}a'(\xi)(3\tilde{\lambda}(\xi)+2\tilde{\mu}(\xi))\Big[R(\tilde{H}(\xi,t+y-\xi))$$

$$-R(\tilde{H}(\xi,t-y+\xi))\Big]\mathrm{d}\xi.$$

Therefore, from relations (1.102)–(1.105), we obtain the system of integral equations (1.106), (1.112), and (1.115). Conversely, under fulfillment of the matching conditions (1.108) and (1.91), from these equations, one can readily derive the differential relations of the inverse problem (1.102)–(1.105). The above argument justifies the following assertion.

Lemma 1.1 *Let*

$$g_0(t)\in C^2\left[0,T\right],\ \rho(z)\in C^3\left[0,\tau^{-1}(T/2)\right],\ \lambda(z)\in C^3\left[0,\tau^{-1}(T/2)\right],$$

$$\mu(z)\in C^3\left[0,\tau^{-1}(T/2)\right],\ \alpha(y)\in C\left[0,Y\right],$$

$Y := \tilde{H}\,(T/2, T/2)$ and let the matching conditions (1.108) and (1.111) be satisfied. Then solving the inverse problem (1.102)–(1.105) in the class of functions $h(t) \in C\,[0, T]$ and $(w, w_t) \in C\,(D_T)\,,\ D_T := \{(y, t)|0 \le y \le t \le T - y\}\,,$ is equivalent to solving system of Eqs. (1.106), (1.112), and (1.115).

For system (1.106), (1.112), and (1.115), we have the following local unique solvability theorem.

Theorem 1.7 *Let the assumptions of the Lemma 1.1 be valid, and let $T, T_0, T_1, k,$ and γ be given positive numbers; moreover, let $T_1 - T_0 > 0$. Then there exists $T^* \in (0, T]$ such that the system of integral equations (1.106), (1.112), and (1.115) has a unique solution in the domain D_{T^*}.*

Proof The function $H(z, t)$ occurring on the right-hand sides in the integral equations (1.112) and (1.115) is a regular solution of the heat equation (1.86) in the domain $z > 0, t > 0$ with the conditions $(\partial H/\partial z - \gamma H)\big|_{z=+0} = -\gamma(T_1 - T_0)t$ and $H|_{t=+0} = 0$. Therefore, the functions $H(z, t)$, $H_z(z, t)$, and $H_t(z, t)$ are continuous in the domain $D_T \backslash \{z = 0\}$. From the closed-form expressions for these functions, one can readily see that they are also continuous for $z = 0$. Under the assumptions of Theorem 1.7, this implies that relations (1.106), (1.112), and (1.115) form a closed system of Volterra integral equations of the second kind with continuous free terms and kernels for the unknown functions $w(z, t)$, $h(t)$, and $w_t(z, t)$. The small parameter in this system is the integration interval, whose length does not exceed T. Therefore, the Banach contraction mapping principle can be used for small T, which provides the existence of a unique solution of system (1.106), (1.112), and (1.115). This solution can be found by the successive approximation method. Each iteration leads to continuous functions. Hence the limit functions are continuous solutions of Eqs. (1.106), (1.112), and (1.115).

Remark 1.1 The number T^* in Theorem 1.7 will be chosen from the condition that the integral operators in Eqs. (1.106), (1.112), and (1.115) are contraction operators on some set of continuous functions and satisfy the inequality

$$\tilde{H}(T^*/2, T^*/2) < T^*.$$

Now let us obtain a stability estimate for the solution and prove the uniqueness theorem for arbitrary $T > 0$. Let $K(h_0)$ be the set of functions $h(t) \in C[0, T]$ satisfying the inequality $|h(t)| \le h_0$ with a given constant h_0 for $t \in [0, T]$, and let

$$\begin{aligned} R_0 &= \max\{\|\rho(z)\|_{C^3[0,\tau^{-1}(T/2]}, \|\lambda(z)\|_{C^3[0,\tau^{-1}(T/2]}, \|\mu(z)\|_{C^3[0,\tau^{-1}(T/2]}, \\ &\|\alpha(y)\|_{C[0,Y]}, \|g_0(t)\|_{C^2[0,T]}, \|\rho^*(z)\|_{C^3[0,\tau^{-1}(T/2]}, \|\lambda^*(z)\|_{C^3[0,\tau^{-1}(T/2]}, \\ &\|\mu^*(z)\|_{C^3[0,\tau^{-1}(T/2]}, \|\alpha^*(y)\|_{C[0,Y]}, \|g_0^*(t)\|_{C^2[0,T]}\}, \end{aligned}$$

Theorem 1.8 *Let $h(t) \in K(h_0)$ and $h^*(t) \in K(h_0)$ be the solutions of the inverse problem (1.102)–(1.105) with data*

$$\{\rho(z), \lambda(z), \mu(z), \alpha(y), g_0(t)\},$$

and

$$\{\rho^*(z), \lambda^*(z), \mu^*(z), \alpha^*(y), g_0^*(t)\},$$

respectively. Then there exists a positive number $C = C(h_0, R_0, \gamma, T, T_0, T_1, h_0)$ such that the following stability estimate holds:

$$\|h(t) - h^*(t)\|_{C[0,T]} \le C\Big[\|\rho - \rho^*\|_{C^3[0,\tau^{-1}(T/2]} + \|\lambda - \lambda^*\|_{C^3[0,\tau^{-1}(T/2]}$$

$$+\|\mu - \mu^*\|_{C^3[0,\tau^{-1}(T/2]} + \|\alpha - \alpha^*\|_{C[0,Y]} + \|g_0 - g_0^*\|_{C^2[0,T]}\Big]. \tag{1.116}$$

Proof First, by considering Eq. (1.106) for $h(t) \in K(h_0)$ as an integral equation for $w(y, t)$ and by forming a series of successive approximations by the usual scheme, we note that the absolute value of the function $w(y, t)$ in the domain D_T is bounded by some constant $c_0 = c_0(h_0, R_0, \gamma, T, T_0, T_1, h_0)$, i.e.,

$$|w(y, t)| \le c_0, \quad (y, t) \in D_T. \tag{1.117}$$

Let us introduce the differences

$$\bar{w} = w - w^*, \quad \bar{h} = h - h^*, \quad \bar{w}_t = w_t - w_t^*, \quad \bar{\rho} = \rho - \rho^*, \quad \bar{\lambda} = \lambda - \lambda^*,$$
$$\bar{\mu} = \mu - \mu^*, \quad \bar{\alpha} = \alpha - \alpha^*, \quad \bar{g}_0 = g_0 - g_0^*.$$

The domain D_T admits the equivalent description

$$D_T = \left\{(y, t)|0 \le y \le \frac{T}{2} - \left|t - \frac{T}{2}\right|, 0 \le t \le T\right\}.$$

Let

$$\varphi(t) = \max\left[\left|\bar{h}(t)\right|, \max_{0\le y\le T/2-|t-T/2|}|\bar{w}(y, t)|, \max_{0\le y\le T/2-|t-T/2|}|\bar{w}_t(y, t)|\right],$$

$t \in [0, T]$. By forming equations for $\bar{w}$, $\bar{h}$, and $\bar{w}_t$ with the use of relations (1.106), (1.112), and (1.115) and by making simple estimates, we obtain the integral inequality

$$\varphi(t) \le c_*\Big[\|\rho - \rho^*\|_{C^3[0,\tau^{-1}(T/2]} + \|\lambda - \lambda^*\|_{C^3[0,\tau^{-1}(T/2]} + \|\mu - \mu^*\|_{C^3[0,\tau^{-1}(T/2]}$$

$$+ \|\alpha - \alpha^*\|_{C[0,Y]} + \|g_0 - g_0^*\|_{C^2[0,T]}\Big] + c_{**}\int_0^t \varphi(\tau)d\tau, \tag{1.118}$$

for $(y, t) \in D_T$, where c_* and c_{**} are positive constants depending on h_0, R_0, k, γ, T, T_0, and T_1. By using the Gronwall inequality, from (1.118) we obtain the estimate (1.116) with some constant C.

The uniqueness theorem for each $T > 0$ is a consequence of Theorem 1.8.

Theorem 1.9 *Let the functions* $(h(t), h^*(t)) \in C\,[0, T]$, $\rho(z), \lambda(z), \mu(z), \alpha(y)$, $g_0(t)$, *and* $\rho^*(z)$, $\lambda^*(z)$, $\mu^*(z), \alpha^*(y)$, *and* $g_0^*(t)$ *have the same meaning as in Theorem 1.7. If, in addition* $\rho(z) = \rho^*(z)$, $\lambda(z) = \lambda^*(z)$, $\mu(z) = \mu^*(z)$ *for* $z \in \left[0,\ \tau^{-1}(T/2\right]$, $\alpha(y) = \alpha^*(y)$ *for* $y \in [0, Y]$, *and* $g_0(t) = g_0^*(t)$ *for* $t \in [0, T]$, *then* $h(t) = h^*(t)$ *on the segment* $[0, T]$.

1.4 Inverse Problem for an Integro-differential Equation of Acoustics

We consider the initial boundary-value problem for the one-dimensional integro-differential equation of acoustics

$$\frac{1}{c^2(z)}\frac{\partial^2 v}{\partial t^2} = \varrho_z - \frac{\partial \ln \rho(z)}{\partial z}\varrho,\ z > 0,\ t > 0, \tag{1.119}$$

$$v\mid_{t\le 0} \equiv 0, \quad \varrho(+0, t) = \delta'(t), \tag{1.120}$$

where $c(z) > 0$ is the wave velocity, $\rho(z)$ is the medium density, $v(z, t)$ is the acoustic pressure; $\varrho(z, t)$ is the stress; and the functions $\varrho(z, t)$ and $v(z, t)$ are related as

$$\varrho(z, t) = \frac{\partial v}{\partial z}(z, t) + \int_0^t k(t - \tau)\frac{\partial v}{\partial z}(z, \tau)d\tau; \tag{1.121}$$

$\delta'(t)$ is as before the derivative of the Dirac delta function $\delta(t)$. Note that the solution to problem (1.119)–(1.121) is the limit (in the sense of generalized functions) of the solutions to problem (1.119)–(1.121), corresponding to $\delta'(t - t_0)$ instead of $\delta'(t)$ in the boundary condition (1.120) at t_0 tending to $+0$.

In this section the inverse problem is posed as follows: to determine the kernel $k(t), t > 0$, that enters Eq. (1.119) via relation (1.121), given that for the solution of the problem we have

$$v(+0, t) = g(t), \quad t > 0. \tag{1.122}$$

Equation (1.119), which allows for absorption by an imperfectly elastic medium by means of relation (1.121), is encountered in geophysics when the properties of the medium are studied by seismic waves. In fact, the system of Boltzmann equations (one of the most common for linear inelastic media) reduces to Eq. (1.119) in the one-dimensional case under appropriate assumptions about smoothness.

Let us introduce a new variable x by the formula

$$x = \psi(z) = \int\limits_0^z \frac{d\xi}{c(\xi)},$$

and adopt the notation

$$\widetilde{v}(x,t) := v(\psi^{-1}(x), t),\ a(x) := c(\psi^{-1}(x)),\ b(x) := \rho(\psi^{-1}(x)).$$

where $\psi^{-1}(x)$ is an inverse of the function $\psi(z)$. In what follows, we assume that $c(z) > 0$ and $\rho(z) > 0$. The main result in the section is the following theorem on the local unique solvability of the inverse problem.

Theorem 1.10 *Assume that a function $g(t)$ can be represented as*

$$g(t) = -c(+0)\delta(t) + \theta(t)g_0(t),$$

where the function $g_0(t)$ belongs to the space $C^2\,[0,\ T]$, the functions $c(z)$ and $\rho(z)$ belong to the space $C^3\left[0, \psi^{-1}(T)\right]$. Then, for sufficiently small $T_0 \in (0, T)$, the solution of the inverse problem in Eqs. (1.119)–(1.122) exists and is unique in the class of kernels $k(t)$ belonging to the space $C^2\,[0,\ T_0]$.

Relations (1.119)–(1.122) for the new functions $\widetilde{v}$, λ and new variable x become

$$\frac{\partial^2 \widetilde{v}}{\partial t^2} = \left(\frac{\partial^2}{\partial x^2} - \frac{\lambda'(x)}{\lambda(x)}\frac{\partial}{\partial x}\right)\left[\widetilde{v}(x,t) + \int\limits_0^t k(t-\tau)\widetilde{v}(x,\tau)d\tau\right],$$
$$x > 0, t > 0, \tag{1.123}$$

$$\widetilde{v}\,|_{t\le 0} \equiv 0,\ \frac{\partial \widetilde{v}}{\partial x}(+0,t) + \int\limits_0^t k(t-\tau)\frac{\partial \widetilde{v}}{\partial x}(+0,\tau)d\tau = c(+0)\delta'(t), \tag{1.124}$$

$$\widetilde{v}(+0,t) = g(t), \quad t > 0, \tag{1.125}$$

where $\lambda(x) := a(x)b(x)$. Further, $c(0), h(0), \lambda(0)$, etc. are understood to be the respective values of these functions for x tending to zero from the right. We transform the integro-differential equation (1.123) so that, first, the integrand does not contain the derivatives of the function $\widetilde{v}$ with respect to x and, second, the coefficient in front of $\frac{\partial \widetilde{v}}{\partial x}$ in the terms outside the integral sign is zero. These requirements can be fulfilled by introducing a new function u with the formula

$$\left[\widetilde{v}(x,t) + \int\limits_0^t k(t-\tau)\widetilde{v}(x,\tau)d\tau\right]\exp\left(-\frac{k(0)t}{2}\right)\sqrt{\frac{\lambda(0)}{\lambda(x)}} = u(x,t).$$

Then it is easy to verify by direct calculations that the function $\widetilde{v}$ can be expressed in terms of the function u using the formula

$$\widetilde{v}(x,t) = \left[\exp\left(k(0)t/2\right)u(x,t) + \int_0^t h(t-\tau)\exp\left(k(0)\tau/2\right)u(x,\tau)\mathrm{d}\tau\right]\sqrt{\frac{\lambda(x)}{\lambda(0)}},$$

where

$$h(t) = -k(t) - \int_0^t k(t-\tau)h(\tau)\mathrm{d}\tau. \tag{1.126}$$

Let us introduce the notation

$$\Lambda(x) := \frac{h^2(0)}{4} - h'(0) + \frac{2\lambda(x)\lambda''(x) - 3(\lambda'(x))^2}{4\lambda^2(x)},$$
$$\tilde{k}(t) := h''(t)\exp\left(\frac{h(0)t}{2}\right), \quad \tilde{g}_0(t) := g(t)\exp\left(\frac{h(0)t}{2}\right),$$
$$h_0(t) := k(t)\exp\left(\frac{h(0)t}{2}\right).$$

Then Eqs. (1.123)–(1.125) in the new functions $u(x,t)$ and $h(t)$ become

$$\frac{\partial^2 u}{\partial t^2} = \frac{\partial^2 u}{\partial x^2} + \Lambda(x)u + \int_0^t \tilde{k}(t-\tau)u(x,\tau)\mathrm{d}\tau, \ x > 0, \ t > 0, \tag{1.127}$$

$$u|_{t\leq 0} \equiv 0, \tag{1.128}$$

$$\left.\frac{\partial u}{\partial x}\right|_{x=+0} = c(0)\delta'(t) + \frac{1}{2}c(0)k(0)\delta(t) - \frac{\lambda'(0)}{2\lambda(0)}u(0,t), \tag{1.129}$$

$$u|_{x=+0} = \tilde{g}_0(t) + \int_0^t h_0(t-\tau)\tilde{g}_0(\tau)\mathrm{d}\tau, \tag{1.130}$$

It follows from the theory of hyperbolic equations that the function $u(x,t)$ as a solution to the direct problem in Eqs. (1.127)–(1.130) possesses the property $u \equiv 0, t < x, x > 0$ and, in a neighborhood of the characteristic line $t = x$, has the structure:

$$u(x,t) = \alpha(x)\delta(t-x) + \theta(t-x)\tilde{u}(x,t), \tag{1.131}$$

where $\tilde{u}(x,t)$ is a regular function.

Let us denote $\beta(x) := \tilde{u}(x, x+0)$. Substituting the function (1.131) into Eqs. (1.127)–(1.130) and using the method of separation of singularities [7], we find $\alpha'(x) = 0$, $\alpha(0) = -c(0)$, $2\beta'(x) - \Lambda(x)\alpha(x) = 0$, $\alpha'(0) - \beta(0) + \alpha(0)\frac{\lambda'(0)}{2\lambda(0)} = \frac{1}{2}$ $c(0)k(0)$. We solve these ordinary differential equations to obtain

$$\alpha(x) = -c(0), \quad \beta(x) = -\frac{c(0)}{2}\left(k(0) + \frac{\lambda'(0)}{\lambda(0)} + \int_0^x \Lambda(\xi)\mathrm{d}\xi\right).$$

It follows from the above that the function $\tilde{u}(x,t)$ in the domain $D := \{(x,t) : t > x > 0\}$ satisfies the equations

$$\frac{\partial^2 \tilde{u}}{\partial t^2} = \frac{\partial^2 \tilde{u}}{\partial x^2} + \Lambda(x)\tilde{u} - c(+0)\tilde{k}(t-x) + \int_x^t \tilde{k}(t-\tau)\tilde{u}(x,\tau)\mathrm{d}\tau, \tag{1.132}$$

$$\tilde{u}|_{t=x+0} = \beta(x), \tag{1.133}$$

$$\left[\frac{\partial \tilde{u}}{\partial x} + \frac{\lambda'(0)}{2\lambda(0)}\tilde{u}(x,t)\right]_{x=0} = 0, \tag{1.134}$$

$$\tilde{u}|_{x=0} = \tilde{g}_0(t) - c(0)h_0(t) + \int_0^t h_0(\tau)\tilde{g}_0(t-\tau)\mathrm{d}\tau, \quad t > 0. \tag{1.135}$$

By requiring that the functions $\tilde{u}(x,t)$ and $(\partial\tilde{u}/\partial x)(x,t)$ for $x = t = 0$, from relations (1.133)–(1.135) we obtain the following expressions for the values $h(0)$ and $h'(0)$ in terms of the known values:

$$h(0) = -\frac{2}{c(0)}\tilde{g}_0(0) - \frac{\lambda'(0)}{\lambda(0)},$$

$$h'(0) = \frac{1}{2}(2h^2(0) - \Lambda(0)) - \frac{1}{c(0)}(\tilde{g}_0'(0) + \tilde{g}_0(0)h(0)).$$

When deriving the last expression, we used the following relations, which follow from Eq. (1.126):

$$h'(t) = -k'(t) - k(0)h(t) - \int_0^t k'(t-\tau)h(\tau)\mathrm{d}\tau,$$

$$h'(0) = -k'(0) + k^2(0).$$

In what follows, we assume that, in the relations for $\Lambda(x)$, $h(0)$ and $h'(0)$ have been replaced by their already established values.

The proof of the theorem is based on the following lemma.

Lemma 1.2 *If the conditions of Theorem 1.10 are satisfied, then problem (1.132)–(1.135) for* $(x,t) \in D_T$, $D_T = ((x,t)|0 \le x \le t \le T - x)$ *is equivalent to the problem of determining functions* $\tilde{u}(x,t)$, $(\partial\tilde{u}/\partial t)(x,t)$, $\tilde{k}(t)$, $h_0(t)$, $h_0'(t)$, $h_0'(t)$ *from the system of equations:*

$$\tilde{u}(x,t) = \beta(x) + \int_x^t \frac{\partial \tilde{u}}{\partial \tau}(x,\tau)\mathrm{d}\tau, \tag{1.136}$$

$$\begin{aligned}\frac{\partial \tilde{u}}{\partial t}(x,t) &= \frac{1}{2}\Bigg(\tilde{g}_0'(t-x) - c(0)h_0'(t-x) - h(0)\tilde{g}_0(t-x) - c(0)\tilde{k}(t-x)x\\ &+\frac{\lambda'(0)}{2\lambda(0)}\tilde{u}(0,t-x)\Bigg) - \frac{c(0)}{4}\Lambda\left(\frac{x+t}{2}\right) + \frac{1}{2}\int_0^{t-x} h_0'(t-y-\tau)\tilde{g}_0(\tau)\mathrm{d}\tau\\ &+\frac{1}{2}\int_0^x \left[\Lambda(\xi)\tilde{u}(\xi,t-x+\xi) - \int_0^{t-x}\tilde{k}(\tau)\tilde{u}(\xi,t-x+\xi-\tau)\mathrm{d}\tau\right]\mathrm{d}\xi\\ &+\frac{1}{2}\int_x^{\frac{t+x}{2}}\Bigg[\Lambda(\xi)\tilde{u}(\xi,t+x-\xi) - c(0)\tilde{k}(t+x-2\xi)\\ &+\int_0^{t+x-2\xi}\tilde{k}(\tau)\tilde{u}(\xi,t+x-\xi-\tau)\mathrm{d}\tau\Bigg]\mathrm{d}\xi,\end{aligned} \tag{1.137}$$

$$\begin{aligned}\tilde{k}(t) &= -\frac{1}{2}\Lambda'\left(\frac{t}{2}\right) - \frac{2}{c(0)}\Bigg[\tilde{g}_0''(t) - c(0)h_0''(t) + \tilde{g}_0(0)h_0'(t) - \frac{\lambda'(0)}{2\lambda(0)}\frac{\partial\tilde{u}}{\partial t}(0,t)\\ &-\frac{1}{2}\Lambda\left(\frac{t}{2}\right)\beta\left(\frac{t}{2}\right)\Bigg] - \frac{2}{c(0)}\int_0^t \tilde{g}_0''(t-\tau)h_0(\tau)\mathrm{d}\tau - \frac{2}{c(0)}\int_0^t \tilde{k}(\tau)\beta\left(\frac{t-\tau}{2}\right)\mathrm{d}\tau\\ &+\frac{2}{c(0)}\int_0^{\frac{t}{2}}\left[\Lambda(\xi)\frac{\partial\tilde{u}}{\partial t}(\xi,t-\xi) + \int_0^{t-2\xi}\tilde{k}(\tau)\frac{\partial\tilde{u}}{\partial t}(\xi,t-\xi-\tau)\mathrm{d}\tau\right]\mathrm{d}\xi,\end{aligned} \tag{1.138}$$

$$h_0(t) = -h(0) + \left(\frac{h^2(0)}{2} - h'(0)\right)t + \int_0^t (t-\tau)h_0''(\tau)\mathrm{d}\tau, \tag{1.139}$$

$$h_0'(t) = \frac{h^2(0)}{2} - h'(0) + \int_0^t h_0''(\tau)\mathrm{d}\tau, \tag{1.140}$$

$$h_0''(t) = -\tilde{k}(t) + \left(\frac{h^2(0)}{4} - h'(0)\right) h_0(t) - \int_0^t \tilde{k}(t-\tau)h_0(\tau)\mathrm{d}\tau. \tag{1.141}$$

Proof Note that the following factorization identities hold:

$$\frac{\partial^2 \tilde{u}}{\partial t^2} - \frac{\partial^2 \tilde{u}}{\partial x^2} = \left(\frac{\partial}{\partial t} - \frac{\partial}{\partial x}\right)\left(\frac{\partial}{\partial t} + \frac{\partial}{\partial x}\right)\tilde{u} = \left(\frac{\partial}{\partial t} + \frac{\partial}{\partial x}\right)\left(\frac{\partial}{\partial t} - \frac{\partial}{\partial x}\right)\tilde{u}.$$

Taking these identities into account, we integrate Eq. (1.132) along the corresponding characteristics of the first-order differential operators in the domain $(x, t) \in D_T$. We integrate along the characteristic of the operator $\partial/\partial t - \partial/\partial x$ from the point (x, t) to the point $((x+t)/2, (x+t)/2)$ in the plane of variables (ξ, τ). Using the relation $(\partial/\partial t + \partial/\partial x)\,\tilde{u}((x+t)/2, (x+t)/2) = -\frac{c(0)}{2}\Lambda\left(\frac{x+t}{2}\right)$, which follows from condition (1.133) after differentiation with respect to x, we obtain

$$\left(\frac{\partial}{\partial t} + \frac{\partial}{\partial x}\right)\tilde{u}(x,t) = -\frac{c(0)}{2}\Lambda\left(\frac{x+t}{2}\right) + \int_x^{(x+t)/2} \Bigg[\Lambda(\xi)\tilde{u}(\xi, t+x-\xi)$$

$$-c(0)\tilde{k}(t+x-2\xi) - \int_0^{t+x-2\xi} \tilde{k}(\tau)\tilde{u}(\xi, t+x-\xi-\tau)\mathrm{d}\tau\Bigg]\mathrm{d}\xi. \tag{1.142}$$

Let us perform integration along the characteristic of the operator $\partial/\partial t + \partial/\partial x$ from the point $(0, t-x)$ to the point (x, t). We use formulas (1.134) and (1.135) to obtain

$$\left(\frac{\partial}{\partial t} - \frac{\partial}{\partial x}\right)\tilde{u}(x,t) = \tilde{g}_0'(t-x) - h(0)\tilde{g}_0(t-x) - c(+0)\tilde{k}(t-x)x$$

$$-c(0)h_0'(t-x) + \frac{\lambda'(0)}{2\lambda(0)}\tilde{u}(t-x, 0) + \int_0^{t-x} h_0'(t-x-\tau)\tilde{g}_0(\tau)\mathrm{d}\tau$$

$$+\int_0^x \left[\Lambda(\xi)\tilde{u}(\xi, t-x+\xi) - \int_0^{t-x} \tilde{k}(\tau)\tilde{u}(\xi, t-x+\xi-\tau)\mathrm{d}\tau\right]\mathrm{d}\xi. \tag{1.143}$$

Summing (1.142) and (1.143), we arrive at Eq. (1.137). By setting $x = 0$ in Eq. (1.142) and using conditions (1.134) and (1.135), we find

$$\tilde{g}_0'(t) + \tilde{g}_0(0)h_0(t) - c(0)h_0'(t) - \frac{\lambda'(0)}{2\lambda(0)}\tilde{u}(0,t) + \int_0^t \tilde{g}_0'(t-\tau)h_0(\tau)\mathrm{d}\tau$$

$$= -\frac{c(0)}{2}\Lambda\left(\frac{t}{2}\right) + \int_0^{t/2}\big[\Lambda(\xi)\tilde{u}(\xi, t-\xi) - c(0)\tilde{k}(t-2\xi)$$

$$- \int_0^{t-2\xi}\tilde{k}(\tau)\tilde{u}(\xi, t-\xi-\tau)\mathrm{d}\tau\big]\mathrm{d}\xi.$$

Differentiating this relation with respect to t, after simple transformations, we obtain Eq. (1.138).

To complete the system of integral equations (1.136)–(1.138), we use formulas (1.139)–(1.141), which follow from the definition of the function $h_0(t)$ and relation (1.126). Under the condition of Theorem 1.10, the equivalence of the system of integral equations (1.136)–(1.141) and the inverse problem in Eqs. (1.132)–(1.135) can be proved in the standard way [8]. Lemma 1.2 is proved.

Let us introduce the notation

$$\frac{h^2(0)}{4} - h'(0) =: r_{00}, \quad \frac{\lambda'(0)}{2\lambda(0)} =: \lambda_0, \quad \frac{c(0)}{2} =: c_0,$$

$$\frac{1}{3}\left(\frac{2}{c(0)}\tilde{g}_0(0) + \frac{\lambda'(0)}{\lambda(0)}\right) =: L_0, \quad \frac{\lambda'(0)\tilde{g}_0(0)}{3\lambda(0)c(0)} =: M_0.$$

We write the system of Eqs. (1.136)–(1.141) as the operator equation

$$\varphi = A\varphi, \tag{1.144}$$

where

$$\varphi = \Big[\varphi_1(x,t), \varphi_2(x,t), \varphi_3(t), \varphi_4(t), \varphi_5(t), \varphi_6(t)\Big]$$

$$= \Big[\tilde{u}(x,t),\ \frac{\partial \tilde{u}}{\partial t}(x,t) + c_0(\tilde{k}(t-x)x + h_0'(t-x) + \lambda_0 h_0(t-x)),\ \tilde{k}(t) - 2h_0''(t)$$

$$+3L_0h_0'(t) - 3M_0h_0(t),\ h_0(t),\ h_0'(t), h_0''(t) + \tilde{k}(t) - r_{00}h_0(t)\Big]$$

is a vector function with its first two components belonging to the space $C(D_T)$ and the other components to the space $C[0, T]$ (we denote the linear space of such vector functions by C_6), while the operator A acts in the space C_6 and has the form

$$A = (A_1,\ A_2,\ A_3,\ A_4,\ A_5,\ A_6)^T.$$

If we introduce the vector function

$$\begin{aligned}
&\varphi_0(x,t) = (\varphi_{01},\ \varphi_{02},\ \varphi_{03},\ \varphi_{04},\ \varphi_{05},\ \varphi_{06}) \\
&:= \Bigg[\beta(x),\ \frac{1}{2}(\tilde{g}_0'(t-x) + (\lambda_0 - h(0))\tilde{g}_0(t-x)) - \frac{c_0}{2}\Lambda\left(\frac{x+t}{2}\right), \\
&-\frac{1}{2}\Lambda'\left(\frac{t}{2}\right) - \frac{1}{c_0}\left[\tilde{g}_0''(t) - \lambda_0\tilde{g}_0'(t) - \frac{1}{2}\Lambda\left(\frac{t}{2}\right)\beta\left(\frac{t}{2}\right)\right], \\
&-h(0) + \left(\frac{h^2(0)}{2} - h'(0)\right)t,\ \frac{h^2(0)}{2} - h'(0),\ 0\Bigg],
\end{aligned}$$

then, according to Eqs. (1.136)–(1.141), the components of the operator A in Eq. (1.144) become

$$\begin{aligned}
A_1\varphi = \varphi_{01} + \int_x^t \Bigg[&\varphi_2(x,\tau) - c_0\varphi_5(\tau - x) - c_0 x\Bigg(\frac{1}{3}(2\varphi_6(\tau - x) + \varphi_3(\tau - x)) \\
&-L_0\varphi_5(\tau - x) + (M_0 + \frac{2r_{00}}{3})\varphi_4(\tau - x)\Bigg) - \lambda_0 c_0\varphi_4(\tau - x)\Bigg]\mathrm{d}\tau,
\end{aligned}$$

$$\begin{aligned}
A_2\varphi = \varphi_{02} &+ \frac{\lambda_0}{2}\int_0^{t-x}\varphi_4(\tau)\tilde{g}_0(t-x-\tau)\mathrm{d}\tau + \frac{1}{2}\int_0^{t-x}\varphi_5(t-x-\tau)\tilde{g}_0(\tau)\mathrm{d}\tau \\
&+\frac{1}{2}\int_0^x\Bigg[\Lambda(\xi)\varphi_1(\xi, t-x+\xi) \\
&-\int_0^{t-x}\left(\frac{1}{3}(2\varphi_6(\tau) + \varphi_3(\tau)) - L_0\varphi_5(\tau) + (M_0 + \frac{2r_{00}}{3})\varphi_4(\tau)\right) \\
&\times\varphi_1(\xi, t-x+\xi-\tau)\mathrm{d}\tau\Bigg]\mathrm{d}\xi + \frac{1}{2}\int_x^{\frac{t+x}{2}}\Bigg[\Lambda(\xi)\varphi_1(\xi, t+x-\xi) \\
&-2c_0\Bigg(\frac{1}{3}(2\varphi_6(t+x-2\xi) + \varphi_3(t+x-2\xi)) - L_0\varphi_5(t+x-2\xi) \\
&+(M_0 + \frac{2r_{00}}{3})\varphi_4(t+x-2\xi)\Bigg) \\
&+\int_0^{t+x-2\xi}\left(\frac{1}{3}(2\varphi_6(\tau) + \varphi_3(\tau)) - L_0\varphi_5(\tau) + (M_0 + \frac{2r_{00}}{3})\varphi_4(\tau)\right) \\
&\times\varphi_1(\xi, t+x-\xi-\tau)\mathrm{d}\tau\Bigg]\mathrm{d}\xi,
\end{aligned}$$

$$
\begin{aligned}
A_3\varphi &= \varphi_{03} + \frac{\lambda_0}{c_0}\int_0^t \varphi_4(\tau)\tilde{g}_0'(t-\tau)\mathrm{d}\tau - \frac{1}{c_0}\int_0^t \varphi_4(\tau)\tilde{g}_0''(t-\tau)\mathrm{d}\tau \\
&-\frac{1}{c_0}\int_0^t \left(\frac{1}{3}(2\varphi_6(\tau)+\varphi_3(\tau)) - L_0\varphi_5(\tau) + (M_0 + \frac{2r_{00}}{3})\varphi_4(\tau)\right) \\
&\times\beta\left(\frac{t-\tau}{2}\right)\mathrm{d}\tau + \frac{1}{c_0}\int_0^{t/2}\Bigg[\Lambda(\xi)\Bigg(\varphi_2(\xi, t-\xi) + c_0\xi\Bigg(\frac{1}{3}(2\varphi_6(t-\xi) \\
&+\varphi_3(t-\xi)) - L_0\varphi_5(\tau) + (M_0 + \frac{2r_{00}}{3})\varphi_4(t-\xi)\Bigg) - c_0\varphi_5(t-\xi)\Bigg) \\
&+\int_0^{t-2\xi}\left(\frac{1}{3}(2\varphi_6(\tau)+\varphi_3(\tau)) - L_0\varphi_5(\tau) + (M_0 + \frac{2r_{00}}{3})\varphi_4(\tau)\right) \\
&\times\Bigg(\varphi_2(\xi, t-\xi-\tau) + c_0\xi\Bigg(\frac{1}{3}(2\varphi_6(t-\xi-\tau) + \varphi_3(t-\xi-\tau)) \\
&-L_0\varphi_5(t-\xi-\tau) + (M_0 + \frac{2r_{00}}{3})\varphi_4(t-\xi-\tau)\Bigg) \\
&-c_0\varphi_5(t-\xi-\tau)\Bigg)\mathrm{d}\tau\Bigg]\mathrm{d}\xi,
\end{aligned}
$$

$$
A_4\varphi = \varphi_{04} + \int_0^t (t-\tau)\left(\frac{1}{3}(\varphi_6(\tau) - \varphi_3(\tau)) + L_0\varphi_5(\tau) - (M_0 - \frac{1}{3}r_{00})\varphi_4(\tau)\right)\mathrm{d}\tau,
$$

$$
A_5\varphi = \varphi_{05} + \int_0^t \left(\frac{1}{3}(\varphi_6(\tau) - \varphi_3(\tau)) + L_0\varphi_5(\tau) - (M_0 - \frac{1}{3}r_{00})\varphi_4(\tau)\right)\mathrm{d}\tau,
$$

$$
\begin{aligned}
A_6\varphi = \varphi_{06} - &\int_0^t \Bigg(\frac{1}{3}(2\varphi_6(t-\tau) + \varphi_3(t-\tau)) - L_0\varphi_5(t-\tau) \\
&+(M_0 + \frac{2}{3}r_{00})\varphi_4(t-\tau)\Bigg)\varphi_4(\tau)\mathrm{d}\tau,
\end{aligned}
$$

Let us consider the above-introduced linear space C_6 of vector functions φ. In this space, we introduce the norm by the formula:

$$
\|\varphi\| = \max_{1\le i\le 6}\|\varphi_i\|,
$$

where $\|\varphi_i\| = \max_{(x,t)\in D_T}|\varphi_i(x,t)|$, $i = 1, 2$; $\|\varphi_i\| = \max_{t\in[0,T]}|\varphi_i(t)|$, $i = 3, 4, 5, 6$.

By $B(\varphi_0)$ we denote the set of vector functions $\varphi(x,t)$ that satisfy the inequality

$$||\varphi - \varphi_0|| \le ||\varphi_0||.$$

Obviously, $\|\varphi\| \le 2\|\varphi_0\|$ for $\varphi(x,t) \in B(\varphi_0)$. Let us prove that A is a contraction operator in the metric space $B(\varphi_0)$ provided the number T is sufficiently small.

Let us verify the first condition of contraction operator. To simplify calculations, we denote

$$P_0 := 1 + L_0 + M_0 + \frac{2r_{00}}{3}, \quad \Lambda_0 := \max_{x\in[0,T/2]} |\Lambda(x)|, \quad B_0 := \max_{x\in[0,T/2]} |\beta(x)|,$$
$$G_0 := \max_{t\in[0,T]} |\tilde{g}_0(t)|, \quad G_1 := \max_{t\in[0,T]} \left|\tilde{g}_0'(t)\right|, \quad G_2 := \max_{t\in[0,T]} \left|\tilde{g}_0''(t)\right|.$$

Let $\varphi \in B(\varphi_0)$. Then, for $(x,t) \in D_T$ we have the estimates

$$|A_1\varphi - \varphi_{01}| \le \left|\int_x^t \left[\varphi_2(x,\tau) - c_0\varphi_5(\tau - x) - c_0 x\left(\frac{1}{3}(2\varphi_6(\tau - x) + \varphi_3(\tau - x))\right.\right.\right.$$
$$\left.\left.\left. -L_0\varphi_5(\tau - x) + (M_0 + \frac{2r_{00}}{3})\varphi_4(\tau - x)\right) - \lambda_0 c_0 \varphi_4(\tau - x)\right] d\tau\right|$$
$$\le ||\varphi_0||T\left[2 + c_0\left(\frac{P_0 T}{2} + 2(1+\lambda_0)\right)\right] =: \alpha_1\|\varphi_0\|,$$

$$|A_2\varphi - \varphi_{02}| \le ||\varphi_0||T\left[(\lambda_0 + 1)G_0 + \Lambda_0 + P_0\left(\frac{3}{2}||\varphi_0||T + c_0\right)\right] =: \alpha_2\|\varphi_0\|,$$

$$|A_3\varphi - \varphi_{03}| \le ||\varphi_0||T\left[\frac{2}{c_0}\left(\lambda_0 G_1 + G_2 + P_0 B_0 + \frac{1}{2}\Lambda_0 + \frac{1}{2}P_0 T||\varphi_0||\right)\right.$$
$$\left. + \Lambda_0 + P_0 T\left(\frac{\Lambda_0}{4} + \frac{P_0 T||\varphi_0||}{6}\right)\right] =: \alpha_3\|\varphi_0\|,$$

$$|A_4\varphi - \varphi_{04}| \le ||\varphi_0||T^2\left[\frac{2}{3} + L_0 + M_0 + \frac{r_{00}}{3}\right] =: \alpha_4\|\varphi_0\|,$$

$$|A_5\varphi - \varphi_{05}| \le ||\varphi_0||T\left[\frac{4}{3} + 2L_0 + 2M_0 + \frac{2r_{00}}{3}\right] =: \alpha_5\|\varphi_0\|,$$

$$|A_6\varphi - \varphi_{06}| \le 4||\varphi_0||^2 T P_0 =: \alpha_6\|\varphi_0\|,$$

It follows from the above estimates that the operator A maps the set $B(\varphi_0)$ to itself if the number T satisfies the condition

$$\max_{1\le i\le 6} \alpha_i < 1. \tag{1.145}$$

Now we verify the second condition of contraction operator. Let $\varphi^k := (\varphi_1^k, \varphi_2^k, \varphi_3^k, \varphi_4^k, \varphi_5^k, \varphi_6^k)$, and let $\varphi^k \in B(\varphi_0)$, $k = 1, 2$. We use the obvious inequalities

$$|\varphi_k^1\varphi_s^1 - \varphi_k^2\varphi_s^2| \le |\varphi_k^1 - \varphi_k^2||\varphi_s^1| + |\varphi_k^2||\varphi_s^1 - \varphi_s^2| \le 4\|\varphi_0\|\|\varphi^1 - \varphi^2\|$$

to obtain

$$\begin{aligned}
||(A\varphi^1 - A\varphi^2)_1|| = \int_x^t \Big[&\|\varphi_2^1(x,\tau) - \varphi_2^2(x,\tau)\| - c_0\|\varphi_5^1(\tau - x) - \varphi_5^2(\tau - x)\| \\
&-c_0 x\left(\frac{1}{3}(2\|\varphi_6^1(\tau - x) - \varphi_6^2(\tau - x)\| + \|\varphi_3^1(\tau - x) - \varphi_3^2(\tau - x)\|)\right. \\
&\left.-L_0\|\varphi_5^1(\tau - x) - \varphi_5^2(\tau - x)\| + M_0\|\varphi_4^1(\tau - x) - \varphi_4^2(\tau - x)\|\right) \\
&+\lambda_0 c_0\|\varphi_4^1(\tau - x) - \varphi_4^2(\tau - x)\|\Big]d\tau \\
\le ||\varphi^1 - \varphi^2||T\Big[1 + c_0\Big(&\frac{P_0 T}{4} + 1 + \lambda_0\Big)\Big] =: ||\varphi^1 - \varphi^2||\beta_0,
\end{aligned}$$

$$||(A\varphi^1 - A\varphi^2)_2|| \le ||\varphi^1 - \varphi^2||T\left[\frac{1}{2}(\lambda_0 + 1)G_0 + \frac{1}{2}\Lambda_0 + P_0\left(\frac{3}{2}||\varphi_0||T + \frac{1}{2}c_0\right)\right]$$

$$=: ||\varphi^1 - \varphi^2||\beta_2,$$

$$||(A\varphi^1 - A\varphi^2)_3|| \le ||\varphi^1 - \varphi^2||T\left[\frac{1}{c_0}\left(\lambda_0 G_1 + G_2 + P_0 B_0 + \frac{1}{2}\Lambda_0 + P_0 T||\varphi_0||\right)\right.$$

$$\left.+\frac{1}{2}\Lambda_0 + P_0 T\left(\frac{\Lambda_0}{8} + \frac{P_0 T||\varphi_0||}{6}\right)\right] =: ||\varphi^1 - \varphi^2||\beta_3,$$

$$||(A\varphi^1 - A\varphi^2)_4|| \le ||\varphi^1 - \varphi^2||T^2\left[\frac{1}{3} + \frac{1}{2}(L_0 + M_0) + \frac{r_{00}}{6}\right] =: ||\varphi^1 - \varphi^2||\beta_4,$$

$$||(A\varphi^1 - A\varphi^2)_5|| \le ||\varphi^1 - \varphi^2||T\left[\frac{2}{3} + L_0 + M_0 + \frac{r_{00}}{3}\right] =: ||\varphi^1 - \varphi^2||\beta_5,$$

$$||(A\varphi^1 - A\varphi^2)_6|| \le 4||\varphi^1 - \varphi^2||T P_0||\varphi_0|| =: ||\varphi^1 - \varphi^2||\beta_6.$$

Therefore, $||A\varphi^1 - A\varphi^2|| \le \rho||\varphi^1 - \varphi^2||$, where $\rho < 1$, if $T > 0$ satisfies condition (1.145) Thus, if the number T is small enough for condition (1.145) to be satisfied, then A is a contraction operator in the metric space $B(\varphi_0)$. Then, according to the Banach principle, there exists a unique solution of Eq. (1.144) in the space $B(\varphi_0)$ [9] (pp. 87–97).

Since $h_0(t) = k(t)\exp(h(0)t/2)$, the obtained function $h_0(t)$ can be used to determine the function $k(t)$ using the formula

$$k(t) = h_0(t)\exp(-h(0)t/2).$$

1.5 Problem of Determining the Memory in a Two-Dimensional System of Maxwell Equations with Distributed Data

The problems of finding the functions $\epsilon(x)$, $\mu(x)$, $\sigma(x)$ which describe electrodynamic characteristics of the medium from the system of Maxwell equations are the subject of a large number of works (see, e.g., the bibliography in [10]). In electromagnetic fields with a greater frequency compared to the frequencies characteristic of establishing electrical and magnetic polarization of medium, the most general form of the dependence of D and B (inductions of electric and magnetic fields, respectively) on the values of E and H (stresses of the corresponding fields) can be written in the form of integral relations [11]:

$$D(x,t) = \varepsilon E + \int_0^t \varphi(t-\tau)E(x,\tau)d\tau,$$

$$B(x,t) = \mu H + \int_0^t \psi(t-\tau)H(x,\tau)d\tau, \tag{1.146}$$

$$E = (E_1, E_2, E_3),\ H = (H_1, H_2, H_3),\ D = (D_1, D_2, D_3),$$
$$B = (B_1, B_1, B_3),\ x = (x_1, x_2, x_3),$$

$\varphi(t) = \mathrm{diag}(\varphi_1, \varphi_2, \varphi_3)$, $\psi(t) = \mathrm{diag}(\psi_1, \psi_2, \psi_3)$ are diagonal matrices functions representing memory of the medium.

In this section, the problem of determining the functions $\varphi(t)$, $\psi(t)$ is investigated. Wherein ϵ, μ, σ are considered as constants. Also assume that $\varphi_1 = \varphi_2 = \varphi_3$, $\psi_1 = \psi_2 = \psi_3$. In view of (1.146), we consider two-dimensional system of Maxwell equations [10]

$$\epsilon\frac{\partial E_2}{\partial t} - \frac{\partial H_1}{\partial z} + \frac{\partial H_3}{\partial x} + (\sigma + \varphi_0)E_2 + \int_0^t \varphi'(t-\tau)E_2(\tau,x,z)d\tau = 0,$$

$$\mu\frac{\partial H_1}{\partial t} - \frac{\partial E_2}{\partial z} + \psi_0 H_1 + \int_0^t \psi'(t-\tau)H_1(\tau,x,z)d\tau = 0, \tag{1.147}$$

$$\mu\frac{\partial H_3}{\partial t} + \frac{\partial E_2}{\partial x} + \psi_0 H_3 + \int_0^t \psi'(t-\tau)H_3(\tau,x,z)d\tau = 0,$$

where $\varphi_0 = \varphi(0)$, $\psi_0 = \psi(0)$. Further in (1.147) for simplicity of calculations, we put $\epsilon = \mu = \sigma = 1$ and introduce the new functions

$$U_1 = E_2 + H_1, \; U_2 = E_2 - H_1, \; U_3 = H_3.$$

Then, the system of Eq. (1.147) is written in the following canonical form:

$$\left(I_3 \frac{\partial}{\partial t} + K \frac{\partial}{\partial z} + E \frac{\partial}{\partial x} + N\right) U + \int_0^t \Phi(\tau) U(t - \tau, x, z) \mathrm{d}\tau = 0. \tag{1.148}$$

Here I_3 is the unit three-dimensional matrix, $K = \mathrm{diag}(-1, 1, 0)$, $U = (U_1, \; U_2, \; U_3)^*$, $*$ is the symbol of transposition,

$$E = \begin{pmatrix} 0 & 0 & 1 \\ 0 & 0 & 1 \\ \frac{1}{2} & \frac{1}{2} & 0 \end{pmatrix}, \quad N = \frac{1}{2} \begin{pmatrix} 1 + \varphi_0 + \psi_0 & 1 + \varphi_0 - \psi_0 & 0 \\ 1 + \varphi_0 - \psi_0 & 1 + \varphi_0 + \psi_0 & 0 \\ 0 & 0 & \psi_0 \end{pmatrix},$$

$$\Phi = \frac{1}{2} \begin{pmatrix} \varphi' + \psi' & \varphi' - \psi' & 0 \\ \varphi' - \psi' & \varphi' + \psi' & 0 \\ 0 & 0 & \psi' \end{pmatrix} (t).$$

Consider for the system (1.148) in the domain

$$G = \left\{(t, x, z) \middle| \, t > 0, \; -\infty < x < \infty, \; 0 < z < H\right\}$$

initial–boundary value problem

$$\begin{aligned} U\big|_{t=0} &= f(x, z), \; -\infty < x < \infty, \; 0 < z < H, \\ U_1\big|_{z=H} &= g_1(t, x), \; t > 0, \; -\infty < x < \infty, \\ U_2\big|_{z=0} &= g_2(t, x), \; t > 0, \; -\infty < x < \infty. \end{aligned} \tag{1.149}$$

We set the inverse problem: determine the functions $\varphi(t)$, $\psi(t)$, if the following conditions with respect to the solutions of problem (1.148) and (1.149) are known:

$$h_1(t) = \int_{-\infty}^{\infty} U_1(t, x, 0) \exp(i\lambda x) \mathrm{d}x,$$

$$h_2(t) = \int_{-\infty}^{\infty} U_2\,(t, x, H) \exp(i\lambda x) \mathrm{d}x, \; t > 0 \tag{1.150}$$

for some fixed λ. First, we assume that the functions $f(x, z)$, $g_1(t, x)$, $g_2(t, x)$ are finite ones in x for fixed (t, z), consequently, this condition guarantees the existence of Fourier transforms:

$$\tilde{f}(\lambda, z) = \int_{-\infty}^{\infty} f(x, z) \exp(i\lambda x) dx,$$

$$\tilde{g}_i(\lambda, t) = \int_{-\infty}^{\infty} g_i(t, x) \exp(i\lambda x) dx, \quad i = 1, 2.$$

Then, due to hyperbolicity of the system (1.148) the solution of initial–boundary problem (1.148) and (1.149) has the same property. We write relations (1.148) and (1.149) in terms of the Fourier transformation with respect to variable x

$$\left(I_3 \frac{\partial}{\partial t} + K \frac{\partial}{\partial z} + \tilde{N} \right) V + \int_0^t \Phi(\tau) V(t - \tau, \lambda, z) d\tau = 0, \tag{1.151}$$

$$V\big|_{t=0} = \tilde{f}(\lambda, z), \ 0 < z < H,$$
$$V_1\big|_{z=H} = \tilde{g}_1(\lambda, t), \ V_2\big|_{z=0} = \tilde{g}_2(\lambda, t), \ t > 0, \tag{1.152}$$

where

$$V(t, \lambda, z) = \int_{-\infty}^{\infty} U(t, x, z) \exp(i\lambda x) dx,$$

$$N = \frac{1}{2} \begin{pmatrix} 1 + \varphi_0 + \psi_0 & 1 + \varphi_0 - \psi_0 & -i\lambda \\ 1 + \varphi_0 - \psi_0 & 1 + \varphi_0 + \psi_0 & -i\lambda \\ -\frac{i\lambda}{2} & -\frac{i\lambda}{2} & \psi_0 \end{pmatrix}.$$

We introduce a new function W by the formula

$$W(t, \lambda, z) = V_t(t, \lambda, z).$$

To obtain relations for W, we differentiate with respect to t Eq. (1.151) and the boundary conditions in (1.152), and the value of W at $t = 0$ will be found using Eq. (1.151) and initial conditions. Then, the function W is the solution of the following problem:

$$\left(I_3 \frac{\partial}{\partial t} + K \frac{\partial}{\partial z} + \tilde{N} \right) W + \Phi(t) \tilde{f}(\lambda, x) + \int_0^t \Phi(\tau) W(t - \tau, x, z) d\tau = 0,$$

$$W\big|_{t=0} = -\left(K \frac{\partial}{\partial z} + \tilde{N} \right) \tilde{f}(\lambda, z), \quad 0 < z < H, \tag{1.153}$$

$$W_1\big|_{z=H} = \tilde{g}_{1t}(\lambda, t), \quad W_2\big|_{z=0} = \tilde{g}_{2t}(\lambda, t), \ t > 0.$$

Let (t, z) be an arbitrary point of the domain $G_H = \{(t, z) | \ t > 0, \ 0 < z < H\}$. On the plane of variables (τ, ξ) we draw the characteristics of differential operators through this point

$$\frac{\partial}{\partial t}-\frac{\partial}{\partial z},\ \frac{\partial}{\partial t}+\frac{\partial}{\partial z},\ \frac{\partial}{\partial t},$$

and continue them to the intersection in the domain $\tau \le t$ with the boundary of G_H. The intersection point is denoted by $\left(t_0^i, z_0^i\right), i = 1, 2, 3$. Integrating the ith component of equality (1.153) along the corresponding characteristic from $\left(t_0^i, z_0^i\right)$ to the point (t, z). Then one gets relations

$$\begin{aligned} W_i(t,\lambda,z) = W\left(t_0^i,\lambda,z_0^i\right) &- \int_{t_0^i}^{t} \sum_{j=1}^{3} \varphi_{ij} \tilde{f}_j(\lambda, z+\lambda_i(\tau-t))\mathrm{d}\tau \\ &- \int_{t_0^i}^{t} \Bigg[\sum_{j=1}^{3} n_{ij}(\tau) W_j(\tau,\lambda,\xi) \\ &+ \int_0^{\tau} \sum_{j=1}^{3} \varphi_{ij}(\tau) W_j(\tau-\gamma,\lambda,\xi)\mathrm{d}\gamma \Bigg]_{\xi=z+\lambda_i(\tau-t)} \mathrm{d}\tau, \quad (1.154) \end{aligned}$$

$i = 1, 2, 3$. Here

$$t_0^1(t,z) = \begin{cases} t+z-H,\ i=1,\ t>H-z \\ 0,\ i=1,\ t\le H-z \\ t-z,\ i=2,\ t>z \\ 0,\ i=2,\ t\le 0, i=3,\ (t,z)\in G_H, \end{cases}$$

$$W_i\left(t_0^i,\lambda,z_0^i\right) = \left\{ \begin{array}{l} \tilde{g}_{1t}(t+z-H,\lambda),\ i=1, t>H-z, \\ \tilde{f}_{1z}(\lambda,t+z)-\sum_{j=1}^{3} n_{1j}\tilde{f}_j(\lambda,t+z),\ i=1, t\le H-z, \\ \tilde{g}_{2t}(t-z,\lambda),\ i=2, t>z, \\ -\tilde{f}_{2z}(\lambda,t+z)-\sum_{j=1}^{3} n_{2j}\tilde{f}_j(\lambda,t-z),\ i=2, t\le z, \\ -\sum_{j=1}^{3} n_{3j}\tilde{f}_j(\lambda,z),\ i=3,\ (t,z)\in G_H, \end{array} \right\},$$

n_{ij}, φ_{ij} $(i, j = 1, 2, 3)$ denote the elements matrices $\tilde{N}$, Φ, respectively.

Consider now an arbitrary point $(0, z) \in \Gamma_0 = ((t, z)\ t = 0, 0 \le z \le H)$, and through this point, we draw the characteristic of differential operators

$$\frac{\partial}{\partial t}-\frac{\partial}{\partial z},\ \frac{\partial}{\partial t}+\frac{\partial}{\partial z}$$

up to their crossing the points of the lateral boundary of domain G_H. Integrating the first two equations of the system (1.153) and using the initial data and the data (1.150), we find

$$\lambda_i \tilde{f}_{iz}(\lambda, z) - \sum_{j=1}^{3} n_{ij} \tilde{f}_j(\lambda, z) - h_{it}(t_i(z)) + \int\limits_0^{t_i(z)} \sum_{j=1}^{3} \varphi_{ij}(\tau) \tilde{f}_j(\lambda, z + \lambda_i(\tau - t))d\tau$$

$$+ \int\limits_0^{t_i(z)} \left[\sum_{j=1}^{3} n_{ij} W_j(\tau, \lambda, \xi) + \int\limits_0^{\tau} \sum_{j=1}^{3} \varphi_{ij}(\tau) W_j(\tau - \gamma, \lambda, \xi) d\gamma \right]_{\xi = z + \lambda_i \tau} d\tau,$$

$$i = 1, 2, \qquad (1.155)$$

Assuming in equality (1.156) $z = 0$ for $i = 1$, and $z = H$ for $i = 2$ we obtain

$$n_{11} \tilde{f}_1(\lambda, 0) + n_{12} \tilde{f}_2(\lambda, 0) = -\tilde{f}_{1z}(\lambda, 0) - h_1'(0) + i\lambda \tilde{f}_3(\lambda, 0),$$
$$n_{21} \tilde{f}_1(\lambda, H) + n_{22} \tilde{f}_2(\lambda, H) = -\tilde{f}_{2z}(\lambda, H) - h_2'(0) + i\lambda \tilde{f}_3(\lambda, H).$$

Since $n_{11} = n_{11}$, $n_{12} = n_{21}$, from the last equalities with the fulfilling the condition

$$F(\lambda) = \begin{vmatrix} \tilde{f}_1(\lambda, 0) & \tilde{f}_2(\lambda, 0) \\ \tilde{f}_2(\lambda, H) & \tilde{f}_1(\lambda, H) \end{vmatrix} \neq 0, \qquad (1.156)$$

we uniquely find

$$n_{11} = \frac{1}{F(\lambda)} \Big\{ \tilde{f}_1(\lambda, H) \Big[-\tilde{f}_{1z}(\lambda, 0) - h_1'(0) + i\lambda \tilde{f}_3(\lambda, 0) \Big]$$
$$- \tilde{f}_2(\lambda, 0) \Big[-\tilde{f}_{2z}(\lambda, H) - h_2'(0) + i\lambda \tilde{f}_3(\lambda, H) \Big] \Big\},$$
$$n_{12} = \frac{1}{F(\lambda)} \Big\{ \tilde{f}_1(\lambda, 0) \Big[-\tilde{f}_{2z}(\lambda, H) - h_2'(0) + i\lambda \tilde{f}_3(\lambda, H) \Big]$$
$$- \tilde{f}_2(\lambda, H) \Big[-\tilde{f}_{1z}(\lambda, H) - h_1'(0) + i\lambda \tilde{f}_3(\lambda, H) \Big] \Big\}.$$

From these equalities, one can find numbers

$$\varphi_0 = n_{11} + n_{12} - 1, \;\; \psi_0 = n_{11} - n_{12}. \qquad (1.157)$$

Thus, the matrix $\tilde{N}$ becomes known.

Lemma 1.3 *Let $\varphi \in C^1[0, H]$, $\psi \in C^1[0, H]$, $\tilde{f}_i \in C^2[0, H]$, $i = 1, 2, 3$; $\tilde{g}_i \in C^1[0, H]$, $i = 1, 2$ and λ is some fixed number. Then, the functions $\frac{\partial W_i}{\partial z}$ $(i = 1, 2)$ are piecewise-continuous, and the function $\frac{\partial W_3}{\partial z}$ is continuous in the domain G_H, moreover, $\frac{\partial W_2}{\partial z}$ has a finite discontinuity along the characteristic $z = H - t$ and $\frac{\partial W_1}{\partial z}$ has a finite discontinuity along the characteristic $z = t$.*

To *prove* Lemma 1.3, we differentiate the equalities (1.154) with respect to z:

$$W_{iz}(t,\lambda,z) = W_z\left(t_0^i,\lambda,z_0^i\right) + \sum_{j=1}^{3}\Big[\varphi_{ij}(t_0^i)\tilde{f}_j\left(\lambda,z+\lambda_i(t_0^i-t)\right)$$

$$+n_{ij}W_j\left(t_0^i,\lambda,z+\lambda_i(t_0^i-t)\right)$$

$$+\int_0^{t_0^i}\sum_{j=1}^{3}\varphi_{ij}(\gamma)W_j\left(t_0^i-\gamma,\lambda,z+\lambda_i(t_0^i-t)\right)\mathrm{d}\gamma\Big]\frac{\mathrm{d}}{\mathrm{d}z}t_0^i$$

$$-\int_{t_0^i}^{t}\sum_{j=1}^{3}\varphi_{ij}(\tau)\tilde{f}_{jz}(\lambda,z+\lambda_i(\tau-t))\mathrm{d}\tau - \int_{t_0^i}^{t}\Big[\sum_{j=1}^{3}n_{ij}(\tau)W_{jz}(\tau,\lambda,\xi)$$

$$+\int_0^{\tau}\sum_{j=1}^{3}\varphi_{ij}(\gamma)W_{jz}(\tau-\gamma,\lambda,\xi)\mathrm{d}\gamma\Big]_{\xi=z+\lambda_i(\tau-t)}\mathrm{d}\tau,\quad i=1,2,3. \quad (1.158)$$

It is easy to see that the system of integral equations (1.158) has continuous kernels and piecewise continuous free members (first, second, third terms). Therefore, this equation has a unique solution and it can be obtained via the method of successive approximations. It is not difficult to believe that each iteration leads to piecewise-continuous functions $\frac{\partial W_1}{\partial z}$, $\frac{\partial W_2}{\partial z}$, with discontinuities along the corresponding characteristics, outgoing from the point $(0,0)$, $(H,0)$ and function $\frac{\partial W_3}{\partial z}$ which is continuous in the domain G_H. Note that for the solvability of the inverse problem (1.148)–(1.150) when the conditions of the lemma 1.3 are satisfied, it is necessary that the functions

$$h_i(t)\in C^1[0,H],\quad i=1,2,$$

satisfies the matching conditions with the initial data:

$$h_1(0)=\tilde{f}_1(\lambda,0),\quad h_2(0)=\tilde{f}_2(\lambda,H),$$

had piecewise-continuous second derivatives on $[0,H]$, moreover, for $h_i'(t)$ finite discontinuities are admissible only at the point $t=H$.

Theorem 1.11 *Let the conditions of the lemma 1.3, inequalities (1.156) and necessary conditions on the data of the inverse problem be satisfied. Then there exists such number H^*, that for $H\in(0,H^*)$ solution of inverse problem (1.148)–(1.150) exists, is unique and is determined by setting $h_i(t)$ for $t\in(0,H)$, $i=1,2$.*

For each i, differentiating (1.155) with respect to z, and taking into account the conditions from (1.153), we obtain

$$\varphi_{11}(z)\tilde{f}_1(\lambda,0)+\varphi_{12}(z)\tilde{f}_2(\lambda,0)=\tilde{f}_{1zz}(\lambda,z)+h_1'(\lambda,z)+\sum_{j=1}^{3}n_{1j}\tilde{f}_{jz}(\lambda,z)$$
$$-\int_0^z\sum_{j=1}^{3}\varphi_{1j}(\tau)\tilde{f}_{jz}(\lambda,z-\tau)d\tau$$
$$-\sum_{j=1}^{3}n_{1j}W_j(z,\lambda,0)-\int_0^z\sum_{j=1}^{3}\varphi_{1j}(\tau)W_j(z-\tau,\lambda,0)d\tau$$
$$-\int_0^z\sum_{j=1}^{3}\left[n_{1j}w_{jz}(\tau,\lambda,\xi)+\int_0^\tau\varphi_{1j}(\nu)W_{jz}(\tau-\nu,\lambda,\xi)d\nu\right]_{\xi=z-\tau}d\tau$$
$$=0, \tag{1.159}$$

$$\varphi_{11}(z)\tilde{f}_1(\lambda,H)+\varphi_{12}(z)\tilde{f}_2(\lambda,H)$$
$$=\tilde{f}_{2zz}(\lambda,H+z)+h_2'(\lambda,z)-\sum_{j=1}^{3}n_{2j}\tilde{f}_{jz}(\lambda,H+z)$$
$$-\int_0^z\sum_{j=1}^{3}\varphi_{1j}(\tau)\tilde{f}_{jz}(\lambda,H+z-\tau)d\tau-\sum_{j=1}^{3}n_{2j}w_j(z,\lambda,H)$$
$$+\int_0^z\sum_{j=1}^{3}\varphi_{1j}(\tau)W_j(z-\tau,\lambda,H)d\tau+\int_0^z\sum_{j=1}^{3}$$
$$\times\left[n_{2j}W_{jz}(\tau,\lambda,\xi)+\int_0^\tau\varphi_{1j}(\nu)W_{jz}(\tau-\nu,\lambda,\xi)d\nu\right]_{\xi=H+z+\tau}d\tau$$
$$=0. \tag{1.160}$$

The following equalities are also used here:

$$\varphi_{11}=\varphi_{22},\quad \varphi_{12}=\varphi_{21},\quad \varphi_{13}=\varphi_{23}=0.$$

The system of Eqs. (1.159) and (1.160) by virtue of condition (1.156) can be solved with respect to the functions $\varphi_{11},\varphi_{12}$. Then,

$$\varphi'(t)=\varphi_{11}+\varphi_{12},\quad \psi'(t)=\varphi_{11}-\varphi_{12}.$$

We add to relations (1.154), (1.158)–(1.160) the obvious equalities

$$\varphi(z)=\varphi_0+\int_0^z\varphi'(\tau)d\tau,\ \psi(z)=\psi_0+\int_0^z\psi'(\tau)d\tau, \tag{1.161}$$

where the numbers φ_0, ψ_0 are determined by formulas (1.157). Consider the square $G_{H,H} = \{(t, z)|\ 0 \le t \le H, 0 \le z \le H\}$. The system of Eqs. (1.154), (1.158)–(1.161) in this square is closed with respect to functions W W_i, W_{iz}, $i = 1, 2, 3$, $\varphi(z)$, $\varphi'(z)$, $\psi(z)$, $\psi'(z)$. To obtain from this system the normal form of nonlinear integral equations of the second kind, it is sufficient for beginning, to express the nonintegral terms $W_i(z, \lambda, 0)$, $W_i(z, \lambda, H)$, $i = 1, 2$, in (1.159) and (1.160) using equalities (1.150), the boundary conditions from (1.153) and $W_3(z, \lambda, 0)$, $W_3(z, \lambda, H)$ with the corresponding values using Eq. (1.154) for $i = 3$. Then, the nonintegral terms φ_{1i}, $i = 1, 2$ in (1.158) can be expressed using the solved system of Eqs. (1.159) and (1.160) with respect to φ_{1i}, $i = 1, 2$. The system of Eqs. (1.154), (1.158)–(1.160) contains the integration interval as a small parameter, which does not exceed H. Therefore, for small H, to this system is applicable the contracted mapping principle. It is easy to check that each iteration leads to continuous functions W_3, W_{3z}, $\varphi(z)$, $\varphi'(z)$, $\psi(z)$, $\psi'(z)$ in the domain $G_{H,H}$ and to piecewise-continuous functions W_i, W_{iz}, $i = 1, 2$ with discontinuities along the corresponding characteristics outgoing points $(0, 0)$, $(H, 0)$. Theorem 1.11 is proved.

1.6 Conclusions

In this chapter, the one-dimensional inverse problems of determining the memory of a medium entering into an integral term of convolution type in the second-order hyperbolic integro-differential equations are investigated. Section 1.1 discussed the problem of finding a kernel from an integro-differential wave equation in a half-space $\mathbb{R}^3_+ = \left\{(x_1, x_2, x_3) \in \mathbb{R}^3 \middle|\ x_3 > 0\right\}$. Further, in Sect. 1.2, the results obtained in the previous section are generalized to the case of a second-order hyperbolic-type equation with variable coefficients at lower derivatives with respect to $x = (x_1, x_2, x_3)$, the main part of which coincides with the wave operator. Based on the method of S. L. Sobolev in this section there have been constructed Volterra type integral equations which are equivalent to the inverse problem. The fixed point theorem (Banach's theorem) was applied to these integral equations. In both sections, the properties of direct problems solution, which are well-posed boundary–initial value problems, are also studied. In subsequent sections, by the method of the previous sections there have also been investigated three one-dimensional inverse problems for a system of thermoelasticity equations in a vertically inhomogeneous disconnected medium, an integro-differential equation of acoustics, and a two-dimensional system of Maxwell equations with memory.

The main results of this chapter are local existence theorems and conditional stability estimates for the solution of inverse problems.

References

1. Romanov, V.G. 1972. *Some inverse problems for hyperbolic-type equations*. Novosibirsk: Nauka (Russian).
2. Sobolev, S.L. 1930. Wave equation for an inhomogeneous medium, *Tr. Seysm. ins.* 6: 1–57 (in Russian).
3. Sobolev S.L. 1934. To the question on integration of wave equation in an inhomogeneous medium, *Tr. Seysm. ins.* 1 (42) (in Russian).
4. Lavrent'ev, M.M., V.G. Romanov, and S.P. Shishatskii. 1980. *Ill-posed problems of mathematical physics and analyses*. Moscow: Nauka (in Russian).
5. Kovalenko, A.D. 1975. *Thermoelasticity*. Kiev: Vish. shk (in Russian).
6. Yakhno, V.G. 1990. *Inverse problems for differential equations of elasticity*. Novosibirsk: Nauka (in Russian).
7. Courant, R., and D. Hilbert. 1962. *Methods of mathematical physics, II*, 830. New York/London: Interscience Publication.
8. Durdiev, D.K. 2009. Global solvability of one inverse problem for integro-differential equation of electrodynamics, *Uzb. Mat. J.* 3: 36–45 (in Russian).
9. Karchevsky, A.L., and A.G. Fat'yanova. 2001. Numerical solution of the inverse problem for an elasticity system with after effect for an upright nonuniform medium. *Sibir. Zh. Vychisl. Mat.* 4 (3): 259–268 (Russian).
10. Romanov, V.G. 1994. *Kabanikhin Inverse problems for Maxwell's equations*. Utrecht: VSP.
11. Landau, L.D., and E.M. Lifshits. 1959. *Electrodynamics of continuous media*. Moscow: Nauka (Russian).

Chapter 2
The Solvability of Multidimensional Inverse Problems of a Memory Kernel Determination

For multidimensional inverse problems there are only special cases for which solvability is established. One of such classes of functions in which solvability takes place is the class of analytic functions. The technique used here is based on the scale method of Banach spaces of analytic functions, developed in the works of Ovsyannikov [1, 2] and Nirenberg [3]. This method was first applied to the problem of solvability of multidimensional inverse problems by Romanov [4] (Sect. 3).

In this chapter, we consider inverse problems for the case when spatial variables belong to the half space, i.e.,

$$(x, z) \in \mathbb{R}_+^{m+1} := \mathbb{R}^n \times \mathbb{R}_+, \ \mathbb{R}_+ = \{z \in \mathbb{R}|_{z>0}\},$$

and the inverse problem is to determine the memory kernel $k(x, t)$, which is included in the integral term of the hyperbolic equation. In this case, it is assumed that the initial data in all the considered problems are equal to zero, and the Neumann boundary condition is given on

$$\partial\mathbb{R}_+^{m+1} \times [0, T], \ \partial\mathbb{R}_+^{m+1} := \left\{(x, z) \in \mathbb{R}_+^{m+1} | z = 0\right\}, \quad T > 0,$$

and is a smooth function in t and analytical in $x \in \mathbb{R}^n$. The inverse problems are investigated under the condition that solutions of the corresponding initial–boundary value problems are given for $(x, z, t) \in \partial\mathbb{R}_+^{m+1} \times [0, T]$, and they are analytic functions of a real variable x. A scale of functions, which are analytic in x, is introduced, and local existence theorems for the solution of the inverse problems are established.

© The Author(s), under exclusive license to Springer Nature Singapore Pte Ltd. 2023
D. K. Durdiev and Z. D. Totieva, *Kernel Determination Problems in Hyperbolic Integro-Differential Equations*, Infosys Science Foundation Series in Mathematical Sciences, https://doi.org/10.1007/978-981-99-2260-4_2

2.1 Definition of a Multidimensional Memory Kernel in the Integro-differential Wave Equation

Consider the problem of determining a pair of functions u and k satisfying the equations

$$u_{tt} - u_{zz} - \Delta u = \int_0^t k(x, \tau)u(x, z, t - \tau)d\tau, \ (x, t) \in \mathbb{R}^{m+1}, \ z > 0, \tag{2.1}$$

$$u\big|_{t<0} \equiv 0, \ u_z\big|_{z=0} = \delta'(t) + h(x, t)\theta(t), \ (x, t) \in \mathbb{R}^{m+1}, \tag{2.2}$$

$$u\big|_{z=0} = -\delta(t) + f(x, t)\theta(t), \ (x, t) \in \mathbb{R}^{m+1}. \tag{2.3}$$

Here Δ is the Laplace operator in the variables $(x_1, x_2, \ldots, x_m) = x$, $\theta(t)$ is the Heaviside function, h, f are given smooth functions.

In this section it will be shown that the inverse problem (2.1)–(2.3) is locally uniquely solvable in the class of functions analytic with respect to the variable x, and for the solution, a stability estimate is written and a global uniqueness theorem is established.

Following the monograph [4] (p. 92), we introduce into consideration the Banach space $A_s(r)$, $s > 0$, of functions $\varphi(x)$, $x \in \mathbb{R}^m$, which are analytic in the neighborhood of the origin and satisfy the relation

$$\|\varphi\|_s(r) := \sup_{|x|<r} \sum_{|\alpha|=0}^{\infty} \frac{s^{|\alpha|}}{\alpha!} |D^\alpha \varphi(x)| < \infty.$$

Here $r > 0$, $s > 0$ and

$$D^\alpha := \frac{\partial^{|\alpha|}}{\partial x_1^{\alpha_1} \ldots \partial x_m^{\alpha_m}}, \ \alpha := (\alpha_1, \ldots, \alpha_m),$$

$$|\alpha| := \alpha_1 + \cdots + \alpha_m, \ \alpha! := (\alpha_1)! \ldots (\alpha_m)!.$$

In the following, the parameter r will be considered fixed, while a parameter s is variable. Then, it is formed a scale of Banach spaces $A_s(r)$, $s > 0$ of analytic functions. The following property is obvious: if $\varphi(x) \in A_s(r)$, then $\varphi(x) \in A_{s'}(r)$ for all $s' \in (0, s)$, consequently, $A_s(r) \subset A_{s'}(r)$, if $s' \in (0, s)$ and the following inequality is valid:

$$\|D^\alpha \varphi\|_{s'} \leq c_\alpha \frac{\|\varphi\|_s(r)}{(s - s')^{|\alpha|}} \tag{2.4}$$

for any α with constant c_α, which depends only on α.

The estimate (2.4) can be shown as follows [4] (p. 93):

$$\|D^\alpha\varphi\|_{s'}(r) = \sup_{|x|<r} \sum_{|\beta|=0}^{\infty} \frac{(s')^{|\beta|}}{\beta!} \left|D^{\alpha+\beta}\varphi(x)\right|$$
$$\le \|\varphi\|_s(r)(s')^{-|\alpha|} \sup_{\beta} \left[\frac{(\alpha+\beta)!}{\beta!}\left(\frac{s'}{s}\right)^{|\alpha+\beta|}\right]$$
$$\le \|\varphi\|_s(r)(s')^{-|\alpha|} \sup_{k\ge|\alpha|} \left[k^{|\alpha|}\left(\frac{s'}{s}\right)^{k}\right] \le c_\alpha \frac{\|\varphi\|_s}{(s-s')^{|\alpha|}},$$

where $c_\alpha = |\alpha|^{|\alpha|}$. Since the parameter r is fixed, we will further omit it and use $\|\varphi\|_s$, A_s instead $\|\varphi\|_s(r)$ and $A_s(r)$.

Equalities (2.1) and (2.2) imply that $u \equiv 0,\ 0 < t < z,\ x \in \mathbb{R}^m$. The function $u(x, z, t)$ as a solution to problem (2.1) and (2.2) has the following structure in a neighborhood of the characteristic hyperplane $t = z$:

$$u(x, z, t) = -\delta(t-z) + \tilde{u}(x, z, t)\theta(t-z), \tag{2.5}$$

where $\tilde{u}(x, z, t)$ is a function which remains continuous in passage through the hyperplane $t = z$. The last equation implies the equality

$$\tilde{u}(x, z, z+0) = 0. \tag{2.6}$$

In particular, here arises one of the necessary conditions for solvability of the posed problem:

$$f(x, +0) = 0, \tag{2.7}$$

which is assumed to be satisfied in the sequel. In view of the above and the equality $u(x, z, t) = \tilde{u}(x, z, t)$, which is valid in the domain $D = \{(x, z, t)|\ x \in \mathbb{R}^m,\ 0 < z < t\}$, we rewrite the problem (2.1)–(2.3) in the form (we omit the symbol ˜ over u to simplify notation):

$$u_{tt} - u_{zz} = \Delta u - k(x, t-z) + \int_0^{t-z} k(x,\tau)u(x, z, t-\tau)d\tau,\ (x, z, t) \in D, \tag{2.8}$$

$$u\big|_{z=0} = f(x, t),\ u_z\big|_{z=0} = h(x, t),\ t > 0,\ x \in \mathbb{R}^m, \tag{2.9}$$

$$u\big|_{t=z+0} \equiv 0,\ x \in \mathbb{R}^m. \tag{2.10}$$

Starting with these equalities, we now construct a system of integro-differential equations for the functions u and k. With the help of d'Alembert's formula, from equalities (2.8) and (2.9) we obtain the equation

$$u(x,z,t)=u_0(x,z,t)+\frac{1}{2}\int\limits_{\Delta(z,t)}\Big[\Delta u(x,\xi,\tau)-k(x,\tau-\xi)$$

$$+\int\limits_0^{\tau-\xi}k(x,\gamma)u(x,\xi,\tau-\gamma)\mathrm{d}\gamma\Big]\mathrm{d}\tau\,\mathrm{d}\xi,\ (x,z,t)\in D. \tag{2.11}$$

Here

$$\Delta(z,t)=\{(\tau,\xi)|\,0\le\xi\le z,\ t-z+\xi\le\tau\le t+z-\xi\},$$

$$u_0(x,z,t)=\frac{1}{2}\left[f(x,t+z)+f(x,t-z)\right]+\frac{1}{2}\int\limits_{t-z}^{t+z}h(x,\tau)\mathrm{d}\tau. \tag{2.12}$$

Passing to the limit in the formula (2.11) as $t\to z+0$, taking conditions (2.7) and (2.10) into account we come to equality

$$\int\limits_0^z\int\limits_\xi^{2z-\xi}\Big[\Delta u(x,\xi,\tau)-k(x,\tau-\xi)$$

$$+\int\limits_0^{\tau-\xi}k(x,\gamma)u(x,\xi,\tau-\gamma)\mathrm{d}\gamma\Big]\mathrm{d}\tau\,\mathrm{d}\xi=-2u_0(x,z,z+0).$$

Differentiating this equality with respect to z, we obtain

$$\int\limits_0^z\Big[\Delta u(x,\xi,2z-\xi)-k\left[x,2(z-\xi)\right]$$

$$+\int\limits_0^{z-\xi}k(x,\gamma)u(x,\xi,2z-\xi-\gamma)\mathrm{d}\gamma\Big]\mathrm{d}\xi=-\frac{\mathrm{d}}{\mathrm{d}z}u_0(x,z,z+0). \tag{2.13}$$

From formulas (2.12) and (2.13), it is easy to derive the following matching condition:

$$f_t(x,+0)=-h(x,+0),$$

which, together with condition (2.7), is necessary for the solvability of the problem (2.1)–(2.4).

The equation for k can be obtained by differentiating the equality (2.13) with respect to z after replacing the variable $2z-\xi$ in the second integral by ξ':

$$k(x,z) = k_0(x,z) + 2\int_0^{z/2}\left[\Delta v(x,\xi,z-\xi) + \int_0^{z-2\xi} k(x,\gamma)v(x,\xi,z-\xi-\gamma)\mathrm{d}\gamma\right]\mathrm{d}\xi, \quad (2.14)$$

where

$$k_0(x,z) = 8\,[f_{tt}(x,t) + h_t(x,t)]_{t=2z}\,, \quad v(x,z,t) \equiv u_t(x,z,t).$$

Now we use equality (2.11) for calculating $v(x,z,t)$. Applying the equivalent description of the domain $\Delta(z,t)$ in the form

$$\Delta(z,t) = \{(\xi,\tau)|\; t-z \le \tau \le t+z,\; 0 \le \xi \le z - |t-\tau|\}\,,$$

we differentiate (2.11) with respect to t. Making the change of the integration variable $\tau - t = \xi'$, we find

$$v(x,z,t) = v_0(x,z,t) + \frac{1}{2}\int_z^{-z}\Bigg[\Delta u(x,z-|\xi|,\xi+t) - k(x,\xi+|\xi|+t-z) + \int_0^{\xi+|\xi|+t-z} k(x,\gamma)u(x,z-|\xi|,t+\xi-\gamma)\mathrm{d}\gamma\Bigg]\operatorname{sgn}\xi\,\mathrm{d}\xi, \quad (2.15)$$

where $v_0(x,z,t) = (\partial/\partial t)u_0(x,z,t)$.

The system of Eqs. (2.11), (2.14), and (2.15) is a closed integro-differential equations for the functions u, k, v. Note that the operator Δ in the functions u and v appears in the system only under the integral sign. We will henceforth consider system (2.11), (2.14), and (2.15) in the domain

$$D_T = G_T \times \mathbb{R}^m,\; G_T = \{(z,t)|\; 0 \le z \le t \le T - z\}\,,\; T > 0.$$

Let $C_{(z,t)}\left(G_T;\; A_{s_0}\right)$ denote the class of functions with values in A_{s_0} which are continuous in the variables (z,t) in the domain G_T. For fixed (z,t), the norm of a function $g(x,z,t)$ in A_{s_0} will be denoted by $\|g\|_{s_0}(z,t)$. The norm of a function g in $C_{(z,t)}\left(G_T;\; A_{s_0}\right)$ is defined by the equality

$$\|g\|_{C_{(z,t)}(G_T;\; A_{s_0})} = \sup_{(z,t)\in G_T} \|g\|_{s_0}(z,t). \quad (2.16)$$

Theorem 2.1 *Suppose that the above assumptions on the functions $f(x,t)$ and $h(x,t)$ are satisfied. Moreover, assume*

$$(f(x,t),\; h(x,t),\; f_t(x,t),\; h_t(x,t),\; f_{tt}(x,t)) \in C_t\left([0,T];\; A_{s_0}\right),$$

$$\max\left\{\|f\|_{s_0}(t),\ \max\left(1,\ \frac{T}{2}\right)\|h\|_{s_0}(t),\ \|f_t\|_{s_0}(t)\right\} \le \frac{R}{2},$$

$$\max\left\{\|f_{tt}\|_{s_0}(t),\ \|h_t\|_{s_0}(t)\right\} \le \frac{R}{16},\quad t\in[0,T],\ R>0.$$

Then, there exists such $a \in \left(0,\ \frac{T}{2s_0}\right)$ *that for every* $s \in (0, s_0)$ *in the domain* $\Gamma_{sT} = D_T \bigcap \{(x,z,t)|\ 0 \le z \le a(s_0 - s)\}$ *there is a unique solution to the system of Eqs. (2.11), (2.14), and (2.15) in the class of functions*

$$(u(x,z,t),\ v(x,z,t)) \in C_{(z,t)}\left(P_{sT};\ A_{s_0}\right)$$

$$k(x,t) \in C_t\left([0,\ a(s_0-s)];\ A_{s_0}\right),$$

$$P_{sT} = G_T \bigcap \{(z,t)|\ 0 \le z \le a(s_0 - s)\}$$

exists and it is unique, for which the estimates

$$\|u-u_0\|_s(z,t) \le R,\ \|k-k_0\|_s(z) \le \frac{R}{(s_0-s)^2},$$

$$\|v-v_0\|_s(z,t) \le \frac{R}{s_0-s},\quad (z,t)\in P_{sT}$$

are valid.

Proof Under the conditions of Theorem 2.1 we have

$$(u_0, k_0, v_0) \in C_{(z,t)}\left(G_T;\ A_{s_0}\right),\ \|u_0\|_s(z,t) \le R,$$

$$\|k_0\|_s(z) \le R,\ \|v_0\|_s(z,t) \le R,\ (z,t)\in G_T,\ 0<s<s_0.$$

Let a_n denote the members of the monotone decreasing sequence that is defined by the equalities

$$a_{n+1} = \frac{a_n}{1+1/(n+1)^2},\quad n=0,\ 1,\ 2,\ldots \tag{2.17}$$

Denote

$$a = \lim_{n\to\infty} a_n = a_0 \prod_{n=0}^{\infty}\left(1+1/(n+1)^2\right)^{-1}. \tag{2.18}$$

The number $a_0 \in \left(0,\ \frac{T}{2s_0}\right)$ will be chosen in an appropriate way. For the system of Eqs. (2.11), (2.14), and (2.15) we construct the process of successive approximations according to the following scheme:

$$u_{n+1}(x,z,t) = u_0(x,z,t) + \frac{1}{2}\int\limits_{\Delta(z,t)} \Bigg[\Delta u_n(x,\xi,\tau) - k_n(x,\tau-\xi) + \int\limits_0^{\tau-\xi} k_n(x,\gamma)u_n(x,\xi,\tau-\gamma)\mathrm{d}\gamma\Bigg]\mathrm{d}\tau\,\mathrm{d}\xi,$$

$$k_{n+1}(x,z) = k_0(x,z) + 2\int\limits_0^{z/2}\Bigg[\Delta v_n(x,\xi,z-\xi) + \int\limits_0^{z-2\xi} k_n(x,\gamma)v_n(x,\xi,z-\xi-\gamma)\mathrm{d}\gamma\Bigg]\mathrm{d}\xi, \qquad (2.19)$$

$$v_{n+1}(x,z,t) = v_0(x,z,t) + \frac{1}{2}\int\limits_z^{-z}\Bigg[\Delta u_n(x,z-|\xi|,\xi+t) - k_n(x,\xi+|\xi|+t-z) + \int\limits_0^{\xi+|\xi|+t-z} k_n(x,\gamma)u_n(x,z-|\xi|,t+\xi-\gamma)\mathrm{d}\gamma\Bigg]\operatorname{sgn}\xi\,\mathrm{d}\xi.$$

Define the function $s'(z)$ by the formula

$$s'(z) = \frac{s+\nu^n(z)}{2}, \quad \nu^n(z) = s_0 - \frac{z}{a_n}. \qquad (2.20)$$

Introduce the notations $p_n = u_{n+1} - u_n$, $q_n = k_{n+1} - k_n$, $r_n = v_{n+1} - v_n$, $n = 0, 1, 2, \ldots$. Then, p_n, q_n, r_n satisfy the relations

$$p_0(x,z,t) = \frac{1}{2}\int\limits_{\Delta(z,t)} \Bigg[\Delta u_0(x,\xi,\tau) - k_0(x,\tau-\xi) + \int\limits_0^{\tau-\xi} k_0(x,\gamma)u_0(x,\xi,\tau-\gamma)\mathrm{d}\gamma\Bigg]\mathrm{d}\tau\,\mathrm{d}\xi,$$

$$q_0(x,z) = 2\int\limits_0^{z/2}\Bigg[\Delta v_0(x,\xi,z-\xi) + \int\limits_0^{z-2\xi} k_0(x,\gamma)v_0(x,\xi,z-\xi-\gamma)\mathrm{d}\gamma\Bigg]\mathrm{d}\xi,$$

$$
\begin{aligned}
&r_0(x,z,t)\\
&=\frac{1}{2}\int_z^{-z}\Bigg[\Delta u_0(x,z-|\xi|,\xi+t)-k_0(x,\xi+|\xi|+t-z)\\
&+\int_0^{\xi+|\xi|+t-z}k_0(x,\gamma)u_0(x,z-|\xi|,t+\xi-\gamma)\mathrm{d}\gamma\Bigg]\operatorname{sgn}\xi\,\mathrm{d}\xi,
\end{aligned}
$$

$$
\begin{aligned}
p_{n+1}(x,z,t)&=u_0(x,z,t)\\
&+\frac{1}{2}\int_{\Delta(z,t)}\Bigg\{\Delta p_n(x,\xi,\tau)-q_n(x,\tau-\xi)\\
&+\int_0^{\tau-\xi}\Big[q_n(x,\gamma)u_{n+1}(x,\xi,\tau-\gamma)\\
&+k_n(x,\gamma)p_n(x,\xi,\tau-\gamma)\Big]\mathrm{d}\gamma\Bigg\}\mathrm{d}\tau\,\mathrm{d}\xi,
\end{aligned}
$$

$$
\begin{aligned}
&q_{n+1}(x,z)\\
&=2\int_0^{z/2}\Bigg\{\Delta r_n(x,\xi,z-\xi)\\
&+\int_0^{z-2\xi}\Big[q_n(x,\gamma)r_{n+1}(x,\xi,z-\xi-\gamma)\\
&+k_n(x,\gamma)r_n(x,\xi,z-\xi-\gamma)\Big]\mathrm{d}\gamma\Bigg\}\mathrm{d}\xi,
\end{aligned}
$$

$$
\begin{aligned}
&r_{n+1}(x,z,t)\\
&=\frac{1}{2}\int_z^{-z}\Bigg\{\Delta p_n(x,z-|\xi|,\xi+t)-q_n(x,\xi+|\xi|+t-z)\\
&+\int_0^{\xi+|\xi|+t-z}\Big[q_n(x,\gamma)u_{n+1}(x,z-|\xi|,t+\xi-\gamma)\\
&+k_n(x,\gamma)p_n(x,z-|\xi|,t+\xi-\gamma)\Big]\mathrm{d}\gamma\Bigg\}\operatorname{sgn}\xi\,\mathrm{d}\xi
\end{aligned}
$$

for $n=0,1,2,\ldots$.

Show that $a_0 \in (0, \frac{T}{2s_0})$ can be chosen so that the following inequalities be valid for all $n = 0, 1, 2, \ldots$:

$$\lambda_n = \max\left\{ \sup_{(z,t,s)\in F_n} \left[\|p_n\|_s(z,t)\frac{\nu^n(z) - s}{z} \right], \right.$$

$$\sup_{(z,t,s)\in F_n} \left[\|q_n\|_s(z)\frac{(\nu^n(z) - s)^3}{z} \right],$$

$$\left. \sup_{(z,t,s)\in F_n} \left[\|r_n\|_s(z,t)\frac{(\nu^n(z) - s)^2}{z} \right] \right\} < \infty, \tag{2.21}$$

$$\|u_{n+1} - u_0\|_s(z,t) \le R, \ \|k_{n+1} - k_0\|_s(z) \le \frac{R}{(s_0 - s)^2},$$

$$\|v_{n+1} - v_0\|_s(z,t) \le \frac{R}{s_0 - s}, \ (z,t,s) \in F_{n+1}, \tag{2.22}$$

where

$$F_n = \{(z,t,s) |\ (z,t) \in G_T, \ 0 \le z \le a_n(s_0 - s), \ 0 < s < s_0\}.$$

Indeed, using the relations for p_n, q_n, r_n, we find

$$\|p_0\|_s(z,t) \le \frac{1}{2} \int\limits_{\Delta(z,t)} \left[\|\Delta u_0\|_{s'(\xi)}(\xi,\tau) + \|k_0\|_{s'(\xi)}(\tau - \xi) \right.$$

$$\left. + \int\limits_0^{\tau-\xi} \|k_0\|_{s'(\xi)}(\gamma)\|u_0\|_{s'(\xi)}(\xi, \tau - \gamma)d\gamma \right] d\tau\, d\xi$$

$$\le \frac{1}{2} \int\limits_{\Delta(z,t)} \left[\frac{Rc_0}{(s'(\xi) - s)^2} + R(1 + Rt) \right] d\tau\, d\xi.$$

Here c_0 is a positive constant, such that

$$\|\ \Delta\varphi\|_s \le c_0\frac{\|\varphi\|_{s'}}{(s' - s)^2}, \quad s' > s > 0.$$

It is easy to check $c_0 = 4m$.

Taking the function $s'(\xi)$ from (2.20), for $n = 0$ we have

$$\|p_0\|_s(z,t)$$
$$\leq \int_0^z (z-\xi)\left[\frac{4Rc_0}{(\nu^0(\xi)-s)^2} + R(1+Rt)\mathrm{d}\xi\right] \leq R\left[4c_0 + s_0^2(1+RT)\right]$$
$$\times \int_0^z \frac{(z-\xi)\mathrm{d}\xi}{(\nu^0(\xi)-s)^2} \leq a_0 R\left[4c_0 + s_0^2(1+RT)\right]\frac{z}{\nu^0(z)-s}, \quad (z,t,s) \in F_0.$$

Proceeding analogously, we obtain

$$\|q_0\|_s(z) \leq 2\int_0^{z/2}\left[\frac{4Rc_0}{(\nu^0(\xi)-s)^2} + R^2 z\right]\mathrm{d}\xi$$
$$\leq R\left(4c_0 + s_0^2 RT\right)\frac{z}{(\nu^0(z)-s)^2} \leq s_0 R\left(4c_0 + s_0^2 RT\right)\frac{z}{(\nu^0(z)-s)^3},$$

$$\|r_0\|_s(z,t) \leq \frac{1}{2}\int_{-z}^{z}\left[\frac{4Rc_0}{(\nu^0(\xi)-s)^2} + R(1+3Rt)\right]\mathrm{d}\xi$$
$$\leq R\left[4c_0 + s_0^2(1+3RT)\right]\frac{z}{(\nu^0(z)-s)^2}, \quad (z,t,s) \in F_0.$$

These estimates imply validity of inequality (2.21) for $n = 0$. Moreover, for $(z,t,s) \in F_1$ we find

$$\|u_1 - u_0\|_s(z,t) = \|p_0\|_s(z,t) \leq \frac{\lambda_0 z}{\nu^0(z)-s} \leq \frac{\lambda_0 a_1}{1 - a_1/a_0} = a_0\lambda_0,$$

$$\|k_1 - k_0\|_s(z,t) = \|q_0\|_s(z) \leq \frac{\lambda_0 z}{(\nu^0(z)-s)^3} \leq \frac{4a_0\lambda_0}{(s_0-s)^2},$$

$$\|v_1 - v_0\|_s(z,t) = \|r_0\|_s(z,t) \leq \frac{\lambda_0 z}{(\nu^0(z)-s)^2} \leq \frac{2a_0\lambda_0}{s_0-s}.$$

Choosing a_0 so as to have $4a_0\lambda_0 \leq R$, we conclude that inequalities (2.22) are satisfied for $n = 0$.

By way of induction, we show that inequalities (2.21) and (2.22) are also valid for the other values of n if a_0 is chosen suitably. Assume that inequalities (2.21) and (2.22) hold for $n = 0, 1, 2, \ldots, i$. Then, for $(z,t,s) \in F_{i+1}$ we have

$$\|p_{i+1}\|_s(z,t) \le \frac{1}{2}\int\limits_{\Delta(z,t)} \Bigg[\|\Delta p_i\|_{s'(\xi)}(\xi,\tau) + \|q_i\|_{s'(\xi)}(\tau-\xi)$$

$$+ \int\limits_0^{\tau-\xi} \Big(\|q_i\|_{s'(\xi)}(\gamma)\|u_{i+1}\|_{s'(\xi)}(\xi,\tau-\gamma)$$

$$+ \|k_i\|_{s'(\xi)}(\gamma)\|p_i\|_{s'(\xi)}(\xi,\tau-\gamma)\Big)\mathrm{d}\gamma\Bigg]\mathrm{d}\tau\,\mathrm{d}\xi$$

$$\le \frac{1}{2}\int\limits_{\Delta(z,t)} \Bigg[\frac{c_0\lambda_i\xi}{(s'(\xi)-s)^2(\nu^i(\xi)-s)} + \frac{\lambda_i\xi}{(\nu^i(\xi)-s)^3}$$

$$+ \frac{4R\lambda_i t\xi}{(\nu^i(\xi)-s)^3} + \frac{2\lambda_i t\xi}{(\nu^i(\xi)-s)}\frac{R(1+s_0^2)}{(s_0-s)^2}\Bigg]\mathrm{d}\tau\,\mathrm{d}\xi$$

$$\le \lambda_i\left(4c_0 + 1 + 6Rt + 2Rs_0^2 t\right)\int\limits_0^z \frac{(z-\xi)\xi\,\mathrm{d}\xi}{(\nu^{i+1}(\xi)-s)^3}$$

$$\le \lambda_i a_0^2\left[2c_0 + RT(3+s_0^2)\right]\frac{z}{\nu^{i+1}(z)-s}.$$

Here in the intermediate calculations, the function s' is defined by equality (2.20) with $n=i$ and the inequalities

$$\|u_i\|_s(z,t) \le 2R, \quad \|k_i\|_s(z) \le R\frac{1+s_0^2}{(s_0-s)^2}$$

are used, the latter valid by the induction hypothesis, together with the obvious inequalities $a_i \le a_0$ and $\nu^{i+1}(z) \le \nu^i(z)$. Similar arguments for q_{i+1} and r_{i+1} lead to inequalities

$$\|q_{i+1}\|_s(z)$$

$$\le \frac{1}{2}\int\limits_0^{z/2}\Bigg\{\frac{c_0\lambda_i\xi}{(s'(\xi)-s)^2(\nu^i(\xi)-s)^2} + z\Bigg[\frac{\lambda_i R(1+s_0)\xi}{(\nu^i(\xi)-s)^3}$$

$$+ \frac{4R\lambda_i t\xi}{(\nu^i(\xi)-s)^3(s_0-s)} + \frac{\lambda_i R(1+s_0^2)\xi}{(\nu^i(\xi)-s)^2(s_0-s)}\Bigg]\Bigg\}\mathrm{d}\xi$$

$$\le 2\lambda_i\left[4c_0 + RT(2+s_0+s_0^2)\right]\int\limits_0^{z/2}\frac{\xi\,\mathrm{d}\xi}{(\nu^{i+1}(\xi)-s)^4}$$

$$\le \lambda_i a_0\left[4c_0 + RT(2+s_0+s_0^2)\right]\frac{z}{(\nu^{i+1}(z)-s)^3}.$$

$$\|r_{i+1}\|_s(z,t) \leq \frac{1}{2}\int_{-z}^{z}\Bigg[\frac{c_0\lambda_i\xi}{(s'(\xi)-s)^2(\nu^i(\xi)-s)}$$

$$+\frac{\lambda_i(1+6Rt)\xi}{(\nu^i(\xi)-s)^3}+\frac{3\lambda_i R(1+s_0^2)t\xi}{(\nu^i(\xi)-s)(s_0-s)^2}\Bigg\}d\xi$$

$$\leq \frac{1}{2}\lambda_i\left[4c_0+1+3RT(3+s_0^2)\right]\int_{-z}^{z}\frac{\xi\, d\xi}{(\nu^{i+1}(\xi)-s)^3}$$

$$\leq \lambda_i a_0\left[4c_0+1+3RT(3+s_0^2)\right]\frac{z}{(\nu^{i+1}(z)-s)^2},\quad (z,t,s)\in F_{i+1}.$$

The obtained estimates yield

$$\lambda_{i+1}\leq \lambda_i\rho,\ \lambda_{i+1}<\infty,\ \rho = a_0\left[4c_0+1+3RT(3+s_0+s_0^2)\right]\max(a_0,1).$$

Moreover, for $(z,t,s)\in F_{i+2}$, we have

$$\|u_{i+2}-u_0\|_s(z,t)\leq \sum_{n=0}^{i+1}\|p_n\|_s(z,t)$$

$$\leq \sum_{n=0}^{i+1}\frac{\lambda_n z}{\nu^n(z)-s}\leq \sum_{n=0}^{i+1}\frac{\lambda_n a_{i+2}}{1-a_{i+2}/a_n}$$

$$\leq \sum_{n=0}^{i+1}\lambda_n a_n(n+1)^2 \leq \lambda_0 a_0\sum_{n=0}^{i+1}\rho^n(n+1)^2,$$

$$\|k_{i+2}-k_0\|_s(z)\leq \sum_{n=0}^{i+1}\|q_n\|_s(z)$$

$$\leq \sum_{n=0}^{i+1}\frac{\lambda_n z}{(\nu^n(z)-s)^3}\leq \frac{1}{(s_0-s)^2}\sum_{n=0}^{i+1}\frac{\lambda_n a_{i+2}}{(1-a_{i+2}/a_n)^3}$$

$$\leq \frac{\lambda_0 a_0}{(s_0-s)^2}\sum_{n=0}^{i+1}\rho^n(n+1)^6,$$

$$\|v_{i+2} - v_0\|_s(z,t) \le \sum_{n=0}^{i+1} \|r_n\|_s(z,t)$$

$$\le \sum_{n=0}^{i+1} \frac{\lambda_n z}{(\nu^n(z) - s)^2} \le \frac{1}{s_0 - s} \sum_{n=0}^{i+1} \frac{\lambda_n a_{i+2}}{(1 - a_{i+2}/a_n)^2}$$

$$\le \frac{\lambda_0 a_0}{s_0 - s} \sum_{n=0}^{i+1} \rho^n (n+1)^4.$$

Now we choose $a_0 \in \left(0, \frac{T}{2s_0}\right)$ so as to obtain

$$\rho < 1, \ \lambda_0 a_0 \sum_{n=0}^{\infty} \rho^n (n+1)^6 \le R.$$

Then,

$$\|u_{i+2} - u_0\|_s(z,t) \le R, \ \|k_{i+2} - k_0\|_s(z) \le \frac{R}{(s_0 - s)^2},$$

$$\|v_{i+2} - v_0\|_s(z,t) \le \frac{R}{s_0 - s}, \quad (z,t,s) \in F_{i+2}.$$

Since the choice of a_0 which is independent of the number of the approximation, all the successive approximations $(u_n, \ k_n, \ v_n)$ belong to

$$C_{(z,t)}\left(F; \ A_s\right), \ F = \bigcap_{n=0}^{\infty} F_n,$$

and satisfy the inequalities

$$\|u_n - u_0\|_s(z,t) \le R, \ \|k_n - k_0\|_s(z) \le \frac{R}{(s_0 - s)^2},$$

$$\|v_n - v_0\|_s(z,t) \le \frac{R}{s_0 - s}, \quad (z,t,s) \in F.$$

For $s \in (0, s_0)$ the series

$$\sum_{n=0}^{\infty}(u_n - u_{n-1}), \ \sum_{n=0}^{\infty}(k_n - k_{n-1}), \ \sum_{n=0}^{\infty}(v_n - v_{n-1})$$

converge uniformly in the norm of the space

$$C_{(z,t)}\left(P_{sT};\ A_s\right),\ P_{sT}=G_T\bigcap\{(z,t)|\ 0\le z\le a(s_0-s)\},$$

therefore $u_n\to u$, $v_n\to v$, $k_n\to k$, and the limit functions u, k, v are elements of $C_{(z,t)}\left(P_{sT};\ A_s\right)$ and satisfy Eqs. (2.11), (2.14), and (2.15).

Now we prove that this solution is unique. Let $(u,\ v,\ k)$ and $(\hat{u},\ \hat{v},\ \hat{k})$ be any two solutions satisfying the inequalities

$$\|u-u_0\|_s(z,t)\le R,\ \|k-k_0\|_s(z)\le\frac{R}{(s_0-s)^2},$$

$$\|v-v_0\|_s(z,t)\le\frac{R}{s_0-s},\ (z,t,s)\in F.$$

Denote $\tilde{p}=u-\hat{u}$, $\tilde{r}=r-\hat{r}$, $\tilde{q}=q-\hat{q}$,

$$\lambda:=\max\left\{\sup_{(z,t,s)\in F}\left[\|\tilde{p}\|_s(z,t)\frac{\nu(z)-s}{z}\right],\ \sup_{(z,t,s)\in F}\left[\|\tilde{r}\|_s(z,t)\frac{(\nu(z)-s)^2}{z}\right],\right.$$
$$\left.\sup_{(z,t,s)\in F}\left[\|\tilde{q}\|_s(z)\frac{(\nu(z)-s)^3}{z}\right]\right\}<\infty,$$

where $\nu(z)=s_0-z/a$, $a=a_0\prod_{n=0}^{\infty}\left(1+1/(n+1)^2\right)^{-1}$. Then, for the functions $\tilde{p},\tilde{r},\tilde{q}$ can be obtained the relations

$$\tilde{p}(x,z,t)=\frac{1}{2}\int_{\Delta(z,t)}\left\{\Delta\tilde{p}(x,\xi,\tau)-\tilde{q}(x,\tau-\xi)\right.$$
$$\left.+\int_0^{\tau-\xi}\left[\tilde{q}(x,\gamma)\hat{u}(x,\xi,\tau-\gamma)+k(x,\gamma)\tilde{p}(x,\xi,\tau-\gamma)\right]\mathrm{d}\gamma\right\}\mathrm{d}\tau\,\mathrm{d}\xi,$$

$$\tilde{r}(x,z,t)=\frac{1}{2}\int_z^{-z}\left\{\Delta\tilde{p}(x,z-|\xi|,\xi+t)-\tilde{q}(x,\xi+|\xi|+t-z)\right.$$
$$+\int_0^{\xi+|\xi|+t-z}\left[\tilde{q}(x,\gamma)\hat{u}(x,z-|\xi|,t+\xi-\gamma)\right.$$
$$\left.\left.+k(x,\gamma)\tilde{p}(x,z-|\xi|,t+\xi-\gamma)\right]\mathrm{d}\gamma\right\}\operatorname{sgn}\xi\,\mathrm{d}\xi,$$

$$\tilde{q}(x,z) = 2\int_0^{z/2}\Bigg\{\Delta\tilde{r}(x,\xi,z-\xi)$$

$$+\int_0^{z-2\xi}\left[\tilde{q}(x,\gamma)\hat{r}(x,\xi,z-\xi-\gamma)+k(x,\gamma)\tilde{r}(x,\xi,z-\xi-\gamma)\right]\mathrm{d}\gamma\Bigg\}\mathrm{d}\xi.$$

Applying to these equations the estimates given above, we find the inequality in the form

$$\lambda \le \lambda\rho',\ \ \rho' := a\left[4c_0 + 1 + 3RT(3 + s_0 + s_0^2)\right]\max(a,1) < \rho < 1.$$

Consequently, $\lambda = 0$. Therefore, $u = \hat{u}$, $r = \hat{r}$, $q = \hat{q}$. Theorem 2.1 is proved. □

Remark 2.1 The above-constructed functions u and k represent a weak (= generalized) solution to problem (2.8)–(2.10) in the domain $\mathbb{R}^m \times P'_{sT}$ and the functions u' and k, where

$$u'(x,z,t) = -\delta(t-z) + u(x,z,t)\theta(t-z),$$

a generalized solution to problem (2.1)–(2.3).

Consider the set Υ consisting of all pairs of functions $(f,\ h)$ that are the elements of $C_t\left([0,T];\ A_{s_0}\right)$, $s_0 > 0$, for which the conditions of Theorem 2.1 are valid with fixed R, T, and s_0. The following stability theorem holds:

Theorem 2.2 *Let $(f,\ h) \in \Upsilon$ and $(\bar{f},\ \bar{h}) \in \Upsilon$. Then, the corresponding solutions $(u,\ k,\ v)$ and $(\bar{u},\ \bar{k},\ \bar{v})$ to the system of Eqs. (2.11), (2.14), and (2.15) satisfy the following estimates:*

$$\|u - u\|_s(z,t) \le c\epsilon,\ \ \|k - \bar{k}\|_s(z) \le \frac{c\epsilon}{(s_0 - s)^2},$$

$$\|v - \bar{v}\|_s(z,t) \le \frac{c\epsilon}{s_0 - s},\ \ (z,t) \in P_{sT},\ 0 < s < s_0, \tag{2.23}$$

where

$$\epsilon = \max\Bigg[\max\|f - \bar{f}\|_{s_0}(t),\ \max\|h - \bar{h}\|_{s_0}(t)\max\left(1,\ \frac{T}{2}\right),$$

$$\max\|f_t - \bar{f}_t\|_{s_0}(t),\ \max\|h_t - \bar{h}_t\|_{s_0}(t),\ \max\|f_{tt} - \bar{f}_{tt}\|_{s_0}(t)\Bigg],$$

$t \in [0,T]$, and the constant c depends only on R, T, and s_0.

Proof For the differences $u - \bar{u} = \tilde{u}$, $k - \bar{k} = \tilde{k}$, $v - \bar{v} = \tilde{v}$, $f - \bar{f} = \tilde{f}$, and $h - \bar{h} = \tilde{h}$, from relations (2.11), (2.14), and (2.15) we derive the equalities

$$\tilde{u}(x,z,t) = \tilde{u}_0(x,z,t) + \frac{1}{2}\int_{\Delta(z,t)} \Big\{ \Delta\tilde{u}(x,\xi,\tau) - \tilde{k}(x,\tau-\xi)$$

$$+ \int_0^{\tau-\xi} \Big[\tilde{k}(x,\gamma)u(x,\xi,\tau-\gamma) + \bar{k}(x,\gamma)\tilde{u}(x,\xi,\tau-\gamma) \Big] d\gamma \Big\} d\tau\, d\xi,$$

$$\tilde{k}(x,z) = \tilde{k}_0(x,z) + 2\int_0^{z/2} \Big\{ \Delta\tilde{v}(x,\xi,z-\xi)$$

$$+ \int_0^{z-2\xi} \Big[\tilde{k}(x,\gamma)v(x,\xi,z-\xi-\gamma) + \bar{k}(x,\gamma)\tilde{v}(x,\xi,z-\xi-\gamma) \Big] d\gamma \Big\} d\xi, \quad (2.24)$$

$$\tilde{v}(x,z,t) = \tilde{v}_0(x,z,t) + \frac{1}{2}\int_z^{-z} \Big\{ \Delta\tilde{u}(x,z-|\xi|,\xi+t) - \tilde{k}(x,\xi+|\xi|+t-z)$$

$$+ \int_0^{\xi+|\xi|+t-z} \Big[\tilde{k}(x,\gamma)u(x,z-|\xi|,t+\xi-\gamma)$$

$$+ \bar{k}(x,\gamma)\tilde{u}(x,z-|\xi|,t+\xi-\gamma) \Big] d\gamma \Big\} \operatorname{sgn}\xi\, d\xi,$$

in which $\tilde{u}$ is calculated by the formula (2.12) with f, h replaced by $\tilde{f}, \tilde{h}$,

$$\tilde{k}_0(x,z) = 8\left(\tilde{f}_{tt}(x,t) + \tilde{h}_t(x,t)\right)\Big|_{t=2z}; \quad \tilde{v}(x,z,t) = \tilde{u}_{0t}(x,z,t).$$

It is obvious that

$$\|\tilde{u}_0\|_{s_0}(z,t) \le 2\epsilon, \quad \|\tilde{k}_0\|_{s_0}(z) \le 16\epsilon, \quad \|\tilde{v}_0\|_{s_0}(z,t) \le 2\epsilon, \quad (z,t) \in P_{sT}. \quad (2.25)$$

Theorem 2.1 yields the following estimates:

$$\|u\|_s(z,t) \le 2R, \quad \|\bar{k}\|_s(z) \le \frac{R(1+s_0^2)}{(s_0-s)^2}, \quad \|v\|_s(z,t) \le \frac{R(1+s_0)}{s_0-s}.$$

Applying to the system of Eq. (2.24) which is linear in $\tilde{u}$, $\tilde{k}$, $\tilde{v}$ the method of successive approximation that was used in the proof of Theorem 2.1, we find that, for the same choice of a_0, solutions to system (2.24) satisfy the inequalities

$$\|\tilde{u}-\tilde{u}_0\|_s(z,t) \leq c_1\epsilon, \quad \|\tilde{k}-\tilde{k}_0\|_s(z) \leq \frac{c_1\epsilon}{(s_0-s)^2},$$

$$\|\tilde{v}-\tilde{v}_0\|_s(z,t) \leq \frac{c_1\epsilon}{s_0-s}, \quad (z,t) \in P_{sT}, \quad 0<s<s_0,$$

in which c_1 depends only on R, T, and s_0. Hence, taking (2.25) into account, we obtain inequalities (2.23).

Denote by H_{s_0} the set of pairs of functions $(u,\ k)$ such that u, u_t, u_z and k belong to A_{s_0} and are continuous for $(z,t) \in G_T,\ T>0$.

Theorem 2.3 *There exists at most one solution to inverse problem (2.8)–(2.10) in the set* H_{s_0}.

Proof The proof is carried out by way of contradiction. Suppose that there are two solutions $(u_1,k_1) \in H_{s_0}$ and $(u_2,k_2) \in H_{s_0}$ to the inverse problem (2.8)–(2.10). Fix any $s \in (0,s_0)$. Since $(u_i,k_i) \in H_{s_0}$, $i=1,2$, by virtue of Theorem 2.1, there exists such $a>0$ that the following relations are valid:

$$u_1(x,z,t) = u_2(x,z,t),\ k_1(x,z) = k_2(x,z),$$
$$x \in \mathbb{R}^n,\ (z,t) \in G_T,\ 0 \leq z \leq a. \tag{2.26}$$

Denote by a^* the least upper bound of all possible numbers $a \in (0,T/2)$ for which equalities (2.26) hold. Since $(u_1,\ k_1) \neq (u_2,\ k_2)$, we have $a^* \in (0,T/2)$. Now consider the problem similar to (2.8)–(2.10), in which the plane $z=a^*$ plays the role of the support of the Cauchy data. Since on this plane the traces of the functions u_1, u_2 and their derivatives u_{1z}, u_{2z} coincide, then, in view of Theorem 2.1 there exists a number $a^{**} \in (a^*,\ T/2)$ such that $(u_1,\ k_1) = (u_2,\ k_2)$, $x \in \mathbb{R}^n$, $(z,t) \in G_T$, $a^* \leq z \leq a^{**}$, which fact contradicts the definition of a^*. The contradiction proves the validity of Theorem 2.3.

2.2 Definition of the Memory Kernel Standing at Some Derivatives of a Direct Problem Solution

In this section, we indicate the statements of several inverse problems, the study of which can be carried out according to the method of Sect. 2.1. Unlike the problem that was considered in the previous section, where the memory kernel is multiplied by the solution of the direct problem, the equations are investigated here, in which the memory kernel is multiplied by the second derivative in time of direct problem solution and the first-order differential operator with analytical coefficients.

At the beginning, we consider an initial–boundary problem for the wave equation with memory

$$u_{tt} - u_{zz} - \Delta u = \int_0^t k(x,\tau)u_{tt}(x,z,t-\tau)\mathrm{d}\tau,$$

$$(x,t) \in \mathbb{R}^{n+1}, \ z \in \mathbb{R}_+, \tag{2.27}$$

$$u|_{t<0} \equiv 0, u_z|_{z=0} = -\delta'(t) + g(x,t)\theta(t), \ (x,t) \in \mathbb{R}^{n+1}. \tag{2.28}$$

The inverse problem is to find $k(x,t)$ via the trace of solution of problem (2.27) and (2.28) on the hyperplane $z = 0$ for all $x \in \mathbb{R}^n, \ t < T, \ T > 0$, i.e.,

$$u|_{z=0} = F(x,t), \ x \in \mathbb{R}^n, \quad t < T. \tag{2.29}$$

It is assumed that unknown function $k(x,t)$ at any fixed x belongs to class $C^2[0, \ T]$.

In Eq. (2.27) by integrating by parts we “relocate” time derivative and by introducing a new function $v(x,z,t)$ according to the formula

$$u = \rho(x,t)v(x,z,t), \ \rho(x,t) := \exp[k_0(x)t/2], \ k_0(x) := k(x,0),$$

from equalities (2.27) and (2.28), we have

$$v_{tt} - v_{zz} = \Delta v + t\,(\nabla k_0, \nabla v) + H(x,t)v + \int_0^t h(x,t-\tau)v(x,z,\tau)\mathrm{d}\tau,$$

$$(x,t) \in \mathbb{R}^{n+1}, \quad z > 0, \tag{2.30}$$

$$v|_{t<0} \equiv 0, \ v_z|_{z=0} = -\delta'(t) - [k_0(x)/2]\delta(t) + G(x,t)\theta(t),$$

$$(x,t) \in \mathbb{R}^{n+1}, \tag{2.31}$$

where $\nabla = (\partial/\partial x_1, \ldots, \partial/\partial x_n)$, $(\cdot,\cdot)$ is scalar product as well as the notations are introduced

$$H(x,t) := k_{0t}(x) + k_0^2(x)/4 + t\Delta k_0(x)/2 + (t^2/2)\sum_{i=1}^n k_{0x_i}^2(x),$$

$$h(x,t) := \exp[-k_0(x)t/2]k_{tt}(x,t),$$

$$G(x,t) = \exp[-k_0(x)t/2]g(x,t). \tag{2.32}$$

To obtain the second condition from (2.31), the equality

$$\exp[-k_0(x)t/2]\delta'(t) = \delta'(t) + [k_0(x)/2]\delta(t),$$

is used, which is valid in the generalized sense.

From the theory of hyperbolic equations we conclude that $v \equiv 0,\ 0 < t < z$, $x \in \mathbb{R}^n$. We represent the solution of (2.30) and (2.31) as

$$v(x,z,t) = \delta(t-z) + \tilde{v}(x,z,t)\theta(t-z).$$

Using the method of separation of variables, it is not difficult to find

$$v(x,z,z+0) = (1/2)\left[k_0(x) + \int_0^z H(x,\xi)\mathrm{d}\xi\right] =: \beta(x,z).$$

That is why the function $F(x,t)$ in (2.29) should be represented as

$$F(x,t) = \delta(t) + f(x,t)\theta(t),\ (x,t) \in \mathbb{R}^{n+1}. \tag{2.33}$$

It is clear that $\tilde{v} = v$ for $t > z$. For the regular part of function $v(x,z,t)$ in the domain $t > z > 0$, $x \in \mathbb{R}^n$, the inverse problem (2.30), (2.31), and (2.29) is equivalent to the problem

$$\begin{aligned} v_{tt} - v_{zz} &= \Delta v + t\,(\nabla k_0, \nabla v) + H(x,t)v \\ &\quad + \int_0^{t-z} h(x,t-\tau)v(x,z,\tau)\mathrm{d}\tau, \end{aligned} \tag{2.34}$$

$$v|_{z=0} = f(x,t),\ v_z|_{z=0} = G(x,t), \tag{2.35}$$

$$v|_{t=z+0} = \beta(x,z). \tag{2.36}$$

In Eqs. (2.34)–(2.36) we replace the variables z, t by z_1, t_1 by the formulas

$$z_1 = t + z,\ t_1 = t - z.$$

Then, $v(x,z,t) = v\Big(x, (z_1 - t_1)/2, (z_1 + t_1)/2\Big) := v_1\,(x, z_1, t_1)$. The problem (2.34)–(2.36) in new variables is rewritten as the problem of finding the functions $v,\ k$ from the equations

$$\frac{\partial^2 v_1(x, z_1, t_1)}{\partial t_1 \partial z_1}$$
$$= -\frac{1}{4}\Bigg[\Delta v_1 + (t_1 + z_1)(\nabla k_0, \nabla v_1)/2 + H(x, (t_1 + z_1)/2)v_1 + h(x, t_1)$$
$$+ \int_0^{t_1} h(x, \tau)v_1(x, z_1 - \tau, t_1 - \tau)\mathrm{d}\tau\Bigg],$$

$$(x, z_1, t_1) \in \{(x, z_1, t_1) \mid x \in \mathbb{R}^n, 0 \le t_1 \le z_1\} := D, \qquad (2.37)$$

$$v_1 \mid_{t_1=z_1} = f(x, z_1), \quad \frac{\partial}{\partial z_1} v_1 \mid_{t_1=z_1} = \frac{1}{2} f_t(x, t) \mid_{t=z_1} + \frac{1}{2} G(x, z_1),$$
$$z_1 > 0, \quad x \in \mathbb{R}^n, \qquad (2.38)$$

$$v_1 \mid_{t_1=0} = \beta[x, (1/2)z_1], \quad z_1 > 0, \ x \in \mathbb{R}^n. \qquad (2.39)$$

We recall that h is related to k according to the second formula of (2.32).

We introduce the function

$$w(x, z_1, t_1) = \frac{\partial}{\partial z_1} v_1(x, z_1, t_1), \quad t_1 < z_1. \qquad (2.40)$$

Demanding the continuity of functions $v(x, z, t)$, $w(x, z, t)$ when $z_1 = t_1 = 0$, $x \in \mathbb{R}^n$, from (2.38) and (2.39) it is not difficult to express $k_0(x)$, $k_{0t}(x)$ by the known functions:

$$k_0(x) = 2f(x, 0), \quad k_{0t}(x) = 2f_t(x, t)|_{t=0} - f^2(x, 0) + 2g(x, 0).$$

Further we will assume that in the equalities for $H(x, z)$, $G(x, t)$, $\beta(x, z)$ instead of functions $k_0(x)$, $k_{0t}(x)$ their expressions by means of the latter equations are used. For simplicity we will omit index 1 in z_1, t_1, v_1.

We denote by $C_z^i(A_s, G)$ a class of functions which are continuously differentiable i-times in z and continuous in t in the domain G. For fixed (z, t) the norm of the function $\omega(x, z, t)$ in A_{s_0}, as earlier, we denote by $\|\omega\|_{s_0}(z, t)$. The norm of the function ω in $C(A_s, G)$ is defined by the equality

$$\|\omega\|_{C(A_s, G)} = \sup_{(z,t)\in G} \|\omega\|_{s_0}(z, t).$$

Theorem 2.4 *Let* $f(x, 0)$, $f_t(x, t)|_{t=0}$, $f_{x_i}(x, 0)$, $i = 1, \ldots, n$, $\Delta f(x, 0)$, $g(x, 0)$ *belong to* A_{s_0}, $s_o > 0$, *and* $f(x, t)$, $f_t(x, t)$, $f_{tt}(x, t)$, $g(x, t)$, $g_t(x, t)$ *belong to* $C(A_{s_0}, [0, T])$, *and*

$$\max\left[\|f(x,t)\|_{s_0}(t),\ \|w_0\|_{s_0}(z),\ \|h_0(x,z)\|_{s_0}(z),\ \|k_0(x)\|_{s_0}\right] = R,$$

for

$$(z,t) \in G_T := \{(z,t)|0 \le t \le z \le T\}.$$

Then, for any $\chi > 0$ *we can find the number* $a = a(s_0, T, R, n)$, $as_0 < T$ *so that for any* $s \in (0, s_0)$ *there exists a unique solution of the problem (2.37)–(2.39)*

$$v(x,z,t) \in C^1_z\left(A_{s_0},\ D_s\right),\ k(x,t) \in C^2_t\left(A_{s_0},\ [0, a(s_0 - s)]\right),$$

D_s *is the domain on the plane* z, t : $D_s := \{(z,t)|0 \le t \le z < a(s_0 - s)\}$, *and solution satisfies the following inequalities:*

$$\|v - v_0\|_s(z,t) \le \chi,\ \|k - k^0\|_s(z) \le \frac{2\chi}{(s_0 - s)}, \tag{2.41}$$

where

$$\begin{aligned} h_0(x,z) &:= 2f(x,0)\exp[-f(x,0)z]g(x,z) \\ &- 2\exp[-f(x,0)z]g_t(x,t)|_{t=z} + (1/2)\Delta f(x,0) - H(x,z)f(x,z) \\ &- \Delta f(x,z) - 2z\,(\nabla f(x,0), \nabla f(x,z)) \\ &- 2f_{tt}(x,t)|_{t=z} + (z/2)\sum_{i=1}^{n} f^2_{x_i}(x,0), \end{aligned}$$

$$v_0 := f(x,z),\ k^0 := k_0(x) + zk_{0t},\ (z,t) \in D_s.$$

Proof At the beginning, the problem (2.37)–(2.39) is reduced to the closed system of the Volterra-type integro-differential equations in the domain $(x, z, t) \in D$. The equation for v is rewritten with the help of Eq. (2.40), while the equation for k is rewritten with the help of the second equation of (2.32):

$$v(x,z,t) = v_0(x,z) + \int_t^z w(x,\xi,t)\mathrm{d}\xi, \tag{2.42}$$

$$k(x,z) = k^0(x) + \int_0^z (z-\xi)\exp[f(x,0)\xi]\,h(x,\xi)\mathrm{d}\xi. \tag{2.43}$$

For fixed $x \in \mathbb{R}^n$ by integrating equality (2.37) on the plane (τ, ξ) along the line $\xi = z$ from the point (t, z) to the point (z, z) and using the second condition in (2.38), we can get the equation for $w(x, z, t)$:

$$w(x,z,t) = w_0(x,z) + \frac{1}{4}\int_t^z [\Delta v(x,z,\tau) + (z+\tau)\nabla f(x,0)\nabla v(x,z,\tau) + H[x,(z+\tau)/2]v(x,z,\tau) + h(x,\tau) + \int_0^\tau h(x,\eta)v(x,z-\eta,\tau-\eta)\mathrm{d}\eta]\mathrm{d}\tau, \tag{2.44}$$

where $w_0(x,z) = (1/2)f_t(x,t)|_{t=z} + (1/2)G(x,z)$. Using Eq. (2.39) differentiated by z, when $t=0$ from (2.44), we find

$$\int_0^z \Big[\Delta v(x,z,\tau) + (z+\tau)\left(\nabla f(x,0), \nabla v(x,z,\tau)\right) + H[x,(z+\tau)/2]v(x,z,\tau) + h(x,\tau) + \int_0^\tau h(x,\eta)v(x,z-\eta,\tau-\eta)\mathrm{d}\eta\Big]\mathrm{d}\tau = 2f_t(x,0) + 2g(x,0) + (z/2)\Delta f(x,0) + (z^2/4)\sum_{i=1}^n f_{x_i}^2(x,0) - 2f_t(x,z) - 2\exp[-f(x,0)z]g(x,z).$$

The equation for $h(x,t)$ can be easily found by differentiation of the latter equation on z and by using the first condition from (2.38):

$$h(x,t) = h_0(x,t) - \int_0^z \Big[\Delta w(x,z,\tau) + (z+\tau)\left(\nabla f(x,0), \nabla w(x,z,\tau)\right) + \left(\nabla f(x,0), \nabla v(x,z,\tau)\right) + H_z[x,(z+\tau)/2]v(x,z,\tau) + H[x,(z+\tau)/2]w(x,z,\tau) + h(x,\tau)f(x,z-\tau) + \int_0^\tau h(x,\eta)w(x,z-\eta,\tau-\eta)\mathrm{d}\eta\Big]\mathrm{d}\tau, \tag{2.45}$$

where the function $h_0(x,t)$ is determined in Theorem 2.4.

Equations (2.42)–(2.45) represent a closed system of integro-differential equations for the functions v, k, w, h in the domain $D_T = G_T \times \mathbb{R}^n$. For convenience we introduce the function vector

$$\phi(x,z,t) = (\phi_1, \phi_2, \phi_3, \phi_4) := (v, k, w, h).$$

We rewrite the system (2.42)–(2.45) as an operator equation

$$\phi = M\phi, \tag{2.46}$$

where $M = (M_1, \ldots, M_4)$ is defined by the right sides of Eqs. (2.42)–(2.45), and

$$\phi^0 = M(0), \;\; \phi^0 = (\phi_1^0, \phi_2^0, \phi_3^0, \phi_4^0) := (v_0, k^0, w_0, h_0).$$

We define the iterations for Eq. (2.46):

$$\phi^{i+1} = M\phi^i, i = 0, 1, 2, \ldots, \;\; \phi^i = (\phi_1^i, \ldots, \phi_4^i) \tag{2.47}$$

and

$$\phi^{i+1} - \phi^i := \psi^i, \;\; i = 0, 1, 2, \ldots, \;\; \psi^i = (\psi_1^i, \ldots, \psi_4^i).$$

Let the sequence of numbers $a_0, \; a_1, \ldots, a_i, \ldots$ be determined by the expressions $a_{i+1} = a_i/(1 + (i + 1)^{-2})$, $i = 0, 1, 2, \ldots$. Here a_0 is a fixed positive number. The number $a_0 < T/s_0$ will be chosen later. With the numerical sequence a_σ we link the sequence of enclosed fields $F_i = \{(z, t, s)|0 < s < s_0, 0 \le t \le z < a_i(s_0 - s)\}$.

The following lemma is valid.

Lemma 2.1 *If the conditions of Theorem 2.4 for any fixed* $\chi > 0$ *and any* $i = 0, 1, 2, \ldots$ *are fulfilled, then there exist* $a_0 \in (0, T/s_0)$, $a_0 = a_0(R, \; s_0, \; \chi, \; n)$ *and* $\lambda_i = \lambda_i(R, \; s_0, \; \chi, \; n) > 0$, *such that for each* $s \in (0, \; s_0)$

$$\left(\psi_1^i, \; \psi_3^i\right) \in C(A_s, \; D_{si}), \; \left(\psi_2^i, \; \psi_4^i\right) \in C\left(A_s, \; [0, \; a_i(s_0 - s)]\right),$$

$D_{si} := \{(z, t)|0 \le t \le z < a_i(s_0 - s)\}$, *and the following inequalities are valid:*

$$\|\psi_1^i\|_s(z, t) \le \frac{\lambda_i z}{a_i(s_0 - s) - z}, \;\; \|\psi_j^\sigma\|_s(z, t) \le \frac{\lambda_i a_i z}{[a_i(s_0 - s) - z]^2}, \;\; j = 2, 3,$$

$$\|\psi_4^i\|_s(z) \le \frac{\lambda_i a_i^2 z}{[a_i(s_0 - s) - z]^3}, \;\; (z, t, s) \in F_i; \tag{2.48}$$

$$\|\phi_1^{i+1} - \phi_1^0\|_s(z, t) \le \chi, \;\; \|\phi_j^{i+1} - \phi_j^0\|_s(z, t) \le \frac{2\chi}{s_0 - s}, \;\; j = 2, 3,$$

$$\|\phi_4^{i+1} - \phi_4^0\|_s(z) \le \frac{4\chi}{(s_0 - s)^2}, \;\; (z, t, s) \in F_{i+1}. \tag{2.49}$$

As in Sect. 2.1, we use the Nirenberg method and its modification to prove the lemma. Using estimate (2.4) it is not difficult to check that inequalities (2.48), (2.49) are satisfied when $i = 0$, besides λ_0 is proportional to a_0. We determine the validity of these inequalities for any i using the induction method. We assume that if the statement of the lemma is valid for $i \le \sigma$ and prove that, then it is valid for $i =$

$\sigma+1$, as well. Using the inductive assumption, we find that $\psi^{\sigma+1}\in C(A_s,\ D_{s(\sigma+1)})$. Besides, from (2.46), we find

$$\|\psi_1^{\sigma+1}\|_s(z,t)\le\int\limits_t^z\|\psi_3^{\sigma}\|_s(\xi,t)\mathrm{d}\xi$$

$$\le\int\limits_t^z\frac{\lambda_\sigma a_\sigma\xi\ \mathrm{d}\xi}{[a_\sigma(s_0-s)-\xi]^2}\le\frac{\lambda_\sigma a_\sigma z}{[a_\sigma(s_0-s)-z]}\le a_0\frac{\lambda_\sigma z}{a_{\sigma+1}(s_0-s)-z},$$

$$\|\psi_2^{\sigma+1}\|_s(z)\le\int\limits_0^z\|\psi_4^{\sigma}\|_s(\xi)\mathrm{d}\xi\le a_0\frac{\lambda_\sigma a_{\sigma+1}z}{[a_{\sigma+1}(s_0-s)-z]^2},$$

$$\|\psi_3^{\sigma+1}\|_s(z,t)\le\frac{1}{4}\int\limits_t^z\{\|\Delta\psi_1^{\sigma}\|_s(z,\xi)+2TR\sum_{j=1}^{n}\|\frac{\partial}{\partial x_j}\psi_1^{\sigma}\|_s(z,\xi)$$

$$+R\|\psi_1^{\sigma}\|_s(z,\xi)+\|\psi_4^{\sigma}\|_s(\xi)+\int\limits_0^\xi[\|\psi_4^{\sigma}\|_s(\tau)\|\phi_1^{\sigma+1}|_s(z-\tau,\xi-\tau)$$

$$+\|\phi_4^{\sigma}\|_s(\tau)\|\psi_1^{\sigma}\|_s(z-\tau,\xi-\tau)]\mathrm{d}\tau\}\mathrm{d}\xi.$$

Using inequality (2.4) to estimate

$$\|\Delta\psi_1^{\sigma}\|_s,\ \|\frac{\partial}{\partial x_j}\psi_1^{\sigma}\|_s,\ \ j=1,2,\dots,n,$$

and assuming that $s'=s'(\xi)=(s+s_0-\xi/a_n)/2$, we get

$$\|\psi_3^{\sigma+1}\|_s(z,t)\le\frac{1}{4}\int\limits_t^z\Bigg\{\frac{16na_\sigma^2}{[a_\sigma(s_0-s)-\xi]^2}\cdot\frac{2\lambda_\sigma z}{a_\sigma(s_0-s)-z}$$

$$+\frac{8TRna_\sigma}{a_\sigma(s_0-s)-\xi}\cdot\frac{2\lambda_\sigma z}{a_\sigma(s_0-s)-z}+\frac{2R\lambda_\sigma z}{a_\sigma(s_0-s)-z}$$

$$+\frac{\lambda_\sigma a_\sigma^2\xi}{[a_\sigma(s_0-s)-\xi]^3}+\int\limits_0^\xi\left[\frac{\lambda_\sigma a_\sigma^2\tau(\chi+R)}{(a_\sigma(s_0-s)-\tau)^3}+\frac{4\chi+Rs_0^2}{(s_0-s)^2}\cdot\frac{2\lambda_\sigma z}{a_\sigma(s_0-s)-z}\right]\mathrm{d}\tau\Bigg\}\mathrm{d}\xi$$

$$\le c_1a_0\frac{a_{\sigma+1}\lambda_\sigma z}{[a_{\sigma+1}(s_0-s)-z]^2},$$

where $c_1 = c_1(s_0, n, a_0, \chi, R, T)$. In the latter inequality it is used that $a_\sigma^2 \le a_{\sigma+1}a_0$ are valid for any $\sigma \ge 1$. Moreover, $a_\sigma^3 \le a_{\sigma+1}^2 a_0$ for $\sigma \ge 1$. These inequalities can be easily checked by the method of mathematical induction. The latter inequalities are used to estimate the function $\psi_4^{\sigma+1}$. Similarly, for $\psi_4^{\sigma+1}$ we obtain

$$\|\psi_4^{\sigma+1}\|_s(z) \le c_2 a_0 \frac{a_{\sigma+1}^2 \lambda_\sigma z}{[a_{\sigma+1}(s_0 - s) - z]^3},$$

where $c_2 = c_2(s_0,\ n,\ a_0,\ \chi,\ R,\ T)$. It is obvious that $c_i,\ i = 1, 2$ in the latter estimates are monotonically nondecreasing functions of the $s_0,\ a_0$ parameters. From the estimations made above, it follows that (2.48) is valid when $i = \sigma + 1$, if we assume that

$$\lambda_{\sigma+1} = a_0 c \lambda_\sigma,$$

where

$$c = \max(1,\ s_0,\ c_1,\ c_2).$$

Now we show that inequalities (2.49) are also satisfied when $i = \sigma + 1$, if the number a_0 is chosen properly. For $(x, z, t) \in F_{\sigma+2}$

$$\|\phi_1^{\sigma+2} - \phi_1^0\|_s(z,t) \le \sum_{j=0}^{\sigma+1} \|\psi_1^j\|_s(z,t) \le \sum_{j=0}^{\sigma+1} \frac{\lambda_j z}{a_j(s_0 - s) - z}$$

$$\le \lambda_0 \sum_{j=0}^{\sigma+1} (a_0 c)^j \left(\frac{a_j}{a_{j+1}} - 1 \right) \le \lambda_0 \sum_{j=0}^{\sigma+1} (a_0 c)^j (j+1)^2,$$

$$\|\phi_\gamma^{\sigma+2} - \phi_\gamma^0\|_s(z,t) \le \sum_{j=0}^{\sigma+1} \|\psi_\gamma^j\|_s(z,t) \le \frac{2\lambda_0}{s_0 - s} \sum_{j=0}^{\sigma+1} (a_0 c)^j (j+1)^4,\ \gamma = 2, 3,$$

$$\|\phi_4^{\sigma+2} - \phi_4^0\|_s(z) \le \sum_{j=0}^{\sigma+1} \|\psi_4^j\|_s(z) \le \frac{4\lambda_0}{(s_0 - s)^2} \sum_{j=0}^{\sigma+1} (a_0 c)^j (j+1)^6.$$

That is why inequalities (2.49) are satisfied when $i = \sigma + 1$, if the number a_0 is chosen so that

$$a_0 c < 1,\ \lambda_0 \sum_{j=0}^{\infty} (a_0 c)^j (j+1)^6 \le \chi. \tag{2.50}$$

It is clear that a_0 can be always chosen to be so small that inequalities (2.50) are satisfied. Thus, the validity of the lemma is proven. Further, we assume that the number a_0 is chosen from the conditions (2.50).

In order to complete the proof of Theorem 2.4, we should note that under the chosen value of a_0 the ϕ^i sequence uniformly converges in the norm of space $C(A_s,\ D_s)$, $a = \lim a_i$, the limiting function belongs to $C(A_s,\ D_s)$ and provides the solution of the operator equation (2.46). The limit transition in the first two inequalities (2.49) when $(x, z, t) \in F = \{(z, t, s)|0 < s < s_0, 0 \le t \le z < a(s_0 - s)\}$ leads to the values coinciding with (2.41). The uniqueness of the constructed solution is established by using the described above technics.

We indicate a few settings of inverse problems of memory determination which could be studied by the above methods.

Problem 1 Find the functions $u(x, z, t),\ k(x, t)$ that satisfy the following equalities:

$$u_{tt} - u_{zz} - \Delta u = \int_0^t k(x, \tau) u_t(x, z, t - \tau) d\tau,$$

$$(x, t) \in \mathbb{R}^{n+1}, z > 0, \tag{2.51}$$

$$u|_{t<0} \equiv 0,\ u_z|_{z=0} = -g(x)\delta'(t),\ (x, t) \in \mathbb{R}^{n+1}, \tag{2.52}$$

$$u|_{z=0} = g(x)\delta(t) + f(x, t)\theta(t),\ (x, t) \in \mathbb{R}^{n+1}, \tag{2.53}$$

where $g(x),\ f(x, t)$ are given smooth functions.

The solution of problem (2.51) and (2.52) is represented as

$$u(x, z, t) = g(x)\delta(t - z) + \tilde{u}(x, z, t)\theta(t - z). \tag{2.54}$$

We introduce the function

$$\tilde{u}(x, (z - t)/2,\ (z + t)/2) := v(x, z, t).$$

As in the above investigated problem (2.51)–(2.52) can be replaced by the equivalent problem of finding the functions $v(x, z, t),\ k(x, t)$ in the domain $z > t > 0,\ x \in \mathbb{R}^n$ from the equations

$$\frac{\partial^2 v(x, z, t)}{\partial t \partial z} = -\frac{1}{4}\left[\Delta v + k_0(x) v - g(x) h(x, t) - \int_0^t h(x, \tau) v(x, z - \tau, t - \tau) d\tau\right],$$

$$(x, z, t) \in \{(x, z, t) \mid x \in \mathbb{R}^n, 0 \le t \le z\} := D, \tag{2.55}$$

$$v|_{t=z} = f(x, z),\ \frac{\partial}{\partial z} v|_{t=z} = \frac{1}{2} f_t(x, t)|_{t=z},\ z > 0,\ x \in \mathbb{R}^n, \tag{2.56}$$

$$v|_{t=0} = \frac{1}{2} k_0(x) z,\ z > 0,\ x \in \mathbb{R}^n, \tag{2.57}$$

in which

$$k_0 := k(x,0) = 1/g(x)[f_t(x,t)|_{t=0} + \Delta g(x)],\ /h(x,t) := k_t(x,t),\ t > 0.$$

With respect to problem (2.5)–(2.7) the analogue of Theorem 2.4 is valid.

Theorem 2.5 *Let*

$$\left\{g(x),\ 1/g(x),\ (\partial/\partial z)f(x,z)|_{z=0},\ (\partial^3/\partial^2 x_i \partial z)f(x,z)|_{z=0}\right\} \in A_{s_0},$$

$s_0 > 0,\ i = 0, 1, \ldots, n,$

$$\left\{f(x,z),\ (\partial/\partial z)f(x,z),\ (\partial^2/\partial^2 z)f(x,z)\right\} \in C\left(A_{s_0},\ [0,T]\right),$$

when some fixed $s_0 > 0,\ T > 0$, *whereas*

$$\max\left[\|g(x)\|_{s_0}, \|f(x,z)\|_{s_0}, \|f_z(x,z)\|_{s_0}, 1/\|g(x)\|_{s_0}, \|\Delta g(x)\|_{s_0}, \|f_{zz}(x,z)\|_{s_0}\right] = R,$$

for $t \in [0,T]$ *and the consistency condition is satisfied:*

$$f_z(x,z)|_{z=0} + \Delta g(x) = g(x) f_z(x,z)|_{z=0}.$$

Then for any $\chi > 0$ *the number* $a \in (0, T/s_0)$ *can be found such that for any* $s \in (0, s_0)$ *there exists the unique solution of problem (2.55)–(2.57)*

$$v(x,z,t) \in C_z^1\left(A_{s_0},\ D_s\right),\ k(x,t) \in C_t^1\left(A_{s_0},\ [0, a(s_0 - s)]\right),$$

where the domain D_s *is determined in Theorem 2.4, and for the solution inequalities (2.41) are valid with the functions*

$$v_0 = f(x,z),\ k_0(x) = \frac{1}{g(x)}[f_t(x,t)|_{t=0} + \Delta g(x)].$$

Problem 2 Find the functions $u(x,z,t),\ k(x,t)$ that satisfy the following equalities:

$$u_{tt} - u_{zz} - Lu = \int_0^t k(x,\tau) L_0 u(x,z,t-\tau)\mathrm{d}\tau,$$

$$(x,t) \in \mathbb{R}^{n+1},\ z > 0, \tag{2.58}$$

$$u|_{t<0} \equiv 0,\ u_z|_{z=0} = -g(x)\delta'(t),\ (x,t) \in \mathbb{R}^{n+1}, \tag{2.59}$$

$$u|_{z=0} = g(x)\delta(t) + f(x,t)\theta(t),\ (x,t) \in \mathbb{R}^{n+1}. \tag{2.60}$$

Here

$$L = \sum_{i=1,j=1}^{n} a_{ij}(x)\frac{\partial^2}{\partial x_i \partial x_j} + L_0, \; L_0 = \sum_{i=1}^{n} b_i(x)\frac{\partial}{\partial x_i} + c(x),$$

$a_{ij}, \; b_i, \; c, \; g, \; f(1 \le i, j \le n)$ are given smooth functions.

In the case of inequality (2.4), let us estimate the differential expressions $Lu, \; L_0u$ with analytical coefficients. We assume that functions $a_{ij}(x), \; b_i(x), \; c(x), \; 1 \le i, j \le n$, are the elements of $A_{s_0}, \; s_0 > 0$, and

$$d := \max_{1\le i,j\le n} \|a_{ij}\|_{s_0}, \; b_0 := \max_{1\le i,j\le n} \|b_i\|_{s_0}, \; c_0 := \|c\|_{s_0}.$$

For the operators $L_0, \; L$ from (2.19) it follows:

$$\|L_0 v\|_{s'} \le \frac{c_1}{s-s'}\|v\|_s, \; \|Lv\|_{s'} \le \frac{c_2}{(s-s')^2}\|v\|_s, \; s' \in (0,s),$$

$$c_1 := nb_0 + s_0c_0, \; c_2 := 4n^2 d + c_1 s_0.$$

The solution of problem (2) and (2.59) is represented as

$$u(x,z,t) = \alpha(x,z)\delta(t-z) + \tilde{u}(x,z,t)\theta(t-z) \tag{2.61}$$

and we use the method of separation of singularities. Denoting $\tilde{u}(x,z,z+0) =: \beta(x,z)$, we put (2.61) in (2) and (2.59) and find

$$\alpha(x,z) \equiv \alpha(x) = g(x), \; \beta(x,z) = (z/2)Lg(x), \; x \in \mathbb{R}^n, \; z > 0. \tag{2.62}$$

It is not difficult, using equalities (2.61) and (2.62), to replace the inverse problem (2)–(2.60) by initial-characteristic problem with Cauchy data on $z = 0$. It should be noted that when $t > z > 0$ the equality $u = \tilde{u}$ takes place, then the functions $u(x,z,t), \; k(x,z)$ satisfy the equations

$$u_{tt} - u_{zz} - Lu = L_0 g(x)k(x,t-z) + \int_0^{t-z} k(x,\tau)L_0 u(x,z,t-\tau)d\tau,$$

$$(x,z,t) \in G := \{(x,z,t) | x \in \mathbb{R}^n, t > z > 0\}, \tag{2.63}$$

$$u|_{z=0} = f(x,t), \; u_z|_{z=0} = 0, \; x \in \mathbb{R}^n, t > 0, \tag{2.64}$$

$$u|_{z=0} = \frac{1}{2}zLg(x), \; x \in \mathbb{R}^n, \; t > 0. \tag{2.65}$$

Further, the problem (2.63)–(2.64) is reduced to the following system of integro-differential equations for $u,\ u_t,\ k$:

$$u(x,z,t) = u_0(x,z,t) + \frac{1}{2}\int_{\Delta(z,t)} [Lu(x,\xi,\tau) - L_0 g(x)k(x,\tau-\xi) + \int_0^{\tau-\xi} k(x,\gamma)L_0 u(x,\xi,\tau-\gamma)d\gamma]d\tau\, d\xi, \tag{2.66}$$

$$u_t(x,z,t) = u_{0t}(x,z,t) + \frac{1}{2}\int_{-z}^{z} [Lu(x,z-|\xi|,t+\xi) - L_0 g(x)k(x,t - z + |\xi| + \xi) + \int_0^{t-z+|\xi|+\xi} k(x,\gamma)L_0 u(x,z-|\xi|,t+\xi-\gamma)d\gamma]\mathrm{sgn}\xi\, d\xi, \tag{2.67}$$

$$k(x,z) = k_0(x,z) + \frac{2}{L_0 g(x)}\int_0^{z/2} [Lu_t(x,\xi,z-\xi) + \frac{1}{2}\xi L_0 g(x)k(x,z-2\xi) + \int_0^{z-2\xi} k(x,\gamma)L_0 u_t(x,\xi,z-\xi-\gamma)d\gamma]d\xi, \tag{2.68}$$

where

$$u_0(x,z,t) = \frac{1}{2}[f(x,t+z) + f(x,t-z)],$$

$$k_0(x,z) = \frac{1}{L_0 g(x)}\left[(1/4)zL^2 g(x) - 2f_{tt}(x,t)|_{t=z}\right].$$

For the system (2.66)–(2.68) the following theorem is valid.

Theorem 2.6 *Assume that the consistency conditions*

$$f(x,0) = 0,\ \ f_t(x,t)|_{t=0} = (1/2)Lg(x)$$

are satisfied, besides

$$(a_{ij}, b_i, c, g)(x) \in A_{s_0},\ \ 1 \le i,j \le n,$$

$$[f(x,t),\ f_t(x,t),\ f_{tt}(x,t)] \in C\left(A_{s_0},\ [0,T]\right),$$

$$\max[\|f\|_{s_0}(t),\ \|f_t\|_{s_0}(t),\ \|f_{tt}\|_{s_0}(t),\ \|L_0 g\|_{s_0},\ 1/\|L_0 g\|_{s_0},\ \|L^2 g\|_{s_0}] = R,$$

for $t \in [0, T],\ R > 0$. *Then,* $a \in (0, T/2),\ as_0 < (T/2)$ *can be found such that for any* $s \in (0,\ s_0)$ *in the domain* $G_T \cap \{(x, z, t)|x \in \mathbb{R}^n, 0 \le z \le a(s_0 - s)\}$ *there exists the unique solution of the system (2.66)–(2.68), for which*

$$(u(x, z, t), u_t(x, z, t)) \in C(A_{s_0}, P_{sT})$$

$$k(x, z) \in C(A_{s_0},\ [0, a(s_0 - s)]),\ P_{sT} := G_T \cap \{(z, t)|0 \le z \le a(s_0 - s)\},$$

moreover,

$$\|u - u_0\|_s(z, t) \le R_0,\ \|k - k_0\|_s(t) \le \frac{R_0}{(s_0 - s)^2},$$

$$\|u_t - u_{0t}\|_s(z, t) \le \frac{R_0}{(s_0 - s)},$$

$(z, t) \in P_{sT},\ R_0 = R_0(R,\ T,\ s_0,\ n)$ *is a constant.*

Theorems 2.5 and 2.6 are proved similarly to Theorem 2.4.

2.3 Definition of Memory Kernel in a Two-Dimensional System Maxwell Equations

Consider the Maxwell system of integro-differential equations:

$$\nabla \times H = (\partial/\partial t)D(x, t) + \sigma E + j,$$

$$\nabla \times E = -(\partial/\partial t)B(x, t),\ (x, t) \in \mathbb{R}^4, \tag{2.69}$$

where

$$D(x, t) = \varepsilon E + \int_0^t \varphi(x, t - \tau)E(x, \tau)d\tau,$$

$$B(x, t) = \mu H + \int_0^t \psi(x, t - \tau)H(x, \tau)d\tau, \tag{2.70}$$

$$E = (E_1, E_2, E_3),\ H = (H_1, H_2, H_3),\ D = (D_1, D_2, D_3),$$

$$B = (B_1, B_1, B_3),\ x = (x_1, x_2, x_3),$$

$\varphi(x,t)$, $\psi(x,t)$ are the scalar functions characterizing memory of the medium, the density of the external electric current has the form

$$j = j^0 g(x_1)\delta(x_3)\delta(t), \; j^0 = \begin{pmatrix} 0 \\ 1 \\ 0 \end{pmatrix}. \tag{2.71}$$

Add to the equations (2.69)–(2.71) the following zero initial conditions:

$$E|_{t<0} \equiv 0, \; H|_{t<0} \equiv 0. \tag{2.72}$$

As a direct problem for system (2.69) and (2.70), we consider the problem of the excitation of an electromagnetic field (initially absent) by a pulse current (2.71).

Suppose that in Eqs. (2.69) and (2.70) the functions ε, μ, σ, φ, ψ do not depend on the coordinate x_2. Then, the solution of the Cauchy problem (2.69)–(2.72) also does not depend on x_2 and $E_1 = E_3 = H_2$. The inverse problem is to define the function $\varphi = \varphi(x_1, t)$, if the following information about the solution of problem (2.69)–(2.72) is known:

$$E|_{x_3=0} = f(x_1, t), \; t \geq 0, \; x_1 \in \mathbb{R}. \tag{2.73}$$

Assume that ε, μ are constants in $\mathbb{R}^2$ and σ is piecewise-constant: $\sigma = \sigma_0$ for $x_3 \geq 0$; $\sigma = 0$ for $x_3 < 0$. Also suppose that $\psi = 0$ (the absence of a magnetic history of the medium). Without loss of generality, we set $\varepsilon = \mu = 1$. By virtue of the assumptions made for nonzero field components, problem (2.69)–(2.71) can be written as

$$\frac{\partial E_2}{\partial t} - \frac{\partial H_1}{\partial x_3} + \frac{\partial H_3}{\partial x_1} + \int_0^t \varphi(x_1, \tau) \frac{\partial E_2}{\partial t}(x_1, x_3, t-\tau)d\tau = 0,$$

$$\frac{\partial H_1}{\partial t} - \frac{\partial E_2}{\partial x_3} = 0, \; \frac{\partial H_3}{\partial t} + \frac{\partial E_2}{\partial x_1} = 0, \; x_3 \neq 0;$$

$$(E_2, \; H_1, \; H_3)_{t<0} \equiv 0, \tag{2.74}$$

$$[H_1]_{x_3=0} = g(x_1)\delta(t), \; [E_2]_{x_3=0} = 0,$$

where $[H_1]_{x_3=0} = H_1\big|_{x_3=+0} - H_1\big|_{x_3=-0}$. In fact, its solution is reduced to solving the Cauchy problem for component E_2:

$$\frac{\partial^2 E_2}{\partial t^2} - \frac{\partial^2 E_2}{\partial x_1^2} - \frac{\partial^2 E_2}{\partial x_3^2} + \sigma \frac{\partial E_2}{\partial t}$$

$$+ \int_0^t \varphi(x_1, \tau) \frac{\partial^2 E_2}{\partial t^2}(x_1, x_3, t - \tau) d\tau = 0, \ x_3 \neq 0, \tag{2.75}$$

$$E_2\big|_{t<0} \equiv 0, \ [E_2]_{x_3=0} = 0, \ \left[\frac{\partial E_2}{\partial x_3}\right]_{x_3=0} = g(x_1)\delta'(t),$$

and the subsequent calculation of H_1, H_3 by integration over t of the corresponding equalities from (2.74).

Thus, the inverse problem is reduced to the problem of determining $\varphi(x_1, t)$ from relations (2.75), if the information (2.73) is known. Since the unknown function φ does not depend on x_3, the inverse problem (2.75) and (2.73) is sufficient to consider in the half-space $x_3 > 0$. In this case, the problem (2.75) has the form

$$\frac{\partial^2 E_2}{\partial t^2} - \frac{\partial^2 E_2}{\partial x_3^2} = \frac{\partial^2 E_2}{\partial x_1^2} - \sigma \frac{\partial E_2}{\partial t}$$

$$- \int_0^t \varphi(x_1, \tau) \frac{\partial^2 E_2}{\partial t^2}(x_1, x_3, t - \tau) d\tau = 0,$$

$$x_3 \neq 0, \ x_3 > 0, \ (x, t) \in \mathbb{R}^2, \tag{2.76}$$

$$E_2\big|_{t<0} \equiv 0, \ \frac{\partial E_2}{\partial x_3}\Big|_{x_3=0} = g(x_1)\delta'(t), \ (x, t) \in \mathbb{R}^2. \tag{2.77}$$

In view of

$$\frac{\partial^2 E_2}{\partial t^2}(x_1, x_3, t - \tau) = \frac{\partial^2 E_2}{\partial \tau^2}(x_1, x_3, t - \tau),$$

we integrate twice by parts in (2.76). By introducing a new function u using the formula

$$E_2 = \beta(x_1, t) u(x_1, z, t), \ \beta(x_1, t) := \exp\left[-(\sigma_0 + \varphi_0(x_1))\frac{t}{2}\right],$$

$$\varphi_0(x_1) := \varphi(x_1, 0), \ z \equiv x_3,$$

from equalities (2.76) and (2.77) we have

$$u_{tt} - u_{zz} = u_{x_1 x_1} + t\varphi_0'(x_1) u_{x_1} + H(x_1, t) u + \int_0^t h(x_1, \tau) u(x_1, z, t - \tau) d\tau,$$

$$z > 0,\ (x, t) \in \mathbb{R}^2, \tag{2.78}$$

$$u\big|_{t<0} = 0,\ u_z\big|_{z=0} = g(x_1)\delta'(t) + \frac{\sigma_0 + \varphi_0(x_1)}{2} g(x_1)\delta(t),\ (x, t) \in R^2. \tag{2.79}$$

Here the following relation was used

$$\exp\left[(\sigma_0 + \varphi_0(x_1))\frac{t}{2}\right] = g(x_1)\delta'(t) + \frac{\sigma_0 + \varphi_0(x_1)}{2} g(x_1)\delta(t),$$

which is fulfilled in the sense of generalized functions, and the following notations are also introduced

$$H(x_1, t) := -\varphi_t(0, x_1) + \frac{(\sigma_0 + \varphi_0(x_1))^2}{4} + \frac{t^2\left[\varphi_0'(x_1)\right]^2}{4} + \frac{t\varphi_0''(x_1)}{2},$$

$$h(x_1, \tau) := \exp\left[(\sigma_0 + \varphi_0(x_1))\frac{t}{2}\right]\varphi_{tt}(x_1, \tau).$$

Additional condition (2.73) in terms of the function u and the z variable looks like

$$u\big|_{z=0} = F(x_1, t),\ t \ge 0,\ x_1 \in \mathbb{R},$$

where $F(x_1, t) := \exp\left[(\sigma_0 + \varphi_0(x_1))t/2\right] f(x_1, t)$.

Let $C_t^2(A_{s_0}, [0, T])$, $T > 0$ be a class of twice continuously differentiable with respect to t on $[0, T]$ functions, with values in A_{s_0}. Suppose that $g(x_1) \in A_{s_0}$ and $\varphi(t, x_1) \in C_t^2(A_{s_0}, [0, T])$. Then, $h(t, x_1) \in C_t(A_{s_0}, [0, T])$ ($C_t(A_{s_0}, [0, T])$ is the class of continuous with respect to t on $[0, T]$ functions, with values in A_{s_0}), and we get the inverse problem, which is very close to that considered in the previous section (with $n = 1$). It can be proved that the solution of the direct problem (2.78) and (2.79) can be represented as

$$u(x_1, z, t) = -g(x_1)\delta(t - z) + \tilde{F}(t, x_1)\theta(t - z).$$

Therefore, for the solvability of the inverse problem, $F(x_1, \tau)$ should be representable in the form:

$$F(x_1, \tau) = -g(x_1)\delta(t) + \tilde{F}(x_1, \tau)\theta(t),\ t \le T,\ x_1 \in \mathbb{R},$$

where $F(x_1, \tau) \in C_t^2(A_{s_0}, [0, T])$.

An analogue of Theorem 2.4, established in Sect. 2.2, holds in this case as well.

2.4 Differential Properties of the Solution of the Memory Kernel Determination Problem in a Two-Dimensional Case

Consider the problem (2.1)–(2.3) about the definition of the function $k(x, t)$, $(x, t) \in \mathbb{R}^2$. For simplicity, we assume that $h(x, t) = 0$. Then, relations (2.8)–(2.10) take the form

$$u_{tt} - u_{zz} - u_{xx} + k(x, t - z) - \int_0^{t-z} k(x, \tau)u(x, z, t - \tau)d\tau = 0,$$

$$t > z > 0, \quad x \in \mathbb{R}, \tag{2.80}$$

$$u|_{z=0} = f(x, t), \quad u_z|_{z=0} = 0, \quad t > 0, \quad x \in \mathbb{R}, \tag{2.81}$$

$$u|_{t=z+0} = 0, \quad x \in \mathbb{R}. \tag{2.82}$$

In Eqs. (2.80)–(2.82), we make the change of variables z, t to z_1, t_1 by the formulas:

$$z_1 = t + z, \ t_1 = t - z.$$

Then, $u(x, z, t) = u(x, (z_1 - t_1)/2, (z_1 + t_1)/2) := u_1(x, z_1, t_1)$. Relations (2.80)–(2.82) in the new variables are rewritten in the form

$$\frac{\partial^2 u_1(x, z_1, t_1)}{\partial t_1 \partial z_1} + \frac{1}{4}\left[u_{1xx} - k(x, t_1) + \int_0^{t_1} k(x, \tau)u_1(x, z_1 - \tau, t_1 - \tau)d\tau \right] = 0,$$

$$0 < t_1 < z_1, \quad x \in \mathbb{R}, \tag{2.83}$$

$$u_1 \,|_{t_1=z_1} = f(x, z_1), \quad \frac{\partial}{\partial z_1} u_1 \,|_{t_1=z_1} = \frac{1}{2} f_{z_1}(x, z_1), \quad z_1 > 0, \quad x \in \mathbb{R}, \tag{2.84}$$

$$u_1 \,|_{t_1=0} = 0, \quad x \in \mathbb{R}. \tag{2.85}$$

Introduce a function

$$\vartheta(x, z_1, t_1) = \frac{\partial}{\partial z_1} u_1(x, z_1, t_1), \quad t_1 < z_1. \tag{2.86}$$

Lemma 2.2 *If the conditions $f(x,0) = f_z(x,0) = 0$ are fulfilled, then, the inverse problem (2.83)–(2.85) in $G_T = \{(x,z,t) \mid x \in \mathbb{R},\ 0 \le t \le z \le T\}$ is equivalent to the following system of equations (for simplicity, further, we will omit index 1 in functions and variables)*

$$u(x,z,t) - \int_t^z \vartheta(x,\eta,t)\mathrm{d}\eta = u^0(x,t),$$

$$\vartheta(x,z,t) - \frac{1}{4}\int_t^z \Bigg[u_{xx}(x,z,\xi) + k(x,\xi) - \int_0^\xi k(x,\tau)u(x,z-\tau,\xi-\tau)\mathrm{d}\tau \Bigg]\mathrm{d}\xi = \upsilon^0(x,z), \tag{2.87}$$

$$k(x,z) + \int_0^z \Bigg[\vartheta_{xx}(x,z,\xi) - k(x,\xi)f(x,z-\xi) - \int_0^\xi k(x,\tau)\vartheta(x,z-\tau,\xi-\tau)\mathrm{d}\tau \Bigg]\mathrm{d}\xi = k^0(x,z),$$

where

$$u^0(x,t) = f(x,t),\quad \vartheta^0(x,z) = \frac{1}{2}\frac{\partial}{\partial z} f(x,z),$$
$$k^0(x,z) = -2\frac{\partial^2}{\partial z^2} f(x,z) - \frac{\partial^2}{\partial x^2} f(x,z),\quad \frac{\partial^i}{\partial z^i} f(x,z) = \frac{\partial^i}{\partial t^i} f(x,t)\,|_{t=z},$$
$$i = 0,1,2. \tag{2.88}$$

Proof The first equation (2.87) is easily obtained by integrating the equality (2.86) over z from the point (z,t) to the point (z,z) on the plane of the variables (η,ξ) and using the first condition from (2.84). To obtain the second equation (2.87), it is necessary to integrate Eq. (2.83) with respect to t from the point (z,t) to the point (z,z) and use the second condition from (2.84). Further, pass to limit when $t \to 0$ in the second equation of (2.87). To find the value of $\vartheta(x,z,0)$ we use the differentiated equality (2.85) with respect to z. Then, one gets

$$\int_0^z [u_{xx}(x,z,\xi) + k(x,\xi) - \int_0^\xi k(x,\tau)u(x,z-\tau,\xi-\tau)d\tau]d\xi = -4v^0(x,z).$$

Differentiating this equality with respect to z and solving relatively $k(x,z)$ we get the third equation of (2.87).

It is easy to show that inverse transformations also take place. Indeed, integrating the third equation (2.87) over z into limits from t to z, changing the order integration, where it is necessary, and taking into account the conditions $f(x,0) = f_z(x,0) = 0$, we derive the second equation (2.87). Differentiating with respect to the corresponding variables the first and second equations of the system (2.87), we easily arrive at equalities (2.83)–(2.85). Lemma 2.2 is proved.

Definition 2.1 We say that the function $h(x)$ belongs to A'_s, $s > 0$, if it is representable by Fourier series

$$h(x) = \sum_{n=-\infty}^{\infty} h_n \exp(-inx)$$

and its s-norm $\|h\|_s = \sum_{n=-\infty}^{\infty} \mid h_n \mid \exp(sn)$ is finite.

Definition 2.2 If some function $u(x,z,t)$ belongs to A'_s, $s > 0$, for fixed $(z,t) \in G$ and is continuous in domain G, as an element of A'_s, then, we say that it belongs to $C(A'_s, G)$, and by the symbol $\| u \|_s (z,t)$ denote its s-norm for fixed $(z,t) \in G$.

Theorem 2.7 *Let*

$$\left((\partial^i/\partial t^i)f, (\partial^2/\partial x^2)f\right) \in C(A_{s_0}, [0,T]), \quad i = 0, 1, 2$$

for some fixed $s_0 > 0$, $T > 0$, *moreover*

$$\max\left\{\|f\|_{s_0}, \left\|\frac{\partial f}{\partial z}\right\|_{s_0}, 2\left\|\frac{\partial^2 f}{\partial z^2}\right\|_{s_0} + \left\|\frac{\partial^2 f}{\partial x^2}\right\|_{s_0}\right\} \le R,$$

$R > 0$, $t \in [0,T]$ *and*

$$f(x,+0) = (\partial/\partial t)f(x,+0) = 0.$$

Then, for an arbitrary χ *there exists a number such* $a \in (0, T/s_0)$ *that for any* $s \in (0, s_0)$ *and* $(z,t) \in G = \{(z,t) | 0 \le t \le z < a(s_0 - s)\}$ *the solution of the system (2.87) exists, is uniquely and* $(u, \vartheta) \in C(A'_{s_0}, G)$, $k \in C(A'_{s_0}, [0, a(s_0 - s)])$, *besides, for solution the following estimates are fulfilled:*

$$\|u - u^0\|_s(z,t) \le \chi, \quad \|\vartheta - \vartheta^0\|_s(z,t) \le \frac{2\chi}{s_0 - s},$$

$$\|k - k^0\|_s(z) \leq \frac{4\chi}{(s_0 - s)^2}, \quad (z, t) \in G. \tag{2.89}$$

Proof For the system of Eq. (2.87) we consider successive approximations

$$u^\sigma(x, z, t), \quad \vartheta^\sigma(x, z, t), \quad k^\sigma(x, z),$$

defining them for $\sigma = 0, 1, 2, \ldots$ by the formulas

$$u^{\sigma+1}(x, z, t) - \int_t^z \vartheta^\sigma(x, \eta, t)\mathrm{d}\eta = u^0(x, t),$$

$$\vartheta^{\sigma+1}(x, z, t) - \frac{1}{4}\int_t^z \Big[u^\sigma_{xx}(x, z, \xi)$$
$$+ k^\sigma(x, \xi) - \int_0^\xi k^\sigma(x, \tau)u^\sigma(x, z - \tau, \xi - \tau)\mathrm{d}\tau\Big]\mathrm{d}\xi = \vartheta^0(x, z), \tag{2.90}$$

$$k^{\sigma+1}(x, z) + \int_0^z \Big[\vartheta^\sigma_{xx}(x, z, \xi)$$
$$- k^\sigma(x, \xi)f(x, z - \xi) - \int_0^\xi k^\sigma(x, \tau)\vartheta^\sigma(x, z - \tau, \xi - \tau)\mathrm{d}\tau\Big]\mathrm{d}\xi = k^0(x, z).$$

Here $u^0(x, t)$, $\vartheta^0(x, z)$, $k^0(x, z)$ are defined by the formulas (2.88). In addition, we introduce the differences between two successive approximations:

$$u^{\sigma+1} - u^\sigma = \tilde{u}^\sigma, \ \vartheta^{\sigma+1} - \vartheta^\sigma = \tilde{\vartheta}^\sigma, \ k^{\sigma+1} - k^\sigma = \tilde{k}^\sigma, \ \sigma = 0, 1, 2, \ldots \tag{2.91}$$

The functions $\tilde{u}^\sigma$, $\tilde{\vartheta}^\sigma$, $\tilde{k}^\sigma$, defined by formula (2.91), as follows from (2.90), satisfy the system of equalities

$$\tilde{u}^\sigma(x, z, t) = \int_t^z \tilde{\vartheta}^{\sigma-1}(x, \eta, t)\mathrm{d}\eta,$$

$$\tilde{\vartheta}^{\sigma}(x,z,t)=\frac{1}{4}\int\limits_{t}^{z}\Big\{\tilde{u}_{xx}^{\sigma-1}(x,z,\xi)+\tilde{k}^{\sigma-1}(x,\xi)$$

$$-\int\limits_{0}^{\xi}\Big[\tilde{k}^{\sigma-1}(x,\tau)u^{\sigma}(x,z-\tau,\xi-\tau)+k^{\sigma-1}(x,\tau)\tilde{u}^{\sigma-1}(x,z-\tau,\xi-\tau)\Big]d\tau\Big\}d\xi,$$

$$\tilde{k}^{\sigma}(x,z)=-\int\limits_{0}^{z}\Big\{\tilde{\vartheta}_{xx}^{\sigma-1}(x,z,\xi)-\tilde{k}^{\sigma-1}(x,\xi)f(x,z-\xi)$$

$$-\int\limits_{0}^{\xi}\Big[\tilde{k}^{\sigma-1}(x,\tau)\vartheta^{\sigma}(x,z-\tau,\xi-\tau)+k^{\sigma-1}(x,\tau)\tilde{\vartheta}^{\sigma-1}(x,z-\tau,\xi-\tau)\Big]d\tau\Big\}d\xi,$$

$$\sigma=0,1,2,\ldots \tag{2.92}$$

For $\sigma=0$, in these equations it is necessary to formally set $\tilde{u}^{-1}=u^{0}$, $\tilde{\vartheta}^{-1}=\vartheta^{0}$, $\tilde{k}^{-1}=k^{0}$, $k^{-1}=0$.

Let the sequence of numbers $a_0, a_1, \ldots, a_\sigma, \ldots$ be defined by relations $a_{\sigma+1}=a_\sigma/(1+(\sigma+1)^{-2})$, $\sigma=0,1,2,\ldots$. Here a_0 is some fixed positive number generating the sequence a_σ. The number $a_0<T/s_0$ will be chosen later. We associate with numerical sequence a_σ the following embedded domains

$$F_\sigma=\Big\{(z,t,s)|0<s<s_0,\ 0\le t\le z<a_\sigma(s_0-s)\Big\}.$$

The following lemma holds:

Lemma 2.3 *If the conditions of Theorem 2.7 are fulfilled, for any fixed $\chi>0$ and for any $\sigma=0,1,2,\ldots$ there exist such numbers $a_0\in(0,T/s_0)$, $a_0=(R,s_0,\chi)$ and $\lambda_k=\lambda_k(R,\ s_0,\ \chi)>0$ that for $s\in(0,s_0)$*

$$\left(\tilde{u}^{\sigma},\ \tilde{\vartheta}^{\sigma}\right)\in C\left(A_s,G_\sigma\right),\ \tilde{k}^{\sigma}\in C\left(A_s,[0,a_\sigma(s_0-s)]\right),$$

$$G_\sigma:=\{(z,t)|0\le t\le z<a_\sigma(s_0-s)\}$$

and the following inequalities are valid:

$$\|\tilde{u}^{\sigma}\|_s(x,t)\le\frac{\lambda_\sigma z}{a_\sigma(s_0-s)-z},\quad \|\tilde{\vartheta}^{\sigma}\|_s(x,t)\le\frac{\lambda_\sigma a_\sigma z}{[a_\sigma(s_0-s)-z]^2},$$

$$\|\tilde{k}^{\sigma}\|_s(x)\le\frac{\lambda_\sigma a_\sigma^2 z}{[a_\sigma(s_0-s)-z]^3},\quad (z,t,s)\in F_\sigma; \tag{2.93}$$

$$\|u^{\sigma+1} - u^0\|_s(x,t) \le \chi, \ \|\vartheta^{\sigma+1} - \vartheta^0\|_s(x,t) \le \frac{2\chi}{s_0 - s},$$

$$\|k^{\sigma+1} - k^0\|_s(z) \le \frac{4\chi}{(s_0 - s)^2}, \ (z,t,s) \in F_{\sigma+1}. \tag{2.94}$$

Proof We use the method of mathematical induction. Check, first of all, that the statement of Lemma 2.3 is true at $\sigma = 0$. Under the conditions of Theorem 2.7, we have $\left(\tilde{u}^0, \ \tilde{v}^0, \ \tilde{k}^0\right) \in C\left(A'_s, \ G_\sigma\right)$ for each $s \in (0, \ s_0)$, if $a_0 s_0 \le T/2$. Using equalities (2.92), we find that for $(z,t,s) \in F_0$ the following estimates are valid:

$$\|\tilde{u}^o\|_s(z,t) \le \int_t^z \|v^o\|_s(\eta)\mathrm{d}\eta \le \int_t^z R\eta \le \frac{z}{a_0(s_0 - s) - z} a_0 s_0 R,$$

$$\|\tilde{v}^o\|_s(z,t) \le \frac{1}{4}\int_t^z \Big[\|u^o_{xx}\|_s(\xi) + \|k^o\|_s(\xi)$$

$$+ \int_0^\xi \|k^o\|_s(\tau)\|u^o\|_s(\xi - \tau)\mathrm{d}\tau\Big]\mathrm{d}\xi \le \frac{1}{4}\int_t^z \left[\frac{4\|u^0\|_{s_0}(\xi)}{e^2(s_0 - s)^2} + R + R^2\xi\right]\mathrm{d}\xi$$

$$\le \frac{1}{4}\int_t^z \left[\frac{4R}{e^2(s_0 - s)^2} + R + R^2\xi\right]\mathrm{d}\xi \le \frac{1}{4}\left[\frac{4e^2 Rz}{(s_0 - s)^2} + (R + R^2 z)z\right]$$

$$\le \frac{a_0 z}{\left[a_0(s_0 - s) - z\right]^2} \frac{a_0 R}{4}\left(4e^2 + s_0^2 + a_0 s_0^3 R\right).$$

Here, in intermediate inequalities, the estimate of s-norm of the second derivative of an analytic function through s_0-norm of the function itself is used. This has the form:

$$\|h_{xx}\|_s \le \frac{4\|h\|_{s_0}}{e^2(s_0 - s)^2)}, \quad s < s_0.$$

In fact, according to Definition 2.1, we have

$$
\begin{aligned}
\|h_{xx}\|_s &= \sum_{-\infty}^{\infty} |h_n| n^2 \exp(|n|s) \\
&\le \sup_n \left[n^2 \exp\left(-|n|(s_0 - s)\right) \right] \sum_{-\infty}^{\infty} |h_n| \exp(|n|s) \\
&\le \frac{\|h\|_{s_0}}{(s_0 - s)^2} \sup_{t>0} \left[t^2 \exp(-t) \right] = \frac{4\|h\|_{s_0}}{e^2(s_0 - s)^2)}.
\end{aligned}
$$

Similarly to the previous one, the estimate for $\|\tilde{k}^0\|_s$ is obtained:

$$
\begin{aligned}
\|\tilde{k}^0\|_s(z) &\le \int_0^z \Bigg[\|v^0_{xx}\|_s(\xi) + \|k^0\|_s(\xi)\|f\|_s(z - \xi) \\
&+ \int_0^{\xi} \|k^0\|_s(\tau)\|v^0\|_s(z - \tau)\mathrm{d}\tau \Bigg] \mathrm{d}\xi \le \int_0^z \left[\frac{4\|v^0\|_{s_0}(\xi)}{e^2(s_0 - s)^2} + (1 + \xi)R^2 \right] \mathrm{d}\xi \\
&\le \frac{a_0^2 z}{\left[a_0(s_0 - s) - z \right]^3} a_0 s_0 R \left(4e^2 + s_0^2 R + a_0 s_0^3 R \right).
\end{aligned}
$$

Therefore, estimates (2.93) are true for $\sigma = 0$, if

$$
\lambda_0 = a_0 R \left[4 + s_0 \max(1, R) + a_0 s_0 3R \right] \max(1, s_0). \tag{2.95}
$$

At the same time for $(z, t, s) \in F_1$ we have

$$
\|u_1 - u_0\|_s(z, t) = \|\tilde{u}_0\|_s(z, t) \le \frac{\lambda_0 z}{a_0(s_0 - s)} \le \frac{\lambda_0 a_1}{a_0 - a_1} = \lambda_0,
$$

$$
\begin{aligned}
\|v_1 - v_0\|_s(z, t) &= \|\tilde{v}_0\|_s(z, t) \le \frac{\lambda_0 a_0 z}{[a_0(s_0 - s) - z]^2} \\
&\le \frac{\lambda_0 a_0 a_1}{(a_0 - a_1)^2(s_0 - s)} = \frac{2\lambda_0}{s_0 - s}, \\
\|k_1 - k_0\|_s(z, t) &= \|k^0\|_s(z) \le \frac{\lambda_0 a_0^2 z}{[a_0(s_0 - s) - z]^3} \\
&\le \frac{\lambda_0 a_0^2 a_1}{(a_0 - a_1)^3(s_0 - s)^2} = \frac{4\lambda_0}{(s_0 - s)^2}.
\end{aligned}
$$

Therefore, inequalities (2.91) will be satisfied for $\sigma = 0$, if a_0 is chosen from the condition $\lambda_0 < \chi$.

Thus, the first step of the induction method is justified. Suppose that the statement of the lemma holds for $\sigma \leq n$, and prove that then it is also true for $\sigma = n+1$. Using the inductive hypothesis, we find that

$$(\tilde{u}^{n+1}, \tilde{\vartheta}^{n+1}, \tilde{k}^{n+1}) \in C(A'_s, G_{n+1}).$$

In addition, from (2.92) we obtain

$$\|\tilde{u}^{n+1}\|_s(z,t) \leq \int_t^z \|\tilde{\vartheta}\|_s(\eta,t)\mathrm{d}\eta \leq \int_t^z \frac{\lambda_n a_n \eta}{[a_n(s_0-s)-\eta]^2}\mathrm{d}\eta$$

$$\leq \frac{\lambda_n a_n z}{a_n(s_0-s)-z} \leq \lambda_n a_0 \frac{z}{a_{n+1}(s_0-s)-z},$$

$$\|\tilde{\vartheta}^{n+1}\|_s(z,t) \leq \frac{1}{4}\int_t^z \{\frac{4e^{-2}}{(s'-s)^2}\|\tilde{u}^n\|_s(z,\xi) + \|\tilde{k}^n\|_s(\xi)$$

$$+ \int_0^\xi [\|\tilde{k}^n\|_s(\tau)\|u^{n+1}\|_s(z-\tau,\xi-\tau) + \|k^n\|_s(\tau)\|\tilde{u}^n\|_s(z-\tau,\xi-\tau)]\mathrm{d}\tau\}\mathrm{d}\xi.$$

Putting $s' = s'(\xi) = (s+s_0-\xi/a_n)/2$ in this inequality, we get

$$\|\tilde{\vartheta}^{n+1}\|_s(z,t)$$

$$\leq \frac{1}{4}\int_t^z \left\{\frac{16e^{-2}a_n^2}{[a_n(s_0-s)-\xi]^2}\frac{2\lambda_n z}{a_n(s_0-s)-z} + \frac{\lambda_n a_n^2 \xi}{[a_n(s_0-s)-\xi]^3}\right.$$

$$\left. + \int_0^\xi \left[\frac{\lambda_n a_n^2 \tau(\chi+R)}{(a_n(s_0-s)-\tau)^3} + \frac{4\chi+Rs_0^2}{(s_0-s)^2}\frac{2\lambda_n z}{a_n(s_0-s)-z}\right]\mathrm{d}\tau\right\}\mathrm{d}\xi$$

$$\leq \frac{a_{n+1}\lambda_n z}{[a_{n+1}(s_0-s)-z]^2}a_0(8e^{-2}+1+2\chi+(1+s_0^2)R).$$

Similar calculations for $\tilde{k}^{n+1}$ lead to the inequality

$$\|\tilde{k}^{n+1}\|_s(z) \leq \frac{a_{n+1}^2\lambda_n z}{(a_n(s_0-s)-z)^3}a_0(32e^{-2}+Rs_0(1+2s_0)+\chi(1+8s_0)).$$

From the above estimates it follows that inequalities (2.93) are fulfilled for $\sigma = n+1$, if one puts

$$\lambda_{n+1} = \lambda_n \rho,$$

$$\rho = a_0 \max[8e^{-2} + 1 + 2\chi + (1 + s_0^2)R, 32e^{-2} + Rs_0(1 + 2s_0^2) + \chi(1 + 8s_0)]. \tag{2.96}$$

Now show that inequalities (2.94) also hold for $\sigma = n + 1$, if the number a_0 is chosen appropriately. For $(z, t, s) \in F_{n+2}$ the following inequalities are valid:

$$\|u^{n+2} - u^0\|_s(z, t) \leq \sum_{\sigma=0}^{n+1} \|\tilde{u}^\sigma\|_s(z, t) \leq \sum_{\sigma=0}^{n+1} \frac{\lambda_\sigma z}{a_\sigma(s_0 - s) - z}$$

$$\leq \lambda_0 \sum_{\sigma=0}^{n+1} \frac{\rho^\sigma a_{\sigma+1}}{a_\sigma - a_{\sigma+1}} \leq \lambda_0 \sum_{\sigma=0}^{n+1} \rho^\sigma \left(\frac{a_\sigma}{a_{\sigma+1}} - 1\right) \leq \lambda_0 \sum_{\sigma=0}^{n+1} \rho^\sigma (\sigma + 1)^2,$$

$$\|\vartheta^{n+2} - \vartheta^0\|_s(z, t) \leq \frac{2\lambda_0}{s_0 - s} \sum_{\sigma=0}^{n+1} \rho^\sigma (\sigma + 1)^4, \; \|k^{n+2} - k^0\|_s(z)$$

$$\leq \frac{4\lambda_0}{(s_0 - s)^2} \sum_{\sigma=0}^{n+1} \rho^\sigma (\sigma + 1)^6.$$

Therefore, inequalities (2.94) are satisfied for $\sigma = n + 1$, if the number a_0 is chosen so that

$$\rho < 1, \; \lambda_0 \sum_{\sigma=0}^{\infty} \rho^\sigma (\sigma + 1)^6 \leq \chi, \tag{2.97}$$

where numbers λ_0 and ρ are defined by the formulas (2.95) and (2.96), respectively. It is clear that one can always choose the parameter a_0 so small that inequalities (2.97) hold. Thus, the validity of Lemma 2.3 is established.

Further, we assume that the number a_0 is chosen from the condition (2.97). Continuing the proof of Theorem 2.7, we note that for the chosen number a_0 the sequence $(u^\sigma, \vartheta^\sigma, k^\sigma)$ converges uniformly in the norm of the space $C(A_s', G)$, $a = \lim a_\sigma$ and determines the unique solution of the system of Eq. (2.87), and at the same time the solution of the inverse problem. Limit transition in inequalities (2.94) for

$$(z, t, s) \in F = \{(z, t, s) | 0 < s < s_0, 0 \leq t \leq z < a(s_0 - s)\}$$

leads to estimates that coincide with (2.91).

The uniqueness of the found solution is established using the above described technique for evaluating by the standard technique (see Sect. 2.1).

Theorem 2.8 *Suppose that for $t \in [0, T]$*

$$\left\{\frac{\partial^i f}{\partial t^i}, \frac{\partial^2 f}{\partial x^2}, \frac{\partial^3 f}{\partial z \partial x^2}\right\} \in C(A_{s_0}, [0, T]), \quad i = 0, 1, 2;$$

$$\max\left\{\|f\|_{s_0}, \left\|\frac{\partial f}{\partial z}\right\|_{s_0}, 2\left\|\frac{\partial^2 f}{\partial z^2}\right\|_{s_0} + \left\|\frac{\partial^2 f}{\partial x^2}\right\|_{s_0}, \left\|\frac{\partial^3 f}{\partial z \partial x^2}\right\|_{s_0}\right\} \leq R,$$

$$f(x,+0)=\frac{\partial f(x,+0)}{\partial t}=0$$

and $\gamma\in(0,s_0)$ is an arbitrary fixed number. Then, for any $\chi>0$ there exists a positive number $C=C(R,\ s_0,\ T,\ \chi)$ such that for any $s\in(0,s_0-\gamma)$ the solution defined by Theorem 2.7 is continuously differentiable with respect to (z,t) in domain

$$G_\gamma\subseteq G,\ G_\gamma=G_\gamma(R,\ s_0,\ T,\ \chi,\ s)=\{(z,t)|0\le t\le z<a(s_0-\gamma-s)\}$$

and the following estimates hold:

$$\|u\|_s(z,t)\le c,\ \ \|\vartheta\|_s(z,t)=\left\|\frac{\partial u}{\partial z}\right\|_s(z,t)\le\frac{c}{\gamma},$$

$$\max\left[\left\|\frac{\partial u}{\partial t}\right\|_s(z,t),\ \|k\|_s(z),\ \left\|\frac{\partial\vartheta}{\partial t}\right\|_s(z,t)\right]\le\frac{c}{\gamma^2},$$

$$\max\left[\left\|\frac{\partial\vartheta}{\partial z}\right\|_s(z,t),\ \left\|\frac{\partial u_{xx}}{\partial z}\right\|_s(z,t)\right]\le\frac{c}{\gamma^3},\tag{2.98}$$

$$\max\left[\left\|\frac{\partial u_{xx}}{\partial t}\right\|_s(z,t),\ \left\|\frac{\partial\vartheta_{xx}}{\partial t}\right\|_s(z,t)\right]\le\frac{c}{\gamma^4},$$

$$\max\left[\left\|\frac{\partial k}{\partial z}\right\|_s(z),\ \left\|\frac{\partial\vartheta_{xx}}{\partial z}\right\|_s(z,t)\right]\le\frac{c}{\gamma^5},\ \ (z,t)\in G_\gamma.$$

Proof Let $(u,\ \vartheta,\ k)$ be the solution of the system of Eq. (2.87) defined by Theorem 2.7. Below we will propose that the numbers $s_0,\ a,\ \chi,\ R$ and the domain G have the same meaning as in the Theorem 2.7. First, we find the expressions for the first derivatives of functions ϑ and k with respect to variable z. For this, we differentiate the second and third equations of the system (2.87) with respect to z:

$$\begin{aligned}\vartheta_z(x,z,t)&=(1/2)f_{zz}(x,z)+(1/4)f_{xx}(x,z)+(1/4)k(x,z)\\&+(1/4)\int_t^z\Big[\vartheta_{xx}(x,z,\xi)-k(x,\xi)f(x,z-\xi)\\&-\int_0^\xi k(x,\tau)\vartheta(x,z-\tau,\xi-\tau)d\tau\Big]d\xi,\end{aligned}\tag{2.99}$$

$$k_z(x,z) = -2f_{zz}(x,z) - f_{xx}(x,z) - (1/2)f_{zxx} - \int_0^z \Big[(\vartheta_z)_{xx}(x,z,\xi) - (3/2)k(x,\xi)f_z(x,z-\xi) - \int_0^\xi k(x,\tau)\vartheta_z(x,z-\tau,\xi-\tau)\mathrm{d}\tau\Big]\mathrm{d}\xi. \tag{2.100}$$

According to (2.89), the following inequalities hold:

$$\|u\|_s(z,t) \le R + \chi := c_1,\ \|\vartheta\|_s(z,t) \le \frac{Rs_0 + 2\chi}{s_0 - s} := \frac{c_2}{s_0 - s},$$

$$\|k\|_s(z) \le \frac{Rs_0^2 + 4\chi}{(s_0 - s)^2} := \frac{c_3}{(s_0 - s)^2}. \tag{2.101}$$

Using these relations, we estimate the functions ϑ_z and k_z, defined by equalities (2.99) and (2.100):

$$\|\vartheta_z\|_s(z,t) \le \frac{s_0^3 R + s_0 c_3 + (4c_2 + s_0 c_3 + T s_0 c_2 c_3)\,T}{(s_0 - s)^3} := \frac{c_4}{(s_0 - s)^3}, \tag{2.102}$$

$$\|k_z\|_s(z) \le \frac{2Rs_0^5 + \left(16c_4 + 2s_0^2 R c_3 + c_3 c_4 T\right) T}{(s_0 - s)^5} := \frac{c_5}{(s_0 - s)^5}. \tag{2.103}$$

Now let $s \in (0, s_0 - \gamma)$ for some $\gamma \in (0, s_0) - (z,t) \in G_\gamma$. Then, from (2.101)–(2.102) the following estimates follow

$$\|u_z\|_s = \|\vartheta\|_s(z,t) \le c_2\gamma^{-1},\ \|k\|_s(z) \le c_3\gamma^{-2},\ \|\vartheta_z\|_s(z,t) \le c_4\gamma^{-3},$$

$$\|k_z\|_s(z) \le c_5\gamma^{-5}.$$

Since u_t satisfies the following relation:

$$u_t(x,z,t) = (1/2)f_t(x,z) + \int_t^z \vartheta_t(x,\eta,z)\mathrm{d}\eta,$$

then

$$\|u_t\|_s(z,t) \le (Rs_0 + c_4 T)\gamma^{-2} := c_6\gamma^{-2}.$$

To estimate s-norm of function u_{xx}, we use the inequality

$$\|u_{xx}\|_s(z,t) \le 4e^{-2}(s'-s)^{-2}\|u\|_{s'}(z,t),$$

putting in it $s' = s + \gamma/2$. It is clear that $s' < s_0 - z/a$, and then

$$\|u_{xx}\|_s(z,t) \le 16e^{-2}(R+\chi)\gamma^{-2} := c_7\gamma^{-2}.$$

The estimate for ϑ_t follows from (2.86) and previous inequalities

$$\|\vartheta_t\|_s(z,t) \le \frac{1}{4}(c_7 + c_3(1 + c_1T))\gamma^{-2} := c_8\gamma^{-2}.$$

Similar inequalities, with the same choice s', we use to evaluate the first derivatives of functions u_{xx}, ϑ_{xx} with respect to the variables t, z:

$$\left\|\frac{\partial u_{xx}}{\partial t}\right\|_s(z,t) \le 4e^{-2}(s'-s)^{-2}\|u_t\|_{s'}(z,t) \le 16e^{-2}c_6\gamma^{-4} := c_9\gamma^{-4},$$

$$\left\|\frac{\partial u_{xx}}{\partial z}\right\|_s(z,t) \le 4e^{-2}(s'-s)^{-2}\|u_z\|_{s'}(z,t) \le 16e^{-2}c_2\gamma^{-3} := c_{10}\gamma^{-3},$$

$$\left\|\frac{\partial \vartheta_{xx}}{\partial t}\right\|_s(z,t) \le 4e^{-2}(s'-s)\|\vartheta_t\|_{s'}(z,t) \le 16e^{-2}c_6\gamma^{-4} := c_{11}\gamma^{-4},$$

$$\left\|\frac{\partial u_{xx}}{\partial z}\right\|_s(z,t) \le 4e^{-2}(s'-s)\|\vartheta_z\|(z,t) \le 16e^{-2}c_4\gamma^{-5} := c_{12}\gamma^{-5}.$$

From the above obtained inequalities it follows the estimates (2.98) at a suitable choice of constant c. Theorem 2.8 is proved.

The estimates obtained above for the derivatives of solution of the system of Volterra-type integro-differential equations (2.87) can be successfully used in the numerical solving of the inverse problem.

2.5 Conclusions

This chapter studied the questions of solvability of multidimensional kernel determination inverse problems for the second-order hyperbolic integro-differential equations by the method of the Banach's scale of analytic functions spaces, developed in the works of L. V. Ovsyannikov and L. Nirenberg. There were shown that problems of this chapter are uniquely solvable in the class of functions having finite smoothness in the time variable t and analytical with respect to the spatial variable x, stability estimates are obtained for the solutions, and uniqueness theorems are formulated. Problems are investigated in the cases of equations, and when in integral terms the memory is multiplied by the second time derivative and the first-order differential operator in variable x with real analytical coefficients of the direct problem solution. Here, it also considered the inverse problem for a two-dimensional system of Maxwell equations with memory. It is shown that this problem, reformulated in several other terms, can be reduced to the already investigated problem.

References

1. Ovsyannikov, L.V. 1965. Singular operator in the scale of the Banach spaces. *Dokl. AN SSSR* 163 (4): 819–822 (Russian).
2. Ovsyannikov, L.V. 1971. Nonlinear Cauchy problem in the scales of the Banach spaces. *Dokl. AN SSSR* 200 (4): 789–792 (Russian).
3. Nirenberg, L. 1974. *Topics in nonlinear functional analysis*. New York: Courant Institute Math. Sci., New York Univ.
4. Romanov, V.G. 2005. *Stability in inverse problems*. Moscow: Nauchniy Mir (Russian).

Chapter 3
Global Solvability of Memory Reconstruction Problems

There are the existence theorems for sufficiently small domains of definition of unknown functions in the inverse problems, that is, solvability for such problems is local. An example of the absence of a global solution of the inverse problem is given in [1]. Even with smooth data, the solution of the inverse problem may not exist on any preassigned interval. This phenomenon is connected with the nonlinearity of the problem. However, in problems of determining the kernel of an integral term in second-order hyperbolic equations, where the nonlinearity is convolutional, it is possible to obtain global existence theorems in the space of continuous functions with exponential weight. In this direction, in addition to the works of the authors, we also note [2–7].

In the Sect. 3.1 the problem of finding a memory kernel from a one-dimensional wave equation for the half-line is investigated. In the Sect. 3.2 the results obtained in the previous section were generalized to the case of the n-dimensional wave equation in a half-space. In the Sect. 3.3 the unknown kernel has the form of a diagonal matrix and is determined from the three-dimensional wave equation in $\mathbb{R}^3$. In the Sect. 3.4 the problem of determining two one-dimensional functions that are included in a initial–boundary problem for the integro-differential equation, main part of which coincides with $n+1-$ dimensional wave operator, is considered. This is one of the versions for setting the inverse problem of determining the kernel of the integral operator in a situation where the shape of the signal from a point source of disturbances causing the medium to oscillate is unknown and to be determined together with the properties of the medium. In the Sect. 3.5 under study is the multidimensional inverse problem of determining the convolutional kernel of the integral term in an integro-differential wave equation. We apply the methods of scales of Banach spaces of real analytic functions of spatial variable and weight norms in the class of continuous functions in time.

In this chapter the theorems of global existence, uniqueness, and the stability estimates are obtained.

© The Author(s), under exclusive license to Springer Nature Singapore Pte Ltd. 2023
D. K. Durdiev and Z. D. Totieva, *Kernel Determination Problems in Hyperbolic Integro-Differential Equations*, Infosys Science Foundation Series in Mathematical Sciences, https://doi.org/10.1007/978-981-99-2260-4_3

3.1 The Problem of Determining the Memory Kernel from the Integro-Differential Equation of String Vibrations

Consider the initial–boundary value problem

$$u_{tt} - u_{xx} - \int_0^t k(\tau)u(x, t-\tau)d\tau = 0,\ (x,t) \in \mathbb{R}^2_+, \tag{3.1}$$

$$u\big|_{t<0} \equiv 0,\ u_x\big|_{x=0} = \delta'(t) + f(t)\theta(t), \tag{3.2}$$

in which $\mathbb{R}^2_+ = \left\{(x,t) \in \mathbb{R}^2 |\ x > 0\right\}$.

For given functions $k(t)$, $f(t)$, the problem of finding a function $u(x,t)$ that satisfies (in a generalized sense) the equalities (3.1), (3.2) is called a direct problem.

It follows from the theory of hyperbolic equations, the function $u(x,t)$ as a solution to the direct problem has the property $u = 0$, $t < x$, $x > 0$ and in the neighborhood of the characteristic line $t = x$ has following structure:

$$u(x,t) = -\delta(t-x) + \theta(t-x)v(x,t). \tag{3.3}$$

We set the inverse problem: determine the function $k(t)$, included in the integral of Eq. (3.1) according to the given the regular part of the solution of the direct problem at any time $t \geq 0$ at the point $x = 0$:

$$v(0,t) = F(t). \tag{3.4}$$

The norm in $C[0,\ T]$ and $C^n[0,\ T]$, $n = 1, 2$ is defined in the usual ways:

$$\|g(t)\|_{C[0,\ T]} = \max_{t\in[0,T]} |g(t)|,$$

$$\|g(t)\|_{C^n[0,\ T]} = \sum_{i=0}^{n} \max_{t\in[0,T]} \left|\frac{d^i g(t)}{dt}\right|.$$

The main results of this section are the following theorems:

Theorem 3.1 *Let $f(t) \in C^1[0,T]$, $F(t) \in C^2[0,T]$ and the matching condition*

$$f(0) = -F'(0)$$

is fulfilled. Then, the inverse problem (3.1)–(3.4) has a unique solution $k(t) \in C[0,T]$, $T > 0$.

Denote by $A(k_0)$ the set of functions $k(t) \in C[0, T]$, satisfying for some $T > 0$ the condition

$$\|k\|_{C[0,T]} \leq k_0$$

with a constant $k_0 > 0$.

Theorem 3.2 *Let $k^{(i)} \in A(k_0)$, $i = 1, 2$ be the solutions of the inverse problem (3.1)–(3.4) with data $f^{(i)}(t)$, $F^{(i)}(t)$, $i = 1, 2$, respectively. Then, there exists the positive constant C, depending on T, k_0, $\|f^{(1)}\|_{C^1[0,T]}$, $\|F^{(1)}\|_{C^2[0,T]}$ that the following estimate*

$$\|k^{(1)} - k^{(2)}\|_{C[0,T]} \leq C \left(\|f^{(1)}(t) - f^{(2)}(t)\|_{C^1[0,T]} + \|F^{(1)} - F^{(2)}\|_{C^2[0,T]}\right) \quad (3.5)$$

is valid.

At the beginning, we prove Theorem 3.1. Substituting the function (3.3) into Eq. (3.1), we find that the function $v(x, t)$ in the domain $t > x > 0$ satisfies the equations

$$v_{tt} - v_{xx} + k(t - x) - \int_0^{t-x} k(\tau)v(x, t - \tau)d\tau = 0, \quad (3.6)$$

$$v|_{t=x+0} = 0, \quad (3.7)$$

$$v_x|_{x=0} = f(t). \quad (3.8)$$

Since the research method allows to simultaneously find functions $v(x, t)$, $k(t)$, it is convenient to interpret the direct and inverse problems as the problem of determining functions $v(x, t)$, $k(t)$ from the equations (3.6)–(3.8), (3.4).

Lemma 3.1 *Under fulfillment of the conditions of Theorem 3.1, the problem (3.4), (3.6)–(3.8) for $(x, t) \in D_T$, $D_T = \left\{(x, t)|0 \leq x \leq t \leq \leq T - x\right\}$ is equivalent to the problem of finding the vector of functions $v(x, t)$, $v_t(x, t)$, $k(t)$ from the following system of equations:*

$$v(x, t) = \int_x^t v_t(x, \tau)d\tau, \quad (3.9)$$

$$
\begin{aligned}
v_t(x,t) &= \frac{1}{2}F'(t-x) \\
&- \frac{1}{2}f(t-x) - \frac{x}{2}k(t-x) \\
&+ \frac{1}{2}\int_0^x \int_0^{t-x} k(\tau)v(\xi, t-x+\xi-\tau)\mathrm{d}\tau\mathrm{d}\xi \\
&+ \frac{1}{2}\int_x^{(x+t)/2}\left[-k(t+x-2\xi) + \int_0^{t+x-2\xi} k(\tau)v(\xi, t+x-\xi-\tau)\mathrm{d}\tau\right]\mathrm{d}\xi,
\end{aligned} \tag{3.10}
$$

$$
k(t) = 2F''(t) + 2f'(t) - \int_0^{t/2}\int_0^{t-2\xi} k(\tau)v_t(\xi, t-\xi-\tau)\mathrm{d}\tau\mathrm{d}\xi. \tag{3.11}
$$

In order to prove the lemma 3.1, it is sufficient in (3.6), after substituting the wave operator $v_{tt} - v_{xx}$ with the product

$$
\left(\frac{\partial}{\partial t} - \frac{\partial}{\partial x}\right)(v_t + v_x) = \left(\frac{\partial}{\partial t} + \frac{\partial}{\partial x}\right)(v_t - v_x)
$$

to integrate equality (3.6) along the corresponding characteristics of first-order differential operators for $(x,t) \in D_T$. Integration along characteristics of the operator $\partial/\partial t - \partial/\partial x$ performs from point (x,t) to point $((x+t)/2,\ (x+t)/2)$ on the plane of variables $(\xi,\ \theta)$.

Using the differentiation with respect to x equality (3.7) to calculate the expression in the form $(v_t + v_x)((x+t)/2,\ (x+t)/2) = 0$, we obtain

$$
\begin{aligned}
&(v_t + v_x)(x,t) \\
&= \int_x^{(x+t)/2}\left[-k(t+x-2\xi) + \int_0^{t+x-2\xi} k(\tau)v(\xi, t+x-\xi-\tau)\mathrm{d}\tau\right]\mathrm{d}\xi.
\end{aligned} \tag{3.12}
$$

The integration along the characteristic of the operator $\partial/\partial t + \partial/\partial x$ is performed from the point $(0,\ t-x)$ to the point (x,t). Using the condition (3.8) and differentiation with respect to t equality (3.4), we find

$$(v_t - v_x)(x,t)$$
$$= F'(t-x) - f(t-x) - k(t-x)x + \int_0^x \int_0^{t-x} k(\tau)v(\xi, t-x+\xi-\tau)\mathrm{d}\tau \mathrm{d}\xi. \tag{3.13}$$

From equations (3.12) and (3.13) it follows (3.10). In Eq. (3.10), setting $x = 0$ and using the condition (3.4) we find

$$F'(t) + f(t)$$
$$= \int_0^{t/2} \left[-k(t-2\xi) + \int_0^{t-2\xi} k(\tau)v(\xi, t-\xi-\tau)\mathrm{d}\tau \right] \mathrm{d}\xi. \tag{3.14}$$

Differentiating this equality, after simple calculations we arrive at the Eq. (3.11).

If the conditions of agreement in Theorem 3.1 are fulfilled, then it is easy to show that the inverse transformations also take place. Indeed, replacing the right and left parts (3.11) t by $t - 2\eta$, multiply both sides by $\mathrm{d}\eta$ and integrate over η in the limits from 0 to t. In the repeated integrals of the resulting equality, changing the order of integration and taking into account the relation (3.7), we obtain Eq. (3.14), which, in turn, is equivalent to equality (3.9) with $x = 0$. Equivalence of equality (3.9) to (3.12), (3.13) and the derivation of relations (3.6)–(3.8) from the last equations are obvious. The obvious equality (3.9) is given for the closure of the system of equations (3.10), (3.11). Thus, Lemma 3.1 is proved.

Continue the proof of Theorem 3.1. We write the system of the equations (3.9)–(3.11) in the form of an operator equation

$$\varphi = A\varphi, \tag{3.15}$$

where

$$\varphi = [\varphi_1(x,t), \varphi_2(x,t), \varphi_3(t)]^* = [v(x,t), v_t(x,t) + (x/2)k(t-x), k(t)]^*$$

is a vector function with components $\varphi_i (i = 1, 2, 3)$, $*$ is as before the sign of transposition, the operator A is defined on the set of functions $\varphi \in C[D_T]$ and in accordance with equalities (3.9)–(3.11) has the form

$$A = (A_1,\ A_2,\ A_3):$$

$$A_1\varphi = \int_x^t \Big[\varphi_2(x,\tau) - \frac{x}{2}\varphi_3(\tau - x)\Big]\mathrm{d}\tau,$$

$$A_2\varphi = \frac{1}{2}F'(t-x) - \frac{1}{2}f(t-x) + \frac{1}{2}\int_0^x\int_0^{t-x} \varphi_3(\tau)\varphi_1(\xi, t - x + \xi - \tau)\mathrm{d}\tau\mathrm{d}\xi$$

$$+\frac{1}{2}\int_x^{(x+t)/2} \big[-\varphi_3(t + x - 2\xi)$$

$$+\int_0^{t+x-2\xi} \varphi_3(\tau)\varphi_3(\xi, t + x - \xi - \tau)\mathrm{d}\tau\big]\mathrm{d}\xi,$$

$$A_3\varphi = 2F''(t) + 2f'(t)$$

$$-\int_0^{t/2}\int_0^{t-2\xi} \varphi_3(\tau)\Big[\varphi_2(\xi, t - \xi - \tau) - \frac{\xi}{2}\varphi_3(t - 2\xi - \tau)\Big]\mathrm{d}\tau\mathrm{d}\xi. \tag{3.16}$$

Denote by C_ρ the Banach space of continuous functions generated by the family of weight norms:

$$\|\varphi\|_\rho$$

$$= \max\left\{\sup_{(x,t)\in D_T}\left|\varphi_i(x,t)\mathrm{e}^{-\rho t}\right|,\ i = 1, 2,\ \sup_{t\in[0,T]}\left|\varphi_3(t)\mathrm{e}^{-\rho t}\right|\right\},\quad \rho \geq 0.$$

It is obviously, when $\rho = 0$ this space coincides with the space of continuous functions with the usual norm. This norm will be denoted below as $\|\psi\|$. Due to inequality

$$\mathrm{e}^{-\rho T}\|\varphi\| \leq \|\varphi\|_\rho \leq \|\varphi\| \tag{3.17}$$

the norms $\|\varphi\|_\rho$ and $\|\varphi\|$ are equivalent for any fixed $T \in (0, \infty)$. The number ρ will be chosen later. Let $Q_\rho(\varphi_0, R)$ be a ball of radius R with center at the point φ_0 of some weight space $C_\rho(\rho \geq 0)$, where

$$\varphi_0(x,t) \quad = (\varphi_{01},\ \varphi_{02},\ \varphi_{03})^* := \Big[0,\ \frac{1}{2}\left(F'(t-x) - f(t-x)\right),\ 2\left(F''(t) + f'(t)\right)\Big]^*.$$

It is easy to see that for $\varphi \in Q_\rho(\varphi_0, R)$ the estimate $\|\varphi\|_\rho \leq \|\varphi_0\|_\rho + R \leq \|\varphi_0\| + R := R_0$ (R_0 is a known number) is valid.

Let $\varphi(x,t) \in Q_\rho(\varphi_0, R)$. We prove that with a suitable choice of $\rho > 0$ the operator A maps the ball $Q_\rho(\varphi_0, R)$ to the same ball, i.e., $A\varphi \in Q_\rho(\varphi_0, R)$. Using equalities (3.16) and making up the norm of differences, for $(x, t) \in D_T$ we have

$$\|A_1\varphi - \varphi_{01}\|_\rho = \sup_{(x,t)\in D_T} \left|(A_1\varphi - \varphi_{01})e^{-\rho t}\right|$$

$$= \sup_{(x,t)\in D_T} \Big|\int_x^t \big[\varphi_2(x,\tau)e^{-\rho\tau}e^{-\rho(t-\tau)}$$

$$- \frac{x}{2}\varphi_3(\tau - x)e^{-\rho(\tau-x)}e^{-\rho(t-\tau+x)}\big]d\tau\Big|$$

$$\leq \left(\|\varphi_2\|_\rho + \frac{T}{4}\|\varphi_3\|_\rho\right)\frac{1}{\rho} \leq R_0\left(1 + \frac{T}{4}\right)\frac{1}{\rho} := \alpha_1\frac{1}{\rho}.$$

Similarly, we find

$$\|A_2\varphi - \varphi_{02}\|_\rho = \frac{1}{2}\sup_{(x,t)\in D_T} \left|(A_2\varphi - \varphi_{02})e^{-\rho t)}\right|$$

$$= \frac{1}{2}\sup_{(x,t)\in D_T} \Big|\int_0^x\int_0^{t-x} \varphi_3(\tau)e^{-\rho\tau}\varphi_1(\xi, t - x + \xi - \tau)e^{-\rho(t-x-\tau+\xi)}$$

$$\times e^{-\rho(x-\xi)}d\tau d\xi + \int_x^{(t+x)/2} \big[-\varphi_3(t + x - 2\xi)e^{-\rho(t+x-2\xi)}e^{-\rho(2\xi-x)}$$

$$+ \int_0^{t+x-2\xi} \varphi_3(\tau)e^{-\rho\tau}\varphi_1(\xi, t + x - \xi - \tau)$$

$$\times e^{-\rho(t+x-\xi-\tau)}e^{-\rho(\xi-x)}d\tau\big]d\xi\Big|$$

$$\leq \frac{1}{2}\left[\frac{T}{2}\|\varphi_3\|_\rho\|\varphi_1\|_\rho + \frac{1}{4}\|\varphi_3\|_\rho + \frac{3T}{4}\|\varphi_3\|_\rho\|\varphi_1\|_\rho\right]\frac{1}{\rho}$$

$$\leq R_0\left[\frac{R_0T}{2} + \frac{1}{4} + \frac{3R_0T}{4}\right]\frac{1}{\rho} := \alpha_2\frac{1}{\rho},$$

$$\|A_3\varphi - \varphi_{03}\|_\rho = \sup_{t\in[0,T]} \left|(A_3\varphi - \varphi_{03})e^{-\rho t}\right|$$

$$= \sup_{t\in[0,T]} \Big|-\int_0^{t/2}\int_0^{t-2\xi} [\varphi_3(\tau)e^{-\rho\tau}\varphi_2(\xi, t - \xi - \tau)e^{-\rho(t-\tau-\xi)}e^{-\rho\xi}$$

$$- \frac{\xi}{2}\varphi_3(t - 2\xi - \tau)e^{-\rho(t-2\xi-\tau)}\varphi_3(\tau)e^{-\rho\tau}e^{-2\rho\xi}]d\tau d\xi\Big|$$

$$\leq \left[\frac{T}{1}\|\varphi_3\|_\rho\|\varphi_2\|_\rho + \frac{T^2}{8}\|\varphi_3\|_\rho^2\right]\frac{1}{\rho} \leq R_0^2T\left(1 + \frac{T}{8}\right)\frac{1}{\rho} := \alpha_3\frac{1}{\rho}.$$

Choosing $\rho \geq \alpha_0 =: (1/R)\max(\alpha_1, \alpha_2, \alpha_3)$, we obtain that A transfers the ball $Q_\rho(\varphi_0, R)$ to the same ball $Q_\rho(\varphi_0, R)$. Now let φ^1, φ^2 be any two elements of $Q_\rho(\varphi_0, R)$. Then, using auxiliary inequalities of the form

$$\begin{aligned}&\left|\varphi_i^1\varphi_j^1 - \varphi_i^2\varphi_j^2\right| \mathrm{e}^{-\rho t)}\\ &\quad \leq \left|\varphi_i^1\right| \left|\varphi_j^1 - \varphi_j^2\right| \mathrm{e}^{-\rho t)} + \left|\varphi_j^2\right| \left|\varphi_i^1 - \varphi_i^2\right| \mathrm{e}^{-\rho t)} \leq 2R_0 \left\|\varphi^1 - \varphi^2\right\|_\rho,\\ &(x,t) \in D_T,\end{aligned}$$

we have

$$\begin{aligned}\left\|(A\varphi^1 - A\varphi^2)_1\right\|_\rho &= \sup_{(x,t)\in D_T} \left|(A\varphi^1 - A\varphi^2)_1 \mathrm{e}^{-\rho t}\right|\\ &= \sup_{(x,t)\in D_T} \Big| \int_x^t [(\varphi_2^1 - \varphi_2^2)(x,\tau)\mathrm{e}^{-\rho\tau}\mathrm{e}^{-\rho(t-\tau)} - \frac{x}{2}(\varphi_3^1\\ &\quad - \varphi_3^2)(\tau - x)\mathrm{e}^{-\rho(\tau-x)}\mathrm{e}^{-\rho(t-\tau+x)}]\mathrm{d}\tau\Big|\\ &\leq \left(1 + \frac{T}{4}\right)\left\|\varphi^1 - \varphi^2\right\|_\rho \frac{1}{\rho} =: \beta_1 \left\|\varphi^1 - \varphi^2\right\|_\rho \frac{1}{\rho}.\end{aligned}$$

Similarly, we find

$$\begin{aligned}&\left\|(A\varphi^1 - A\varphi^2)_2\right\|_\rho = \sup_{(x,t)\in D_T} \left|(A\varphi^1 - A\varphi^2)_2 \mathrm{e}^{-\rho t}\right|\\ &= \frac{1}{2}\sup_{(x,t)\in D_T} \Big| \int_0^x \int_0^{t-x} [\varphi_3^1(\tau)\mathrm{e}^{-\rho\tau}(\varphi_1^1 - \varphi_1^2)(\xi, t - x + \xi - \tau)\mathrm{e}^{-\rho(t-x-\tau+\xi)}\\ &+ \varphi_1^2(\xi, t - x + \xi - \tau)\mathrm{e}^{-\rho(t-x-\tau+\xi)}(\varphi_3^1 - \varphi_3^2)(\tau)\mathrm{e}^{-\rho\tau}]\mathrm{e}^{-\rho(x-\xi)}\mathrm{d}\tau\mathrm{d}\xi\\ &+ \int_x^{(x+t)/2} [(\varphi_3^2 - \varphi_3^1)(t + x - 2\xi)\mathrm{e}^{-\rho(t+x-2\xi)}\mathrm{e}^{-\rho(2\xi-x)}\\ &+ \int_0^{t+x-2\xi} (\varphi_3^1(\tau)\mathrm{e}^{-\rho\tau}(\varphi_1^1 - \varphi_1^2)(\xi, t + x - \xi - \tau)\mathrm{e}^{-\rho(t+x-\xi-\tau)}\\ &+ \varphi_1^2(\xi, t + x - \xi - \tau)\mathrm{e}^{-\rho(t+x-\xi-\tau)}\\ &\times (\varphi_1^1 - \varphi_1^2)(\tau)\mathrm{e}^{-\rho\tau})\mathrm{e}^{-\rho(\xi-x)}\mathrm{d}\tau]\mathrm{d}\xi\Big|\\ &\leq \frac{1}{2}\left[\frac{1}{4} + R_0T + \frac{3R_0T}{2}\right]\left\|\varphi^1 - \varphi^2\right\|_\rho \frac{1}{\rho} =: \beta_2 \left\|\varphi^1 - \varphi^2\right\|_\rho \frac{1}{\rho},\\ &\left\|(A\varphi^1 - A\varphi^2)_3\right\|_\rho = \sup_{t\in[0,T]} \left|(A\varphi^1 - A\varphi^2)_3 \mathrm{e}^{-\rho t}\right|\end{aligned}$$

$$= \sup_{t\in[0,T]} \Big| - \int_0^{t/2}\int_0^{t-2\xi} [(\varphi_3^1(\tau)e^{-\rho\tau}(\varphi_2^1 - \psi_2^2)(\xi, t-\xi-\tau)e^{-\rho(t-\tau-\xi)}$$
$$+ \varphi_2^2(\xi, t-\xi-\tau)e^{-\rho(t-\tau-\xi)}(\varphi_3^1 - \varphi_3^2)(\tau)e^{-\rho\tau})e^{-\rho\xi}$$
$$- \frac{\xi}{2}(\varphi_3^1(t-2\xi-\tau)e^{-\rho(t-2\xi-\tau)}(\varphi_3^1 - \varphi_3^2)(\tau)e^{-\rho\tau})$$
$$+ \varphi_3^2 e^{-\rho\tau}(\varphi_3^1 - \varphi_3^2)(t-2\xi-\tau)e^{-\rho(t-2\xi-\tau)})e^{-2\rho\xi}]d\tau d\xi \Big|$$
$$\leq R_0 T\left[2 + \frac{T}{4}\right]\left\|\varphi^1 - \varphi^2\right\|_\rho \frac{1}{\rho} =: \beta_3 \left\|\varphi^1 - \varphi^2\right\|_\rho \frac{1}{\rho}.$$

Let $\beta_0 := \max(\beta_1, \beta_2, \beta_3)$. As follows from the above estimates, if the number ρ is chosen from the condition $\rho > \max(\alpha_0, \beta_0)$, then operator A is contraction one on $Q_\rho(\varphi_0, R)$. Consequently, according to the Banach principle, Eq. (3.15) has, and moreover, the only solution in $Q_\rho(\varphi_0, R)$ for any fixed $T > 0$. The Theorem 3.1 is proved.

Remark 3.1 The pair of functions $v(x,t), k(t)$, constructed above, is the generalized solution of the problem (3.6)–(3.8), (3.4) in the domain D_T, the pair of functions $u(x,t)$, $k(t)$, where $u(x,t) = -\delta(t-x) + \theta(t-x)v(x,t)$ is the generalized solution to the problem (3.1)–(3.4).

Example 3.1 Let $f(t) = 0, \ \ F(t) = -t^2$. One can easily show that in this case the solution to the inverse problem (3.6)–(3.8), (3.4) uniquely exists:

$$k(t) = -\frac{1}{2}e^{\sqrt[3]{3/2}\cdot t} - \frac{1}{2}e^{-1/2\sqrt[3]{3/2}\cdot t}\left[\cos\left(\sqrt[6]{(3/2)^5}\cdot t\right) - \sin\left(\sqrt[6]{(3/2)^5}\cdot t\right)\right], \tag{3.18}$$

$v(x,t) = x^2 - t^2$, the pair of functions $u(x,t)$, $k(t)$, where $u(x,t) = -\delta(t-x) + \theta(t-x)v(x,t)$ and $k(t)$ is determined by (3.18), is the generalized solution to the problem (3.1)–(3.4). In the case under consideration, Theorem 3.1 holds.

Now we prove the conditional stability theorem 3.2. First of all, we note that substituting the expression for $v_t(x,t)$ by formula (3.10) into (3.9) for the known function $k(t) \in A(k_0)$ we obtain an integral equation of the second kind of Volterra-type with respect to $v(x,t)$:

$$v(x,t) = G(x,t) + \frac{1}{2}\int_0^t \Big[\int_0^x \int_0^{\tau-x} k(\alpha)v(\xi, \tau - x + \xi - \alpha)d\alpha d\xi$$
$$+ \int_x^{(x+\tau)/2}\int_0^{\tau+x-2\xi} k(\alpha)v(\xi, \tau + x - \xi - \alpha)d\alpha d\xi\Big]d\tau,$$

where

$$G(x,t) = \frac{1}{2}\int\limits_0^{t-x}\big[F'(\tau) - f(\tau) - xk(\tau) - \frac{1}{2}\int\limits_0^{\tau} k(\alpha)\mathrm{d}\alpha\big]\mathrm{d}\tau.$$

We represent $v(x,t)$ in the form

$$v(x,t) = v_0(x,t) + \sum_{n=1}^{\infty} v_n(x,t),$$

in which $v_n(x,t)$ are given by formulas:

$$v(x,t) = G(x,t),$$

$$v_n(x,t) = \frac{1}{2}\int\limits_0^{t}\Big[\int\limits_0^{x}\int\limits_0^{\tau-x} k(\alpha)v_{n-1}(\xi,\tau - x + \xi - \alpha)\mathrm{d}\alpha\mathrm{d}\xi$$

$$+ \int\limits_x^{(x+\tau)/2}\int\limits_0^{\tau+x-2\xi} k(\alpha)v_{n-1}(\xi,\tau + x - \xi - \alpha)\mathrm{d}\alpha\mathrm{d}\xi\Big]\mathrm{d}\tau, \ n = 1, 2, \ldots.$$

It is clear, that v_n are continuous functions in D_T together with the first derivatives with respect to x, t for all $n \geq 0$. Doing the same as at the obtaining the estimate (1.17) from Sect. 1.1 we find that in the domain D_T the function $v(x,t)$ is modulo bounded with some constant c_0 depending on k_0, T, $\|f\|_{C[0,T]}$, $\|F\|_{C^1[0,T]}$, i.e.,

$$|v(x,t)| \leq c_0, \ (x,t) \in D_T.$$

Suppose that the functions $f^{(i)}$, $F^{(i)}(t)$, $k^{(i)}$, $i = 1, 2$ are such that they are defined in Theorem 3.1. The regular part of the solution of the direct problem (3.1)–(3.3) for $k = k^{(i)}$, $f = f^{(i)}$, we denote by $v^{(i)}(x,t)$. And also introduce the differences

$$\tilde{v} = v^{(1)} - v^{(2)}, \ \tilde{v_t} = v_t^{(1)} - v_t^{(2)}, \ \tilde{k} = k^{(1)} - k^{(2)},$$
$$\tilde{f} = f^{(1)} - f^{(2)}, \ \tilde{F} = F^{(1)} - F^{(2)}.$$

Then from relations (3.9)–(3.11) we obtain

$$\tilde{v}(x,t) = \int_x^t \tilde{v}_t(x,\tau)\mathrm{d}\tau,$$

$$\tilde{v}_t(x,t) = \frac{1}{2}\tilde{F}'(t-x) - \frac{1}{2}\tilde{f}(t-x) - \frac{x}{2}\tilde{k}(t-x)$$
$$+\frac{1}{2}\int_0^x\int_0^{t-x}\big[\tilde{k}(\tau)v^{(1)}(\xi,t-x+\xi-\tau)$$
$$+k^{(2)}(\tau)\tilde{v}(\xi,t-x+\xi-\tau)\big]\mathrm{d}\tau\mathrm{d}\xi + \frac{1}{2}\int_x^{(x+t)/2}\{-\tilde{k}(t+x-2\xi) \tag{3.19}$$

$$+\int_0^{t+x-2\xi}\big[\tilde{k}(\tau)v^{(1)}(\xi,t+x-\xi-\tau)+k^{(2)}(\tau)\tilde{v}(\xi,t+x-\xi-\tau)\big]\mathrm{d}\tau\}\mathrm{d}\xi, \tag{3.20}$$

$$\tilde{k}(t) = 2\tilde{F}''(t) + 2\tilde{f}'(t)$$
$$-\int_0^{t/2}\int_0^{t-2\xi}\left[\tilde{k}(\tau)v_t^{(1)}(\xi,t-\xi-\tau)+k^{(2)}(\tau)\tilde{v}_t(\xi,t-\xi-\tau)\right]\mathrm{d}\tau\mathrm{d}\xi. \tag{3.21}$$

The domain D_T admits an equivalent description in the form

$$D_T = \big\{(\xi,\tau)|\ |\tau-t+x| \le \xi \le (t+x)/2 - |(t+x)/2-\tau|,$$
$$(t-x)/2 \le \tau \le t\big\}.$$

Let

$$\phi(t) = \max\left[\max_{0\le x\le T/2-|t-T/2|}|\tilde{v}(x,t)|,\ \max_{0\le x\le T/2-|t-T/2|}|\tilde{v}_t(x,t)|,\ |\tilde{k}(t)|\right],$$
$$t\in[0,T].$$

Conducting simple estimates in equations (3.19)–(3.21), we find that for $(x,t)\in D_T$ the integral inequality

$$\phi(t) \quad \le c_1\left(\|f^{(1)}(t)-f^{(2)}(t)\|_{C^1[0,T]} + \|F^{(1)}-F^{(2)}\|_{C^2[0,T]}\right) + c_2\int_0^t \phi(\tau)\mathrm{d}\tau$$

is valid, where $c_1 and$ c_2 are positive constants depending on k_0, T, $\|f^{(1)}\|_{C^1[0,T]}$, $\|F^{(1)}\|_{C^2[0,T]}$. From this, using the Gronwall inequality, we obtain estimate (3.5) with some constant C.

Here we note that similar results can be obtained for the problem of determining the function $k(t) \in C^2[0, T]$ from the equation

$$u_{tt} - u_{xx} - \int_0^t k(\tau) u_{tt}(x, t - \tau) d\tau = 0, \, (x, t) \in R_+^2, \tag{3.22}$$

with initial–boundary conditions (3.2) for a given information (3.4) with respect to the regular part of the solution to the direct problem (3.22), (3.2). In fact, in Eq. (3.22) integrating by parts and introducing new unknown function $\tilde{u}(x, t)$ by formula

$$u(x, t) = \exp\left(\frac{k_0 t}{2}\right) \tilde{u}(x, t), \, k_0 := k(0)$$

from (3.22), (3.2) we get

$$\tilde{u}_{tt} - \tilde{u}_{xx} - h_0 \tilde{u} - \int_0^t h(\tau) \tilde{u}(x, t - \tau) d\tau = 0, \;\; (x, t) \in R_+^2, \tag{3.23}$$

$$\tilde{u}|_{t<0} \equiv 0, \;\; \tilde{u}_x|_{x=0} = \delta'(t) - \frac{k_0}{2} \delta(t) + g(t)\theta(t), \tag{3.24}$$

where the relation $\exp(-k_0 t/2)\delta'(t) = \delta'(t) - (k_0/2)\delta(t)$ is used, as well as the following notations were introduced

$$h_0 : = k'(0) + k_0^2/4, \;\; h(t) := \exp(-k_0 t/2) k''(t),$$
$$g(t) : = \exp(-k_0 t/2) f(t).$$

Alike equality (3.3), the solution of problem (3.22), (3.23), we present in the form

$$\tilde{u}(x, t) = -\delta(t - x) + \theta(t - x) v(x, t).$$

It is easy to show that the function $v(x, t)$ in the domain $t \geq x \geq 0$ satisfies the relations

$$v_{tt} - v_{xx} - h_0 v + h(t - x) - \int_0^{t-x} h(\tau) v(x, t - \tau) d\tau = 0, \tag{3.25}$$

$$v|_{t=x+0} = \frac{1}{2} \left(-k_0 + h_0 x\right), \tag{3.26}$$

$$v_x|_{x=0} = g(t). \tag{3.27}$$

Unknown coefficients $k_0, k'(0)$ can be expressed in terms of the known ones. Indeed, requiring the continuity of the function $v(x,t)$ and its first-order derivatives at $x = t = 0$, we find

$$k_0 = -2F(0), k'(0) = 2F'(0) - F^2(0) + 2f(0).$$

With respect to the inverse problem (3.25)–(3.27), (3.4), an analog of Theorem 3.1 is valid.

Theorem 3.3 *Let $f(t) \in C^1[0,T]$, $F(t) \in C^2[0,T]$. Then, the inverse problem (3.25)–(3.27), (3.4) has a unique solution $h(t) \in C[0,T]$, $T > 0$.*

Using the constructed function $h(t)$ the unknown function $k(t)$ is found from the formula

$$k(t) = k_0 + k'(0)t + \int_0^t (t-\tau)\exp[k(0)\tau/2]h(\tau)d\tau.$$

Denote by $B(k_{00})$ the set of functions $k(t) \in C^2[0,T]$, satisfying for some $T > 0$ the condition

$$\|k\|_{C^2[0,T]} \leq k_{00}$$

with constant $k_{00} > 0$.

The conditional stability theorem for solution of the inverse problem (3.25)–(3.27), (3.4) is also valid.

Theorem 3.4 *Let $k^{(i)} \in B(k_{00})$, $i = 1, 2$ be solutions to the inverse problem (3.25)–(3.27), (3.4) with data $f^{(i)}(t)$, $F^{(i)}(t)$, $i = 1, 2$, respectively. Then, there is a positive constant C, depending on T, k_{00}, $\|f^{(1)}\|_{C^1[0,T]}$, $\|F^{(1)}\|_{C^2[0,T]}$ such that the following estimate*

$$\|k^{(1)} - k^{(2)}\|_{C^2[0,T]} \leq C\left(\|f^{(1)}(t) - f^{(2)}(t)\|_{C^1[0,T]} + \|F^{(1)} - F^{(2)}\|_{C^2[0,T]}\right)$$

is true.

Remark 3.2 The Neumann boundary condition in (3.2) can be replaced by condition $u_x|_{x=0} = \delta(t) + \eta(t)\theta(t)$. Then, considering the function $w = u_t$ as a solution to the direct problem and assuming that $\eta(t)$ satisfies conditions $\eta(0) = 0$, $\eta'(t) = f(t)$, we arrive at above the investigated problem.

3.2 Memory Identification Problem from Integro-Differential Wave Equation

Consider n-dimensional wave equation with memory

$$u_{tt} - u_{xx} - \Delta u - \int_0^t k(\tau)u(x, y, t-\tau)d\tau = 0,\ (y,t) \in \mathbb{R}^{n+1},\ x > 0, \quad (3.28)$$

in which Δ is Laplace operator with respect to variables $y = (y_1, \ldots, y_n)$, with the following initial and boundary conditions

$$u\big|_{t<0} \equiv 0,\ u_x\big|_{x=0} = \delta'(t)\delta(y) + f(t)\theta(t)\delta(y). \quad (3.29)$$

For given functions $k(t)$ and $f(t)$ the problem (3.28)–(3.29) is well posed and defines a function $u(x, y, t)$ that has a compact support for any finite t.

Denote by $\tilde{u}(x, \lambda, t)$, $\lambda = (\lambda_1, \ldots, \lambda_n)$ the Fourier transform of the function $u(x, y, t)$:

$$\tilde{u}(x, \lambda, t) = \frac{1}{\sqrt{(2\pi)^n}} \int_{\mathbb{R}^n} u(x, y, t)e^{i(\lambda,y)}dx,\ (\lambda, y) = \lambda_1 y_1 + \ldots + \lambda_n y_n.$$

Then the problem (3.28), (3.29) is transformed to the form with respect to $\tilde{u}$:

$$\tilde{u}_{tt} - \tilde{u}_{xx} + |\lambda|^2\tilde{u} - |\lambda|^2 \int_0^t k(\tau)\tilde{u}(x, \lambda, t-\tau)d\tau = 0,\ (\lambda, t) \in \mathbb{R}^{n+1},\ x > 0, \quad (3.30)$$

$$\tilde{u}\big|_{t<0} \equiv 0,\ \tilde{u}_x\big|_{y=0} = \delta'(t) + f(t)\theta(t), \quad (3.31)$$

where $|\lambda|^2 = \lambda_1^2 + \ldots + \lambda_n^2$.

Similarly to (3.3) we can conclude that for fixed λ, the solution of problem (3.30), (3.31) has the structure

$$u(x, \lambda, t) = -\delta(t-x) + \theta(t-x)v(x, \lambda, t).$$

Here $v(x, \lambda, t)$ is a regular function. From relations (3.30)–(3.31) we deduce that for fixed λ, this function in domain $t > x > 0$ satisfy relations

$$v_{tt} - v_{xx} + k(t-x) + |\lambda|^2 v - |\lambda|^2 \int_0^{t-x} k(\tau)v(x, \lambda, t-\tau)d\tau = 0, \quad (3.32)$$

$$v\big|_{t=x+0} = 0,\ v_x\big|_{x=0} = f(t). \tag{3.33}$$

We pose the inverse problem: to determine $k(t)$, $t > 0$ from the given trace of the solution of problem (3.32), (3.33) on the line $x = 0$, for fixed $\lambda = \lambda_0 \in \mathbb{R}^n$ and $t < T, T > 0$, i.e.,

$$v(0, \lambda_0, t) = F(t),\ \lambda_0 \in \mathbb{R}^n,\ t < T. \tag{3.34}$$

Note that the inverse problem (3.32)–(3.34) for $\lambda = 0$ is identical to the problem explored in the previous section. We give here typical theorems of unique global solvability and stability for the inverse problem (3.32)–(3.34). Their proofs can easily be carried out according to the scheme described in the previous section and are not given here.

Theorem 3.5 *Let $f(t) \in C^1[0, T]$, $F(t) \in C^2[0, T]$ and the matching conditions*

$$F(\lambda_0, 0) = 0,\ f(0) = \frac{|\lambda_0|^2}{2} - F'(0)$$

are valid. Then, the inverse problem (3.32)–(3.34) has a unique solution $k(t) \in C[0, T]$, $T > 0$.

We denote, as in Sect. 3.1, by $A(k_0)$ the set of functions $k(t) \in C[0, T]$, satisfying the condition

$$\|k\|_{C[0,T]} \leq k_0$$

for some $T > 0$ with constant $k_0 > 0$.

Theorem 3.6 *Let $k^{(i)} \in A(k_0)$, $i = 1, 2$ be solutions to the inverse problem (3.32)–(3.34) with data $f^{(i)}(t)$, $F^{(i)}(t)$, $i = 1, 2$, respectively. Then, there exists a positive constant C, depending on T, k_0, λ_0, $\|f^{(1)}\|_{C^1[0,T]}$, $\|F^{(1)}\|_{C^2[0,T]}$ such that the estimate*

$$\|k^{(1)} - k^{(2)}\|_{C[0,T]} \leq C\left(\|f^{(1)}(t) - f^{(2)}(t)\|_{C^1[0,T]} + \|F^{(1)} - F^{(2)}\|_{C^2[0,T]}\right)$$

is valid.

3.3 Inverse Problem for an Integro-Differential Equation of Electrodynamics

Let the vectors E and H form a solution of the Cauchy problem for the system of Maxwell equations with zero initial data:

$$\nabla \times H = (\partial/\partial t)D(x, t) + \sigma E + j,\ \nabla \times E = -(\partial/\partial t)B(x, t),\ (x, t) \in \mathbb{R}^4,$$

$$(E, H)|_{t<0} \equiv 0. \tag{3.35}$$

Here $j = j(x, t)$ is a given function characterizing the external current density. The values of D and B (the electric induction and the magnetic induction, respectively) at the moment depend not only on E and H (the intensities of the respective fields) but also on the entire time history of these fields:

$$
\begin{aligned}
D(x,t) &= \varepsilon E + \int_0^t \varphi(t-\tau)E(x,\tau)\mathrm{d}\tau,\\
B(x,t) &= \mu H + \int_0^t \psi(t-\tau)H(x,\tau)\mathrm{d}\tau,\\
&E = (E_1, E_2, E_3),\ H = (H_1, H_2, H_3),\ D = (D_1, D_2, D_3),\\
&B = (B_1, B_1, B_3),\ x = (x_1, x_2, x_3),
\end{aligned} \tag{3.36}
$$

where $\varphi(t) = diag(\varphi_1, \varphi_2, \varphi_3)$ and $\psi(t) = diag(\psi_1, \psi_2, \psi_)$ are diagonal matrices representing the memory.

In what follows, we assume that the electrodynamic parameters ε, μ, and σ are given positive constants and $\psi = 0$. (The latter condition means that the magnetic field has no prehistory.)

For further studies, it is convenient to derive a second-order equation for H from the Maxwell system (3.35), (3.36). To this end, under our assumptions, we apply the operation $\nabla\times$ to the first equations and the operation $\partial/\partial t$ to the second equations in system (3.35). Since $\nabla \times \nabla \times H = \nabla(\mathrm{div} H) - \Delta H$, we see that simple computations give

$$
\mu(\varepsilon H_{tt} + \sigma H_t) - \Delta H + \int_0^t \mu\varphi(\tau)H_{tt}(x, t-\tau)\mathrm{d}\tau = \nabla \times j. \tag{3.37}
$$

Here we have used the relation $\mathrm{div} H = 0$, which follows from relations (3.35). Indeed, by applying div to the second equations in (3.35) and by using the relation $\mathrm{div}(\nabla \times E) = 0$, we obtain $\partial/\partial t(\mathrm{div} H) = 0$. From this, by integrating the last relation under the initial conditions in (3.35), we arrive at $\mathrm{div} H = 0$.

In what follows, we assume that $\varepsilon = \mu = 1$. Otherwise, the change of variables $t_1 = t/\sqrt{\varepsilon\mu}$ provides the desired relations. We integrate twice by parts the integral in (3.37) and introduce the new vector function $u(x,t) = (u_1, u_2, u_3)$, where $u_i = H_i \exp[-0.5(\sigma + \varphi_i(0))t]$, $i = 1, 2, 3$. Then equation (3.37) and the initial condition can be written as

$$
u_{tt} - \Delta u + cu + \int_0^t k(t-\tau)u(x,\tau)\mathrm{d}\tau = F(x,t), \tag{3.38}
$$

$$u|_{t<0} = 0, \tag{3.39}$$

where $c = diag(c_1,\ c_2,\ c_3),\ c_i := \varphi_{it} - 0.25(\sigma + \varphi_i(0))^2,$

$$k(t) = diag(k_1,\ k_2,\ k_3),\ k_i := \varphi'' \exp[-0.5(\sigma + \varphi_i(0)t)],$$
$$F(x,t) = (F_1,\ F_2,\ F_3),\ F_i := \exp[-0.5(\sigma + \varphi_i(0)t)](\nabla \times j)_i,\ i = 1, 2, 3,$$

and $(\nabla \times j)_i$ is the ith component of the vector $(\nabla \times j)$.

We shall solve problem (3.38), (3.39) for the case in which the function $F(x,t)$ has the form

$$F(x,t) = \nu\delta(x,t). \tag{3.40}$$

Here $\nu = (\nu_1,\ \nu_2,\ \nu_3)$ is a unit vector, $\nu_i \neq 0,\ i = 1, 2, 3$, and $\delta(x,t)$ is the Dirac delta function. The matrix c in Eq. (3.38) is assumed to be a known constant matrix.

The problem in which the vector $u(x,t)$ should be determined from (3.38)–(3.40) for a given matrix function $k(t)$ will be called the *direct problem.* We seek the solution of this Cauchy problem in the form

$$u(x,t) = \frac{\nu\theta(t)}{4\pi|x|}\delta(t - |x|) + v(x,t), \tag{3.41}$$

where $v(x,t)$ is the scattered wave, $\theta(t) = 1$ for $t \geq 0$, and $\theta(t) = 0$ for $t < 0$. We pose the following *inverse problem:* find the function $k(t) = diag(k_1,\ k_2,\ k_3)$ occurring in the integral in Eq. (3.38) from the information on the scattered wave $v(x,t)$ at an arbitrary time $t \geq 0$ at the point $x = 0$:

$$v(0,t) = f(t),\ f(t) = (f_1,\ f_2,\ f_3). \tag{3.42}$$

Theorem 3.7 *Suppose that $f(t) \in C^1[0,T]$ and the matching condition*

$$f(0) = -\frac{c\nu}{8\pi}.$$

holds. Then, the inverse problem (3.38)–(3.42) has a unique solution $k(t) \in C[0,T]$, $T > 0$.

Proof By substituting the functions (3.40), (3.41) into equation (3.38) and by taking into account the fact that $(\theta(t)/4\pi|x|)\delta(t - |x|)$ is the fundamental solution of the wave equation, we obtain

$$v_{tt} - \Delta v + cv + \frac{c\nu\theta(t)}{4\pi|x|}\delta(t-|x|)$$
$$+ \frac{\nu\theta(t-|x|)}{4\pi|x|}k(t-|x|) + \theta(t-|x|)\int_{|x|}^{t} k(t-\tau)v(x,\tau)\mathrm{d}\tau = 0,\ (x,t) \in R^4,$$
$$v|_{t<0} = 0.$$

Hence it follows that $v|_{t<|x|} = 0$. By using the Kirchhoff formula, we obtain the following integral equation in the domain $t \geq |x|$:

$$v(x,t) = -\frac{c\nu}{16\pi^2}\int\int\int_{|\xi|+|x-\xi|\leq t} \frac{\delta(t-|\xi|-|x-\xi|)}{|\xi||x-\xi|}\mathrm{d}\xi$$
$$-\frac{\nu}{8\pi^2}\int\int\int_{|\xi|+|x-\xi|\leq t} \frac{k(t-|\xi|-|x-\xi|)}{|\xi||x-\xi|}\mathrm{d}\xi - \frac{1}{4\pi}\int\int\int_{|\xi|+|x-\xi|\leq t} \frac{1}{|x-\xi|}$$
$$\times\left[cv(\xi,t-|x-\xi|) + \int_0^{t-|\xi|-|x-\xi|} k(\alpha)v(\xi,t-|x-\xi|-\alpha)\mathrm{d}\alpha\right]\mathrm{d}\xi. \tag{3.43}$$

Thus, the inverse problem is reduced to finding the vector function $k(t)$ in vector integral Eq. (3.43) on the given condition (3.42). Let us transform the triple integrals occurring in (3.43) into repeated integrals. To this end, for a given x, consider the family of ellipsoids

$$G(x,t) = \{\xi :\ |\xi| + |x-\xi| = \alpha,\ |x| \leq \alpha \leq t\},$$

and the ordinary spherical coordinate system r, θ, φ, with its pole located at the point 0 and the polar axis running through the points $(0,\ x)$. We characterize a point $\xi \in G(x,t)$ by the curvilinear coordinates $r,\ \theta,\ \varphi$. Let us use the polar equation of the ellipsoid $G(x,t)$ in the form (1.8) and for points $\xi \in G(x,t)$ the relations (1.9), (1.10). By passing to the variables θ, φ, α in the triple integrals in equation (3.43) and using the main property of the delta function and the relation

$$\frac{\alpha^2-|x|^2}{2}\int_0^{\pi} \frac{\sin\theta\mathrm{d}\theta}{(\alpha-|x|\cos\theta)^2} = 1,$$

we obtain the equation

$$v(x,t) = -\frac{c\nu}{8\pi} - \frac{\nu}{8\pi}\int_{|x|}^{t} k(t-\alpha)\mathrm{d}\alpha - \frac{1}{2\pi}\int_{|x|}^{t}\frac{\mathrm{d}\alpha}{\alpha^2-|x|^2}\int\int\int_{G(x,\alpha)} r^3(\theta,\alpha)$$

$$\times\left[cv(\xi,t-\alpha+r) + \int_0^{t-\alpha} k(\tau)v(\xi,t-\alpha+r-\tau)\mathrm{d}\tau\right]\mathrm{d}\omega,\ t \geq |x|, \quad (3.44)$$

where $\mathrm{d}\mathrm{d}\omega = \sin\theta \mathrm{d}\theta \mathrm{d}\varphi$ is a solid angle differential. In particular, it follows that

$$v(x,|x|+0) = -\frac{c\nu}{8\pi}. \quad (3.45)$$

Whence, by using the relation

$$\frac{1}{\alpha^2-|x|^2}\int\int r^3(\theta,\alpha)\mathrm{d}\omega = \frac{\pi\alpha}{2},$$

and by differentiating equation (3.44) and using (3.45) , we obtain

$$v_t(x,t) = -\frac{\nu}{8\pi}k(t-|x|) + \frac{c\nu}{32\pi}(t^2-|x|^2) + \frac{c\nu}{32\pi}\int_{|x|}^{t} k(t-\alpha)\alpha\mathrm{d}\alpha$$

$$-\frac{1}{2\pi}\int_{|x|}^{t}\frac{\mathrm{d}\alpha}{\alpha^2-|x|^2}\int\int_{G(x,\alpha)} r^3(\theta,\alpha)$$

$$\times\left[cv_t(\xi,t-\alpha+r) + \int_0^{t-\alpha} k(\tau)v_t(\xi,t-\alpha+r-\tau)\mathrm{d}\tau\right]\mathrm{d}\omega,\ t \geq |x|$$

(3.46)

We set in (3.46) $x = 0$ and, in view of the additional information (3.42), obtain equation with respect to vector function $\tilde{k}(t) =: (k_1,\ k_2,\ k_3)$

$$\tilde{k}(t) = g(t) - c\int_0^t k(t-\alpha)\alpha\mathrm{d}\alpha - \gamma\int_0^t\frac{\mathrm{d}\alpha}{\alpha}\int\int_{G(0,\alpha)}[cv_t(\xi,t-\alpha+r)$$

$$+\int_0^{t-\alpha} k(\tau)v_t(\xi,t-\alpha+r-\tau)\mathrm{d}\tau]\mathrm{d}\omega, t \geq 0, \quad (3.47)$$

where $\gamma =: 2(1/\nu_1,\ 1/\nu_2,\ 1/\nu_3)$,

$$g(t) = (g_1,\ g_2,\ g_3) =: \big((-8\pi/\nu_1) f_1' + (c_1 t^2)/4,$$
$$(-8\pi/\nu_2) f_2' + (c_2 t^2)/4,\ (-8\pi/\nu_3) f_3' + (c_3 t^2)/4\big).$$

Consider the sphere $S(T) = \{\xi : |\xi| = T\}$ and take pairs (x, t) such that the corresponding spheres $|x - \xi| = t$ are contained inside the sphere $S(T)$. For this, it is sufficient to require that the set of points (x, t) lies in the four-dimensional domain

$$\Omega = \{(x, t) : |x| \leq t \leq T - |x|\}.$$

□

The system of vector equations (3.46), (3.47) is a closed system of integral equations for the vector functions $v_t(x, t)$ and $\tilde{k}(t)$. We rewrite it in the form of the operator integral equation

$$\psi = L\psi, \tag{3.48}$$

where

$$\psi = \begin{bmatrix} \psi_1(x, t) \\ \psi_2(t) \end{bmatrix} = \begin{bmatrix} v_t(x, t) + \frac{\nu}{8\pi} k(t - |x|) \\ \tilde{k}(t) \end{bmatrix}$$

is a vector function with components $\psi_i (i = 1, 2)$, each of which is in turn a vector function itself and the operator L is defined on the set of functions $\psi \in C[\Omega]$ and, according to (3.46) and (3.47), has the form

$$L = (L_1,\ L_2),$$

where

$$L_1\psi = \frac{c\nu}{32\pi}(t^2 - |x|^2) - \frac{c\nu}{32\pi} \int_{|x|}^{t} \tilde{\psi}_2(t - \alpha)\alpha \mathrm{d}\alpha$$
$$- \frac{1}{2\pi} \int_{|x|}^{t} \frac{\mathrm{d}\alpha}{\alpha^2 - |x|^2} \int\int_{G(x,\alpha)} r^3(\theta, \alpha)\big[c\psi_1(\xi, t - \alpha + r) - \frac{c\nu}{8\pi} \tilde{\psi}_2(t - \alpha)$$
$$+ \int_0^{t-\alpha} \tilde{\psi}_2(\tau)\big(\psi_1(\xi, t - \alpha + r - \tau) - \frac{\nu}{8\pi} \tilde{\psi}_2(t - \alpha - \tau)\big)\mathrm{d}\tau\big]\mathrm{d}\omega, \tag{3.49}$$

$$L_2\psi = g(t) - c\int\limits_0^t \tilde{\psi}_2(t-\alpha)\alpha \mathrm{d}\alpha$$

$$-\gamma\int\limits_0^t \frac{\mathrm{d}\alpha}{\alpha}\int\int_{G(0,\alpha)} \Big[c\psi_1(\xi, t-\alpha+r) - \frac{c\nu}{8\pi}\tilde{\psi}_2(t-\alpha)$$

$$+\int\limits_0^{t-\alpha} \tilde{\psi}_2(\tau)\big(\psi_1(\xi, t-\alpha+r-\tau) - \frac{\nu}{8\pi}\tilde{\psi}_2(t-\alpha-\tau)\big)\mathrm{d}\tau\Big]\mathrm{d}\omega.$$

Here $\tilde{\psi}_2 = diag(\psi_{21}, \psi_{22}, \psi_{23})$.

Let C_μ be the Banach space of continuous functions generated by the family of weighted norms

$$\|\psi\|_\mu = \max\Big\{\sup_{(x,t)\in\Omega} \left|\psi_{1i}(x,t)\mathrm{e}^{-\mu(t-|x|)}\right|,$$
$$\sup_{t\in[0,T]} \left|\psi_{2i}(t)\mathrm{e}^{-\mu t}\right|,\ 1 \le i \le 3\Big\},\ \mu \ge 0.$$

Obviously, for $\mu = 0$ this space is the space of continuous functions with the ordinary norm, which will be denoted by $\|\psi\|$ and

$$\|\psi\| = \max\left\{\sup_{(x,t)\in\Omega} |\psi_{1i}(x,t)|,\ \sup_{t\in[0,T]} |\psi_{2i}(t)|,\ 1 \le i \le 3\right\},\ \mu \ge 0.$$

By virtue of the inequality

$$\mathrm{e}^{-\mu T}\|\psi\| \le \|\psi\|_\mu \le \|\psi\|$$

the norms $\|\psi\|_\mu$ and $\|\psi\|$ are equivalent for any given $T \in (0, \infty)$. The number μ will be chosen later. Let $Q_\mu(\psi_0, R)$ be the ball of radius R centered at the point ψ_0 of some weighted space $C_\mu(\mu \ge 0)$, where

$$\psi_0 = \begin{bmatrix}\psi_{01}\\ \psi_{02}\end{bmatrix} := \begin{bmatrix}\frac{c\nu}{32\pi}(t^2 - |x|^2)\\ g(t)\end{bmatrix}.$$

One can readily see that

$$\|\psi\|_\mu \le \|\psi_0\|_\mu + R \le \|\psi_0\| + R := R_0$$

for $\psi \in Q_\mu(\psi_0, R)$, where R_0 is a known number.

Let $\psi(x,t) \in Q_\mu(\psi_0, R)$. Let us show that, for an appropriate choice of $\mu > 0$, the operator L maps this ball to the same ball; i.e., $L\psi \in Q_\mu(\psi_0, R)$. Indeed, by computing the norm of the differences with the help of (3.49), we have

$$||L_1\psi - \psi_{01}||_\mu = \sup_{(x,t)\in\Omega} \left|(L_1\psi - \psi_{01})e^{-\mu(t-|x|)}\right|$$

$$= \sup_{(x,t)\in\Omega} \Big| -\frac{c\nu}{32\pi} \int_{|x|}^{t} \tilde{\psi}_2(t-\alpha)e^{-\mu(t-\alpha)}\alpha e^{-\mu(\alpha-|x|)}d\alpha$$

$$-\frac{1}{2\pi}\int_{|x|}^{t} \frac{e^{-\mu(\alpha-|x|)}d\alpha}{\alpha^2 - |x|^2} \int\int_{G(x,\alpha)} r^3(\theta,\alpha)\big[c\psi_1(\xi, t-\alpha+r)e^{-\mu(t-\alpha)}$$

$$-\frac{c\nu}{8\pi}\tilde{\psi}_2(t-\alpha)e^{-\mu(t-\alpha)} + \int_0^{t-\alpha} \tilde{\psi}_2(\tau)e^{-\mu\tau}\left(\psi_1(\xi, t-\alpha+r-\tau)e^{-\mu(t-\alpha-\tau)}\right.$$

$$\left.-\frac{\nu}{8\pi}\tilde{\psi}_2(t-\alpha-\tau)e^{-\mu(t-\alpha-\tau)}\right)d\tau\big]d\omega\Big|$$

$$\le \left(\frac{c_0 T}{32\pi}\|\psi_2\|_\mu + \frac{T}{4}\left(c_0\|\psi_1\|_\mu + \frac{c_0}{8\pi}\right)(1+T)\|\psi_2\|_\mu\right)\frac{1}{\mu} := \alpha_0\frac{1}{\mu},$$

where $c_0 = \max_{1\le i\le 3} c_i$. A similar argument gives the relations

$$||L_2\psi - \psi_{02}||_\mu = \sup_{t\in[0,T]} \left|(L_2\psi - \psi_{02})e^{-\mu t}\right|$$

$$\le c_0 T R_0 \left(1 + \frac{\gamma_0}{2}\left(1 + \left(1 + \frac{1}{8\pi}\right)(1+TR)\right)\right)\frac{1}{\mu} := \alpha_1\frac{1}{\mu}.$$

We choose $\mu \ge \alpha_{00} =: (1/R)\max(\alpha_0, \alpha_1)$ and conclude that L maps the ball $Q_\mu(\psi_0, R)$ into itself.

Now let ψ^1 and ψ^2 be two arbitrary elements of the ball $Q_\mu(\psi_0, R)$. Then, by using auxiliary inequalities of the form

$$\left|\psi_i^1\psi_j^1 - \psi_i^2\psi_j^2\right| e^{-\mu(t-|x|)}$$
$$\le \left|\psi_i^1\right|\left|\psi_j^1 - \psi_j^2\right| e^{-\mu(t-|x|)} + \left|\psi_j^2\right|\left|\psi_j^1 - \psi_j^2\right| e^{-\mu(t-|x|)} \le 2R_0\left\|\psi^1 - \psi^2\right\|_\mu,$$

we obtain the relations

$$\left\| \left(L\psi^1 - L\psi^2\right)_1 \right\|_\mu = \sup_{(x,t)\in\Omega} \left| \left(L\psi^1 - L\psi^2\right)_1 \mathrm{e}^{-\mu(t-|x|)} \right|$$

$$= \sup_{(x,t)\in\Omega} \Big| -\frac{c\nu}{32\pi} \int_{|x|}^{t} (\widetilde{\psi_2^1}(t-\alpha) - \widetilde{\psi_2^2}(t-\alpha))\mathrm{e}^{-\mu(t-\alpha)} \alpha \mathrm{e}^{-\mu(\alpha-|x|)} \mathrm{d}\alpha$$

$$- \frac{1}{2\pi} \int_{|x|}^{t} \frac{\mathrm{e}^{-\mu(\alpha-|x|)}\mathrm{d}\alpha}{\alpha^2 - |x|^2} \int\int_{G(x,\alpha)} r^3(\theta,\alpha)\{c(\psi_1^1(\xi, t-\alpha+r)$$

$$- \psi_1^2(\xi, t-\alpha+r))\mathrm{e}^{-\mu(t-\alpha)} - \frac{c\nu}{8\pi}(\widetilde{\psi_2^1}(t-\alpha) - \widetilde{\psi_2^2}(t-\alpha))\mathrm{e}^{-\mu(t-\alpha)}$$

$$+ \int_0^{t-\alpha} [\widetilde{\psi_2^1}(\tau)\mathrm{e}^{-\mu\tau}(\psi_1^1(\xi, t-\alpha+r-\tau)$$

$$- \frac{\nu}{8\pi}\widetilde{\psi_2^1}(t-\alpha-\tau))\mathrm{e}^{-\mu(t-\alpha-\tau)} - \widetilde{\psi_2^2}(\tau)\mathrm{e}^{-\mu\tau}(\psi_1^2(\xi, t-\alpha+r-\tau)$$

$$- \frac{\nu}{8\pi}\widetilde{\psi_2^2}(t-\alpha-\tau))\mathrm{e}^{-\mu(t-\alpha-\tau)}]\mathrm{d}\tau\}\mathrm{d}\omega\Big| \le$$

$$\le \frac{1}{\mu}\left[\frac{c_0 T^2 R_0}{4\pi}\left(2 + \frac{1}{4\pi}\right) + \frac{c_0 T}{32\pi}\left(9 + \frac{1}{\pi}\right)\right] \psi^1 - \psi^2_\mu$$

$$=: \frac{1}{\mu}\beta_0 \left\|\psi^1 - \psi^2\right\|_\mu .$$

The second component can be estimated in a similar way:

$$\left\| \left(L\psi^1 - L\psi^2\right)_2 \right\|_\mu = \sup_{t\in[0,T]} \left| \left(L\psi^1 - L\psi^2\right)_2 \mathrm{e}^{-\mu t} \right|$$

$$\le \frac{1}{\mu}\left[\frac{c_0 T^2 R_0 \gamma_0}{2}\left(2 + \frac{1}{4\pi}\right) + 2c_0 T\gamma_0 + \frac{c_0 T\gamma_0}{8\pi}\right] \|\psi^1 - \psi^2\|_\mu$$

$$=: \frac{1}{\mu}\beta_1 \left\|\psi^1 - \psi^2\right\|_\mu .$$

Let $\beta_{00} := \max(\beta_0, \beta_1)$. It follows from the preceding estimates that if $\mu > \max(\alpha_{00}, \beta_{00})$, then L is a contraction operator on the ball $Q_\mu(\psi_0, R)$. Then, by the Banach principle, equation (3.48) has a unique solution in the ball $Q_\mu(\psi_0, R)$ for any given $T > 0$. The proof of the theorem is completed.

3.4 Global Solvability of the Problem of Determining Two Unknowns in the Inverse Problem for the Integro-Differential Wave Equation

Consider the initial–boundary problem

$$u_{tt} - u_{yy} - \Delta u - \int_0^t k(\tau)\Delta u(x, y, t-\tau)\mathrm{d}\tau = 0, \ (x, t) \in \mathbb{R}^{n+1}, \ y > 0, \tag{3.50}$$

$$u|_{t<0} \equiv 0, \ u_y|_{y=0} = \delta'(t)\delta(x) + f(t)\theta(t)\delta(x), \tag{3.51}$$

in which Δ is n dimensional Laplacian in the variable $x = (x_1, \ldots, x_n)$.

For the given functions $k(t)$, $f(t)$, the problem of finding a function $u(x, y, t)$, satisfying equalities (3.50), (3.51), is called the direct problem. The boundary condition in (3.51) simulates an instantaneous wave excitation source located at the point $x = 0, \ y = 0$.

In this section it is shown that the method used to prove the global solvability of the definition of the kernel of the integral operator in second-order hyperbolic equations can be applied to the determination of two unknown functions. In this case, one of these unknowns represents the memory kernel, another one determines the regular part of the pulse source.

It follows from the hyperbolic equations theory that the function $u(x, y, t)$ as a solution to problem (3.50), (3.51) has a compact support for any finite $t > 0$. Therefore, to the relations (3.50), (3.51) the Fourier transform in the variable x is applicable.

Denote through $\tilde{u}(\lambda, y, t), \ \lambda = (\lambda_1, \ldots, \lambda_n)$ the Fourier image of the function $u(x, y, t)$:

$$\tilde{u}(\lambda, y, t) = \frac{1}{\sqrt{(2\pi)^n}} \int_{\mathbb{R}^n} u(x, y, t)\mathrm{e}^{i(\lambda,x)}dx, \ (\lambda, x) = \lambda_1 x_1 + \ldots + \lambda_n x_n.$$

Then, the problem (3.50), (3.51) is transformed to the form

$$\tilde{u}_{tt} - \tilde{u}_{yy} + |\lambda|^2\tilde{u} + |\lambda|^2 \int_0^t k(\tau)\tilde{u}(\lambda, y, t-\tau)\mathrm{d}\tau = 0, \ \ (\lambda, t) \in \mathbb{R}^{n+1}, \ \ y > 0, \tag{3.52}$$

$$\tilde{u}|_{t<0} \equiv 0, \ \ \tilde{u}_y|_{y=0} = \delta'(t) + f(t)\theta(t), \tag{3.53}$$

where $|\lambda|^2 = \lambda_1^2 + \ldots + \lambda_n^2$.

From equalities (3.52), (3.53) for a fixed value of λ it follows that $u \equiv 0$, $t < y$, $y > 0$. The function $\tilde{u}(\lambda, y, t)$ as a solution to problem (3.52), (3.53) has the following structure in a neighborhood of the characteristic line $t = y$:

$$\tilde{u}(\lambda, y, t) = -\delta(t - y) + \theta(t - y)v(\lambda, y, t), \tag{3.54}$$

where $v(\lambda, y, t)$ is a regular function.

Now we pose the *inverse problem*: determine the functions $k(t)$, $f(t)$, included in equations (3.52), (3.53), respectively, from information on the regular part of the Fourier transform of the direct problem solution at $y = 0$ for two different values λ of the transformation parameter:

$$v(\lambda^i, 0, t) = F(\lambda^i, t), \; t > 0, \; i = 1, 2. \tag{3.55}$$

Theorem 3.8 *Let* $F(\lambda^i, t) \in C^1[0, T], i = 1, 2$, $F_0(\lambda^1, \lambda^2, t) := F_t(\lambda^1, t) - F_t(\lambda^2, t) \in C^1[0, T]$, *and the matching conditions*

$$F(\lambda^i, 0) = 0, \; i = 1, 2, \; \lambda^1 \neq \lambda^2$$

are fulfilled. Then, inverse problem (3.52)–(3.55) has the unique solution so that $(f(t), \; k(t)) \in C[0, T], \; T > 0$.

Proof Substituting the function (3.54) into equations (3.52), (3.53) and using the singularity separation method, we find that the function $v(\lambda, y, t)$ for a fixed λ in the domain $t > y > 0$ satisfies the equations

$$v_{tt} - v_{yy} + |\lambda|^2 v - |\lambda|^2 k(t - y) + |\lambda|^2 \int_0^{t-y} k(\tau) v(\lambda, y, t - \tau) d\tau = 0, \tag{3.56}$$

$$v|_{t=y+0} = \frac{|\lambda|^2}{2} y, \tag{3.57}$$

$$v_y|_{y=0} = f(t). \tag{3.58}$$

The following statement establishes the equivalence of the inverse problem (3.52)–(3.58) to some closed system of integral equations of Volterra-type. □

Lemma 3.2 *Under the conditions of the theorem 3.8, the problem (3.55)–(3.58) for* $\lambda = \lambda^i = (\lambda_1^i, \ldots, \lambda_n^i)$, $i = 1, 2$ *and* $(y, t) \in D_T$, $D_T = ((y, t)|0 \le y \le t \le T - y)$ *is equivalent to the problem of finding functions* $v(\lambda^i, y, t)$, $v_t(\lambda^i, y, t)$, $k(t)$, $f(t)$, $i = 1, 2$ *from the following system of equations:*

$$v(\lambda^i, y, t) = \frac{|\lambda^i|^2}{2} y + \int\limits_y^t v_t(\lambda^i, y, \tau)\mathrm{d}\tau, i = 1, 2, \tag{3.59}$$

$$\begin{aligned}
v_t(\lambda^i, y, t) &= F_t(\lambda^i, t - y) + \frac{|\lambda^i|^2}{2} yk(t - y) \\
&+ \frac{|\lambda^i|^2}{2} \int\limits_y^{(y+t)/2} \big[k(t + y - 2\xi) - v(\lambda^i, \xi, t + y - \xi) \\
&- \int\limits_0^{t+y-2\xi} k(\tau)v(\lambda^i, \xi, t + y - \xi - \tau)\mathrm{d}\tau\big]\mathrm{d}\xi - \\
&- \frac{|\lambda^i|^2}{2} \int\limits_0^{t-y)/2} \big[k(t - y - 2\xi) - v(\lambda^i, \xi, t - y - \xi) \\
&- \int\limits_0^{t-y-2\xi} k(\tau)v(\lambda^i, \xi, t - y - \xi - \tau)\mathrm{d}\tau\big]\mathrm{d}\xi \\
&- \frac{|\lambda^i|^2}{2} \int\limits_0^y \left[v(\lambda^i, \xi, t - y + \xi) + \int\limits_0^{t-y} k(\tau)v(\lambda^i, \xi, t - y + \xi - \tau)\mathrm{d}\tau \right] \mathrm{d}\xi, i = 1, 2,
\end{aligned} \tag{3.60}$$

$$\begin{aligned}
k(t) &= \frac{2}{|\lambda^1|^2 - |\lambda^2|^2} F_{0t}(\lambda^1, \lambda^2, t) \\
&+ \frac{|\lambda^1|^2 + |\lambda^2|^2}{4} t + \frac{|\lambda^1|^2 + |\lambda^2|^2}{4} \int\limits_0^t k(\tau)(t - \tau)\mathrm{d}\tau \\
&+ \frac{2}{|\lambda^1|^2 - |\lambda^2|^2} \int\limits_0^{t/2} \big[|\lambda^1|^2 v_t(\lambda^1, \xi, t - \xi) - |\lambda^2|^2 v_t(\lambda^2, \xi, t - \xi) \\
&+ \int\limits_0^{t-2\xi} k(\tau)\big(|\lambda^1|^2 v_t(\lambda^1, \xi, t - \xi - \tau) - |\lambda^2|^2 v_t(\lambda^2, \xi, t - \xi - \tau)\big)\mathrm{d}\tau\big]\mathrm{d}\xi,
\end{aligned} \tag{3.61}$$

$$f(t) = -F_t(\lambda^1, t) + \frac{|\lambda|^2}{2} + |\lambda^1|^2 \int_0^{t/2} \Big[k(t-2\xi) - v(\lambda^1, \xi, t-\xi)$$

$$- \int_0^{t-2\xi} k(\tau) v(\lambda^1, \xi, t-\xi-\tau) \mathrm{d}\tau \Big] \mathrm{d}\xi. \tag{3.62}$$

Proof In view of the factorization form of the wave operator

$$v_{tt} - v_{yy} = \left(\frac{\partial}{\partial t} - \frac{\partial}{\partial y}\right)(v_t + v_y) = \left(\frac{\partial}{\partial t} + \frac{\partial}{\partial y}\right)(v_t - v_y), \tag{3.63}$$

integrating (3.56) along the corresponding characteristics of first-order differential operators for $(y, t) \in D_T$ from (3.55)–(3.57) we obtain

$$(v_t + v_y)(\lambda, y, t) = \frac{|\lambda|^2}{2} + |\lambda|^2 \int_y^{(y+t)/2} \Big[k(t+y-2\xi) - v(\lambda, \xi, t+y-\xi)$$

$$- \int_0^{t+y-2\xi} k(\tau) v(\lambda, \xi, t+y-\xi-\tau) \mathrm{d}\tau \Big] \mathrm{d}\xi, \tag{3.64}$$

$$(v_t - v_y)(\lambda, y, t) = 2F_t(\lambda, t) - \frac{|\lambda|^2}{2} + |\lambda|^2 k(t-y) y$$

$$- |\lambda|^2 \int_0^{(t-y)/2} [k(t-y-2\xi) - v(\lambda, \xi, t-y-\xi)$$

$$- \int_0^{t-y-2\xi} k(\tau) v(\lambda, \xi, t-y-\xi-\tau) \mathrm{d}\tau] \mathrm{d}\xi$$

$$- |\lambda|^2 \int_0^y \left[v(\lambda, \xi, t-y+\xi) + \int_0^{t-y} k(\tau) v(\lambda, \xi, t-y+\xi-\tau) \mathrm{d}\tau \right] \mathrm{d}\xi. \tag{3.65}$$

In obtaining equality (3.65), we also used the relations

$$
\begin{aligned}
& v_t(\lambda, 0, t) = F_t(\lambda, t), \\
& v_y(\lambda, 0, t) = \frac{|\lambda|^2}{2} - F_t(\lambda, t) \\
& + |\lambda|^2 \left[\int_0^{t/2} k(t - 2\xi) - v(\lambda, \xi, t - \xi) - \int_0^{t-2\xi} v(\lambda, \xi, t - \xi - \tau) \mathrm{d}\tau \right] \mathrm{d}\xi, \quad (3.66)
\end{aligned}
$$

which follow from (3.55), (3.64). Considering the equation (3.66) for $\lambda = \lambda^1, \ \lambda = \lambda^2$, we compose the difference $v_y(\lambda^1, 0, t) - v_y(\lambda^2, 0, t)$. Taking into account (3.58), we obtain

$$
\begin{aligned}
& \frac{|\lambda^1|^2 - |\lambda^2|^2}{2} + F_t(\lambda^2, t) - F_t(\lambda^1, t) \\
& + \frac{|\lambda^1|^2 - |\lambda^2|^2}{2} \int_0^t k(\tau) \mathrm{d}\tau - \int_0^{t/2} [|\lambda^1|^2 v_t(\lambda^1, \xi, t - \xi) - |\lambda^2|^2 v_t(\lambda^2, \xi, t - \xi) \\
& + \int_0^{t-2\xi} k(\tau)(|\lambda^1|^2 v_t(\lambda^1, \xi, t - \xi - \tau) - |\lambda^2|^2 v_t(\lambda^2, \xi, t - \xi - \tau)) \mathrm{d}\tau] \mathrm{d}\xi = 0.
\end{aligned}
\tag{3.67}
$$

Differentiating the last relation with respect to t, after simple calculations, we arrive at (3.61). Note that the equations (3.59)–(3.61) do not contain the unknown function $f(t)$. After finding the functions v, k the function $f(t)$, according to equality (3.66), is calculated by the formula (3.62). In the formula (3.62) $\lambda = \lambda^1$ or $\lambda = \lambda^2$, the calculation result should not depend on the choice of the parameter λ. For definiteness, in this formula we set $\lambda = \lambda^1$. The obvious equality (3.59) is given for the closure systems of the equations (3.60), (3.61). □

It is easy to see that inverse transforms also take place. Indeed, in the first integral of the right-hand side of (3.61), we change variable τ to τ' by the formula $\tau' = (t - \tau)/2$. Then, on the right-hand and left-hand sides of this equation, replacing t to $t - 2\xi$, we multiply both sides by $\mathrm{d}\xi$ and integrate over ξ from 0 to $t/2$. In the repeated integrals of the resulting equality, we change the order of integration. Using the conditions of the theorem 3.8 after simple calculations, we get (3.67). Equating in this equation functions that depend only on λ^1 and λ^2, we arrive at (3.66) for $\lambda = \lambda^i, i = 1, 2$. Equation (3.66) is obtained after setting $y = 0$ in the equality for $v_y(\lambda, y, t)$, following from (3.64), (3.65). From these equalities we obtain Eq. (3.60). It is clear that after applying the differential operators $(\partial/\partial t - \partial/\partial y)$, $(\partial/\partial t + \partial/\partial y)$ to relations (3.64), (3.65) we arrive at the equations (3.55)–(3.57).

Further proof of Theorem 3.8 can be carried out quite analogously to the proof of Theorem 3.1 from Sect. 3.1.

Remark 3.3 The Neumann boundary condition (3.51) can be replaced by the condition $u_y|_{y=0} = \delta(t)\delta(x) + \omega(t)\theta(t)\delta(x)$. Then, considering the function $w = u_t$ as a solution to the direct problem and assuming that the equalities $\omega(t)$ are true for $\omega(0) = 0$, $\omega'(t) = f(t)$, we arrive at the problem studied in this section.

3.5 About Global Solvability of a Multidimensional Inverse Problem for an Equation with Memory

In this section we obtain the global unique solvability of a multidimensional convolutional kernel determination problem in the class of continuous functions in the time variable and analytic in a part of the space variables.

Consider the problem of determining a pair of functions $u(x, z, t)$ and $k(x, t)$ satisfying the equations

$$u_{tt} - u_{zz} - \Delta u = \int_0^t k(x, \tau)u(x, z, t - \tau)\,d\tau, \ (x, t) \in \mathbb{R}^{m+1}, \ z > 0, \quad (3.68)$$

$$u\Big|_{t<0} \equiv 0, \ u_z\Big|_{z=0} = \delta'(t) + h(x, t)\theta(t), \ (x, t) \in \mathbb{R}^{m+1}, \quad (3.69)$$

$$u\Big|_{z=0} = -\delta(t) + f(x, t)\theta(t), \ (x, t) \in \mathbb{R}^{m+1}. \quad (3.70)$$

Here Δ is the Laplace operator in $(x_1, x_2, \ldots, x_m) = x$; $\delta'(t)$ is the derivative of the Dirac delta function, $\theta(t)$ is the Heaviside function, and h and f are given smooth functions.

We show that problem (3.68)–(3.70) is globally unique solvable in the class of functions continuous in t and analytic in x. We also establish a stability estimate for the solution of the inverse problem

In line with Sect. 2.1 of Chap. 2, introduce the Banach space $A_s(r)$, $s > 0$, of functions $\varphi(x)$, $x \in \mathbb{R}^m$, analytic in a neighborhood of the origin such that

$$\|\varphi\|_s(r) := \sup_{|x|<r} \sum_{|\alpha|=0}^{\infty} \frac{s^{|\alpha|}}{\alpha!} |D^\alpha \varphi(x)| < \infty.$$

Denote by $C\left(A_{s_0};\ G_T\right)$ the class of continuous with respect to (z,t) in domain G_T with values in A_{s_0} functions.

For (z,t) fixed, we denote the norm of $g(x,z,t)$ in A_{s_0} by $\|g\|_{s_0}(z,t)$. The norm of g in $C\left(A_{s_0};\ G_T\right)$ is defined as

$$\|g\|_{C(A_{s_0};\ G_T)} = \sup_{(z,t)\in G_T} \|g\|_{s_0}(z,t). \tag{3.71}$$

From (3.68), (3.69) it follows that $u \equiv 0,\ \ 0 < t < z,\ \ x \in \mathbb{R}^m$. The function $u(x,z,t)$ as a solution to (3.68), (3.69) has the form (2.5) in a neighborhood of the characteristic surface $t = z$.

Using the singularity distinguishing method, from (2.5) we get (2.6). Hence, in particular, there arises the condition (2.7) (one of the necessary conditions solvability for the problem), which we assume valid hereinafter.

Considering the equality $u(x,z,t) = \tilde{u}(x,z,t)$ for $x \in \mathbb{R}^m$ and $0 < z < t$, rewrite (3.68)–(3.70) as

$$u_{tt} - u_{zz} = \Delta u - k(x,t-z) + \int_0^{t-z} k(x,\tau)u(x,z,t-\tau)\,\mathrm{d}\tau,\ \ x \in \mathbb{R}^m,\ \ 0 < z < t, \tag{3.72}$$

$$u\big|_{z=0} = f(x,t),\ u_z\big|_{z=0} = h(x,t),\ t > 0,\ x \in \mathbb{R}^m, \tag{3.73}$$

$$u\big|_{t=z+0} \equiv 0,\ x \in \mathbb{R}^m. \tag{3.74}$$

Thus, (3.68)–(3.70) is reduced to the problem of determining the functions $u(x,z,t)$ and $k(x,t)$ from (3.72)–(3.74).

Let $C(A_{s_0};\ [0,T])$ be the space of continuous with respect to t in $[0,T]$ with values in A_{s_0} functions.

Lemma 3.3 *Let $T > 0$ and $s_0 > 0$ be some fixed numbers, condition (2.7) hold, and*

$$\Big(h(x,t),\ f_t(x,t),\ h_t(x,t),\ f_{tt}(x,t)\Big) \in C\Big(A_{s_0};\ [0,T]\Big).$$

Then, the inverse problem (3.72)–(3.74) for $(x,z,t) \in D_T$ is equivalent to the problem of finding $u(x,z,t),\ u_t(x,z,t)$ and $k(x,t)$ from the system of nonlinear integral Volterra-type equations of the second kind:

$$u(x,z,t) = \int_z^t u_t(x,z,\tau)\,\mathrm{d}\tau, \tag{3.75}$$

$$u_t(x,z,t) = \frac{1}{2} f_t(x,t-z) - \frac{1}{2} h(x,t-z) - \frac{z}{2} k(x,t-z)$$
$$+ \frac{1}{2} \int_0^z \left[\Delta u(x,\xi,t-z+\xi) + \int_0^{t-z} k(x,\tau) u(x,\xi,t-z+\xi-\tau)\,\mathrm{d}\tau \right] \mathrm{d}\xi$$
$$+ \frac{1}{2} \int_z^{(t+z)/2} \left[\Delta u(x,\xi,t+z-\xi) - k(x,t+z-2\xi) \right.$$
$$\left. + \int_0^{t+z-2\xi} k(x,\tau) u(x,\xi,t+z-\xi-\tau)\,\mathrm{d}\tau \right] \mathrm{d}\xi, \tag{3.76}$$

$$k(x,t) = 2 f_{tt}(x,t) + 2h_t(x,t)$$
$$- \int_0^{t/2} \left[\Delta u_t(x,\xi,t-2\xi) + \int_0^{t-2\xi} k(x,\tau) u_t(x,\xi,t-\xi-\tau)\,\mathrm{d}\tau \right] \mathrm{d}\xi. \tag{3.77}$$

Proof Taking (3.63) into account, integrate (3.72) along the corresponding characteristics of the differential operators of the first order for $(z,t) \in G_T$. Integrate along the characteristic of the operator $\partial/\partial t - \partial/\partial z$ from(z,t) to $((z+t)/2, (z+t)/2)$ on the plane of variables (ξ, τ). Using (3.74) and differentiating with respect to z, we obtain

$$(u_t + u_z)(x,z,t) = \int_z^{(z+t)/2} \left[\Delta u(x,\xi,t+z-\xi) - k(x,t+z-2\xi) \right.$$
$$\left. + \int_0^{t+z-2\xi} k(x,\tau) u(x,\xi,t+z-\xi-\tau)\,\mathrm{d}\tau \right] \mathrm{d}\xi. \tag{3.78}$$

Integrate now (3.72) along the characteristic of $\partial/\partial t + \partial/\partial z$ from $(0,\ t-z)$ to (z,t). Using (3.73) and differentiating with respect to t, we get

$$(u_t - u_z)(x,z,t) = f_t(x,t-z) - h(x,t-z) - k(x,t-z)z$$
$$+ \int_0^z \left[\Delta u(x,\xi,t-z+\xi) + \int_0^{t-z} k(x,\tau) u(x,\xi,t-z+\xi-\tau)\,\mathrm{d}\tau \right] \mathrm{d}\xi. \tag{3.79}$$

Equalities (3.78) and (3.79) easily yield (3.76). Putting $z = 0$ in (3.78) and using (3.73), find

$$f_t(x,t) + h(x,t)$$
$$= \int_0^{t/2} \left[\Delta u(x,\xi,t-\xi) - k(x,t-2\xi) + \int_0^{t-2\xi} k(x,\tau)u(x,\xi,t-\xi-\tau)\,d\tau \right] d\xi. \tag{3.80}$$

Differentiating (3.80), we arrive at (3.77) by simple computations. □

Under the agreement condition (2.7) it is easy to show that the inverse transformations are also valid. Indeed, replacing t with $t - 2\tau$ on the right- and left-hand sides of (3.77), multiply both sides by $d\tau$ and integrate with respect to τ from 0 to $t/2$. Changing the order of integration in the iterated integrals of the obtained equality and using (3.74), we obtain (3.80) which, in turn, is equivalent to (3.75) for $z = 0$. Equivalence of (3.75) to (3.78), (3.79) and derivation of (3.72)–(3.74) from the latter equations are obvious. Equality (3.75) is given for the closure of the system of (3.76) and (3.77). Thus, the lemma is proven.

System (3.75)–(3.77) is a closed system of integro-differential equations for u, u_t, and k. Note that the operator Δ in u and u_t enters this system only under the integral sign.

Introduce $v(x,z,t) := u_t + \frac{z}{2}k(x,t-z)$ and rewrite (3.75)–(3.77) as

$$u(x,z,t) = \int_z^t \left(v(x,z,\tau) - \frac{z}{2}k(x,\tau - z) \right) d\tau, \tag{3.81}$$

$$v(x,z,t) = v_0(x,z,t)$$
$$+ \frac{1}{2}\int_0^z \left[\Delta u(x,\xi,t-z+\xi) + \int_0^{t-z} k(x,\tau)u(x,\xi,t-z+\xi-\tau)\,d\tau \right] d\xi$$
$$+ \frac{1}{2}\int_z^{(t+z)/2} \Big[\Delta u(x,\xi,t+z-\xi) - k(x,t+z-2\xi)$$
$$+ \int_0^{t+z-2\xi} k(x,\tau)u(x,\xi,t+z-\xi-\tau)\,d\tau \Big] d\xi, \tag{3.82}$$

$$k(x,t) = k_0(x,t) - \int_0^{t/2} \Bigg[\Delta v(x,\xi,t-2\xi) - \frac{\xi}{2} \Delta k(x,t-2\xi)$$

$$+ \int_0^{t-2\xi} k(x,\tau) \left(v(x,\xi,t-\xi-\tau) - \frac{\xi}{2} k(x,t-\xi-\tau) \right) \mathrm{d}\tau \Bigg] \mathrm{d}\xi, \quad (3.83)$$

where

$$v_0(x,z,t) = \frac{1}{2} f_t(x,t-z) - \frac{1}{2} h(x,t-z),$$
$$k_0(x,t) = 2 f_{tt}(x,t) + 2 h_t(x,t). \quad (3.84)$$

We consider system (3.81)–(3.83) in a domain narrower than D_T, namely in $D_T^\varepsilon = G_T^\varepsilon \times \mathbb{R}^m$, $G_T^\varepsilon = \left\{(z,t) | \; 0 \le z \le t \le \frac{T}{1+\varepsilon} - z\right\}$ for some fixed $\varepsilon > 0$.

Denote by $C_\sigma\left(A_{s_0};\; G_T^\varepsilon\right)$ the space of continuous in (z,t) in G_T^ε with values in A_{s_0} functions which is generated by the family of weight norms

$$\|g\|_{C_\sigma(A_s;\; G_T^\varepsilon)} = \sup_{(z,t)\in G_T^\varepsilon} \|g\|_s(z,t) \mathrm{e}^{-\frac{\sigma t}{T-t-z}}, \quad \sigma \ge 0, \quad 0 < s < s_0. \quad (3.85)$$

It is obvious that (3.85) coincides with (3.71) for $\sigma = 0$. The number σ will be chosen later.

Theorem 3.9 *Let $T > 0$, $s_0 > 0$, and $\varepsilon > 0$ be fixed numbers and the conditions of the Lemma 3.3 be valid. Moreover,*

$$\max_{t\in[0,T]} \left\{ \|h\|_{s_0}(t),\; \|f_t\|_{s_0}(t),\; \frac{1}{4} \|f_{tt}\|_{s_0}(t),\; \frac{1}{4} \|h_t\|_{s_0}(t) \right\} = R,$$

while $R > 0$ is a known number. Then, for a and s such that $a \in (0,\; T/s_0)$, and $s \in (0,\; s_0)$ there is a unique solution to (3.81)–(3.83) in the domain $\Gamma_{sT}^\varepsilon = D_T^\varepsilon \bigcap \left\{(x,z,t) | \; 0 \le t < \frac{1}{1+\varepsilon} a(s_0 - s) - z\right\}$ for which

$$(u(x,z,t),\; v(x,z,t)) \in C\left(A_s;\; P_{sT}^\varepsilon\right),$$
$$k(x,t) \in C\left(A_{s_0};\; \left[0,\; \frac{1}{1+\varepsilon} a(s_0 - s)\right]\right).$$

Moreover,

$$\|u-u_0\|_{C(A_s;\,P^{\varepsilon}_{sT})} \le \tilde{R},\ \|v-v_0\|_{C(A_s;\,P^{\varepsilon}_{sT})} \le \frac{\tilde{R}}{s_0-s},$$

$$\|k-k_0\|_{C(A_s;\,P^{\varepsilon}_{sT})} \le \frac{\tilde{R}}{(s_0-s)^2},$$

$P^{\varepsilon}_{sT} = G^{\varepsilon}_T \bigcap \left\{(z,t)|\ 0 \le z \le t < \frac{1}{1+\varepsilon}a(s_0-s)-z\right\}$, *where* $u_0 = 0$, k_0 *and* v_0 *are determined in (3.84), and* $\tilde{R} = Re^{\sigma/\varepsilon}$, $\sigma > 0$.

Proof Under assumptions of theorem we have

$$(v_0, k_0) \in C\left(A_{s_0};\ G_T\right),$$
$$\|k_0\|_{C_\sigma(A_s;\,P_{sT})} \le \|k_0\|_{C(A_s;\,G_T)} \le R,\ \|v_0\|_{C_\sigma(A_s;\,P_{sT})} \le \|v_0\|_{C(A_s;\,G_T)} \le R,$$

$0 < s < s_0$. Let a_n be the terms of the monotone decreasing sequence

$$a_{n+1} = \frac{a_n}{1+1/(n+1)^2},\ n = 0,\ 1,\ 2, \ldots.$$

Put

$$a = \lim_{n\to\infty} a_n = a_0 \prod_{n=0}^{\infty} \left(1+1/(n+1)^2\right)^{-1},$$

where $a_0 \in (0,\ T/s_0)$. Construct a process of successive approximations to system (3.81)–(3.83) by the scheme

$$u_{n+1}(x,z,t) = \int\limits_z^t \left(v_n(x,z,\tau) - \frac{z}{2}k_n(x,\tau-z)\right) \mathrm{d}\tau,$$

$$v_{n+1}(x,z,t) = v_0(x,z,t)$$
$$+\frac{1}{2}\int\limits_0^z \left[\Delta u_n(x,\xi,t-z+\xi) + \int\limits_0^{t-z} k_n(x,\tau)u_n(x,\xi,t-z+\xi-\tau)\,\mathrm{d}\tau\right]\mathrm{d}\xi$$
$$+\frac{1}{2}\int\limits_z^{(t+z)/2} \Big[\Delta u_n(x,\xi,t+z-\xi) - k_n(x,t+z-2\xi)$$
$$+\int\limits_0^{t+z-2\xi} k_n(x,\tau)u_n(x,\xi,t+z-\xi-\tau)\,\mathrm{d}\tau\Big]\mathrm{d}\xi,$$

$$k_{n+1}(x,t) = k_0(x,t) - \int\limits_0^{t/2}\Big[\Delta v_n(x,\xi,t-2\xi) - \frac{\xi}{2}\Delta k_n(x,t-2\xi)$$

$$+ \int\limits_0^{t-2\xi} k_n(x,\tau)\Big(v_n(x,\xi,t-\xi-\tau) - \frac{\xi}{2}k_n(x,t-\xi-\tau)\Big)\,d\tau\Big]\,d\xi.$$

Put

$$s_n'(z,t) = \frac{s+\nu^n(z,t)}{2}, \quad \nu^n(z,t) = s_0 - \frac{t+z}{a_n}. \tag{3.86}$$

Introduce $p_n = u_{n+1} - u_n$, $r_n = v_{n+1} - v_n$, and $q_n = k_{n+1} - k_n$, $n = 0, 1, 2, \ldots$. □

Then p_n, q_n, and r_n satisfy the relations

$$p_0(x,z,t) = \int\limits_z^t \Big(v_0(x,z,\tau) - \frac{z}{2}k_0(x,\tau-z)\Big)\,d\tau,$$

$$r_0(x,z,t) = -\frac{1}{2}\int\limits_z^{(t+z)/2} k_0(x,t+z-2\xi)\,d\xi,$$

$$q_0(x,t) = -\int\limits_0^{t/2}\Big[\Delta v_0(x,\xi,t-2\xi) - \frac{\xi}{2}\Delta k_0(x,t-2\xi)$$

$$+ \int\limits_0^{t-2\xi} k_0(x,\tau)\Big(v_0(x,\xi,t-\xi-\tau) - \frac{\xi}{2}k_0(x,t-\xi-\tau)\Big)\,d\tau\Big]\,d\xi,$$

$$p_{n+1}(x,z,t) = \int\limits_z^t \Big(r_n(x,z,\tau) - \frac{z}{2}q_n(x,\tau-z)\Big)\,d\tau,$$

$$r_{n+1}(x,z,t) = \frac{1}{2}\int\limits_0^z\Big[\Delta p_n(x,\xi,t-z+\xi)$$

$$+ \int\limits_0^{t-z}\big(k_n(x,\tau)p_n(x,\xi,t-z+\xi-\tau) + u_{n+1}(x,\xi,t-z+\xi-\tau)q_n(x,\tau)\big)\,d\tau\Big]\,d\xi$$

$$+ \frac{1}{2}\int\limits_z^{(t+z)/2}\Big[\Delta p_n(x,\xi,t+z-\xi) - q_n(x,t+z-2\xi)$$

$$+ \int\limits_0^{t+z-2\xi}\big(k_n(x,\tau)p_n(x,\xi,t+z-\xi-\tau) + q_n(x,\tau)u_{n+1}(x,\xi,t+z-\xi-\tau)\big)\,d\tau\Big]\,d\xi,$$

$$q_{n+1}(x,t) = -\int_0^{t/2}\Big[\Delta r_n(x,\xi,t-2\xi) - \frac{\xi}{2}\Delta q_n(x,t-2\xi)$$

$$+\int_0^{t-2\xi}\Big\{q_n(x,\tau)\Big(v_{n+1}(x,\xi,t-\xi-\tau) - \frac{\xi}{2}k_{n+1}(x,t-\xi-\tau)\Big)$$

$$+k_n(x,\tau)\Big(r_n(x,\xi,t-\xi-\tau) - \frac{\xi}{2}q_n(x,t-\xi-\tau)\Big)\Big\}\,\mathrm{d}\tau\Big]\,\mathrm{d}\xi.$$

Prove that we can choose $\sigma > 0$ so that for each $n = 0, 1, 2, \ldots$ the following estimates are valid:

$$\lambda_n(\sigma) = \max\Bigg\{\sup_{(z,t,s)\in F_n}\Big[\|p_n\|_s(z,t)\mathrm{e}^{-\frac{\sigma t}{b_n-t-z}}\Big],\ \sup_{(z,t,s)\in F_n}\Big[\|r_n\|_s(z,t)\times(\nu^n(z,t)-s)\mathrm{e}^{-\frac{\sigma t}{b_n-t-z}}\Big],$$
$$\sup_{(z,t,s)\in F_n}\Big[\|q_n\|_s(t)(\nu^n(z,t)-s)^2\mathrm{e}^{-\frac{\sigma t}{b_n-t-z}}\Big]\Bigg\} < \infty, \tag{3.87}$$

$$\begin{gathered}\|u_{n+1}-u_0\|_s(z,t)\mathrm{e}^{-\frac{\sigma t}{b_n-t-z}} \le R,\ \|v_{n+1}-v_0\|_s(z,t)\mathrm{e}^{-\frac{\sigma t}{b_n-t-z}} \le \frac{R}{s_0-s},\\ \|k_{n+1}-k_0\|_s(t)\mathrm{e}^{-\frac{\sigma t}{b_n-t-z}} \le \frac{R}{(s_0-s)^2},\ (z,t,s)\in F_{n+1},\end{gathered} \tag{3.88}$$

where

$$b_n := a_n(s_0-s),$$
$$F_n = \Big\{(z,t,s)|\ (z,t)\in G_T^{\varepsilon},\ 0<z<t<\frac{b_n}{1+\varepsilon}-z,\ 0<s<s_0\Big\}.$$

Indeed, using the relations for p_n, q_n, and r_n, we find

$$\|p_0\|_s(z,t)\mathrm{e}^{-\frac{\sigma t}{b_0-t-z}} \le \int_z^t\Big(\|v_0(x,z,\tau)\|_s(z,\tau)\mathrm{e}^{-\frac{\sigma\tau}{b_0-\tau-z}}\mathrm{e}^{\frac{\sigma\tau}{b_0-\tau-z}}\mathrm{e}^{-\frac{\sigma t}{b_0-t-z}}$$
$$+\frac{z}{2}\|k_0(x,\tau-z)\|_s(\tau-z)\mathrm{e}^{-\frac{\sigma(\tau-z)}{b_0-\tau}}\mathrm{e}^{\frac{\sigma(\tau-z)}{b_0-\tau}}\mathrm{e}^{-\frac{\sigma t}{b_0-t-z}}\Big)\,\mathrm{d}\tau$$
$$\le \int_z^t\Big(\|v_0\|^{\sigma}_{s,0}\mathrm{e}^{-\frac{\sigma(b_0-z)(t-\tau)}{(b_0-\tau-z)(b_0-t-z)}} + \frac{T}{4}\|k_0\|^{\sigma}_{s,0}\mathrm{e}^{-\frac{\sigma(b_0-z)(t+z-\tau)}{(b_0-\tau)(b_0-t-z)}}\Big)\,\mathrm{d}\tau$$
$$\le \int_z^t\Big(\|v_0\|^{\sigma}_{s,0}\mathrm{e}^{-\frac{\sigma(t-\tau)}{b_0-t-z}} + \frac{T}{4}\|k_0\|^{\sigma}_{s,0}\mathrm{e}^{-\frac{\sigma(t+z-\tau)}{b_0-t-z}}\Big)\,\mathrm{d}\tau$$

$$\leq R\left(1+\frac{T}{4}\right)\frac{b_0-t-z}{\sigma}\leq R\left(1+\frac{T}{4}\right)\frac{a_0 s_0}{\sigma},\quad (z,t,s)\in F_0,$$

where $\|\cdot\|^{\sigma}_{s,i} := \sup\limits_{(z,t,s)\in F_i} \|\cdot\|_s(z,t)\mathrm{e}^{-\frac{\sigma t}{b_i-t-z}}$, $i=0,1,2,\ldots$.

Similarly,

$$\|r_0\|_s(z,t)\mathrm{e}^{-\frac{\sigma t}{b_0-t-z}} \leq \frac{1}{2}\int\limits_z^{(t+z)/2} \|k_0\|^{\sigma}_{s,0}\mathrm{e}^{-\frac{\sigma(2\xi-z)}{b_0-t-z}}\,\mathrm{d}\xi$$

$$\leq \frac{1}{2}R\frac{b_0-t-z}{\sigma}\leq \frac{R}{2(\nu^0(z,t)-s)}\frac{a_0 s_0^2}{\sigma},\ \|q_0\|_s(t)\mathrm{e}^{-\frac{\sigma t}{b_0-t-z}}$$

$$\leq \int\limits_0^{t/2}\Bigg[\|\Delta v_0\|_s(\xi,t-2\xi)\mathrm{e}^{-\frac{\sigma(t-2\xi)}{b_0-(t-\xi)}}\mathrm{e}^{-\frac{2\sigma\xi(b_0-t/2)}{(b_0-(t-\xi))(b_0-t-z)}}$$

$$+\frac{\xi}{2}\|\Delta k_0\|_s(t-\xi)\mathrm{e}^{-\frac{\sigma(t-\xi)}{b_0-(t-\xi)}}\mathrm{e}^{-\frac{\sigma(b_0-t/2)}{(b_0-(t-\xi))(b_0-t-z)}}$$

$$+\int\limits_0^{t-2\xi}\|k_0\|_s(\tau)\mathrm{e}^{-\frac{\sigma\tau}{b_0-\tau-z}}\mathrm{e}^{\frac{\sigma\tau}{b_0-\tau-z}}\left(\|v_0\|_s(\xi,t-\xi-\tau)+\frac{\xi}{2}\|k_0\|_s(t-\xi-\tau)\right)$$

$$\times\,\mathrm{e}^{-\frac{\sigma(t-\xi-\tau)}{b_0-(t-\tau)}}\mathrm{e}^{\frac{\sigma(t-\xi-\tau)}{b_0-(t-\tau)}}\mathrm{e}^{-\frac{\sigma t}{b_0-t-z}}\,\mathrm{d}\tau\Bigg]\,\mathrm{d}\xi$$

$$\leq \int\limits_0^{t/2}\Bigg[\|\Delta v_0\|^{\sigma}_{s,0}(\xi,t-2\xi)\mathrm{e}^{-\frac{2\sigma\xi}{b_0-t-z}}+\frac{\xi}{2}\|\Delta k_0\|^{\sigma}_{s,0}(t-2\xi)\mathrm{e}^{-\frac{2\sigma\xi}{b_0-t-z}}$$

$$+\int\limits_0^{t-2\xi}\|k_0\|^{\sigma}_{s,0}(\tau)\left(\|v_0\|^{\sigma}_{s,0}(\xi,t-\xi-\tau)+\frac{\xi}{2}\|k_0\|^{\upsilon}_{s,0}(t-\xi-\tau)\right)$$

$$\times\,\mathrm{e}^{-\frac{\sigma\xi(b_0-t+z+2\tau)}{(b_0-(t-\tau))(b_0-t-z)}}\,\mathrm{d}\tau\Bigg]\,\mathrm{d}\xi.$$

Let c_0 be a positive constant such that

$$\|\Delta\varphi\|_s\leq c_0\frac{\|\varphi\|_{s'_n}}{\left(s'_n-s\right)^2},\ s'_n>s>0,\ n=0,\ 1,\ 2,\ldots.$$

We can verify that $c_0=4m$.

Taking $s_n'(z,t)$ from (3.86) ($s_n'(z,t) > s$) for $n = 0$ we obtain

$$\|q_0\|_s(t)\mathrm{e}^{-\frac{\sigma t}{b_0-t-z}}$$
$$\leq \int\limits_0^{t/2}\left[\frac{Rc_0\mathrm{e}^{-\frac{2\sigma\xi}{b_0-t-z}}}{(s_0'(\xi,t-2\xi)-s)^2}\left(1+\frac{T}{4}\right)+TR^2\mathrm{e}^{-\frac{\sigma\xi}{b_0-t-z}}\left(1+\frac{T}{4}\right)\right]\mathrm{d}\xi$$
$$\leq \frac{R(1+T/4)(4c_0+s_0^2TR)}{(\nu^0(z,t)-s)^2}\frac{a_0s_0}{\sigma},\quad (z,t,s)\in F_0.$$

Consequently,

$$\lambda_0(\sigma)\leq\frac{1}{\sigma}\left(1+\frac{T}{4}\right)a_0s_0\max\left(1,\ s_0,\ 4c_0+s^2TR\right)R=:\frac{1}{\sigma}\mu_0R.$$

Thus, $\lambda_0(\sigma)$ is finite, i.e., (3.87) is valid for $n = 0$. Moreover, for $(z,t,s)\in F_1$ the following estimates hold:

$$\|u_1-u_0\|_s(z,t)\mathrm{e}^{-\frac{\sigma t}{b_0-t-z}}=\|p_0\|_s(z,t)\mathrm{e}^{-\frac{\sigma t}{b_0-t-z}}\leq\lambda_0(\sigma),$$
$$\|v_1-v_0\|_s(z,t)\mathrm{e}^{-\frac{\sigma t}{b_0-t-z}}=\|r_0\|_s(z,t)\mathrm{e}^{-\frac{\sigma t}{b_0-t-z}}\leq\frac{\lambda_0(\sigma)}{\nu^0(z,t)-s}\leq\frac{2\lambda_0(\sigma)}{s_0-s},$$
$$\|k_1-k_0\|_s(t)\mathrm{e}^{-\frac{\sigma t}{b_0-t-z}}=\|q_0\|_s(t)\mathrm{e}^{-\frac{\sigma t}{b_0-t-z}}\leq\frac{\lambda_0(\sigma)}{(\nu^0(z,t)-s)^2}\leq\frac{4\lambda_0(\sigma)}{(s_0-s)^2}.$$

For $\sigma > 0$ such that $\sigma > 4\mu_0$, inequalities (3.88) hold for $n = 0$.

Using induction, show that (3.87) and (3.88) hold for other n as well if σ is chosen appropriately. Let (3.87) and (3.88) be valid for $n = 0, 1, 2, \ldots, i$. Then, for $(z,t,s)\in F_{i+1}$ we have

$$\|p_{i+1}\|_s(z,t)\mathrm{e}^{-\frac{\sigma t}{b_{i+1}-t-z}}\leq\int\limits_z^t\left(\|r_i\|_s(z,\tau)\mathrm{e}^{-\frac{\sigma\tau}{b_{i+1}-\tau-z}}\mathrm{e}^{-\frac{\sigma(b_{i+1}-z)(t-\tau)}{(b_{i+1}-\tau-z)(b_{i+1}-t-z)}}\right.$$
$$\left.+\frac{z}{2}\|q_i\|_s(\tau-z)\mathrm{e}^{-\frac{\sigma(\tau-z)}{b_{i+1}-\tau}}\mathrm{e}^{-\frac{\sigma(b_{i+1}-z)(t-\tau+z)}{(b_{i+1}-\tau)(b_{i+1}-t-z)}}\right)\mathrm{d}\tau$$
$$\leq\int\limits_z^t\frac{\lambda_i(\sigma)}{\nu^{i+1}(z,\tau)-s}\mathrm{e}^{-\frac{\sigma(t-\tau)}{b_{i+1}-t-z}}\mathrm{d}\tau+\int\limits_0^{t-z}\frac{t-\eta}{2}\frac{\lambda_i(\sigma)}{(\nu^{i+1}(z,\eta)-s)^2}\mathrm{e}^{-\frac{\sigma(t-\eta)}{b_{i+1}-t-z}}\mathrm{d}\eta$$
$$\leq\frac{\lambda_i(\sigma)(1+a_0/2)}{\nu^{i+1}(z,t)-s}\frac{b_{i+1}-t-z}{\sigma}\leq\frac{\lambda_i(\sigma)a_0(1+a_0/2)}{\sigma}.$$

Here we used the obvious inequalities $a_i \le a_0$, $\nu^{i+1}(z,t) \le \nu^i(z,t)$, and $t-\eta < \frac{b_{i+1}}{1+\varepsilon} - (z+\eta) < b_{i+1} - (z+\eta)$. Similar arguments for r_{i+1} and q_{i+1} yield

$$\|r_{i+1}\|_s(z,t)\mathrm{e}^{-\frac{\sigma t}{b_{i+1}-t-z}} \le \frac{1}{2}\int_0^z \Bigg[\frac{c_0\lambda_i(\sigma)}{(s_i'(\xi,t-z+\xi)-s)^2}\mathrm{e}^{-\frac{\sigma(z-\xi)(b_{i+1}+t-z)}{(b_{i+1}-t+z-2\xi)(b_{i+1}-t-z)}}$$

$$+\int_0^{t-z}\left(\frac{R(1+s_0^2)\lambda_i(\sigma)}{(s_0-s)^2}+2R\frac{\lambda_i(\sigma)}{(\nu^i(\xi,\tau)-s)^2}\right)\mathrm{e}^{-\frac{\sigma(z-\xi)(b_{i+1}-t+z+2\tau)}{(b_{i+1}-t+z-2\xi+\tau)(b_{i+1}-t-z)}}\,\mathrm{d}\tau\Bigg]\,\mathrm{d}\xi$$

$$+\frac{1}{2}\int_z^{(t+z)/2}\Bigg[\frac{c_0\lambda_i(\sigma)\mathrm{e}^{-\frac{\sigma(\xi-z)}{b_{i+1}-t-z}}}{(s_i'(t+z-\xi)-s)^2}+\frac{\lambda_i(\sigma)\mathrm{e}^{-\frac{\sigma(2\xi-z)}{b_{i+1}-t-z}}}{(\nu^i(\xi,\tau)-s)^2}$$

$$+\int_0^{t+z-2\xi}\left(\frac{R(1+s_0^2)\lambda_i(\sigma)}{(s_0-s)^2}+2R\frac{\lambda_i(\sigma)}{(\nu^i(\xi,\tau)-s)^2}\right)\mathrm{e}^{-\frac{\sigma(\xi-z)}{b_{i+1}-t-z}}\,\mathrm{d}\tau\Bigg]\,\mathrm{d}\xi$$

$$\le \frac{\lambda_i(\sigma)(b_{i+1}-t-z)}{2\sigma(\nu^{i+1}(z,t)-s)^2}\left[8c_0+1+2RT(3+s_0^2)\right]$$

$$\le \frac{\lambda_i(\sigma)}{\sigma}\frac{a_0[4c_0+1/2+RT(3+s_0^2)]}{\nu^{i+1}(z,t)-s}, \quad (z,t,s)\in F_{i+1}.$$

For the intermediate computations, s_n' is defined by (3.86) for $n=i$. We also used the relations

$$\|\cdot\|^\sigma_{s,i+1} < \|\cdot\|^\sigma_{s,i}, \quad \frac{1}{s_0-s} \le \frac{1}{\nu^i(z,t)-s} \quad i=0,\,1,\,2,\ldots,$$

$$\|u_i\|^\sigma_{s,i+1} < \|u_i\|^\sigma_{s,i-1} \le 2R, \ \|v_i\|^\sigma_{s,i+1} < \|v_i\|^\sigma_{s,i-1} \le R\frac{1+s_0}{s_0-s},$$

$$\|k_i\|^\sigma_{s,i+1} < \|k_i\|^\sigma_{s,i-1} \le R\frac{1+s_0^2}{(s_0-s)^2}, \quad (z,t,s)\in F_{i+1},$$

which are true by the inductive assumption. Similar arguments for q_{i+1} lead to the following inequality (the change of variable $t-2\xi=\eta$ was made beforehand):

$$\|q_{i+1}\|_s(z)\mathrm{e}^{-\frac{\sigma t}{b_{i+1}-t-z}} \le \frac{\lambda_i(\sigma)}{2}\times$$

$$\times\int_0^t\Bigg[\frac{c_0\mathrm{e}^{-\frac{\sigma(t-\eta)}{b_{i+1}-t-z}}}{(s_i'(\frac{t-\eta}{2},\eta)-s)^2}\left(\frac{1}{\nu^{i+1}(\frac{t-\eta}{2},\eta)-s}+\frac{t-\eta}{4(\nu^{i+1}(\frac{t-\eta}{2},\eta)-s)^2}\right)$$

$$+\int_0^\eta\Bigg\{\frac{1}{(\nu^{i+1}(\frac{t-\eta}{2},\tau)-s)^2}\left(\frac{R(1+s_0)}{s_0-s}+\frac{t-\eta}{4}\frac{R(1+s_0^2)}{(s_0-s)^2}\right)$$

$$+\frac{R(1+s_0^2)}{(s_0-s)^2}\left(\frac{1}{\nu^{i+1}(\frac{t-\eta}{2},\frac{t+\eta-2\tau}{2})-s}+\frac{t-\eta}{4(\nu^{i+1}(\frac{t-\eta}{2},\frac{t+\eta-2\tau}{2})-s)^2}\right)\Bigg\}$$

$$\times\, \mathrm{e}^{-\frac{(t-\eta)\sigma}{2(b_{i+1}-t-z)}}\,\mathrm{d}\tau\Bigg]\,\mathrm{d}\xi \le \frac{\lambda_i(\sigma)(b_{i+1}-t-z)}{\sigma}$$

$$\times\left[\frac{4c_0(4+a_0)}{4(\nu^{i+1}(z,t)-s)^3}+RT\frac{4(1+s_0)+a_0(1+s_0^2)}{4(\nu^{i+1}(z,t)-s)^3}+RT\frac{(4+a_0)(1+s_0^2)}{4(\nu^{i+1}(z,t)-s)^3}\right]$$

$$\le \frac{\lambda_i(\sigma)a_0}{\sigma}\frac{c_0(4+a_0)+RT[1+s_0+(2+3a_0)(1+s_0^2)/4]}{(\nu^{i+1}(z,t)-s)^2},\quad (z,t,s)\in F_{i+1}.$$

From the above estimates, it follows that

$$\lambda_{i+1}(\sigma)\le\lambda_i(\sigma)\rho,\ \lambda_{i+1}<\infty,$$
$$\rho=\frac{a_0}{\sigma}\max\Big\{1+a_0/2;\ 1/2+4c_0+RT(3+s_0^2);$$
$$c_0(4+a_0)+RT[1+s_0+(2+3a_0)(1+s_0^2)/4]\Big\}.$$

At the same time, for $(z,t,s)\in F_{i+2}$ we obtain

$$\|u_{i+2}-u_0\|_s(z,t)\mathrm{e}^{-\frac{\sigma t}{b_{i+1}-t-z}}\le\sum_{n=0}^{i+1}\|p_n\|_s\mathrm{e}^{-\frac{\sigma t}{b_n-t-z}}\le\sum_{n=0}^{i+1}\lambda_n(\sigma)\le\lambda_0(\sigma)\sum_{n=0}^{i+1}\rho^n,$$

$$\|v_{i+2}-v_0\|_s(z,t)\mathrm{e}^{-\frac{\sigma t}{b_{i+1}-t-z}}\le\sum_{n=0}^{i+1}\|r_n\|_s\mathrm{e}^{-\frac{\sigma t}{b_n-t-z}}\le\sum_{n=0}^{i+1}\frac{\lambda_n(\sigma)}{\nu^n(z,t)-s}$$

$$\le\frac{1}{s_0-s}\sum_{n=0}^{i+1}\frac{\lambda_n(\sigma)a_n}{a_n-a_{n+1}}\le\frac{\lambda_0(\sigma)}{s_0-s}\sum_{n=0}^{i+1}\rho^n[(n+1)^2+1],$$

$$\|k_{i+2}-k_0\|_s(t)\mathrm{e}^{-\frac{\sigma t}{b_{i+1}-t-z}}\le\sum_{n=0}^{i+1}\|q_n\|_s\mathrm{e}^{-\frac{\sigma t}{b_n-t-z}}$$

$$\le\sum_{n=0}^{i+1}\frac{\lambda_n(\sigma)}{(\nu^n(z,t)-s)^2}\le\frac{1}{(s_0-s)^2}\sum_{n=0}^{i+1}\frac{\lambda_n(\sigma)a_n^2}{(a_n-a_{n+1})^2}$$

$$\le\frac{\lambda_0(\sigma)}{(s_0-s)^2}\sum_{n=0}^{i+1}\rho^n[(n+1)^2+1]^2,$$

Choose $\sigma>0$ such that

$$\rho<1,\ \lambda_0(\sigma)\sum_{n=0}^{\infty}\rho^n[(n+1)^2+1]^2\le R.$$

Then

$$\|u_{i+2} - u_0\|^{\sigma}_{s,i+1} \le R, \ \|v_{i+2} - v_0\|^{\sigma}_{s,i+1} \le \frac{R}{s_0 - s},$$

$$\|k_{i+2} - k_0\|^{\sigma}_{s,i+1} \le \frac{R}{(s_0 - s)^2}, \ (z, t, s) \in F_{i+2}.$$

Note that the norms $\|\cdot\|_s$ and $\|\cdot\|^{\sigma}_{s,i}$ are equivalent for $i = 0, 1, 2, \dots$. Indeed,

$$e^{-\frac{\sigma}{\varepsilon}} \|\cdot\|_s(z, t) \le \|\cdot\|^{\sigma}_{s,i}(z, t) \le \|\cdot\|_s(z, t), \quad (z, t) \in F_{i+1}.$$

Consequently, (3.88) can be rewritten as

$$\begin{aligned} \|u_{n+1} - u_0\|_s \le Re^{\frac{\sigma}{\varepsilon}}, \ \|v_{n+1} - v_0\|_s \le \tfrac{Re^{\frac{\sigma}{\varepsilon}}}{s_0 - s}, \\ \|k_{n+1} - k_0\|_s \le \tfrac{Re^{\frac{\sigma}{\varepsilon}}}{(s_0 - s)^2}, \ (z, t, s) \in F_{n+1}. \end{aligned} \tag{3.89}$$

Since the choice of $\sigma > 0$ is independent of the number of approximations, the successive approximations (u_n, k_n, v_n) belong to

$$C\,(A_s;\ F)\,,\ F = \bigcap_{n=0}^{\infty} F_n,$$

and

$$\|u_n - u_0\|_s(z, t) \le \tilde{R}, \ \|v_n - v_0\|_s(z, t) \le \frac{\tilde{R}}{s_0 - s},$$

$$\|k_n - k_0\|_s(t) \le \frac{\tilde{R}}{(s_0 - s)^2}, \ (z, t, s) \in F.$$

For $s \in (0, s_0)$ the series

$$\sum_{n=0}^{\infty}(u_n - u_{n-1}), \ \sum_{n=0}^{\infty}(v_n - v_{n-1}), \ \sum_{n=0}^{\infty}(k_n - k_{n-1})$$

converge uniformly in the norm of $C\left(A_s;\ P^{\varepsilon}_{sT}\right)$, therefore $u_n \to u$, $v_n \to v$, and $k_n \to k$, and the limit functions u, k, and v are elements of $C\left(A_s;\ P^{\varepsilon}_{sT}\right)$ and satisfy (3.81)–(3.83).

Prove that the found solution is unique. Let $(u,\ v,\ k)$ and $(\hat{u},\ \hat{v},\ \hat{k})$ be two arbitrary solutions satisfying

$$\|u - u_0\|_s(y, t) \le \tilde{R}, \ \|v - v_0\|_s(y, t) \le \frac{\tilde{R}}{s_0 - s}, \ \|k - k_0\|_s(t) \le \frac{\tilde{R}}{(s_0 - s)^2}, \ (z, t, s) \in F.$$

Put $\tilde{p} = u - \hat{u}$, $\tilde{r} = v - \hat{v}$, and $\tilde{q} = k - \hat{k}$ and let

$$\lambda(\sigma) = \max\Bigg\{ \sup_{(z,t,s)\in F} \left[\|\tilde{p}\|_s(z,t) e^{-\frac{\sigma t}{b-t-z}} \right], \sup_{(z,t,s)\in F} \left[\|\tilde{r}\|_s(z,t)(\nu(z,t)-s) e^{-\frac{\sigma t}{b-t-z}} \right],$$

$$\sup_{(z,t,s)\in F} \left[\|\tilde{q}\|_s(t)(\nu(z,t)-s)^2 e^{-\frac{\sigma t}{b-t-z}} \right] \Bigg\} < \infty,$$

where $b := a(s_0 - s)$, $\nu(z,t) = s_0 - \frac{t+z}{a}$, and $a = a_0 \prod_{n=0}^{\infty} \left(1 + 1/(n+1)^2\right)^{-1}$.

Then,

$$\tilde{p}(x,z,t) = \int_z^t \left(\tilde{r}(x,z,\tau) - \frac{z}{2}\tilde{q}(x,\tau - z) \right) d\tau,$$

$$\tilde{r}(x,z,t) = \frac{1}{2} \int_0^z \Bigg[\Delta \tilde{p}(x,\xi,t-z+\xi)$$

$$+ \int_0^{t-z} \left(k(x,\tau)\tilde{p}(x,\xi,t-z+\xi-\tau) + \hat{u}(x,\xi,t-z+\xi-\tau)\tilde{q}(x,\tau) \right) d\tau \Bigg] d\xi$$

$$+ \frac{1}{2} \int_z^{(t+z)/2} \Bigg[\Delta \tilde{p}(x,\xi,t+z-\xi) - \tilde{q}(x,t+z-2\xi)$$

$$+ \int_0^{t+z-2\xi} \left(k(x,\tau)\tilde{p}(x,\xi,t+z-\xi-\tau) + \tilde{q}(x,\tau)\hat{u}(x,\xi,t+z-\xi-\tau) \right) d\tau \Bigg] d\xi,$$

$$\tilde{q}(x,t) = -\int_0^{t/2} \Bigg[\Delta \tilde{r}(x,\xi,t-2\xi) - \frac{\xi}{2}\Delta \tilde{q}(x,t-2\xi)$$

$$+ \int_0^{t-2\xi} \Big\{ \tilde{q}(x,\tau)\Big(\hat{v}(x,\xi,t-\xi-\tau) - \frac{\xi}{2}\hat{k}(x,t-\xi-\tau) \Big)$$

$$+ k(x,\tau)\Big(\tilde{r}(x,\xi,t-\xi-\tau) - \frac{\xi}{2}\tilde{q}(x,t-\xi-\tau) \Big) \Big\} d\tau \Bigg] d\xi.$$

Applying to these equations the estimates presented above, we find

$$\lambda \le \lambda\rho', \ \rho' := \frac{a}{\sigma} \max \Big\{ 1 + a/2;\ 1/2 + 4c_0 + 2RT + RT(1+s_0^2);$$

$$c_0(4+a) + RT[1 + s_0 + (2+3a)(1+s_0^2)/4] \Big\} < \rho < 1.$$

Consequently, $\lambda = 0$; therefore, $u = \hat{u}$, $v = \hat{v}$, and $k = \hat{k}$. The theorem is proven.

Consider the set Υ of all pairs of functions (f, h) representing the elements of $C\left(A_{s_0};\ [0, T]\right)$, $s_0 > 0$, for which the conditions of Theorem 3.9 are valid with fixed R, T, s_0, and ε. Then, we have the stability theorem:

Theorem 3.10 *Let $(f, h) \in \Upsilon$, and $(\bar{f}, \bar{h}) \in \Upsilon$. Then, for the corresponding solutions (u, k, v) and $(\bar{u}, \bar{k}, \bar{v})$ to (3.81)–(3.83), we have*

$$\|u - \bar{u}\|_s \le cM, \ \ \|k - \bar{k}\|_s \le \frac{cM}{(s_0 - s)^2},$$
$$\|v - \bar{v}\|_s \le \frac{cM}{s_0 - s}, \ \ (z, t) \in P^{\varepsilon}_{sT}, \ 0 < s < s_0, \tag{3.90}$$

where

$$M = \max\Big[\max \|h - \bar{h}\|_{s_0}(t), \ \max \|f_t - \bar{f}_t\|_{s_0}(t),$$
$$\max \|h_t - \bar{h}_t\|_{s_0}(t), \max \|f_{tt} - \bar{f}_{tt}\|_{s_0}(t)\Big], \ \ t \in [0, T],$$

and the constant c depends on R, T, s_0, σ, and ε.

Proof For the differences $u - \bar{u} = \tilde{u}$, $k - \bar{k} = \tilde{k}$, $v - \bar{v} = \tilde{v}$, $f - \bar{f} = \tilde{f}$, and $h - \bar{h} = \tilde{h}$ the equalities follow from (3.81)–(3.83)

$$\tilde{u}(x, z, t) = \int\limits_z^t \left(\tilde{v}(x, z, \tau) - \frac{z}{2}\tilde{k}(x, \tau - z)\right) \mathrm{d}\tau, \tag{3.91}$$

$$\tilde{v}(x, z, t) = \frac{1}{2}\int\limits_0^z \Big[\Delta\tilde{u}(x, \xi, t - z + \xi)$$
$$+ \int\limits_0^{t-z} \left(\bar{k}(x, \tau)\tilde{u}(x, \xi, t - z + \xi - \tau) + u(x, \xi, t - z + \xi - \tau)\tilde{k}(x, \tau)\right) \mathrm{d}\tau\Big] \mathrm{d}\xi$$
$$+ \frac{1}{2}\int\limits_z^{(t+z)/2} \Big[\Delta\tilde{u}(x, \xi, t + z - \xi) - \tilde{k}(x, t + z - 2\xi)$$
$$+ \int\limits_0^{t+z-2\xi} \left(\bar{k}(x, \tau)\tilde{u}(x, \xi, t + z - \xi - \tau) + \tilde{k}(x, \tau)u(x, \xi, t + z - \xi - \tau)\right) \mathrm{d}\tau\Big] \mathrm{d}\xi, \tag{3.92}$$

$$\tilde{k}(x,t) = -\int_0^{t/2} \Big[\Delta\tilde{v}(x,\xi,t-2\xi) - \frac{\xi}{2}\Delta\tilde{k}(x,t-2\xi)$$

$$+ \int_0^{t-2\xi} \Big\{ \tilde{k}(x,\tau)\Big(v(x,\xi,t-\xi-\tau) - \frac{\xi}{2}k(x,t-\xi-\tau)\Big)$$

$$+ \bar{k}(x,\tau)\Big(\tilde{v}(x,\xi,t-\xi-\tau) - \frac{\xi}{2}\tilde{k}(x,t-\xi-\tau)\Big)\Big\}\,\mathrm{d}\tau\Big]\,\mathrm{d}\xi. \tag{3.93}$$

where

$$\tilde{v_0}(x,z,t) = \frac{1}{2}\left(\tilde{f}_t(x,t-z) - \tilde{h}(x,t-z)\right),$$
$$\tilde{k}_0(x,t) = 2\left(\tilde{f}_{tt}(x,t) + \tilde{h}_t(x,t)\right).$$

It is obvious that

$$\|\tilde{v}_0\|_{s_0}(z,t) \leq M,\ \|\tilde{k}_0\|_{s_0}(z) \leq 4M, \quad (z,t) \in P^{\varepsilon}_{sT}. \tag{3.94}$$

Theorem 3.9 yields the estimates:

$$\|u\|_s \leq 2\tilde{R},\ \|v\|_s \leq \frac{\tilde{R}(1+s_0)}{s_0-s},\ \|k\|_s \leq \frac{\tilde{R}(1+s_0^2)}{(s_0-s)^2}.$$

□

Applying the method of successive approximations used for the proof of Theorem 3.9 to the system of equations (3.91)–(3.93), which is linear with respect to $\tilde{u}$, $\tilde{k}$, and $\tilde{v}$ we find that, for the same choice of σ, the following inequalities are valid for solution to (3.91)–(3.93):

$$\|\tilde{u} - \tilde{u}_0\|_{\leq} c_1 M,\ \|\tilde{v} - \tilde{v}_0\| \leq \frac{c_1 M}{s_0 - s},$$
$$\|\tilde{k} - \tilde{k}_0\| \leq \frac{c_1 M}{(s_0-s)^2},\ (z,t) \in P^{\varepsilon}_{sT},\ 0 < s < s_0,$$

where c_1 depends on R, T, s_0, σ, and ε. Hence, due to (3.93), it follows (3.89). Theorem 3.10 is proven.

3.6 Conclusions

In this chapter, the global solvability of memory determination problems is investigated. The local nature of existence theorems proved in the previous chapters is due to the fact that the Volterra integral equations obtained in the process of solving the inverse problem are nonlinear. Turns out, since in these equations the nonlinearity is convolutional, it is possible to solve the resulting Volterra equations in space of continuous functions with an exponential weight. This allowed to prove the global solvability of such problems. Here, the problems of determining memory function from initial–boundary value problems for the integro-differential wave equation and the system of equations of electrodynamics are considered. At the end of this chapter, it is studied the multidimensional inverse problem of determining the convolutional kernel of the integral term in an integro-differential wave equation. It is proven the global unique solvability of this problem in the class of continuous functions in the time variable and analytic in a part of the space variables.

References

1. Romanov, V.G. 2005. *Stability in inverse problems,* M.: Nauchniy Mir (Russian).
2. Bukhgeim, A.L. 1993. Inverse problems of memory reconstruction. *Journal of Inverse and Ill-Posed Problems* 1 (3): 193–206.
3. Bukhgeim, A.L., and N.I. Kalinina. 1997. Global convergence of the Newton method in inverse problems of memory reconstruction. *Siberian Mathematical Journal* 38 (5): 881–895.
4. Bukhgeim, A.L., N.I. Kalinina, and V.B. Kardakov. 2000. Two methods for the inverse problem of memory reconstruction. *Siberian Mathematical Journal* 41 (4): 634–642.
5. Janno, J., and L. von Wolfersdorf. 1997. Inverse problems for identification of memory kernels in viscoelasticity. *Mathematical Methods in the Applied Sciences* 20 (4): 291–314.
6. Janno, J., and L. von Wolfersdorf. 1998. Inverse problems for identification of memory kernels in thermo—and poro—viscoelasticity. *Mathematical Methods in the Applied Sciences* 21 (16): 1495–1517.
7. Janno, J., and L. von Wolfersdorf. 2001. An inverse problem for identification of a time- and space-dependent memory kernel in viscoelasticity. *Inverse problems* 17 (1): 13–24.

Chapter 4
Stability in Inverse Problems for Determining of Two Unknowns

This chapter discusses the problems of simultaneously identifying two unknowns. In the Sect. 4.1, wave propagation velocity and the memory of the layered medium will be determined. In the Sect. 4.2, the regular part of the instantaneous source physically simulating a point explosion at the boundary of the domain and the memory of the medium are unknowns. For their determination, two observations are used for how the boundary of the domain fluctuates. Main results are stability estimates and uniqueness theorems for the problems under consideration [1, 2].

4.1 The Problem of Determining the Speed of Sound and Memory of Medium

In this paragraph, we simultaneously determine the wave propagation velocity and the memory of the layered medium from the two-dimensional integro-differential wave equation with variable coefficient.

4.1.1 Statement of the Problem and Preliminary Transformations

Consider for

$$(x, y, t) \in \mathbb{R}_+^2 \times \mathbb{R}$$

the initial–boundary value problem for the equation

© The Author(s), under exclusive license to Springer Nature Singapore Pte Ltd. 2023
D. K. Durdiev and Z. D. Totieva, *Kernel Determination Problems in Hyperbolic Integro-Differential Equations*, Infosys Science Foundation Series in Mathematical Sciences, https://doi.org/10.1007/978-981-99-2260-4_4

$$u_{tt} - \left(a^2 u_x\right)_x - \left(a^2 u_y\right)_y - \int_0^t k(\tau) u_{xx}(x, y, t - \tau) \mathrm{d}\tau = 0, \ t > 0 \tag{4.1}$$

with conditions

$$u\big|_{t<0} \equiv 0, \ u_y\big|_{y=0} = \delta(x)\delta'(t) + f(t)\delta(x)\theta(t), \tag{4.2}$$

in which $\mathbb{R}^2_+ = \left\{(x, y) \in \mathbb{R}^2 | y > 0\right\}$. In these equations, the coefficient $a = a(y)$ is a positive function of the class $C^2\left(\mathbb{R}_+\right)$, $\mathbb{R}_+ = \{y \in \mathbb{R} | y > 0\}$, and $k(t), \ f(t)$ are continuous functions for $t \in \mathbb{R}$.

Given the functions $a(y), k(t), f(t)$ the problem of finding function $u(x, y, t)$, satisfying (in a generalized sense) equalities (4.1), (4.2), we call the direct problem. Suppose that the solution to this problem is given on the boundary of the domain: $\mathbb{R}^2_+ \times \mathbb{R}$

$$u|_{y=0} = F(x, t), \ (x, t) \in \mathbb{R}^2_+. \tag{4.3}$$

The inverse problem is to determine two functions $c(t)$, $k(t)$ for a given function $F(x, t)$.

Denote by $\tilde{u}(\lambda, y, t)$ the Fourier transform of the function $u(x, y, t)$ with respect to the variable x:

$$\tilde{u}(\lambda, y, t) = \frac{1}{\sqrt{2\pi}} \int_{\mathbb{R}} u(x, y, t) e^{i\lambda x} \mathrm{d}x.$$

Given the functions $a(y), k(t), f(t)$ the problem (4.1), (4.2) is correctly set and it has a unique solution $u(x, y, t)$ possessing a compact support for any finite t. Eqs. (4.1), (4.2) with respect to the function $\tilde{u}(\lambda, y, t)$ can be written as

$$\tilde{u}_{tt} - \left(a^2 \tilde{u}_y\right)_y + \lambda^2 a^2 \tilde{u} + \lambda^2 \int_0^t k(\tau) \tilde{u}(\lambda, y, t - \tau) \mathrm{d}\tau = 0, (\lambda, y, t) \in \mathbb{R}^2_+ \times \mathbb{R}, \tag{4.4}$$

$$\tilde{u}\big|_{t<0} \equiv 0, \ \tilde{u}_y\big|_{y=0} = \delta'(t) + f(t)\theta(t). \tag{4.5}$$

Instead of y, we introduce the new variable z by the formula

$$z = \int_0^y \frac{\mathrm{d}\tau}{a(\xi)}. \tag{4.6}$$

The function $y = h(z)$ defined by this equality is monotonic and defines a one-to-one correspondence between y and z. Denote

$$a\left(h(z)\right) =: c(z) \tag{4.7}$$

and introduce the new function $\bar{u}(\lambda, z, t)$ by equality

$$\bar{u}(\lambda, z, t) = \tilde{u}(\lambda, h(z), t)\sqrt{\frac{c(z)}{c(0)}}.$$

Then, the function $\bar{u}(\lambda, z, t)$ satisfies the relations

$$\bar{u}_{tt} - \bar{u}_{zz} - q(\lambda, z)\bar{u} + \lambda^2 \int_0^t k(\tau)\bar{u}(\lambda, y, t-\tau)d\tau = 0,\ (t, y) \in \mathbb{R}^2_+, \lambda \in \mathbb{R}, \tag{4.8}$$

$$\bar{u}\big|_{t<0} \equiv 0,\ (\bar{u}_z + H\bar{u})\big|_{z=0} = \delta'(t) + f(t)\theta(t), \tag{4.9}$$

$$\bar{u}\big|_{z=0} = \bar{F}(\lambda, t),\ t \in \mathbb{R}_+, \tag{4.10}$$

in which

$$q(\lambda, z) = q_0(z) - \lambda^2 q_1(z),\ q_0(z) = \frac{\left[c'(z)\right]^2 - 2c(z)c''(z)}{4c^2(z)},\ q_1 = c^2(z), \tag{4.11}$$

$$H = -\frac{c'(0)}{2c(0)},\ \bar{F}(\lambda, y, t) = \frac{1}{\sqrt{2\pi}} \int_R F(x, t)e^{i\lambda x}dx.$$

Under the above assumption on smoothness of the function $a(y)$, it is obviously that $q_0(z) \in C(R)$, $q_1(z) \in C^2(R)$.

It follows from the theory of hyperbolic equations that $\bar{u}(\lambda, z, t) \equiv 0$ for $0 < t < z,\ z \in R_+$. Therefore, the solution of the problem (4.8), (4.9) is conveniently represented in the form

$$\bar{u}(\lambda, z, t) = -\delta(t-z) + \theta(t-z)v(\lambda, z, t). \tag{4.12}$$

Denote $\beta(\lambda, z) = \bar{u}(\lambda, z, z+0)$. Substituting expression (4.12) into (4.8), (4.9) and equating the coefficients for the same singularities, we find that $\beta(\lambda, z)$ satisfies the ordinary differential equation

$$2\frac{\partial \beta}{\partial z}(\lambda, z) + q(\lambda, z) = 0,$$

with condition

$$\beta(\lambda, 0) = -H.$$

Hence

$$\beta(\lambda, z) = -H - \frac{1}{2}\int_0^z q(\lambda, \xi)\mathrm{d}\xi.$$

Then, $v(\lambda, z, t)$ at fixed λ in domain $t > z > 0$ satisfies equations

$$v_{tt} - v_{zz} - q(\lambda, z)v(\lambda, z, t) - \lambda^2 k(t-z) + \lambda^2 \int_0^{t-z} k(\tau)v(\lambda, z, t-\tau)\mathrm{d}\tau = 0, \tag{4.13}$$

$$v\big|_{t=z+0} = -H - \frac{1}{2}\int_0^z q(\lambda, \xi)\mathrm{d}\xi, \; z \in \mathbb{R}_+ \tag{4.14}$$

$$(v_z + Hv)\big|_{z=0} = f(t), \; t \in \mathbb{R}_+. \tag{4.15}$$

Note that for the solvability of the inverse problem, as follows from the representation (4.12), the function $\bar{F}(\lambda, t)$ must have the following structure:

$$\bar{F}(\lambda, t) = -\delta(t) + F_0(\lambda, t)\theta(t), (\lambda, t) \in \mathbb{R}^2_+, \tag{4.16}$$

where the function $F_0(\lambda, t)$ with respect to the argument t satisfies some smoothness conditions, which will be discussed below. In this regard, the additional information (4.10) for the function v looks like

$$v|_{z=0} = F_0(\lambda, t), \; t \in \mathbb{R}_+. \tag{4.17}$$

Now we narrow down the data of the problem, assuming that the function $F_0(\lambda, t)$ (hence the function $\widetilde{F}(\lambda, t)$) is known only for two values of λ_1, λ_2 such that $\lambda_1^2 \neq \lambda_2^2$. Thus, the inverse problem (4.1)–(4.3) has been reduced to the problem of determining the functions $c(z),\; k(t)$ from the relations (4.13)–(4.15) if the solution of the direct problem is known for $\lambda = \lambda_i,\; i = 1, 2$ and is given by the equalities (4.17). It turns out that with these data the functions $c(z),\; k(t)$ are uniquely determined. After finding $c(z)$, the function $h(z)$, defined the correspondence between the variables y and z, as follows from (4.6), is found by the formula

$$h(z) = \int_0^z c(\xi)\mathrm{d}\xi,$$

and $a(y) = c\left(h^{-1}(y)\right)$. Due to the fact that Eq. (4.13) describes a wave process propagating with a velocity equal to unity, $v(\lambda, 0, t)$ for $t \in [0, T]$ and for a fixed λ depends from function $c(z)$ and its derivatives (through function $q(\lambda, z)$) only on the interval $[0, T/2]$, and from $k(t)$ on $[0, T]$. Therefore, it is natural to expect that the inverse is also true: the function $c(z)$ on each interval $[0, T/2]$ and the function $k(t)$ on $[0, T]$ are determined by $F_0(\lambda, t)$ only on interval $[0, T]$. It turns out that such a local character of the dependence of $c(z)$, $k(t)$ on $F_0(\lambda, t)$ really takes place, and it, of course, is reflected in the results that in what follows, we intend to prove.

4.1.2 Properties of the Direct Problem Solution

We study the properties of the direct problem solution (4.13)–(4.15).

Lemma 4.1 *Let $c(z) \in C^2[0, T/2]$, $k(t) \in C[0, T]$, $f(t) \in C[0, T]$ for some fixed $T > 0$. Then, for each fixed value of the parameter λ the solution of problem (4.13)–(4.15) for*

$$(z, t) \in D_T, \ D_T = \big((z, t)\big|0 \le z \le t \le T - z\big)$$

belongs to the functional class $C^1(D_T)$ and for it the following estimate is true:

$$\|v\|_{C^1(D(T)} \le C\left(\|c(z)\|_{C^2[0,T/2]} + \|k(t)\|_{C[0,T]} + \|f(t)\|_{C[0,T]}\right), \tag{4.18}$$

where C depends only on T, λ, $\|c(z)\|_{C^2[0,T/2]}$ and $\|k(t)\|_{C[0,T]}$. Also function

$$\psi(\lambda_1, \lambda_2, t) = v_t(\lambda_1, 0, t) - v_t(\lambda_2, 0, t)$$

for any fixed λ_j, $j = 1, 2$ is a function of the class $C^1[0, T]$.

Proof Using the formula (3.63) for $v_{tt} - v_{zz}$ from relations (4.13)–(4.15) for $(z, t) \in D_T$ by integration along the corresponding characteristics of the first-order differential operators, we obtain the equalities

$$\begin{aligned}(v_t + v_z)(\lambda, z, t) = &-\frac{1}{2}q\left(\lambda, \frac{z+t}{2}\right)\\ &+ \int\limits_z^{(z+t)/2} \big[q(\lambda, \xi)v(\lambda, \xi, t+z-\xi) + \lambda^2 k(t+z-2\xi)\\ &- \lambda^2 \int\limits_0^{t+z-2\xi} k(\tau)v(\lambda, \xi, t+z-\xi-\tau)d\tau\big]d\xi,\end{aligned} \tag{4.19}$$

$$
\begin{aligned}
v(\lambda,0,t) = {} & -He^{Ht} - \frac{1}{2}\int_0^t e^{H(t-\tau)} q\left(\lambda,\frac{\tau}{2}\right)\mathrm{d}\tau - \int_0^t e^{H(t-\tau)} f(\tau)\mathrm{d}\tau \\
& + \int_0^t e^{H(t-\tau)} \int_0^{\tau/2} \big[q(\lambda,\xi)v(\lambda,\xi,\tau-\xi) + \lambda^2 k(\tau-2\xi) \\
& - \lambda^2 \int_0^{\tau-2\xi} k(\alpha)v(\lambda,\xi,\tau-\xi-\alpha)\mathrm{d}\alpha\big]\mathrm{d}\xi\mathrm{d}\tau,
\end{aligned}
\tag{4.20}
$$

$$
\begin{aligned}
& (v_t - v_z)(\lambda,z,t) \\
& = -2f(t-z) + 2Hv(\lambda,0,t-z) - \frac{1}{2}q\left(\lambda,\frac{t-z}{2}\right) \\
& + \lambda^2 k(t-z)z + \int_0^{(t-z)/2} \big[q(\lambda,\xi)v(\lambda,\xi,t-z-\xi) + \lambda^2 k(t-z-2\xi) \\
& - \lambda^2 \int_0^{t-z-2\xi} k(\tau)v(\lambda,\xi,t-z-\xi-\tau)\mathrm{d}\tau\big]\mathrm{d}\xi \\
& - \int_0^z \big[q(\lambda,\xi)v(\lambda,\xi,t-z+\xi) + \lambda^2 \int_0^{t-z} k(\tau)v(\lambda,\xi,t-z+\xi-\tau)\mathrm{d}\tau\big]\mathrm{d}\xi.
\end{aligned}
\tag{4.21}
$$

From (4.19), (4.21), we find the equations for v_t, v_z, v

$$
\begin{aligned}
v_t(\lambda,z,t) = {} & -f(t-z) + Hv(\lambda,0,t-z) - \frac{1}{4}q\left(\lambda,\frac{t+z}{2}\right) \\
& - \frac{1}{4}q\left(\lambda,\frac{t-z}{2}\right) + \frac{\lambda^2}{2}k(t-z)z \\
& + \frac{1}{2}\int_0^{(t-z)/2} \big[q(\lambda,\xi)v(\lambda,\xi,t-z-\xi) + \lambda^2 k(t-z-2\xi) \\
& - \lambda^2 \int_0^{t-z-2\xi} k(\tau)v(\lambda,\xi,t-z-\xi-\tau)\mathrm{d}\tau\big]\mathrm{d}\xi \\
& + \frac{1}{2}\int_0^z \big[q(\lambda,\xi)v(\lambda,\xi,t-z+\xi) - \lambda^2 \int_0^{t-z} k(\tau)v(\lambda,\xi,t-z+\xi-\tau)\mathrm{d}\tau\big]\mathrm{d}\xi
\end{aligned}
$$

$$+\frac{1}{2}\int\limits_{z}^{(z+t)/2}\Big[q(\lambda,\xi)v(\lambda,\xi,t+z-\xi)+\lambda^2 k(t+z-2\xi)$$

$$-\lambda^2\int\limits_0^{t+z-2\xi}k(\tau)v(\lambda,\xi,t+z-\xi-\tau)\mathrm{d}\tau\Big]\mathrm{d}\xi, \tag{4.22}$$

$$v_z(\lambda,z,t)=f(t-z)-Hv(\lambda,0,t-z)$$

$$-\frac{1}{4}q\left(\lambda,\frac{t+z}{2}\right)+\frac{1}{4}q\left(\lambda,\frac{t-z}{2}\right)-\frac{\lambda^2}{2}k(t-z)z$$

$$-\frac{1}{2}\int\limits_0^{(t-z)/2}\Big[q(\lambda,\xi)v(\lambda,\xi,t-z-\xi)+\lambda^2 k(t-z-2\xi)$$

$$-\lambda^2\int\limits_0^{t-z-2\xi}k(\tau)v(\lambda,\xi,t-z-\xi-\tau)\mathrm{d}\tau\Big]\mathrm{d}\xi$$

$$-\frac{1}{2}\int\limits_0^{z}\Big[q(\lambda,\xi)v(\lambda,\xi,t-z+\xi) \tag{4.23}$$

$$-\lambda^2\int\limits_0^{t-z}k(\tau)v(\lambda,\xi,t-z+\xi-\tau)\mathrm{d}\tau\Big]\mathrm{d}\xi$$

$$+\frac{1}{2}\int\limits_{z}^{(z+t)/2}\Big[q(\lambda,\xi)v(\lambda,\xi,t+z-\xi)+\lambda^2 k(t+z-2\xi)$$

$$-\lambda^2\int\limits_0^{t+z-2\xi}k(\tau)v(\lambda,\xi,t+z-\xi-\tau)\mathrm{d}\tau\Big]\mathrm{d}\xi,$$

$$v(\lambda,z,t)=-H-\frac{1}{2}\int\limits_0^{z}q(\lambda,\xi)\mathrm{d}\xi-\int\limits_0^{t-z}f(\tau)\mathrm{d}\tau+H\int\limits_0^{t-z}v(\lambda,0,\tau)\mathrm{d}\tau$$

$$-\frac{1}{4}\int\limits_z^{t}\left(q\left(\lambda,\frac{\tau+z}{2}\right)+q\left(\lambda,\frac{\tau-z}{2}\right)\right)\mathrm{d}\tau+\frac{\lambda^2 z}{2}\int\limits_0^{t-z}k(\tau)\mathrm{d}\tau$$

$$+\frac{1}{2}\int\limits_z^{t}\int\limits_0^{(\tau-z)/2}\Big[q(\lambda,\xi)v(\lambda,\xi,\tau-z-\xi)+\lambda^2 k(\tau-z-2\xi)$$

$$- \lambda^2 \int\limits_0^{\tau-2\xi-z} k(\alpha)v(\lambda,\xi,\tau-\xi-\alpha-z)\mathrm{d}\alpha\big]\mathrm{d}\xi\mathrm{d}\tau$$

$$+ \frac{1}{2}\int\limits_z^t \int\limits_0^z \big[q(\lambda,\xi)v(\lambda,\xi,\tau-z+\xi)$$

$$- \lambda^2 \int\limits_0^{\tau-z} k(\alpha)v(\lambda,\xi,\tau+\xi-\alpha-z)\mathrm{d}\alpha\big]\mathrm{d}\xi\mathrm{d}\tau$$

$$+ \frac{1}{2}\int\limits_z^t \int\limits_0^{(\tau+z)/2} \big[q(\lambda,\xi)v(\lambda,\xi,\tau+z-\xi) + \lambda^2 k(\tau+z-2\xi)$$

$$- \lambda^2 \int\limits_0^{\tau-2\xi+z} k(\alpha)v(\lambda,\xi,\tau-\xi-\alpha+z)\mathrm{d}\alpha\big]\mathrm{d}\xi\mathrm{d}\tau. \qquad (4.24)$$

Eq. (4.24) is the integral equation of Volterra type in the domain D_T and it has the unique continuous solution. It follows from equalities (4.22), (4.23) that this solution is continuously differentiable in D_T. Substituting the expression for $v(\lambda, 0, t)$, defined by formula (4.20) into Eqs. (4.22)–(4.24) and constructing the method of successive approximations for these equations by the usual scheme, which has factorial convergence by argument t, it is easy to establish the validity of the estimate (4.18) in the domain D_T. Using equality (4.22), we compose the function $\psi(\lambda_1, \lambda_2, t)$

$$\psi(\lambda_1,\lambda_2,t)$$

$$= H[v(\lambda_1,0,t) - v(\lambda_2,0,t)] + \frac{1}{2}\left(\lambda_1^2-\lambda_2^2\right) q_1\left(\frac{t}{2}\right)$$

$$+ \int\limits_0^{t/2} [q(\lambda_1,\xi)v(\lambda_1,\xi,t-\xi) - q(\lambda_2,\xi)v(\lambda_2,\xi,t-\xi) + (\lambda_1^2-\lambda_2^2)k(t-2\xi)$$

$$- \int\limits_0^{t-2\xi} k(\tau)[\lambda_1^2 v(\lambda_1,\xi,t-\xi-\tau) - \lambda_2^2 v(\lambda_2,\xi,t-\xi-\tau)]\mathrm{d}\tau]\mathrm{d}\xi. \qquad (4.25)$$

The right-hand side of this equality belongs to the class $C^1[0,T]$. Therefore $\psi(\lambda_1, \lambda_2, t) \in C^1[0,T]$, for any fixed λ_j, $j = 1, 2$.

As consequence of the Lemma 4.1, we note that under the conditions of lemma , the function $F_0(\lambda, t)$, entering in equality (4.13), for each fixed λ belongs to the class $C^1[0,T]$, and the function $F_{00}(t) \equiv F_0(\lambda_1, t) - F_0(\lambda_2, t)$ belongs to the class $C^2[0,T]$.

4.1.3 Investigation of the Inverse Problem

Lemma 4.2 *Suppose that the function $F(\lambda, t)$ has the structure (4.16) and $F_0(\lambda, t) \in C^1[0, T]$ for any fixed λ, and*

$$F_0(\lambda_1, t) - F_0(\lambda_2, t) \equiv F_{00}(t) \in C^2[0, T].$$

In addition, let the function $F_{0t}(\lambda, 0)$ be increase in λ, $\lambda \in \mathbb{R}$, $f(t) \in C[0, T]$. Then the inverse problem (4.15)–(4.15), (4.17) in the domain D_T is equivalent to the problem of finding the functions v, v_t, $c(z)$, $c'(z)$, $k(t)$ from the following closed system of the following integral equations:

$$\begin{aligned}
v(\lambda_i, z, t) &= -H - \frac{1}{2}\int_0^z q(\lambda_i, \xi)d\xi - \int_0^{t-z} f(\tau)d\tau + H\int_0^{t-z} v(\lambda_i, 0, \tau)d\tau \\
&- \frac{1}{4}\int_z^t \left(q\left(\lambda_i, \frac{\tau+z}{2}\right) + q\left(\lambda_i, \frac{\tau-z}{2}\right)\right)d\tau + \frac{\lambda_i^2 z}{2}\int_0^{t-z} k(\tau)d\tau \\
&+ \frac{1}{2}\int_z^t \int_0^{(\tau-z)/2} \big[q(\lambda_i, \xi)v(\lambda_i, \xi, \tau - z - \xi) + \lambda_i^2 k(\tau - z - 2\xi) \\
&- \lambda_i^2 \int_0^{\tau-2\xi-z} k(\alpha)v(\lambda_i, \xi, \tau - \xi - \alpha - z)d\alpha\big]d\xi d\tau \\
&+ \frac{1}{2}\int_\tau^t \int_0^z \big[q(\lambda_i, \xi)v(\lambda_i, \xi, \tau - z + \xi) \\
&- \lambda_i^2 \int_0^{\tau-z} k(\alpha)v(\lambda_i, \xi, \tau + \xi - \alpha - z)d\alpha\big]d\xi d\tau \\
&+ \frac{1}{2}\int_z^t \int_0^{(\tau+z)/2} \big[q(\lambda_i, \xi)v(\lambda, \xi, \tau + z - \xi) + \lambda_i^2 k(\tau + z - 2\xi) \\
&- \lambda_i^2 \int_0^{\tau-2\xi+z} k(\alpha)v(\lambda_i, \xi, \tau - \xi - \alpha + z)d\alpha\big]d\xi d\tau, \quad i = 1, 2,
\end{aligned} \tag{4.26}$$

$$
\begin{aligned}
&v_t(\lambda_i, z, t)\\
&= -f(t-z) + Hv(\lambda_i, 0, t-z) - \frac{1}{4}q\left(\lambda_i, \frac{t+z}{2}\right) - \frac{1}{4}q\left(\lambda_i, \frac{t-z}{2}\right)\\
&+ \frac{\lambda_i^2}{2}k(t-z)z + \frac{1}{2}\int_0^{(t-z)/2}\big[q(\lambda_i, \xi)v(\lambda_i, \xi, t-z-\xi) + \lambda_i^2 k(t-z-2\xi)\\
&- \lambda_i^2 \int_0^{t-z-2\xi} k(\tau)v(\lambda_i, \xi, t-z-\xi-\tau)d\tau\big]d\xi\\
&+ \frac{1}{2}\int_0^{z}\big[q(\lambda_i, \xi)v(\lambda_i, \xi, t-z+\xi)\\
&- \lambda_i^2 \int_0^{t-z} k(\tau)v(\lambda_i, \xi, t-z+\xi-\tau)d\tau\big]d\xi\\
&+ \frac{1}{2}\int_z^{(z+t)/2}\big[q(\lambda_i, \xi)v(\lambda_i, \xi, t+z-\xi) + \lambda_i^2 k(t+z-2\xi) -\\
&- \lambda_i^2 \int_0^{t+z-2\xi} k(\tau)v(\lambda_i, \xi, t+z-\xi-\tau)d\tau\big]d\xi,\ i = 1, 2,
\end{aligned}
\tag{4.27}
$$

$$
c(z) = c(0) + \int_0^z c'(\xi)d\xi,\ z \in \left[0, \frac{T}{2}\right], \tag{4.28}
$$

$$
c'(z) = c'(0) + \frac{1}{4}\int_0^z \left\{\frac{[c'(\xi)]^2}{c(\xi)} - 2q_0(\xi)c(\xi)\right\}d\xi,\ z \in \left[0, \frac{T}{2}\right], \tag{4.29}
$$

$$
\begin{aligned}
q_0(z) &= -2F_{0t}(\lambda_1, 2z) - 2f(2z) + 2HF_0(\lambda_1, 2z) + \lambda_1^2 q_1(z)\\
&+ 2\int_0^z\big[q(\lambda_1, \xi)v(\lambda_1, \xi, 2z-\xi) + \lambda_1^2 k(2z-2\xi)\\
&- \lambda_1^2 \int_0^{2(z-\xi)} k(\tau)v(\lambda_1, \xi, 2z-\xi-\tau)d\tau\big]d\xi,\ z \in \left[0, \frac{T}{2}\right],
\end{aligned}
\tag{4.30}
$$

$$k(t) = \frac{2}{\lambda_1^2 - \lambda_2^2} F_{00}''(t) - \frac{2H}{\lambda_1^2 - \lambda_2^2} F_{00}'(t)$$
$$- \frac{1}{2} q_1'\left(\frac{t}{2}\right) - \frac{1}{\lambda_1^2 - \lambda_2^2}\left(q\left(\lambda_1, \frac{t}{2}\right) v\left(\lambda_1, \frac{t}{2}, \frac{t}{2}\right) - q\left(\lambda_2, \frac{t}{2}\right) v\left(\lambda_2, \frac{t}{2}, \frac{t}{2}\right)\right)$$
$$- \frac{2}{\lambda_1^2 - \lambda_2^2} \int_0^{t/2} \{q(\lambda_1, \xi) v_t(\lambda_1, \xi, t - \xi) - q(\lambda_2, \xi) v_t(\lambda_2, \xi, t - \xi)$$
$$- k(t - 2\xi)\big(\lambda_1 v(\lambda_1, \xi, \xi) - \lambda_2 v(\lambda_2, \xi, \xi)\big)$$
$$- \int_0^{t-2\xi} k(\tau)\big[\lambda_1^2 v_t(\lambda_1, \xi, t - \xi - \tau) - \lambda_2^2 v_t(\lambda_2, \xi, t - \xi - \tau)\big] d\tau \big\} d\xi, \; t \in [0, T], \tag{4.31}$$

where

$$c(0) = \sqrt{\frac{2F_{00}'(0)}{\lambda_1^2 - \lambda_2^2}}, \; c'(0) = 2F_0(\lambda, 0) c(0). \tag{4.32}$$

Proof At the beginning we establish the validity of equality (4.32). Indeed, setting in Eq. (4.20) $t = 0$ and using conditions (4.17), we find

$$F_0(\lambda_i, 0) = -H, \; i = 1, 2. \tag{4.33}$$

In particular, from this we conclude that $F_0|_{t=0}$ does not depend on λ. Further, from Eq. (4.25) for $t = 0$, we obtain

$$F_{00}'(0) = \frac{1}{2}\left(\lambda_1^2 - \lambda_2^2\right) c^2(0).$$

Hence, by virtue of the positivity of the number $F_{00}'(0) = F_{0t}(\lambda_1, 0) - F_{0t}(\lambda_2, 0)$, we arrive at the first equality (4.32). Taking into account the first equality (4.11), from (4.33) we obtain the second equality (4.32). Since $F_0|_{t=0}$ does not depend on λ, then the second formula from (4.32) uniquely determines $c'(0)$.

Continuing the proof of the Lemma 4.2, we note that Eqs. (4.26) and (4.27) coincide with the equalities (4.22) and (4.24), respectively, for $\lambda = \lambda_i$, $i = 1, 2$. Equation (4.22) is derived from the relations (4.13)–(4.15). Equation (4.24) is obtained from the Eq. (4.22) by integration over t from the point (z, z) to the point (z, t) on the plane of variables (ξ, τ). In turn, equation (4.24) satisfies equalities (4.13)–(4.16). Further, in Eq. (4.22) we set $z = 0$ and use condition (4.17) for $\lambda = \lambda_1$. Then, after simple transformations, we obtain equality (4.30). In this equation, for definiteness, we put $\lambda = \lambda_1$. In fact, the result of the calculation should not depend from the choice of the parameter λ. To obtain equality (4.31), we use relation (4.25), which is derived from Eq. (4.22) using the conditions (4.17). Noting that $\psi(\lambda_1, \lambda_2, t) \equiv F_{00}'(t)$, we differentiate equality (4.25) with respect to the variable t. Then we have

$$F_{00}''(t) = HF_{00}' + \frac{1}{4}(\lambda_1^2 - \lambda_2^2)q_1'\left(\frac{t}{2}\right) + \frac{1}{2}\left[q\left(\lambda_1, \frac{t}{2}\right)v\left(\lambda_1, \frac{t}{2}, \frac{t}{2}\right)\right.$$

$$\left. -q\left(\lambda_2, \frac{t}{2}\right)v\left(\lambda_2, \frac{t}{2}, \frac{t}{2}\right)\right] + \frac{\lambda_1^2 - \lambda_2^2}{2}k(t) + \int\limits_0^{t/2} \{q(\lambda_1, \xi)v_t(\lambda_1, \xi, t-\xi)$$

$$- q(\lambda_2, \xi)v_t(\lambda_2, \xi, t-\xi) - k(t-2\xi)\big(\lambda_1^2 v(\lambda_1, \xi, \xi) - \lambda_2^2 v(\lambda_2, \xi, \xi)\big)$$

$$- \int\limits_0^{t-2\xi} k(\tau)\left[\lambda_1^2 v_t(\lambda_1, \xi, t-\xi-\tau) - \lambda_2^2 v_t(\lambda_2, \xi, t-\xi-\tau)\right] d\tau\big\} d\xi.$$

Solving this equality with respect to $k(t)$, we arrive at the Eq. (4.31).

We verify that inverse transformations take place as well. We show this by the example of the Eq. (4.31). In this equation, replacing t by $t-\tau$, we multiply both sides by $d\tau$ and integrate over τ from 0 to $t/2$. In the repeated integrals of the resulting equality, we change the order of integration. Using condition (4.17) in the opposite direction, after simple calculations, we arrive at equation (4.25), from which follow equalities (4.22) for $\lambda = \lambda_i, i = 1, 2$; $z = 0$ and differentiated with respect to t equality (4.17). Proof of equivalence (4.30) and (4.30) for $\lambda = \lambda_1$ is carried out similarly. The equivalence of equalities (4.28), (4.29) with the last equalities (4.11) is obvious.

The system of integral Eqs. (4.26)–(4.31) is closed in the domain D_T and defines the unique continuous functions $v, v_t, c(z), c'(z), k(t)$ for sufficiently small T. Without dwelling on the theorem of local unique solvability of the problem, we pass to the results characterizing stability estimates and unique solvability for an arbitrary $T > 0$.

Denote by $A(c_0, k_0)$ the set of pairs of functions $\{c(t), k(t)\}$, satisfying for some $T > 0$ the following conditions:

$$0 < c_{00} \le c(z), \ \|c(t)\|_{C^2[0,T/2]} \le c_0, \ \|k(t)\|_{C[0,T]} \le k_0.$$

Besides,

$$\|f(t)\|_{C[0,T]} \le f_0,$$

where $f_0 > 0$ is a known number

Theorem 4.1 *Let* $(c^{(1)}, k^{(1)}) \in A(c_0, k_0)$, $(c^{(2)}, k^{(2)}) \in A(c_0, k_0)$ *be solutions of inverse problem (4.13)–(4.15), (4.17) with data*

$$\left(F_0^{(1)}(\lambda_j, t), f^{(1)}(t)\right), \ \left(F_0^{(2)}(\lambda_j, t), f^{(2)}(t)\right),$$

$j = 1, 2$, respectively. Then, there is a positive constant C, depending on $\lambda_1, \lambda_2, T, c_0, c_{00}, k_0, f_0$, such that the estimate

$$\|k^{(1)} - k^{(2)}\|_{C[0,T]} + \|c^{(1)} - c^{(2)}\|_{C^2[0,T/2]} + \|h^{(1)} - h^{(2)}\|_{C^3[0,T/2]} \leq C\mu \quad (4.34)$$

is fulfilled, where

$$\mu = \sum_{j=1}^{2} \|F_0^{(1)}(\lambda_j, t) - F_0^{(2)}(\lambda_j, t)\|_{C^1[0,T]} + \|F_{00}^{(1)} - F_{00}^{(2)}\|_{C^2[0,T]}$$
$$+ \|f^{(1)} - f^{(2)}\|_{C[0,T]}.$$

Theorem 4.1 obviously implies the following uniqueness theorem for any $T > 0$.

Theorem 4.2 *The functions $c^{(i)} \in C^2[0, T/2]$, $k^{(i)} \in C[0, T]$ and $F_0^{(i)}(\lambda_j, t)$, $f^{(i)}(t)$, $i = 1, 2$, $j = 1, 2$ have the same meaning as in Theorem 4.1. If at the same time*

$$F_0^{(1)}(\lambda_j, t) = F_0^{(2)}(\lambda_j, t),\ j = 1, 2,\ f^{(1)}(t) = f^{(2)}(t)$$

for $t \in [0, T]$, then,

$$c^{(1)}(z) = c^{(2)}(z),\ z \in [0, T/2],\ k^{(1)}(t) = k^{(2)}(t),\ t \in [0, T].$$

Proof Suppose that the functions $f^{(i)}, F^{(i)}(\lambda_j, t), c^{(i)}, k^{(i)}, i = 1, 2, j = 1, 2$ are as defined in Theorem 4.1. The solution of problem (4.13)–(4.15) for $f = f^{(i)}, c = c^{(i)}$, $k = k^{(i)}, \lambda = \lambda_j$ denote by $v^{(ij)}(z, t), i = 1, 2, j = 1, 2$. Introduce also notations

$$\tilde{f}(t) = f^{(1)} - f^{(2)},\ \tilde{c}(z) = c^{(1)} - c^{(2)},\ \tilde{k}(t) = k^{(1)} - k^{(2)},$$

$$\tilde{F}_0(\lambda_j, t) = F_0^{(1)}(\lambda_j, t) - F_0^{(2)}(\lambda_j, t),\ \tilde{h}(z) = h^{(1)} - h^{(2)},\ \tilde{F}_{00} = F_{00}^{(1)} - F_{00}^{(2)},$$

$$\tilde{v}^{(j)}(z, t) = v^{(1j)}(z, t) - v^{(2j)}(z, t),\ j = 1, 2.$$

Let $q_0^{(i)}(z), q_1^{(i)}(z), q^{(ij)}(z) = q_0^{(i)}(z) - \lambda_j^2 q_1^{(i)}(z), \tilde{q}^{(j)}(z) = q^{(1j)} - q^{(2j)}$ be auxiliary functions and $H^{(i)} = -(c^{(i)})'(0)/(2c^{(i)}(0))$ be number corresponding to the functions $c^{(i)}(z)$. Let us write out the corresponding integral relations for the newly introduced functions. From equalities (4.26) and (4.27) it follows that

$$
\begin{aligned}
\tilde{v}^{(j)}(z,t) = -\widetilde{H}& \\
&- \int\limits_0^{t-z} \tilde{f}(\tau)\mathrm{d}\tau + \widetilde{H}\int\limits_0^{t-z} F^{(1)}(\lambda_j,\tau)\mathrm{d}\tau + H^{(2)}\int\limits_0^{t-z} \widetilde{F}(\lambda_j,\tau)\mathrm{d}\tau \\
&- \frac{1}{2}\int\limits_0^{z} \tilde{q}^{(j)}(\xi)\mathrm{d}\xi - \frac{1}{4}\int\limits_z^{t}\left(\tilde{q}^{(j)}\left(\frac{\tau+z}{2}\right) + \tilde{q}^{(j)}\left(\frac{\tau-z}{2}\right)\right)\mathrm{d}\tau \\
&+ \frac{\lambda_j^2 z}{2}\int\limits_0^{t-z}\tilde{k}(\tau)\mathrm{d}\tau + \frac{1}{2}\int\limits_z^{t}\int\limits_0^{(\tau-z)/2}\big[\tilde{q}^{(j)}(\xi)v^{(1j)}(\xi,\tau-z-\xi) \\
&+ q^{(2j)}(\xi)\tilde{v}^{(j)}(\xi,\tau-z-\xi) + \lambda_j^2\tilde{k}(\tau-2\xi-z) \\
&- \lambda_j^2\int\limits_0^{\tau-2\xi-z}\big(\tilde{k}(\alpha)v^{(1j)}(\xi,\tau-\xi-\alpha-z) \\
&+ k^{(2)}(\alpha)\tilde{v}^{(j)}(\xi,\tau-\xi-\alpha-z)\big)\mathrm{d}\alpha\big]\mathrm{d}\xi\mathrm{d}\tau \\
&+ \frac{1}{2}\int\limits_z^{t}\int\limits_0^{z}\big[\tilde{q}^{(j)}(\xi)v^{(1j)}(\xi,\tau-z+\xi) + q^{(2j)}(\xi)\tilde{v}^{(j)}(\xi,\tau-z+\xi) \\
&- \lambda_j^2\int\limits_0^{\tau-z}\big(\tilde{k}(\alpha)v^{(1j)}(\xi,\tau+\xi-\alpha-z) \\
&+ k^{(2)}(\alpha)\tilde{v}^{(j)}(\xi,\tau+\xi-\alpha-z)\big)\mathrm{d}\alpha\big]\mathrm{d}\xi\mathrm{d}\tau \\
&+ \frac{1}{2}\int\limits_z^{t}\int\limits_0^{(\tau+z)/2}\big[\tilde{q}^{(j)}(\xi)v^{(1j)}(\xi,\tau+z-\xi) \\
&+ q^{(2j)}(\xi)\tilde{v}^{(j)}(\xi,\tau+z-\xi) + \lambda_j^2\tilde{k}(\tau+z-2\xi) \\
&- \lambda_j^2\int\limits_0^{\tau-2\xi+z}\big(\tilde{k}(\alpha)v^{(1j)}(\xi,\tau-\xi-\alpha+z) \\
&+ k^{(2)}(\alpha)\tilde{v}^{(j)}(\xi,\tau-\xi-\alpha+z)\big)\mathrm{d}\alpha\big]\mathrm{d}\xi\mathrm{d}\tau,
\end{aligned}
\tag{4.35}
$$

$$
\begin{aligned}
&\tilde{v}_t^j(z,t)\\
&= -\tilde{f}(t-z) + \widetilde{H}F^{(1)}(\lambda_j, t-z) + H^{(2)}\widetilde{F}(\lambda_j, t-z)\\
&\quad -\frac{1}{4}\left(\tilde{q}^{(j)}\left(\frac{t+z}{2}\right) + \tilde{q}^{(j)}\left(\frac{t-z}{2}\right)\right) + \frac{\lambda_j^2}{2}\tilde{k}(t-z)z\\
&\quad +\frac{1}{2}\int_0^{(t-z)/2} [\tilde{q}^{(j)}(\xi)v^{(1j)}(\xi, t-z-\xi)\\
&\quad + q^{(2j)}(\xi)\tilde{v}^{(j)}(\xi, t-z-\xi) + \lambda_j^2\tilde{k}(t-z-2\xi)\\
&\quad -\lambda_j^2 \int_0^{t-z-2\xi} \big(\tilde{k}(\alpha)v^{(1j)}(\xi, t-z-\xi-\alpha)\\
&\quad + k^{(2)}(\alpha)\tilde{v}^{(j)}(\xi, t-z-\xi-\alpha)\big)\mathrm{d}\alpha\big]\mathrm{d}\xi\\
&\quad +\frac{1}{2}\int_0^{z}\big[\tilde{q}^{(j)}(\xi)v^{(1j)}(\xi, t-z+\xi) + q^{(2j)}(\xi)\tilde{v}^{(j)}(\xi, t-z+\xi)\\
&\quad -\lambda_j^2\int_0^{t-z}\big(\tilde{k}(\alpha)v^{(1j)}(\xi, t+\xi-\alpha-z) + k^{(2)}(\alpha)\tilde{v}^{(j)}(\xi, t-z+\xi-\alpha)\big)\mathrm{d}\alpha\big]\mathrm{d}\xi\\
&\quad +\frac{1}{2}\int_z^{(t+z)/2} [\tilde{q}^{(j)}(\xi)v^{(1j)}(\xi, t+z-\xi)\\
&\quad + q^{(2j)}(\xi)\tilde{v}^{(j)}(\xi, t+z-\xi) + \lambda_j^2\tilde{k}(t+z-2\xi)\\
&\quad -\lambda_j^2\int_0^{t+z-2\xi} (\tilde{k}(\alpha)v^{(1j)}(\xi, t+z-\xi-\alpha) + k^{(2)}(\alpha)\tilde{v}^{(j)}(\xi, t+z-\xi-\alpha))\mathrm{d}\alpha]\mathrm{d}\xi,\\
&\quad j = 1, 2.
\end{aligned}
\tag{4.36}
$$

Note that in these equalities

$$
\widetilde{H} = \frac{\tilde{c}(0)\left(c^{(2)}\right)'(0) - \tilde{c}'(0)\left(c^{(2)}\right)(0)}{2c^{(1)}c^{(2)}}. \tag{4.37}
$$

Using equalities (4.28)–(4.31) we find

$$
\tilde{c}(z) = \tilde{c}(0) + \int_0^z \tilde{c}'(\xi)\mathrm{d}\xi,\ z \in \left[0, \frac{T}{2}\right], \tag{4.38}
$$

$$\begin{aligned}\tilde{c}'(z) = \tilde{c}'(0) + \frac{1}{4}\int\limits_0^z \Big\{\tilde{c}'(\xi)\frac{\big[(c^{(1)})'(\xi) + (c^{(2)})'(\xi)\big]}{c^{(1)}(\xi)} - \tilde{c}(\xi)\frac{\big[(c^{(2)})'(\xi)\big]^2}{c^{(1)}(\xi)c^{(2)}(\xi)} \\ - 2\tilde{q}_0(\xi)c^{(1)}(\xi) - 2q_0^{(2)}\tilde{c}(\xi)\Big\}d\xi,\ z \in \left[0, \frac{T}{2}\right],\end{aligned} \tag{4.39}$$

$$\begin{aligned}\tilde{q}_0(z) &= -2\widetilde{F}_{0t}(\lambda_1, 2z) - 2\tilde{f}(2z) \\ &+ 2\widetilde{H}F_0^{(1)}(\lambda_1, 2z) + 2H^{(2)}\widetilde{F}_0(\lambda_1, 2z) + \lambda_1^2\tilde{q}_1(z) \\ &+ 2\int\limits_0^z \Big[\tilde{q}^{(1)}(\xi)v^{(11)}(\xi, 2z-\xi) + q^{(12)}(\xi)\tilde{v}^{(1)}(\xi, 2z-\xi) + \lambda_1^2\tilde{k}(2z-2\xi) \\ &- \lambda_1^2 \int\limits_0^{2(z-\xi)} \big(\tilde{k}(\tau)v^{(11)}(\xi, 2z-\xi-\tau) + k^{(2)}(\tau)\tilde{v}^{(1)}(\xi, 2z-\xi-\tau)\big)d\tau\Big]d\xi, \\ &z \in \left[0, \frac{T}{2}\right],\end{aligned} \tag{4.40}$$

$$\begin{aligned}\tilde{k}(t) &= \frac{2}{\lambda_1^2 - \lambda_2^2}\widetilde{F}'_{00}(t) - \frac{2\widetilde{H}}{\lambda_1^2 - \lambda_2^2}F_{00}^{(1)}(t) - \frac{2H^{(2)}}{\lambda_1^2 - \lambda_2^2}\widetilde{F}_{00}(t) - \frac{1}{2}\tilde{q}'_1\left(\frac{t}{2}\right) \\ &- \frac{1}{\lambda_1^2 - \lambda_2^2}\Big[\tilde{q}^{(1)}\big(\tfrac{t}{2}\big)\big(H^{(1)} - \frac{1}{2}\int\limits_0^{t/2} q^{(11)}(\xi)d\xi\big) \\ &+ q^{(12)}\left(\frac{t}{2}\right)\big(\widetilde{H} - \frac{1}{2}\int\limits_0^{t/2}\tilde{q}^{(1)}(\xi)d\xi\big) \\ &- \tilde{q}^{(2)}\left(\frac{t}{2}\right)\big(H^{(1)} - \frac{1}{2}\int\limits_0^{t/2} q^{(21)}(\xi)d\xi\big) + q^{(22)}\left(\frac{t}{2}\right)\big(\widetilde{H} - \frac{1}{2}\int\limits_0^{t/2}\tilde{q}^{(2)}(\xi)d\xi\big)\Big] \\ &- \frac{2}{\lambda_1^2 - \lambda_2^2}\int\limits_0^{t/2}\Big\{\tilde{q}^{(1)}(\xi)v_t^{(11)}(\xi, t-\xi) \\ &+ q^{(12)}(\xi)\tilde{v}_t^{(1)}(\xi, t-\xi) - \tilde{q}^{(2)}(\xi)v_t^{(12)}(\xi, t-\xi) \\ &+ q^{(22)}(\xi)\tilde{v}_t^{(2)}(\xi, t-\xi) - \tilde{k}(t-2\xi)\big(\lambda_1 v^{(11)}(\xi,\xi) - \lambda_2 v^{(12)}(\xi,\xi)\big) \\ &- k^{(2)}(t-2\xi)\big(\lambda_1\tilde{v}^{(1)}(\xi,\xi) - \lambda_2\tilde{v}^{(2)}(\xi,\xi)\big) \\ &- \int\limits_0^{t-2\xi}\Big[\tilde{k}(\tau)\big(\lambda_1^2 v_t^{(11)}(\xi, t-\xi-\tau) - \lambda_2^2 v_t^{(12)}(\xi, t-\xi-\tau)\big) \\ &+ k^{(2)}(\tau)\big(\lambda_1^2\tilde{v}_t^{(1)}(\xi, t-\xi-\tau) - \lambda_2^2\tilde{v}_t^{(2)}(\xi, t-\xi-\tau)\big)\Big]d\tau\Big\}d\xi,\ t \in [0, T].\end{aligned} \tag{4.41}$$

We estimate the functions of the system (4.35)–(4.41) in domain D_T through value μ, defined in Theorem 4.1. The domain D_T allows equivalent description

$$D_T = \left\{(z,t)\middle|0 \le z \le t \le \frac{T}{2} - \left|\frac{T}{2} - t\right|, 0 \le t \le T\right\}.$$

Let

$$\rho(t) = \max\left\{\max_{0\le z\le T/2-|T/2-t|} \left|\tilde{v}^{(j)}(z,t)\right|, \max_{0\le z\le T/2-|T/2-t|} \left|\tilde{v}_t^{(j)}(z,t)\right|, \left|\tilde{k}(t)\right|,\right.$$

$$\left.\max_{0\le z\le T/2-|T/2-t|} \left|\tilde{c}(z)\right|, \max_{0\le z\le T/2-|T/2-t|} \left|\tilde{c}'(z)\right|, \max_{0\le z\le T/2-|T/2-t|} \left|\tilde{q_0}^{(j)}(z)\right|\right\},$$

$$t \in [0,T],\ j = 1,2.$$

According to Lemma 4.1, the functions v^{ij} are differentiable in the domain D_T and satisfy the estimate

$$\left\|v^{ij}\right\|_{C^1[D_T]} \le M_1,\ i,j = 1,2, \tag{4.42}$$

with some constant M_1, depending only on $\lambda_1,\ \lambda_2,\ T,\ c_0,\ c_{00},\ k_0,\ f_0$. Since the functions $F_{00}^{(i)}(t)$ are traces of the functions $v^{(i1)}(z,t) - v^{(i2)}(z,t)$ for $z = 0,\ i = 1,2$, for each of which, as follows from (4.42), (4.25), an estimate similar to (4.42) holds, then, for $F_{00}^{(i)}(t) = F_0^{(i)}(\lambda_1,t) - F_0^{(i)}(\lambda_2,t)$ we have the inequality

$$\|F_{00}^{(i)}(t)\|_{C^2[0,T]} \le M_2,\ i = 1,2,$$

in which the constant M_2 depends on the same parameters as M_1. From (4.42), (4.17) it follows that the functions $F_0^{(i)}(\lambda_j,t)$ must be bounded by the constant M_1:

$$\|F_0^{(i)}(\lambda_j,t)\|_{C^1[0,T]} \le M_1,\ i,j = 1,2$$

The numbers $c^{(i)}(0),\ (c')^{(i)}(0),\ i{=}1,2$, defined by formulae (4.32) via $\lambda_1,\ \lambda_2,\ F_{00}'$, for similar reasons should be bounded by the constant M_3, which depends from the same parameters as $M_j,\ j = 1,2$. Therefore, for the numbers $H^{(i)},\ \widetilde{H}$, defined by the numbers $c^{(i)}(0),\ (c')^{(i)}(0),\ i = 1,2$, and equality (4.37), the estimates

$$\left|H^{(i)}\right| \le M_4,\ \left|\widetilde{H}\right| \le M_5\mu$$

are valid with some constants.

M_4, M_5, depend on the set of numbers $\lambda_1, \lambda_2, T, c_0, c_{00}, k_0, f_0$. Recall that μ is defined in Theorem 4.1. Next, the functions

$$q_0^{(i)}(z),\ q_1^{(i)}(z),\ q^{(ij)}(z),$$

defined by $c^{(i)}(z)$ and λ_j, $i, j = 1, 2$, for similar reasons satisfy the inequalities

$$\left\|q_0^{(i)}(z)\right\|_{C^2[0,T/2]} \le M_6,\ \left\|q_1^{(i)}(z)\right\|_{C^2[0,T/2]} \le M_7,$$

$$\left\|q^{(ij)}(z)\right\|_{C^2[0,T/2]} \le M_8,\ i, j = 1, 2,$$

where M_6, M_7, M_8 depend on the same parameters as the previous constants.

Given the above, we proceed to estimate in the domain D_T functions $\tilde{v}^{(j)}$, satisfying the integral Eq. (4.35). Note that these equations, like all others, contain terms containing only known and terms with unknown functions. On the right-hand side of Eq. (4.35), the first four terms depend only on known functions and numbers. Therefore, these terms in totality are estimated by the value $K_1\mu$, with constant K_1, depending on M_i, $i = \overline{1, 8}$. As it is easily seen the remaining part is estimated in the domain D_T by an integral of the form

$$L_1 \int_0^t \rho(\tau) d\tau,$$

in which L_1 depends only on M_i, $i = \overline{1, 8}$, which in turn depend on the parameters $\lambda_1, \lambda_2, T, c_0, c_{00}, k_0, f_0$. Thus,

$$|\tilde{v}^{(j)}(z, t)| \le K_1\mu + L_1 \int_0^t \rho(\tau) d\tau,\ (z, t) \in D_T,\ j = 1, 2. \tag{4.43}$$

From the Eqs. (4.38), (4.39) we can see that the functions $\tilde{c}(z)$, $\tilde{c}'(z)$ are estimated in a similar way

$$|\tilde{c}(z)| \le K_2\mu + L_2 \int_0^t \rho(\tau) d\tau, \tag{4.44}$$

$$|\tilde{c}'(z)| \le K_3\mu + L_3 \int_0^t \rho(\tau) d\tau,\ z \in \left[0, \frac{T}{2}\right]. \tag{4.45}$$

The constants K_2, K_3, L_2, L_3 depend on the same parameters as K_1, L_1.
We use the inequalities (4.44), (4.45) to estimate the functions

$$\tilde{q}_1(z) = \tilde{c}(z)(c^{(1)}(z) + c^{(2)}(z)),\ \widetilde{q_1^{(')}}(z) = 2(\tilde{c}(z)(c')^{(1)}(z) + c^{(2)}(z)\widetilde{c^{(')}}(z)),$$

occurring in nonintegral terms in the right-hand sides of Eqs. (4.40), (4.41). Similar reasoning leads to inequalities

$$|\tilde{q}_0(z)| \le K_4\mu + L_4 \int_0^t \rho(\tau)d\tau, \tag{4.46}$$

$$|\tilde{k}(t)| \le K_5\mu + L_5 \int_0^t \rho(\tau)d\tau. \tag{4.47}$$

We use inequalities (4.44)–(4.47) in the estimates of the functions $\tilde{q}^j$, $\tilde{k}$ outside the integral terms of Eq. (4.36) in order to obtain an estimate for the functions $\tilde{v}_t^{(j)}(z, t)$ in the form

$$|\tilde{v}_t^{(j)}(z, t)| \le K_6\mu + L_6 \int_0^t \rho(\tau)d\tau, \ j = 1, 2. \tag{4.48}$$

In the last inequalities, the constants $K_4,\ K_5,\ L_4,\ L_5$ depend by means of the numbers $M_i,\ i = \overline{1, 8}$ on the parameters $\lambda_1,\ \lambda_2,\ T,\ c_0,\ c_{00},\ k_0,\ f_0$. From the relations (4.43)–(4.48) it follows that $\rho(t)$ satisfies integral inequality

$$\rho(t) \le K\mu + L \int_0^t \rho(\tau)d\tau,$$

with the new constants $K,\ L$, depending only on $\lambda_1,\ \lambda_2,\ T,\ c_0, c_{00},\ k_0,\ f_0$. Hence, using the Gronwall inequality, we obtain the estimate (4.34).

4.2 The Problem of Determining the Memory of Medium and Regular Part of Impulsive Source

In this paragraph we suppose the function $a(y)$ as known and consider the initial–boundary value problem (4.1)–(4.3) in which $k(t),\ f(t)$ are unknown functions.

4.2.1 Statement of the Problem and Preliminary Constructions

For simplicity, we set $a(y) = 1,\ y \in \mathbb{R}_+$ and rewrite Eqs. (4.49)–(4.51) in the terms of the Fourier transform

$$\tilde{u}_{tt} - \tilde{u}_{yy} + \lambda^2 \tilde{u} + \lambda^2 \int_0^t k(\tau)\tilde{u}(\lambda, y, t-\tau)\mathrm{d}\tau = 0, \ (\lambda, y, t) \in \mathbb{R}^2_+ \times \mathbb{R}, \quad (4.49)$$

$$\tilde{u}\big|_{t<0} \equiv 0, \ \tilde{u}_y\big|_{y=0} = \delta'(t) + f(t)\theta(t), \quad (4.50)$$

$$\tilde{u}|_{y=0} = \widetilde{F}(\lambda, t), (\lambda, t) \in \mathbb{R}^2_+. \quad (4.51)$$

Here $\tilde{u}(\lambda, y, t)$ also has representation (4.60) instead z with variable y. It can be easily noted, in this case $q(\lambda, y) = -\lambda^2$. Hence, $\beta(\lambda, y) = (1/2)\lambda^2 y$. Then, for a fixed λ in the domain $t > y > 0$ the regular part of solution to the problem (4.49)–(4.51) satisfies the equations

$$v_{tt} - v_{yy} + \lambda^2 v(\lambda, y, t) - \lambda^2 k(t-y) + \lambda^2 \int_0^{t-y} k(\tau) v(\lambda, y, t-\tau)\mathrm{d}\tau = 0, \quad (4.52)$$

$$v\big|_{t=y+0} = \frac{\lambda^2}{2} y, \quad (4.53)$$

$$v_y\big|_{y=0} = f(t). \quad (4.54)$$

Let $F_0(\lambda, t)$ be a regular part of function $\widetilde{F}(\lambda, t)$, having representation (4.16). Narrowing down the problem data we assume that the function $F_0(\lambda, t)$ (therefore, $\widetilde{F}(\lambda, t)$) is known only for two values $\lambda = \lambda_i, \ i = 1, 2$. Thus, the inverse problem (4.1)–(4.3) reduces to the problem of determining the functions $k(t), \ f(t)$ from relations (4.52)–(4.54), provided that the solution of the direct problem is known for $\lambda = \lambda_i, \ i = 1, 2$, and satisfies the following condition:

$$v\big|_{y=0} = F_0(\lambda_i, t), \ t > 0, \ i = 1, 2. \quad (4.55)$$

It turns out that these data uniquely determine the functions $k(t), \ f(t)$.

4.2.2 *Properties of the Direct Problem Solution. Statement and Proof of the Main Results*

Let us study the properties of the solution of the direct problem (4.52)–(4.54). As well in this case it is true the analog of Lemma 4.1.

Lemma 4.3 *Suppose that $k(t) \in C[0,T]$ and $f(t) \in C[0,T]$ for some $T > 0$. Then, for each fixed value of the parameter λ and for $(y,t) \in D_T$, $D_T = \big((y,t)|0 \le y \le t \le T - y\big)$ the solution of problem (4.52)–(4.54) belongs to the functional class $C^1(D_T)$ and the following estimate for the solution holds:*

$$\|v\|_{C^1(D(T))} \le C\left(1 + \|k(t)\|_{C[0,T]} + \|f\|_{C[0,T]}\right), \tag{4.56}$$

where C depends only on T, λ and $\|k(t)\|_{C[0,T]}$. In addition, for any fixed λ_j, $j = 1, 2$, the function

$$\psi(\lambda_1, \lambda_2, t) = v_t(\lambda_1, 0, t) - v_t(\lambda_2, 0, t)$$

is a function of class $C^1[0,T]$.

Proof Proceeding in a similar way, as in the previous paragraph, for $(y,t) \in D_T$ from relations (4.52)–(4.54) one gets

$$\begin{aligned}(v_t + v_y)(\lambda, y, t) = \frac{\lambda^2}{2} + \lambda^2 \int\limits_y^{(y+t)/2} \big[k(t + y - 2\xi) - v(\lambda, \xi, t + y - \xi) \\ - \int\limits_0^{t+y-2\xi} k(\tau) v(\lambda, \xi, t + y - \xi - \tau) d\tau\big] d\xi,\end{aligned} \tag{4.57}$$

$$\begin{aligned}v(\lambda, 0, t) = \frac{\lambda^2}{2} t - \int\limits_0^t f(\tau) d\tau + \lambda^2 \int\limits_0^t \int\limits_0^{\tau/2} \big[k(\tau - 2\xi) - v(\lambda, \xi, \tau - \xi) \\ - \int\limits_0^{\tau-2\xi} k(\alpha) v(\lambda, \xi, \tau - \xi - \alpha) d\alpha\big] d\xi d\tau.\end{aligned} \tag{4.58}$$

$$\begin{aligned}(v_t - v_y)(\lambda, y, t) = \frac{\lambda^2}{2} + \lambda^2 k(t - y) y - 2f(t - y) + \lambda^2 \int\limits_0^{(t-y)/2} \big[k(t - y - 2\xi) \\ - v(\lambda, \xi, t - y - \xi) - \int\limits_0^{t-y-2\xi} k(\tau) v(\lambda, \xi, t - y - \xi - \tau) d\tau\big] d\xi \\ - \lambda^2 \int\limits_0^y \big[v(\lambda, \xi, t - y + \xi) + \int\limits_0^{t-y} k(\tau) v(\lambda, \xi, t - y + \xi - \tau) d\tau\big] d\xi.\end{aligned} \tag{4.59}$$

From (4.57), (4.59), we obtain

$$
\begin{aligned}
v_t(\lambda, y, t) = {} & \frac{\lambda^2}{2} + \frac{\lambda^2}{2} k(t-y)y - f(t-y) + \frac{\lambda^2}{2} \int\limits_0^{(t-y)/2} \Big[k(t-y-2\xi) \\
& - v(\lambda, \xi, t-y-\xi) - \int\limits_0^{t-y-2\xi} k(\tau) v(\lambda, \xi, t-y-\xi-\tau)\mathrm{d}\tau\Big]\mathrm{d}\xi \\
& + \frac{\lambda^2}{2} \int\limits_0^{y} \Big[v(\lambda, \xi, t-y+\xi) + \int\limits_0^{t-y} k(\tau) v(\lambda, \xi, t-y+\xi-\tau)\mathrm{d}\tau\Big]\mathrm{d}\xi \\
& + \frac{\lambda^2}{2} \int\limits_y^{(y+t)/2} \Big[k(t+y-2\xi) - v(\lambda, \xi, t+y-\xi) \\
& - \int\limits_0^{t+y-2\xi} k(\tau) v(\lambda, \xi, t+y-\xi-\tau)\mathrm{d}\tau\Big]\mathrm{d}\xi,
\end{aligned} \tag{4.60}
$$

$$
\begin{aligned}
v_y(\lambda, y, t) = {} & -\frac{\lambda^2}{2} k(t-y)y + f(t-y) - \frac{\lambda^2}{2} \int\limits_0^{(t-y)/2} \Big[k(t-y-2\xi) \\
& - v(\lambda, \xi, t-y-\xi) - \int\limits_0^{t-y-2\xi} k(\tau) v(\lambda, \xi, t-y-\xi-\tau)\mathrm{d}\tau\Big]\mathrm{d}\xi \\
& - \frac{\lambda^2}{2} \int\limits_0^{y} \Big[v(\lambda, \xi, t-y+\xi) + \int\limits_0^{t-y} k(\tau) v(\lambda, \xi, t-y+\xi-\tau)\mathrm{d}\tau\Big]\mathrm{d}\xi \\
& + \frac{\lambda^2}{2} \int\limits_y^{(y+t)/2} \Big[k(t+y-2\xi) - v(\lambda, \xi, t+y-\xi) \\
& - \int\limits_0^{t+y-2\xi} k(\tau) v(\lambda, \xi, t+y-\xi-\tau)\mathrm{d}\tau\Big]\mathrm{d}\xi,
\end{aligned} \tag{4.61}
$$

$$v(\lambda, y, t) = \frac{\lambda^2}{2} y \int_y^t k(\tau - y)\mathrm{d}\tau + \frac{\lambda^2}{2} \int_y^t \int_y^{(\tau+y)/2} k(\tau + y - 2\xi)\mathrm{d}\xi \mathrm{d}\tau$$
$$- \frac{1}{2} \int_y^t f(\tau - y)\mathrm{d}\tau + v(\lambda, 0, t - y) + \frac{\lambda^2}{2} \int_y^t \int_0^y \Big[v(\lambda, \xi, \tau - y + \xi)$$
$$+ \int_0^{\tau - y} k(\alpha) v(\lambda, \xi, \tau - y + \xi - \alpha)\mathrm{d}\alpha\Big]\mathrm{d}\xi \mathrm{d}\tau$$
$$- \frac{\lambda^2}{2} \int_y^t \int_y^{(\tau+y)/2} \Big[+v(\lambda, \xi, \tau + y - \xi)$$
$$+ \int_0^{\tau+y-2\xi} k(\alpha) v(\lambda, \xi, \tau + y - \xi - \alpha)\mathrm{d}\alpha\Big]\mathrm{d}\xi \mathrm{d}\tau. \tag{4.62}$$

In the domain D_T, the equation (4.62) is an integral equation of Volterra type and determines a unique continuous solution. It follows from relations (4.60), (4.61) that this solution is continuously differentiable in D_T. Substituting the expression for $v(\lambda, 0, t)$, given by formula (4.58), into equation (4.62) and constructing, for Eqs. (4.60)–(4.62), the usual scheme of the successive approximation method, which has factorial convergence with respect to the argument t, it is easy to prove that the estimate (4.56) holds in the domain D_T.

Using relation (4.60), we construct the function $\psi(\lambda_1, \lambda_2, t)$ as follows

$$\psi(\lambda_1, \lambda_2, t) = \frac{\lambda_1^2 - \lambda_2^2}{2}$$
$$+ \int_0^{t/2} [(\lambda_1^2 - \lambda_2^2)k(t - 2\xi) - \lambda_1^2 v(\lambda_1, \xi, t - \xi) + \lambda_2^2 v(\lambda_2, \xi, t - \xi)$$
$$- \int_0^{t-2\xi} k(\tau)[\lambda_1^2 v(\lambda_1, \xi, t - \xi - \tau) - \lambda_2^2 v(\lambda_2, \xi, t - \xi - \tau)]\mathrm{d}\tau]\mathrm{d}\xi. \tag{4.63}$$

The right-hand side of (4.63) belongs to the class $C^1[0, T]$. Therefore, $\psi(\lambda_1, \lambda_2, t) \in C^1[0, T]$ for any λ_j, $j = 1, 2$.

As consequence of the Lemma, it is easily noted that under the assumptions of Lemma 4.3 the function $F_0(\lambda_i, t)$ appearing in relation (4.55), belongs, for each fixed λ, to the class $C^1[0, T]$, and the function $F_{00}(t) \equiv F_0(\lambda_1, t) - F_0(\lambda_2, t)$ to the class $C^2[0, T]$. In addition, formulas (4.60), (4.62) imply that the following relations hold:

$$F_0(\lambda_i, 0) = 0,\ F_{0t}(\lambda_i, 0) = \frac{\lambda_i^2}{2} - f(0),\ i = 1, 2.$$

In what follows, we assume that the function $F_0(\lambda, t)$ satisfies these necessary conditions.

The following lemma expresses the equivalence of the inverse problem to some system of nonlinear integral equations.

Lemma 4.4 *Suppose that the assumptions of Lemma 4.3, the necessary conditions on the function F_0 and the relation $\lambda_1^2 \neq \lambda_2^2$ hold. Then, the inverse problem (4.52)–(4.55) in the domain D_T is equivalent to the problem of finding the functions $v,\ v_t,\ k,\ f$ from the following system of integral equations:*

$$\begin{aligned} v(\lambda_j, y, t) = F_0(\lambda_j, t-y) + \frac{\lambda_j^2}{2} y + \frac{\lambda_j^2}{2} \int\limits_{D(y,t)} \Big[k(\tau - \xi) \\ - v(\lambda_j, \xi, \tau) - \int\limits_0^{\tau-\xi} k(\alpha) v(\lambda_j, \xi, \tau - \alpha) d\alpha\Big] d\tau d\xi,\ j = 1, 2, \end{aligned} \tag{4.64}$$

$$\begin{aligned} v_t(\lambda_j, y, t) = F_{0t}(\lambda_j, t-y) + \frac{\lambda_j^2}{2} \int\limits_0^{(t+y)/2} \Big\{k(t - |y-\xi| - \xi) \\ - v(\lambda_j, \xi, t - |y-\xi|) - \int\limits_0^{t-|y-\xi|-\xi} k(\alpha) v(\lambda_j, \xi, t - |y-\xi| - \alpha) d\alpha \\ - \frac{1}{2}\Big[k\left(\frac{t-y}{2} + \left|\xi - \frac{t-y}{2}\right| - \xi\right) - v\left(\lambda_j, \xi, \frac{t-y}{2} + \left|\xi - \frac{t-y}{2}\right|\right) \\ - \int\limits_0^{\frac{t-y}{2} + \left|\xi - \frac{t-y}{2}\right| - \xi} k(\alpha) v\left(\lambda_j, \xi, \frac{t-y}{2} + \left|\xi - \frac{t-y}{2}\right| - \alpha\right) d\alpha\Big] \\ \times \left(1 - \operatorname{sign}\left(\xi - \frac{t-y}{2}\right)\right)\Big\} d\xi, \end{aligned} \tag{4.65}$$

for j=1,2 and

$$k(t) = \frac{2}{\lambda_1^2 - \lambda_2^2} F_{00tt}(t) + \frac{\lambda_1^2 + \lambda_2^2}{2} t + \frac{2}{\lambda_1^2 - \lambda_2^2} \int_0^t \Big\{ \frac{\lambda_1^4 - \lambda_2^4}{4} k(\xi)(t - \xi)$$

$$+ \lambda_1^2 v_t \left(\lambda_1, \frac{t-\xi}{2}, \frac{t+\xi}{2}\right) - \lambda_2^2 v_t \left(\lambda_2, \frac{t-\xi}{2}, \frac{t+\xi}{2}\right) \tag{4.66}$$

$$+ \int_0^\xi k(\tau) \Big[\lambda_1^2 v_t \left(\lambda_1, \frac{t-\xi}{2}, \frac{t+\xi}{2} - \tau\right)$$

$$- \lambda_2^2 v_t \left(\lambda_2, \frac{t-\xi}{2}, \frac{t+\xi}{2} - \tau\right)\Big] d\tau \Big\} d\xi.$$

$$f(t) = -F_{0t}(\lambda, t) + \frac{\lambda^2}{2} + \frac{\lambda^2}{2} \int_0^t \Big[k(\xi) - v\left(\lambda, \frac{t-\xi}{2}, \frac{t+\xi}{2}\right)$$

$$- \int_0^\xi k(\tau) - v\left(\lambda, \frac{t-\xi}{2}, \frac{t+\xi}{2} - \tau\right) d\tau\Big] d\xi, \tag{4.67}$$

where

$$D(y,t) = \left\{ (\xi, \tau) \Big| \frac{t-y}{2} + \left|\xi - \frac{t-y}{2}\right| < \tau < t - |\xi - y|,\ 0 < \xi < \frac{t+y}{2} \right\}.$$

The proof of this lemma differs from the proof of the similar lemma from the previous section. Therefore, here we will give it.

Proof First we note that the system of the integral Eqs. (4.64)–(4.66) is closed in the domain D_T and determines the unique continuous functions $v,\ v_t,\ k$. After determination of these functions, the unknown function $f(t)$ can be calculated by the formula (4.67). In formula (4.67), $\lambda = \lambda_1$ or $\lambda = \lambda_2$ and the result of calculation must not depend on the choice of the parameter λ.

Now we consider the auxiliary problem of determining the function $\varphi(y,t)$, satisfying relations:

$$\varphi_{tt} - \varphi_{yy} = \gamma(y,t),\ 0 < y < t < T - y,$$

$$\varphi(0,t) = \varphi_0(t),\ 0 \le y \le T,\ \varphi(y,y) = \varphi_1(y),\ 0 \le y \le T,$$

in which $\gamma,\ \varphi_0,\ \varphi_1$ are continuous functions and $\varphi_0(0) = \varphi_1(0)$. It is easily verified that the generalized solution of this problem is given by the formula

$$\varphi(y,t)=\varphi_0(t-y)+\varphi_1\left(\frac{t+y}{2}\right)-\varphi_1\left(\frac{t-y}{2}\right)+\frac{1}{2}\int\limits_{D(y,t)}\gamma(\xi,\tau)\mathrm{d}\tau\mathrm{d}\xi, \quad (4.68)$$

which can be verified directly. By formula (4.68) the solution of problem (4.52), (4.53),(4.55) satisfies the integral Eq. (4.64) in the domain D_T. To obtain the Eq. (4.65) we write the integral over the domain $D(y,t)$ as the repeated integral

$$\int\limits_{D(y,t)}\gamma(\xi,\tau)\mathrm{d}\tau\mathrm{d}\xi=\int\limits_0^{(t+y)/2}\int\limits_{(t-y)/2+|\xi-(t-y)/2|}^{t-|y-\xi|}\gamma(\xi,\tau)\mathrm{d}\tau\mathrm{d}\xi$$

and, differentiating relation (4.64) with respect to the variable t, we obtain Eq. (4.65). Differentiating relation (4.64) with respect to the variable y, we find

$$\begin{aligned}v_y(\lambda_j,y,t)=&-F_{0t}\left(\lambda_j,t-y\right)+\frac{1}{2}\lambda_j^2+\frac{1}{2}\lambda_j^2\int\limits_0^{(t+y)/2}\{[k(t-|y-\xi|-\xi)\\&-v\left(\lambda_j,\xi,t-|y-\xi|\right)\\&-\int\limits_0^{t-|y-\xi|-\xi}k(\alpha)v\left(\lambda_j,\xi,t-|y-\xi|-\alpha\right)\mathrm{d}\alpha]\mathrm{sign}(\xi-y)\\&+\frac{1}{2}[k\left(\frac{t-y}{2}+\left|\xi-\frac{t-y}{2}\right|-\xi\right)-v\left(\lambda_j,\xi,\frac{t-y}{2}+\left|\xi-\frac{t-y}{2}\right|\right)\\&-\int\limits_0^{\frac{t-y}{2}+\left|\xi-\frac{t-y}{2}\right|-\xi}k(\alpha)v\left(\lambda_j,\xi,\frac{t-y}{2}+\left|\xi-\frac{t-y}{2}\right|-\alpha\right)\mathrm{d}\alpha]\\&\times\left(1-\mathrm{sign}\left(\xi-\frac{t-y}{2}\right)\right)\}\mathrm{d}\xi,\end{aligned}$$

(4.69)

for $j=1,2$. In this equation, setting $y=0$ and using relation (4.54) we obtain equality (4.67). Considering relation (4.54) for $\lambda=\lambda_j,\ j=1,2$, and forming the difference, we get

$$v_y(\lambda_1,0,t)-v_y(\lambda_2,0,t)=0. \quad (4.70)$$

Using relation (4.70) for $y=0$, from Eq. (4.69) we obtain

$$F_{0t}(\lambda_1,t)-F_{0t}(\lambda_2,t)$$
$$=\frac{\lambda_1^2-\lambda_2^2}{2}+\int\limits_0^{t/2}\{(\lambda_1^2-\lambda_2^2)k(t-2\xi)-\lambda_1^2v(\lambda_1,\xi,t-\xi)+\lambda_2^2v(\lambda_2,\xi,t-\xi)$$
$$-\int\limits_0^{t-2\xi}k(\tau)\left[\lambda_1^2v\left(\lambda_1,\xi,t-\xi-\tau\right)-\lambda_2^2v\left(\lambda_2,\xi,t-\xi-\tau\right)\right]d\tau d\tau\}d\xi. \tag{4.71}$$

Differentiating this equality with respect to the variable t, and carrying out simple calculations, we obtain Eq. (4.66). Note that the unknown function $f(t)$ does not appear in relations (4.64)–(4.66).

It is readily verified that the inverse transformations also hold. Indeed, integrating Eq. (4.65) over t between the limits from y to t and changing, where it is necessary, the order of integration, we obtain Eq. (4.64) from which, in turn, we derive relations (4.52), (4.53), (4.55) for $\lambda=\lambda_j$. Using the formula $\xi'=(t-\xi)/2$, we perform the change of variable ξ to ξ' in the integrals on the right-hand side of (4.66). Further, on both sides of this equality, we replace t by $t-2t$, multiply both sides by dt, and integrate over t between the limits from 0 to $t/2$. Then changing the order of integration in the repeated integrals of the resulting equality, after simple calculations, we obtain Eq. (4.71). The equivalence of relations (4.67) and (4.54) is proved in a similar way. The Lemma 4.4 is proved.

Omitting the theorem on the local unique solvability of our problem, now we pass to formulate the results concerning stability estimates and unique solvability for an arbitrary $T>0$.

Denote by $B(k_0)$ the set of pairs of functions $(k(t),\ f(t))$ satisfying the following condition for some $T>0$:

$$\max\left\{\|k\|_{C[0,T]},\ \|f\|_{C[0,T]}\right\}\le k_0,\ k_0>0.$$

Theorem 4.3 *Suppose that $\left(k^{(1)},\ f^{(1)}\right)\in B(k_0)$, $\left(k^{(2)},\ f^{(2)}\right)\in B(k_0)$ are solutions of the inverse problem (4.52)–(4.55) with given $F_0^{(1)}(\lambda_j,t)$, $F_0^{(2)}(\lambda_j,t)$, $j=1,2$, respectively, $F_{00}^{(i)}\equiv F_0^{(i)}(\lambda_1,t)-F_0^{(i)}(\lambda_2,t)$ and $\lambda_1^2\neq\lambda_2^2$. Then, there exists a positive constant C depending on $\lambda_1,\lambda_2,T,k_0$, such that the following estimate holds:*

$$\|k^{(1)}-k^{(2)}\|_{C[0,T]}+\|f^{(1)}-f^{(2)}\|_{C[0,T]}$$
$$\le C\left(\sum_{j=1}^{2}\|F_0^{(1)}(\lambda_j,t)-F_0^{(2)}(\lambda_j,t)\|_{C^1[0,T]}+\|\widetilde{F}_{00}^{(1)}-\widetilde{F}_{00}^{(2)}\|_{C^2[0,T]}\right). \tag{4.72}$$

Theorem 4.3 readily implies the following uniqueness theorem for any $T>0$.

Theorem 4.4 *Suppose that the functions* $\left(k^{(i)},\ f^{(i)}\right) \in C[0,T]$ *and* $F_0^{(i)}\left(\lambda_j, t\right)$, $i = 1, 2,\ \ j = 1, 2,$ *have the same meaning as in Theorem 4.3. If* $F_0^{(1)}(\lambda_j, t) = F_0^{(2)}(\lambda_j, t),\ j = 1, 2$ *for* $t \in [0, T]$, *then* $k^{(1)} = k^{(2)},\ f^{(1)} = f^{(2)}$ *for* $t \in [0, T]$.

The proof of Theorem 4.3 can be carried out quite similarly to the proof of the corresponding theorem from the previous section. Therefore, we will omit it.

4.3 Conclusions

This chapter investigated two inverse problems of simultaneously determining two unknown functions. In both problems, it is shown that both unknown functions of one variable are uniquely determined by specifying the Fourier transform of the direct problem on the boundary of the half-space with respect to the variable x for two different transformation parameters. The conditional stability estimates and uniqueness theorems of the solution to the problems were obtained.

References

1. Yakhno, V.G. 1990. *Inverse Problems for Differential Equations of Elasticity*. Novosibirsk: Nauka (Russian).
2. Karchevsky, A.L., and V.G. Yakhno. 1999. One-dimensional inverse problems for systems of elasticity with a source of explosive type. *Journal of Investigation Ill - Posed Problems* 7 (4): 347–364.

Chapter 5
Two-Dimensional Special Kernel Determination Problems

This chapter is devoted to the inverse problems of recovering a memory kernel in hyperbolic integro-differential equations, having a finite Fourier series with respect to one of the spatial variables. In the first section it is supposed that the unknown kernel is a trigonometric polynomial with respect to the spatial variables with continuous coefficients with respect to the time variable. In the second section it is determined the two-dimensional spatial part of the three-dimensional kernel and wherein it is supposed that the unknown function is a trigonometric polynomial with respect to the spatial variable y with continuous coefficients with respect to the variable x. The direct problem for a hyperbolic integro-differential equation in both cases is the initial–boundary value problem for the half-space $x > 0$ with the zero initial Cauchy data and a special Neumann's data at $x = 0$. Local solvability theorems and stability estimates for the solution to the inverse problem are obtained [1] and [2].

5.1 A Problem of Identification of a Two-Dimensional Kernel

In this section it is assumed that the time convolution kernel depends on one of the spatial variables, and it has a form of trigonometric polynomial with respect to this variable with continuous coefficients in t-variable.

5.1.1 Setting of the Problem

We consider the integro-differential equation for $(x, y, t) \in \mathbb{R}^3_+$

© The Author(s), under exclusive license to Springer Nature Singapore Pte Ltd. 2023

D. K. Durdiev and Z. D. Totieva, *Kernel Determination Problems in Hyperbolic Integro-Differential Equations*, Infosys Science Foundation Series in Mathematical Sciences, https://doi.org/10.1007/978-981-99-2260-4_5

$$\frac{\partial^2 u}{\partial t^2} - \Delta u = \int\limits_0^t k(y, \tau)u(x, y, t - \tau)\mathrm{d}\tau \tag{5.1}$$

with the initial and boundary conditions

$$u\Big|_{t<0} \equiv 0, \quad \frac{\partial u}{\partial x}\Big|_{x=0} = \delta'(t). \tag{5.2}$$

Here Δ is the two-dimensional Laplacian, i.e., $\Delta = \frac{\partial^2}{\partial x^2} + \frac{\partial^2}{\partial y^2}$, $\mathbb{R}^3_+ = \{(x, y, t) \in \mathbb{R}^3 \mid x > 0\}$.

Boundary condition in (5.2) is generalized function and, that is why the solution to the problem (5.1) and (5.2) is understood as a linear functional over the space of infinitely differentiable functions with compact support (test functions), i.e., $u(x, y, t) \in D'\left(\mathbb{R}^3_+\right)$. Therefore, equalities in equations (1.1), (1.2) are understood as the equality of values of the left and right sides for test functions. We note that the kernel in equation (5.1) does not depend on the spatial variable which is perpendicular to the axis where the boundary condition in (5.2) is given. Moreover, we assume that this kernel can be represented in the form of a finite Fourier series

$$k(y, t) = \sum_{s=-N}^{N} k_s(t)\mathrm{e}^{isy} \tag{5.3}$$

with a fixed integer $N \geq 0$. Denote by $\Omega\,(N, T, K)$ the set of functions $k(y, t)$ for which the coefficients k_s, $|s| \leq N$, are continuous functions on the interval $[0,\ T]$ and satisfy the conditions

$$|k_s(t)| \leq K,\ t \in [0,\ T],\ -N \leq s \leq N. \tag{5.4}$$

For $k(y, t) \in \Omega\,(N, T, K)$ the solution of problem (5.1) and (5.2) is a $2\pi-$periodic function of y and can be represented by the Fourier series

$$u(x, y, t) = \sum_{s=-\infty}^{\infty} u_s(x, t)\mathrm{e}^{isy}, \tag{5.5}$$

whereas it follows from relations (5.1)–(5.3) the coefficients satisfy the following equations

$$\left(\frac{\partial^2}{\partial t^2} - \frac{\partial^2}{\partial x^2} + m^2\right) u_m(x,t) = \int_0^t \sum_{s=-N}^{N} k_s(\tau) u_{m-s}(x, t-\tau) d\tau,$$
$$(x,t) \in \mathbb{R}^2_+ = \left\{(x,t) \in \mathbb{R}^2 \mid x > 0\right\}$$
$$u_m \mid_{t<0} \equiv 0, \ \frac{\partial u_m}{\partial x} \mid_{x=0} = \delta_{0m}\delta'(t), \ m = 0, \pm 1, \pm 2, \ldots. \tag{5.6}$$

In the latter equation δ_{0m} is the Kronecker symbol.

For given functions $k_s(t)$, $s = 0, \pm 1, \pm 2 \pm, \ldots, \pm N$ we call **the direct problem** the problem of finding the functions $u_m(x,t)$, $m = 0, \pm 1, \pm 2, \ldots$ that satisfy (in a generalized meaning) the relations (1.6).

Now let us state the inverse problem that will be studied in the present paper.

The inverse problem. Find the coefficients $k_s(t)$, $s = 0, \pm 1, \pm 2 \pm, \ldots, \pm N$ by using the conditions

$$u_m(0,t) = f_m(t), \ t \in \mathbb{R}_+, \ \mathbb{R}_+ = \{t \in \mathbb{R} \mid t > 0\}, \ m = 0, \pm 1, \pm 2, \ldots, \pm N. \tag{5.7}$$

Definition Functions $k_s(t) \in C\,[0, \infty)$, $s = 0, \pm 1, \pm 2, \ldots, \pm N$ are called a solution to inverse problem (5.6) and (5.7), if the the corresponding solution to direct problem (5.7) $u_m(x,t) \in D'\left(\mathbb{R}^2_+\right)$ (from the class of generalized functions, i.e., distributions) satisfies equalities (1.7) for $f_m(t) \in D'(\mathbb{R}_+)$, $m = 0, \pm 1, \pm 2, \ldots, \pm N$.

Remark It turns out that the solution to direct problem (1.6) at $m = 0$ consists of a singular and a regular generalized function. With that at $m = \pm 1, \pm 2, \ldots$ the solution to this problem are only regular generalized functions. The regular parts of the solutions are smooth in some domain (see Lemma 2.1).

5.1.2 Investigation of the Direct Problem

We begin studying the inverse problem with consideration of some properties of solution to direct problems (5.6). Concerning the solution to direct problems (5.6) the following lemma is valid.

Lemma 5.1 *Let T be an arbitrary positive number, $k_s(y,t) \in \Omega(N, T, K)$ and $D(T) = ((x,t) |\ 0 \le x \le T - t)$. Then the solution to problem (5.6) exists and can be represented in $D(T)$ in the form*

$$u_m(x,t) = -\delta_{0m}\delta(t-x) + v_m(x,t)\theta(t-x), m = 0, \pm 1, \pm 2, \ldots, \tag{5.8}$$

here $\theta(t)$ is the Heaviside step-function: $\theta(t) = 1$ for $t \ge 0$; $\theta(t) = 0$ for $t < 0$ and $v_m(x,t)$ are continuously differentiable functions in the domain $D'(T) = ((x,t) |\ 0 \le x \le t \le T - x)$. Moreover, this solution is unique and there exist *positive constants* $C_1 = C_1(N, T, K) \geq 1$ *and* $C_2 = C_2(N, T, K)$, *continuously depending on* K, T *such that*

$$|v_m(x,t)| \leq C_1 \frac{KT^2}{2}, \quad -N \leq m \leq N,$$

$$|v_m(x,t)| \leq C_1 \frac{\left(KT^2\right)^{n+1} (2N+1)^n \, t^n}{2^{n+1} \cdot n!}, \tag{5.9}$$

if

$$(x,t) \in D'(T), \; Nn < |m| \leq (n+1)N, \; n = 1, 2, \ldots,$$

and

$$\left| \frac{\partial v_m(x,t)}{\partial t} \right| \leq \frac{KT(1+m^2T^2)}{2} [1 + (2N+1)C_2],$$

$$-N \leq m \leq N, \left| \frac{\partial v_m(x,t)}{\partial t} \right| \leq (1+m^2T^2)$$

$$\frac{(KT)^{n+1}(2N+1)^n t^n}{2^{n+1} \cdot n!} \max\{C_1, (2N+1)C_2\}, \tag{5.10}$$

if

$$(x,t) \in D'(T), \; Nn < |m| \leq (n+1)N, \; n = 1, 2, \ldots.$$

Proof It follows from the hyperbolic equation theory that the solution to the problem (5.1) and (5.2) vanishes for all (x, y, t) satisfying the condition $x > t > 0$ because the initial data are zero and the boundary source is located on the axis $x = 0, t = 0$. Hence all $u_m(x,t) = 0$ for $x > t > 0$. To separate the singular part of the solution to direct problem, we represent it in the form

$$u_m(x,t) = \alpha_m \delta(t-x) + v_m(x,t)\theta(t-x), m = 0, \pm 1, \pm 2, \ldots,$$

where α_m are unknown constants, $v_m(x,t)$ are unknown regular functions. Substituting this equality into equations (5.6) and equating the coefficients for the same singularities, we find $\alpha_m = -\delta_{0m}$.

In view of representation (5.8), we rewrite equations (5.6) with respect to $v_m(x,t)$. For this, we use the property of the Dirac function, from which it follows equality

$$\sum_{s=-N}^{N} \delta_{0(m-s)} \int_0^t k_s(\tau)\delta(t-x-\tau)d\tau = \theta(N-|m|)k_m(t-x), \; t > x > 0.$$

Taking this fact into consideration, we conclude that equations (5.6) with respect to $v_m(x,t)$ can be rewritten in the form

$$\left(\frac{\partial^2}{\partial t^2} - \frac{\partial^2}{\partial x^2} + m^2\right) v_m(x,t)$$
$$= -\theta(N - |m|)k_m(t-x) + \sum_{s=-N}^{N} \int_0^{t-x} k_s(\tau) v_{m-s}(x, t-\tau)\mathrm{d}\tau,$$
$$t > x > 0,\ m = 0, \pm 1, \pm 2, \ldots,$$
$$v_m \mid_{t \le x} \equiv 0,\ \frac{\partial v_m}{\partial x} \mid_{x=0} = 0,\ m = 0, \pm 1, \pm 2, \ldots. \tag{5.11}$$

For the sake of convenience, we continue all functions $v_m(x,t)$, for $x < 0$ as even functions: $v_m(-x,t) = v_m(x,t)$. Then problem (5.11) is equivalent to the following integral equation

$$v_m(x,t) = -\frac{\theta(N-|m|)}{2} \iint_{\diamond(x,t)} J_0\left(m\sqrt{(t-\tau)^2 - (x-\xi)^2}\right) k_m(\tau - |\xi|)\mathrm{d}\tau \mathrm{d}\xi$$
$$+ \frac{1}{2} \iint_{\diamond(x,t)} J_0\left(m\sqrt{(t-\tau)^2 - (x-\xi)^2}\right) \int_0^{\tau - |\xi|} \sum_{s=-N}^{N} k_s(\alpha) v_{m-s}(\xi, \tau - \alpha)\mathrm{d}\alpha \mathrm{d}\tau \mathrm{d}\xi,$$
$$(x,t) \in D'(T),\ m = 0, \pm 1, \pm 2, \ldots, \tag{5.12}$$

where $J_0(\zeta)$ is the Bessel function and

$$\diamond(x,t) = \left\{(\xi, \tau) \mid |\xi| \le \tau \le t - |x - \xi|,\ \frac{x-t}{2} \le \xi \le \frac{x+t}{2}\right\}.$$

Recall that the Bessel function $J_\nu(\zeta)$ for a fixed integer $\nu \ge 0$ is defined by the formula

$$J_\nu(\zeta) = \sum_{j=0}^{\infty} \frac{(-1)^j}{j!(j+\nu)!} \left(\frac{\zeta}{2}\right)^{2j+\nu}.$$

From this formula follows the estimate

$$\left|\frac{J_\nu(\zeta)}{\zeta^\nu}\right| \le \frac{1}{2^\nu \nu!},\ |\zeta| \le 2. \tag{5.13}$$

Hereafter we will also use another estimate for J_ν [3]:

$$|J_\nu(\zeta)| \le 1$$

for all $\zeta \in \mathbb{R}$. Consider for equations (5.12) the method of successive approximations. Define

$$v_m(x,t) = \sum_{j=0}^{\infty} v_m^j(x,t), \tag{5.14}$$

where

$$v_m^0(x,t) = -\frac{\theta(N-|m|)}{2} \iint\limits_{\diamond(x,t)} J_0\left(m\sqrt{(t-\tau)^2-(x-\xi)^2}\right) k_m(\tau-|\xi|)\mathrm{d}\tau\mathrm{d}\xi,$$

$$v_m^j(x,t)$$
$$= \frac{1}{2} \iint\limits_{\diamond(x,t)} J_0\left(m\sqrt{(t-\tau)^2-(x-\xi)^2}\right) \int\limits_0^{\tau-|\xi|} \sum_{s=-N}^{N} k_s(\alpha) v_{m-s}^{j-1}(\xi,\tau-\alpha)\mathrm{d}\alpha\mathrm{d}\tau\mathrm{d}\xi,$$
$$(x,t) \in D'(T),\ j = 1,2,\ldots,\ m = 0,\pm 1,\pm 2,\ldots. \tag{5.15}$$

Obviously all functions $v_m^j(x,t)$ are continuous in $D'(T)$. Moreover, estimating these functions, we get the following estimates

$$\left|v_m^0(x,t)\right| \le \frac{\theta(N-|m|)}{2}$$
$$\times \iint\limits_{\diamond(x,t)} \left|J_0\left(m\sqrt{(t-\tau)^2-(x-\xi)^2}\right)\right| |k_m(\tau-|\xi|)|\,\mathrm{d}\tau\mathrm{d}\xi \le \theta(N-|m|)\frac{KT^2}{2},$$

$$\left|v_m^1(x,t)\right| \le \frac{KT^2}{2}\int\limits_0^t \sum_{s=-N}^{N} \max_{|\xi|\le\frac{T}{2}} \left|v_{m-s}^0(\xi,\tau)\right| \mathrm{d}\tau \le \theta(2N-|m|)\frac{K^2T^4(2N+1)t}{2^2},$$

$(x,t) \in D'(T)$.

Continuing these estimates, we easily obtain that

$$\left|v_m^j(x,t)\right| \le \theta\left((j+1)N-|m|\right)\frac{KT^2}{2}\left(\frac{KT^2(2N+1)t}{2}\right)^j \frac{1}{j!},$$
$$(x,t) \in D'(T),\ j = 0,1,2,\ldots.$$

Since $t \le T$ series (5.14) uniformly converges in $(x,t) \in D'(T)$ for all m. Hence, its sum is a continuous function in $D'(T)$. Moreover, the following estimates hold

$$|v_0(x,t)| \le \sum_{j=0}^{\infty} \left|v_0^j(x,t)\right| \le \frac{KT^2}{2}$$

$$\sum_{j=0}^{\infty} \left(\frac{KT^2(2N+1))t}{2}\right)^j \frac{1}{j!} \le \frac{KT^2}{2} C_1,$$

$$|v_m(x,t)| \le \sum_{j=n}^{\infty} \left|v_m^j(x,t)\right| \le \frac{KT^2}{2} \sum_{j=n}^{\infty} \left(\frac{KT^2(2N+1)t}{2}\right)^j \frac{1}{j!}$$

$$\le \frac{\left(KT^2\right)^{n+1} (2N+1)^n t^n}{2^{n+1} n!} C_1, \; (x,t) \in D'(T),$$

$$Nn < |m| \le (n+1)N, \; n = 0, 1, 2, \ldots, \tag{5.16}$$

where $C_1 = \exp\left(KT^2(2N+1)/2\right)$.

Now differentiating equations (5.12) with respect to t and x, we easily check that functions $v_m(x,t)$ are continuously differentiable in $D'(T)$. We check it for the derivatives with respect to t only. The expressions for these derivatives will be useful in the analysis of the inverse problem. From (5.12), we find

$$\frac{\partial v_m(x,t)}{\partial t} = -\frac{\theta(N-|m|)}{2} \int_{\frac{x-t}{2}}^{\frac{x+t}{2}} \Big[k_m(t - |\xi| - |x-\xi|)$$

$$+ \int_{|\xi|}^{t-|x-\xi|} H_m(t-\tau, x-\xi) k_m(\tau - |\xi|) \mathrm{d}\tau \Big] \mathrm{d}\xi$$

$$+ \frac{1}{2} \int_{\frac{x-t}{2}}^{\frac{x+t}{2}} \int_0^{t-|\xi|-|x-\xi|} \sum_{s=-N}^{s=N} k_s(\tau) v_{m-s}(\xi, t - |x-\xi| - \tau) \mathrm{d}\tau \mathrm{d}\xi$$

$$+ \frac{1}{2} \int_{\Diamond(x,t)} H_m(t-\tau, x-\xi) \int_0^{\tau-|\xi|} \sum_{s=-N}^{s=N} k_s(\alpha) v_{m-s}(\xi, \tau - \alpha) \mathrm{d}\alpha \mathrm{d}\tau \mathrm{d}\xi,$$

$$(x,t) \in D'(T), \; m = 0, \pm 1, \pm 2, \ldots, \tag{5.17}$$

where

$$H_m(t,x) = \frac{\partial}{\partial t} J_0\left(m\sqrt{t^2 - x^2}\right) = -m^2 t \frac{J_1(\zeta)}{\zeta}\,|_{\zeta = m\sqrt{t^2-x^2}}\,.$$

It follows from the definition of the Bessel function that $J_1(\zeta)/\zeta$ is a continuous function for all $\zeta \in [0, \infty)$ and

$$\lim_{\zeta\to 0} \frac{J_1(\zeta)}{\zeta} = \frac{1}{2}. \tag{5.18}$$

From relation (5.12) we see that the derivatives $\partial v_m(x,t)/\partial t$ are, indeed, continuous in $D'(T)$ for all m.

Since from (5.13) it follows that $|J_1(\zeta)/\zeta| \leq 1/2$ for all $\zeta \in \mathbb{R}$, we have $|H_m(t,x)| \leq m^2 T/2$. Therefore we can estimate $\partial v_m(x,t)/\partial t$ as follows

$$\left|\frac{\partial v_m(x,t)}{\partial t}\right| \leq \frac{KT(1+m^2T^2)}{2}\left[\theta(N-|m|) + \int_0^t \sum_{s=-N}^{N} \max_{\xi\in\Sigma(x,t,\tau)} |v_{m-s}(\xi,\tau)|\,d\tau\right],$$
$$(x,t) \in D'(T),\ m = 0, \pm 1, \pm 2, \ldots,$$

where $\Sigma(x,t,\tau) = \{\xi : (\xi,\tau) \in \Diamond(x,t)\}$. Using (5.9), we get the following estimates

$$\left|\frac{\partial v_m(x,t)}{\partial t}\right| \leq \frac{KT(1+m^2T^2)}{2}$$
$$\times\left[1 + (2N+1)\int_0^t \max\left\{\max_{\xi\in\Sigma(x,t,\tau),|s|\leq N} |v_s(\xi,\tau)|,\right.\right.$$
$$\left.\left.\max_{\xi\in\Sigma(x,t,\tau),N<|s|\leq 2N} |v_s(\xi,\tau)|\right\}d\tau\right] \leq \frac{KT(1+m^2T^2)}{2}$$
$$\times\left[1 + (2N+1)\max\left\{\frac{KT^2 t}{2\cdot 1!}, \frac{\left(KT^2\right)^2 (2N+1)t^2}{2^2\cdot 2!}C_1\right\}\right],$$
$$|m| \leq N, \tag{5.19}$$

$$\left|\frac{\partial v_m(x,t)}{\partial t}\right| \le \frac{KT(1+m^2T^2)}{2} \int_0^t \sum_{s=-N}^{N} \max_{\xi\in\Sigma(x,t,\tau)} |v_{m-s}(\xi,\tau)|\, d\tau$$

$$\le \frac{KT(1+m^2T^2)}{2}(2N+1) \int_0^t \max_{\xi\in\Sigma(x,t,\tau),N(n-1)<|s|\le(n+2)N} |v_s(\xi,\tau)|\, d\tau$$

$$\le \frac{KT(1+m^2T^2)}{2}(2N+1) \int_0^t \max\Big\{ \max_{\xi\in\Sigma(x,t,\tau),N(n-1)<|s|\le nN} |v_s(\xi,\tau)|,$$

$$\max_{\xi\in\Sigma(x,t,\tau),Nn<|s|\le(n+1)N} |v_s(\xi,\tau)|, \max_{\xi\in\Sigma(x,t,\tau),N(n+1)<|s|\le(n+2)N} |v_s(\xi,\tau)|\Big\} d\tau$$

$$\le \frac{KT(1+m^2T^2)}{2}(2N+1)C_1 \max\Big\{ \frac{(KT^2)^n(2N+1)^{n-1}t^n}{2^n\cdot n!},$$

$$\frac{(KT^2)^{n+1}(2N+1)^n t^{n+1}}{2^{n+1}\cdot(n+1)!}, \frac{(KT^2)^{n+2}(2N+1)^{n+1}t^{n+2}}{2^{n+2}\cdot(n+2)!}\Big\},$$

$$(x,t)\in D'(T),\ Nn<|m|\le(n+1)N,\ n=1,2,\dots. \tag{5.20}$$

Introducing the notation

$$C_2 = C_1 \max\left\{\frac{KT^3}{2}, \frac{(KT^3)^2(2N+1)}{2^3}\right\},$$

we come to the estimates

$$\left|\frac{\partial v_m(x,t)}{\partial t}\right| \le \frac{KT(1+m^2T^2)}{2}[1+(2N+1)C_2],\ -N\le m\le N,$$

$$\left|\frac{\partial v_m(x,t)}{\partial t}\right| \le (1+m^2T^2)\frac{(KT)^{n+1}(2N+1)^n t^n}{2^{n+1}\cdot n!} \max\{C_1, (2N+1)C_2\},$$

$$(x,t)\in \bar{D}'(T),\ Nn<|m|\le(n+1)N,\ n=1,2,\dots,$$

which coincide with (5.10). Similarly one can check that the derivative $\frac{\partial v_m(x,t)}{\partial x}$ is continuous functions in $D'(T)$.

By (5.8), the uniqueness of the solution to problem (5.6) reduces the uniqueness of the solution to (5.11). Since problem (5.11) is equivalent to the integral equation (5.12), the uniqueness of the solution to this equation follows from the general theory of integral equations. The lemma is proven. □

Lemma 5.2 *Let $k(y,t)$, $\hat{k}(y,t)$ be two arbitrary functions of the set $\Omega(N,T,K)$ and $u_m(x,t)$, $\hat{u}_m(x,t)$, $m=0,\pm1,\pm2,\dots$, be the solutions to problem (5.6) which correspond to $k(y,t)$, $\hat{k}(y,t)$ respectively, and $\tilde{v}_m(x,t)=v_m(x,t)-\hat{v}_m(x,t)$, $\tilde{k}_s(t)=k_s(t)-\hat{k}_s(t)$. Then, there exist constants $C_3=C_3(N,T,K)\ge1$ and $C_4=C_4(N,T,K)$, continuously depending on K, T, such that*

$$|\tilde{v}_0(x,t)| \le C_3 \frac{\tilde{K}T^2}{2},$$

$$|\tilde{v}_m(x,t)| \le C_3 \frac{\tilde{K}T^2}{2} \frac{\left(KT^2(2N+1)t\right)^n}{2^n \cdot n!}, \tag{5.21}$$

if

$$(x,t) \in \bar{D}'(T),\ nN < |m| \le (n+1)N,\ n = 0, 1, 2, \ldots.$$

and

$$\left|\frac{\partial \tilde{v}_m(x,t)}{\partial t}\right| \le \tilde{K}\Big[\frac{T(1+m^2T^2)}{2} + C_4 \frac{KT^2}{2}\frac{t}{1!}\Big],\ |m| \le N,$$

$$\left|\frac{\partial \tilde{v}_m(x,t)}{\partial t}\right| \le C_4 \tilde{K} \frac{\left(KT^2(2N+1)\right)^n}{2^2} \frac{t^n}{n!}, \tag{5.22}$$

if

$$(x,t) \in D'(T),\ Nn < |m| \le (n+1)N,,\ n = 1, 2, \ldots,$$

where $\tilde{K} = \max_{-N \le s \le N} \max_{0 \le t \le T} \left|k_s(t) - \hat{k}_s(t)\right|$.

Proof Using equations (5.12) with (v_m, k_s) and $(\hat{v}_m, \hat{k}_s)$ subtracting one from the other, we find

$$\tilde{v}_m(x,t) = \frac{\theta(N-|m|)}{2} \iint\limits_{\Diamond(x,t)} J_0\left(m\sqrt{(t-\tau)^2-(x-\xi)^2}\right) \tilde{k}_m(\tau-|\xi|)\mathrm{d}\tau\mathrm{d}\xi$$

$$+\frac{1}{2} \iint\limits_{\Diamond(x,t)} J_0\left(m\sqrt{(t-\tau)^2-(x-\xi)^2}\right) \int_0^{\tau-|\xi|} \sum_{s=-N}^{N} \Big[k_s(\alpha)\tilde{v}_{m-s}(\xi,\tau-\alpha)$$

$$+\tilde{k}_s(\alpha)\hat{v}_{m-s}(\xi,\tau-\alpha)\Big]\mathrm{d}\alpha\mathrm{d}\tau\mathrm{d}\xi, \tag{5.23}$$

$$(x,t) \in D'(T),\ m = 0, \pm 1, \pm 2, \ldots.$$

Represent $\tilde{v}_m(x,t)$ in the form

$$\tilde{v}_m(x,t) = \sum_{j=0}^{\infty} \tilde{v}_m^j(x,t), \tag{5.24}$$

where

$$\tilde{v}_m^0 = \frac{\theta(N-|m|)}{2} \iint\limits_{\diamond(x,t)} J_0\left(m\sqrt{(t-\tau)^2-(x-\xi)^2}\right) \tilde{k}_m(\tau-|\xi|)d\tau d\xi$$

$$+\frac{1}{2}\iint\limits_{\diamond(x,t)} J_0\left(m\sqrt{(t-\tau)^2-(x-\xi)^2}\right) \int\limits_0^{\tau-|\xi|} \sum_{s=-N}^{N} \tilde{k}_s(\alpha)\hat{v}_{m-s}(\xi,\tau-\alpha)d\alpha d\xi d\tau, \tag{5.25}$$

$$\tilde{v}_m^j(x,t) =$$

$$= \frac{1}{2}\iint\limits_{\diamond(x,t)} J_0\left(m\sqrt{(t-\tau)^2-(x-\xi)^2}\right) \int\limits_0^{\tau-|\xi|} \sum_{s=-N}^{N} k_s(\alpha)\tilde{v}_{m-s}^{j-1}(\xi,\tau-\alpha)d\alpha d\xi d\tau, \tag{5.26}$$

$$(x,t) \in D'(T),\ j=0,1,2\ldots,\ m=0,\pm1,\pm2,\ldots.$$

From here we obtain

$$|\tilde{v}_m^0(x,t)| \le \frac{\tilde{K}T^2}{2}\theta(N-|m|) + \frac{\tilde{K}T^2}{2}\int\limits_0^t \sum_{s=-N}^{N} \max_{|\xi|\le T/2} \left|\hat{v}_{m-s}(\xi,\tau)\right| d\tau,$$

$$|\tilde{v}_m^j(x,t)| \le \frac{KT^2}{2}\int\limits_0^t \sum_{s=-N}^{N} \max_{|\xi|\le T/2} \left|\tilde{v}_{m-s}^{j-1}(\xi,\tau)\right| d\tau, \tag{5.27}$$

$$(x,t) \in D'(T),\ j=1,2,\ldots,\ m=0,\pm1,\pm2,\ldots.$$

Using the estimates (5.9) for functions $\hat{v}_m(x,t)$ here, as result we get

$$|\tilde{v}_0^0(x,t)| \le \frac{\tilde{K}T^2}{2} + \frac{\tilde{K}T^2}{2}(2N+1)\int\limits_0^t \max_{|\xi|\le T/2, |s|\le 2N} \left|\hat{v}_s(\xi,\tau)\right|$$

$$\le \frac{\tilde{K}T^2}{2}\left(1+(2N+1)C_2\right),$$

$$|\tilde{v}_m^0(x,t)| \le \frac{\tilde{K}T^3}{2} + \frac{\tilde{K}T^2}{2}(2N+1)\int\limits_0^t \max_{|\xi|\le T/2, Nn<|s|\le(n+1)N} \left|\hat{v}_s(\xi,\tau)\right|$$

$$\le \frac{\tilde{K}T^2}{2}\left(T+(2N+1)C_2\right),\ |m|\le N,$$

$$|\tilde{v}_m^0(x,t)| \le \frac{\tilde{K}T^2}{2}(2N+1)\int_0^t \max_{|\xi|T/2,|s|\le T} \left|\hat{v}_{m-s}(\xi,\tau)\right|$$

$$\le \tilde{K}\frac{T^{2(n+1)}K^n(2N+1)^n t^n}{2^{n+1}\cdot n!}\max\{C_1,(2N+1)C_2\},$$

$$(x,t)\in D'(T),\ nN<|m|\le(n+1)N,\ n=1,2,\ldots.$$

In view of $C_1 \ge 1$, the last estimates we can write in the more compact form as follows

$$|\tilde{v}_m^0(x,t)| \le (C_1+(2N+1)C_2)\frac{\tilde{K}T^2}{2},\ -N\le m\le N,$$

$$|\tilde{v}_m^0(x,t)| \le (C_1+(2N+1)C_2)\frac{\tilde{K}T^2}{2}\frac{K^nT^{2n}(2N+1)^n t^n}{2^n\cdot n!},$$

$$(x,t)\in D'(T),\ nN<|m|\le(n+1)N,\ n=1,2,\ldots.$$

Further calculations for $j\ge 1$ show that estimates

$$|\tilde{v}_0^j(x,t)| \le (C_1+(2N+1)C_2)\frac{\tilde{K}T^2}{2}\frac{\left(KT^2(2N+1)t\right)^j}{2^j\cdot j!},$$

$$|\tilde{v}_m^j(x,t)| \le$$

$$\le (C_1+(2N+1)C_2)\frac{\tilde{K}T^2}{2}\frac{\left(KT^2(2N+1)t\right)^{n+j}}{2^{n+j}\cdot(n+j)!}\max\left(1,\frac{KT^2t}{2(n+j+1)}\right),$$

$$(x,t)\in D'(T),\ nN<|m|\le(n+1)N,\ n=0,1,2,\ldots$$

hold. Hence,

$$|\tilde{v}_0(x,t)| \le C_1\,(C_1+(2N+1)C_2)\frac{\tilde{K}T^2}{2},$$

$$|\tilde{v}_m(x,t)| \le C_1\,(C_1+(2N+1)C_2)\frac{\tilde{K}T^2}{2}\frac{\left(KT^2(2N+1)t\right)^n}{2^n\cdot n!}\max\left(1,\frac{KT^3}{2}\right),$$

$$(x,t)\in D'(T),\ nN<|m|\le(n+1)N,\ n=0,1,2,\ldots.$$

Thus, estimate (5.21) holds with

$$C_3 = C_1\left(C_1 + (2N+1)C_2\right)\max\left(1, \frac{KT^3}{2}\right).$$

For proving (5.22), we use the relation (5.17) for (v_m, k_s) and $(\hat{v}_m, \hat{k}_s)$. Subtracting one from the other, we get

$$\frac{\partial \tilde{v}_m(x,t)}{\partial t} = \frac{\theta(N-|m|)}{2}\int\limits_{\frac{x-t}{2}}^{\frac{x+t}{2}}\Bigg[\tilde{k}_m(t-|\xi|-|x-\xi|)$$

$$+ \int\limits_{|\xi|}^{t-|x-\xi|} H_m(t-\tau, x-\xi)\tilde{k}_m(\tau-|\xi|)\mathrm{d}\tau\Bigg]\mathrm{d}\xi$$

$$+\frac{1}{2}\int\limits_{\frac{x-t}{2}}^{\frac{x+t}{2}}\int\limits_{0}^{t-|\xi|-|x-\xi|}\sum_{s=-N}^{N}\Big[k_m(\tau)\tilde{v}_{m-s}(\xi, t-|x-\xi|-\tau)$$

$$+\tilde{k}_m(\tau)\hat{v}_{m-s}(\xi, t-|x-\xi|-\tau)\Big]\mathrm{d}\tau\mathrm{d}\xi$$

$$+\frac{1}{2}\int\limits_{\diamond(x,t)} H_m(t-\tau, x-\xi)\int\limits_{0}^{\tau-|\xi|}\sum_{s=-N}^{N}\Big[k_m(\alpha)\tilde{v}_{m-s}(\xi,\tau-\alpha)$$

$$+\tilde{k}_m(\alpha)\ddot{v}_{m-s}(\xi,\tau-\alpha)\Big]\mathrm{d}\alpha\mathrm{d}\tau\mathrm{d}\xi,$$

$$(x,t)\in D'(T),\ m = 0, \pm 1, \pm 2, \ldots.$$

Estimating each term in this equality, we have

$$\left|\frac{\partial \tilde{v}_m(x,t)}{\partial t}\right| \le \frac{\tilde{K}T}{2}\left(1+m^2T^2\right)\theta\left(N-|m|\right)$$

$$+\frac{2T+m^2T^3}{4}\sum_{s=-N}^{N}\int\limits_0^t \max_{\xi\in\Sigma(x,t,\tau)}\Big[K\left|\tilde{v}_{m-s}(\xi,\tau)\right| + \tilde{K}\left|\hat{v}_{m-s}(\xi,\tau)\right|\Big]\mathrm{d}\tau.$$

Using estimates (5.9) and (5.21), we find

$$\left|\frac{\partial \tilde{v}_m(x,t)}{\partial t}\right| \le \tilde{K}\Bigg[\frac{T(1+m^2T^2)}{2}$$
$$+\frac{2T+m^2T^3}{4}\frac{KT^2}{2}(C_1+C_3)\left(1+\frac{KT^3(2N+1)}{2}\right)\frac{t}{1!}\Bigg],\ |m|\le N,$$

$$\left|\frac{\partial \tilde{v}_m(x,t)}{\partial t}\right| \le \tilde{K}\frac{2T+m^2T^3}{4}(C_1+C_3)\frac{\left(KT^2(2N+1)\right)^n}{2^2}\frac{t^n}{n!}$$
$$\times\max\left[1+\frac{(KT^3)(2N+1)}{2},\ \frac{(KT^3)^3(2N+1)^2}{2^3}\right],$$
$$(x,t)\in D'(T),\ Nn<|m|\le(n+1)N,\ n=1,2,\ldots.$$

Hence, estimates (5.22) hold with

$$C_4=\frac{2T+m^2T^3}{4}(C_1+C_3)\max\left[1+\frac{(KT^3)(2N+1)}{2},\ \frac{(KT^3)^3(2N+1)^2}{2^3}\right].$$

The lemma is proven. □

5.1.3 *Inverse Problem—The Existence and Uniqueness Theorem*

Taking into account presentation (5.8) we conclude that for solvability of the inverse problem the functions $f_m(t)$, $|m|\le N$ must be as follows

$$f_m(t)=-\delta_{0m}\delta(t)+\theta(t)g_m(t),\ t\in\mathbb{R},\ |m|\le N, \tag{5.28}$$

where $g_m(t)$ is a given smooth functions. Then, it is obvious that the additional conditions for $v_m(x,t)$, as it follows from (5.7) and (5.8), have the forms

$$v_m(0,t)=g_m(t),\ t>0,\ |m|\le N. \tag{5.29}$$

Using these, from (5.17) after some transformations, one gets

$$g'_m(t) =$$

$$= \frac{\theta(N-|m|)}{2}\int_0^t k_m(\tau)\mathrm{d}\tau + \frac{\theta(N-|m|)}{2}\int_0^{\frac{t}{2}}\int_{\xi}^{t-\xi} H_m(t-\tau,\xi)k_m(\tau-\xi)\mathrm{d}\tau\mathrm{d}\xi$$

$$+\int_0^{\frac{t}{2}}\int_0^{t-2\xi}\sum_{s=-N}^{N} k_s(\tau)v_{m-s}(\xi,t-\xi-\tau)\mathrm{d}\tau\mathrm{d}\xi$$

$$+\int_0^{\frac{t}{2}}\int_{\xi}^{t-\xi} H_m(t-\tau,\xi)\int_0^{\tau-\xi}\sum_{s=-N}^{N} k_s(\alpha)v_{m-s}(\xi,\tau-\alpha)\mathrm{d}\alpha\mathrm{d}\tau\mathrm{d}\xi.$$

By differentiating this equality, we obtain the integral equation for $k_m(t)$:

$$k_m(t) = 2g''_m(t) + \frac{m^2}{2}\int_0^t (t-\tau)k_m(\tau)\mathrm{d}\tau - 2\int_0^{\frac{t}{2}}\int_{\xi}^{t-\xi} H'_m(t-\tau,\xi)k_m(\tau-\xi)\mathrm{d}\tau\mathrm{d}\xi$$

$$-2\int_0^{\frac{t}{2}}\int_0^{t-2\xi}\sum_{s=-N}^{N} k_s(\tau)\left[-\frac{m^2\xi}{2}v_{m-s}(\xi,t-\xi-\tau) + \frac{\partial}{\partial t}v_{m-s}(\xi,t-\xi-\tau)\right]\mathrm{d}\tau\mathrm{d}\xi$$

$$-2\int_0^{\frac{t}{2}}\int_{\xi}^{t-\xi} H'_m(t-\tau,\xi)\int_0^{\tau-\xi}\sum_{s=-N}^{N} k_s(\tau)v_{m-s}(\xi,\tau-\alpha)\mathrm{d}\alpha\mathrm{d}\tau\mathrm{d}\xi,\ |m| \le N. \quad (5.30)$$

For getting the last equation we have used (5.18) and the relation $v_m(x, |x| + 0) = 0$ which follows from (5.12) for all m. In (5.19)

$$H'_m(t,x) = \frac{\partial}{\partial t}H_m(t,x) = -m^2\left[\frac{J_1(\zeta)}{\zeta} - m^2t^2\frac{J_2(\zeta)}{\zeta^2}\right]_{\zeta=\sqrt{t^2-x^2}}. \quad (5.31)$$

Since the functions $J_1(\zeta)/\zeta$ and $J_2(\zeta)/\zeta^2$ are continuous for all $\zeta \in [0,\infty)$, the function $H'_m(t,x)$ is continuous for $(x,t) \in D'(T)$.

Note that integral equations (5.12) and (5.17) determine $v_m(x,t)$, $\frac{\partial v_m(x,t)}{\partial t}$ respectively, as a function of $k_s(t)$, $s = -N,\dots,N$. Hence, introducing the operator $U = (U_{-N},\dots,U_N)$ by the right sides of (5.30), we can rewrite relations (5.30) as the operator equations

$$k_m(t) = U_m\left(k_{-N}(t),\dots,k_N(t)\right), \quad (5.32)$$

$$m = 0,\pm1,\pm2,\dots,\pm N,\ t \in [0,\ T].$$

Let

$$k_m^0(t) = 2g_m''(t),\ m = 0, \pm 1, \pm 2, \dots, \pm N. \tag{5.33}$$

Denote by $\Omega_0\,(N, T, K_0)$ the set of functions $k_s(t),\ -N \le s \le N$, satisfying the conditions

$$\left\|k_m(t) - k_m^0(t)\right\|_{C[0,T]} \le K_0,\ K_0 = \max_{-N\le s\le N} \left\|k_m^0(t)\right\|_{C[0,T]}, \tag{5.34}$$

$$m = 0, \pm 1, \pm 2, \dots, \pm N.$$

If functions $k_s(t) \in \Omega_0\,(N, T, K_0)$ for $-N \le s \le N$ then, obviously, $k_s(t) \in \Omega$ $(N, T, 2K_0)$. For operator equations (5.32) the following theorem holds.

Theorem 5.1 *Let presentation (5.28) be valid. Moreover, let data (5.29) satisfy the following conditions:*

$$g_m(t) \in C^2[0,\ T],\ g_m(0) = g_m'(0) = 0,\ m = 0, \pm 1, \pm 2, \dots, \pm N.$$
$$\text{and} G = \max_{-N\le s\le N} \left\|g_m''(t)\right\|_{C[0,T]}. \tag{5.35}$$

Then, there exists a number $T_0 \in (0,\ T)$ such that operator equations (5.32) have a unique solution, belonging to the set $\Omega_0\,(N, T_0, 2G)$.

Proof It is obvious that $\|k_0\| = 2G$. We prove that operator $U = (U_{-N}, \dots, U_N)$ maps the set $\Omega_0\,(N, T, 2G)$ into itself and is a contraction operator if T is sufficiently small. The theorem of existence and uniqueness then follows immediately from the Contraction mapping principle.

Estimate first the function $H_m'(t-\tau, \xi)$ defined by (5.31)for

$$(\xi, \tau) \in \triangleright(t) = \left\{(\xi, \tau) : 0 \le \xi \le \frac{t}{2},\ \xi \le \tau \le t - \xi\right\}.$$

Since from (5.13) follows that

$$\left|\frac{J_1(\zeta)}{\zeta}\right| \le \frac{1}{2},\ \left|\frac{J_2(\zeta)}{\zeta^2}\right| \le \frac{1}{4}$$

for all $\zeta \in \mathbb{R}$, then the following estimate holds

$$\left|H_m'(t-\tau, \xi)\right| \le \frac{m^2(2 + m^2T^2)}{4},\ (\xi, \tau) \in \triangleright(t),\ t \in [0,\ T]. \tag{5.36}$$

Demonstrate that for sufficiently small T the operator $U = (U_{-N}, \dots, U_N)$ is a contraction mapping of the set $\Omega_0(N, T, 2G)$ into itself. Indeed, for $k_m(t) \in \Omega_0(N, T, 2G), -N \le m \le N$ the inclusion $k_m(t) \in \Omega(N, T, 4G), -N \le m \le N$ is valid. Then, in view of (5.36) from relations (5.30) we obtain

$$\left|k_m(t) - k_m^0(t)\right| \le m^2 G T^2 \left(3 + m^2 T^2\right) + 8G(2N+1)$$

$$\times \iint\limits_{\triangleright(t)} \left[\frac{m^2 \xi}{2} \max_{-N \le s \le N} |v_{m-s}(\xi, \tau - \xi)| + \max_{-N \le s \le N} \left|\frac{\partial}{\partial t} v_{m-s}(\xi, \tau - \xi)\right|\right]$$

$$+ 8G(2N+1) \iint\limits_{\triangleright(t)} \int\limits_0^{\tau - \xi} \max_{-N \le s \le N} |v_{m-s}(\xi, \tau - \alpha)| \, d\alpha d\tau d\xi$$

$$\le T^2 N^2 G \left(3 + N^2 T^2\right) + 2T^2 G(2N+1)$$

$$\times \left[\frac{N^2 T}{4} \max_{-2N \le j \le 2N} \max_{(\xi,\tau) \in \bar{D}'(T)} \left|v_j(\xi, \tau)\right| + \max_{-2N \le j \le 2N} \max_{(\xi,\tau) \in \bar{D}'(T)} \left|\frac{\partial}{\partial t} v_j(\xi, \tau)\right|\right]$$

$$+ \frac{3T^3 G(2N+1)}{4} \max_{-2N \le j \le 2N} \max_{(\xi,\tau) \in \bar{D}'(T)} \left|v_j(\xi, \tau)\right|,$$

$$t \in [0, T], \; -N \le m \le N.$$

For functions $v_m(x, t)$ and $\frac{\partial}{\partial t} v_m(x, t)$ the estimates (5.9) and (5.10) valid with $K = 4G$. Using them, we find

$$\left|k_m(t) - k_m^0(t)\right| \le G N^2 T^2 \left(3 + N^2 T^2\right)$$

$$+ 2T^2 G(2N+1)[\frac{1}{2} G N^2 T^3 C_1 \max\left(1, \; 2GT^3(2N+1)\right)$$

$$+ 2GT(1 + N^2 T^2) \max\left(\max(1, C_1) + (2N+1) C_2\right) \max\left(1, 2GT^2(2N+1)\right)]$$

$$+ \frac{3}{2} G(2N+1) T^2 C_1 \max\left(1, \; 2GT^3(2N+1)\right) \equiv 2GT^2 C_5(N, T, G),$$

$$t \in [0, T], \; -N \le m \le N,$$

where C_5 is positive constant, continuously depending on T. Choosing $T_1 = T_1(N, G)$ as a positive root of the equation

$$T^2 C_5(N, T, G) = \sigma,$$

for some $0 < \sigma < 1$, we obtain that

$$\left|k_m(t) - k_m^0(t)\right| \le 2G,\ t \in [0, T_1],\ -N \le m \le N,$$

i.e., the operator $U = (U_{-N}, \dots, U_N)$ maps the set $\Omega_0\,(N, T_1, 2G)$ into itself. □

Now let $k_m(t),\ \hat{k}_m(t),\ m = -N, \dots, N$ be two solutions of operator equations (5.32) belonging to the set $\Omega_0\,(N, T, 2G)$. Denote corresponding solutions of integral equations (5.12), (5.17) by $v_m,\ \hat{v}_m$ respectively, and its derivatives with respect to t by $\partial v_m/\partial t$ and $\partial \hat{v}_m/\partial t$. Denoting

$$\tilde{v}_m = v_m - \hat{v}_m,\ \frac{\partial \tilde{v}}{\partial t} = \frac{\partial v}{\partial t} - \frac{\partial \hat{v}}{\partial t},$$

$$\tilde{k}_s(t) = k_s(t) - \hat{k}_s(t),\ \tilde{K} = \max_{-N \le s \le N} \max_{0 \le t \le T} \left|\tilde{k}_s(t)\right|$$

from (5.30), we find

$$\tilde{k}_m(t) = \frac{m^2}{2}\int_0^t (t-\tau)\tilde{k}_m(\tau)\mathrm{d}\tau - 2\int_0^{\frac{t}{2}}\int_\xi^{t-\xi} H_m'(t-\tau,\xi)\tilde{k}_m(\tau-\xi)\mathrm{d}\tau\mathrm{d}\xi$$

$$-2\int_0^{\frac{t}{2}}\int_0^{t-2\xi}\sum_{s=-N}^{N}\left\{\tilde{k}_s(\tau)\left[-\frac{m^2\xi}{2}v_{m-s}(\xi, t-\xi-\tau) + \frac{\partial}{\partial t}v_{m-s}(\xi, t-\xi-\tau)\right]\right.$$

$$\left. + \hat{k}_s(\tau)\left[-\frac{m^2\xi}{2}\tilde{v}_{m-s}(\xi, t-\xi-\tau) + \frac{\partial}{\partial t}\tilde{v}_{m-s}(\xi, t-\xi-\tau)\right]\right\}\mathrm{d}\tau\mathrm{d}\xi$$

$$-2\int_0^{\frac{t}{2}}\int_\xi^{t-\xi} H_m'(t-\tau,\xi)\int_0^{\tau-\xi}\sum_{s=-N}^{N}\left\{\tilde{k}_s(\tau)v_{m-s}(\xi, \tau-\alpha)\right.$$

$$\left. + \hat{k}_s(\tau)\tilde{v}_{m-s}(\xi, t-\tau)\right\}\mathrm{d}\alpha\mathrm{d}\tau\mathrm{d}\xi,$$

$$t \in [0, T],\ |m| \le N.$$

Estimating the terms of this equality, we get

$$\left|\tilde{k}_m(t)\right| \leq \frac{T^2 m^2 \tilde{K}}{8}\,(3+m^2T^2)$$
$$+\frac{T^2}{2}(2N+1) \max_{-2N\leq j\leq 2N} \max_{(\xi,\tau)\in \bar{D}'(T)} \left\{ \tilde{K}\left[\frac{m^2T}{4}\left|v_j(\xi,\tau)\right| + \left|\frac{\partial v_j(\xi,\tau)}{\partial t}\right|\right]\right.$$
$$\left. + 4G\left[\frac{m^2T}{4}\left|\tilde{v}_j(\xi,\tau)\right| + \left|\frac{\partial \tilde{v}_j(\xi,\tau)}{\partial t}\right|\right]\right\} + \frac{T^3m^2(2+m^2T^2)}{8}(2N+1)$$
$$\times \max_{-2N\leq j\leq 2N} \max_{(\xi,\tau)\in \bar{D}'(T)} \left[\tilde{K}\left|v_j(\xi,\tau)\right| + 4G\left|\tilde{v}_j(\xi,\tau)\right|\right],$$
$$t\in[0,T],\ m=-N,\ldots,N.$$

Using estimates (5.8), (5.9) and (5.21), (5.22), we find

$$\left|\tilde{k}_m(t)\right| \leq T^2N^2\tilde{K}\frac{3+T^2N^2}{8}$$
$$+\frac{T^2}{2}(2N+1)\tilde{K}\left[\frac{T^2N^2}{2}(C_1+C_3)G\max\left(1, 2G(2N+1)T^3\right)\right.$$
$$+2T\left(1+T^2N^2\right)G\left(C_1+(2N+1)C_2\right)\max\left(1, 2G(2N+1)T^3\right)$$
$$\left.+4TG\max\left(\frac{1+T^2(N^2+4C_4G)}{2}, T^2C_4G(2N+1)\right)\right]$$
$$+\frac{T^5N^2}{4}(2+T^2N^2)(2N+1)\,(C_1+C_3)\max\left(1, 2G(2N+1)T^3\right)$$
$$=\tilde{K}T^2C_6\,(N,T,G),$$

where C_6 is positive constant, continuously depending on T. Set a fixed $\rho\in(0,1)$ and define $T_2 = T_2\,(N, G, \rho)$ as the positive root of the equation

$$T^2C_6\,(N,T,G)=\rho.$$

Then, the estimates

$$\left|\tilde{k}_m(t)\right| \leq \rho\tilde{K},\ m=-N,\ldots,N$$

implies that the operator $U=(U_{-N},\ldots,U_N)$ is contractive on the set Ω_0 $(N, T_2, 2G)$. Taking $T_0=\min\,(T_1,T_2)$ we get that this operator maps the set $\Omega_0\,(N,T_2,2G)$ into itself and is a contraction operator on this set. By the Banach principle operator equation (5.32) has a unique solution belonging to the set $\Omega_0\,(N,T_2,2G)$. The proof of Theorem 5.1 is completed.

5.2 A Problem of Determining a Spatial Part of a Memory Kernel

In this paragraph, we consider the integro-differential equation for $(x,\ y,\ t) \in \mathbb{R}^3_+$

$$\frac{\partial^2 u}{\partial t^2} - \Delta u = \int\limits_0^t k(x, y, \tau) u(x, y, t - \tau) d\tau \tag{5.37}$$

with the initial and boundary conditions (5.2).

Note that the kernel in equation (5.37) depends on all variables as u. We assume that this kernel is represented in the form

$$k(x, y, t) = k_0(t) p(x, y),$$

where $k_0(t)$ is known function and $k_0(t) \in C^1[0, \infty),\ k_0(0) \neq 0$, without loss of generality, we set $k_0(0) = 1$. The function $p(x, y)$ is unknown, and it can be represented in the form of a finite Fourier series

$$p(x, y) = \sum_{s=-N}^{N} p_s(x) e^{isy} \tag{5.38}$$

with a fixed integer $N \geq 0$. Denote by $\Omega\,(N, X, P)$ the set of functions $p(x, y)$ for which the coefficients $p_s,\ |s| \leq N$, are continuous functions on the interval $[0, X],\ X > 0$ and satisfy the conditions

$$|p_s(x)| \leq P,\ x \in [0,\ X],\ -N \leq s \leq N. \tag{5.39}$$

For $p(x, y) \in \Omega\,(N, X, P)$ the solution of problem (5.37), (5.2) is a $2\pi-$periodic function of y and can be represented by the Fourier series

$$u(x, y, t) = \sum_{s=-\infty}^{\infty} u_s(x, t) e^{isy},$$

whereas it follows from relations (5.37), (5.2), (5.39) the coefficients satisfy the following equations

$$\left(\frac{\partial^2}{\partial t^2} - \frac{\partial^2}{\partial x^2} + m^2\right) u_m(x, t) = \sum_{s=-N}^{N} p_s(x) \int\limits_0^t k_0(\tau) u_{m-s}(x, t - \tau) d\tau,$$

$$(x, t) \in \mathbb{R}^2_+ = \left\{(x, t) \in \mathbb{R}^2 \mid x > 0\right\}$$

$$u_m \mid_{t<0} \equiv 0,\ \frac{\partial u_m}{\partial x} \mid_{x=0} = \delta_{0m} \delta'(t),\ m = 0, \pm 1, \pm 2, \ldots. \tag{5.40}$$

Here for given functions $k_0(t)$, $p_s(x)$, $s = 0, \pm1, \pm2\pm, \ldots$ we call **the direct problem** the problem of finding the functions $u_m(x,t)$, $m = 0, \pm1, \pm2, \ldots$ that satisfy (in a generalized meaning) the relations (5.40).

Now we formulate the inverse problem that will be studied in this section.

The inverse problem. Find the coefficients $p_m(x)$, $m = 0, \pm1, \pm2\pm, \ldots, \pm N$ by using the conditions

$$u_m(0,t) = f_m(t),\ t \in \mathbb{R}_+,\ \mathbb{R}_+ = \{t \in \mathbb{R} \mid t > 0\},\ m = 0, \pm1, \pm2, \ldots, \pm N. \tag{5.41}$$

As in the previous section, here concerning the solution to direct problems (5.40) the following lemma is valid.

Lemma 5.3 *Let T be an arbitrary positive number, $p(x,y) \in \Omega(N, X, P)$ and $D(T) = ((x,t)|\ 0 \le x \le T - t)$. Then, the solution to problem (5.40) exists and can be represented in $D(T)$ in the form*

$$u_m(x,t) = -\delta_{0m}\delta(t-x) + v_m(x,t)\theta(t-x), m = 0, \pm1, \pm2, \ldots, \tag{5.42}$$

where $\theta(t)$ is the Heaviside step-function: $\theta(t) = 1$ for $t \ge 0$; $\theta(t) = 0$ for $t < 0$ and $v_m(x,t)$ are continuously differentiable functions in the domain $D'(T) = ((x,t)|\ 0 \le x \le t \le T - x)$. Moreover, this solution is unique and there exist positive constants $C_1 = C_1(N, T, K, P) \ge 1$, continuously depending on $K := \|k_0(t)\|_{C^1[0,T]}$, T such that

$$|v_m(x,t)| \le C_1 \frac{KT^2P}{4},\quad -N \le m \le N,$$

$$|v_m(x,t)| \le C_1 \frac{\left(KT^2P\right)^{n+1} (2N+1)^n t^n}{4^{n+1} \cdot n!},$$

if

$$(x,t) \in \overline{D'}(T),\ Nn < |m| \le (n+1)N,\ n = 1, 2, \ldots.$$

In order to prove this lemma we represent the solution to direct problem (5.40) in the form

$$u_m(x,t) = \alpha_m\delta(t-x) + v_m(x,t)\theta(t-x), m = 0, \pm1, \pm2, \ldots,$$

where α_m are unknown constants, $v_m(x,t)$ are unknown regular functions. Also, as in the previous paragraph, substituting this equality into equations (5.40) and using the method of separation of singularities we find $\alpha_m = -\delta_{0m}$.

In view of representation (5.42), we rewrite equations (5.40) with respect to $v_m(x,t)$ in the form

$$\left(\frac{\partial^2}{\partial t^2} - \frac{\partial^2}{\partial x^2} + m^2\right) v_m(x,t)$$
$$= -\theta(N - |m|)k_0(t-x)p_m(x) + \sum_{s=-N}^{N} p_s(x) \int_0^{t-x} k_0(\tau)v_{m-s}(x, t-\tau)\mathrm{d}\tau,$$
$$t > x > 0, \ m = 0, \pm 1, \pm 2, \dots,$$
$$v_m \mid_{t \le x} \equiv 0, \ \frac{\partial v_m}{\partial x} \mid_{x=0} = 0, \ m = 0, \pm 1, \pm 2, \dots. \tag{5.43}$$

Here we also continue all functions $v_m(x,t)$ and $p_m(x)$ for $x < 0$ as even functions: $v_m(-x,t) = v_m(x,t)$, $p_m(-x) = p_m(x)$. Let $\overline{D'}(T) = \{(x,t) | \ 0 \le |x| \le t \le T - |x|\}$. Then problem (5.43) is equivalent to the following integral equation

$$v_m(x,t)$$
$$= -\frac{\theta(N-|m|)}{2} \iint_{\diamond(x,t)} J_0\left(m\sqrt{(t-\tau)^2 - (x-\xi)^2}\right) k_0(\tau - |\xi|)p_m(\xi)\mathrm{d}\tau\mathrm{d}\xi$$
$$+ \frac{1}{2} \iint_{\diamond(x,t)} J_0\left(m\sqrt{(t-\tau)^2 - (x-\xi)^2}\right) \sum_{s=-N}^{N} p_s(\xi) \int_0^{\tau-|\xi|} k_0(\alpha)v_{m-s}(\xi, \tau - \alpha)$$
$$\times \mathrm{d}\alpha\mathrm{d}\tau\mathrm{d}\xi, \ (x,t) \in \overline{D'}(T), \ m = 0, \pm 1, \pm 2, \dots. \tag{5.44}$$

Further proof of the lemma can be carried out quite analogously to the corresponding one from the previous section.

The following lemma is also true:

Lemma 5.4 . *Let $p(x,y)$, $\hat{p}(x,y)$ be two arbitrary functions of the set $\Omega(N, X, P)$ and $u_m(x,t)$, $\hat{u}_m(x,t)$, $m = 0, \pm 1, \pm 2, \dots$, be the solutions to problem (5.40) which correspond to $p(x,y)$, $\hat{p}(x,y)$ respectively, and $\tilde{v}_m(x,t) = v_m(x,t) - \hat{v}_m(x,t)$, $\tilde{p}_s(x) = p_s(x) - \hat{p}_s(x)$. Then, there exist constants $C_2 = C_2(N, T, K, P) \ge 1$, continuously depending on K, T, such that*

$$|\tilde{v}_0(x,t)| \le C_2 \frac{K\tilde{P}T^2}{2},$$

$$|\tilde{v}_m(x,t)| \le C_2 \frac{K\tilde{P}T^2}{2} \frac{\left(KPT^2(2N+1)t\right)^n}{2^n \cdot n!},$$

if

$$(x,t) \in \overline{D'}(T), \ nN < |m| \le (n+1)N, \ n = 0, 1, 2, \dots,$$

where $\tilde{P} = \max\limits_{-N \le s \le N} \max\limits_{0 \le x \le T/2} |\tilde{p}_s(x)|$.

We omit the proof of this lemma, referring reader to the proof of the corresponding one from the previous section.

In view of the presentation (5.42) we conclude that for solvability of the inverse problem the functions $f_m(t)$, $|m| \leq N$ must have the form

$$f_m(t) = -\delta_{0m}\delta(t) + \theta(t)g_m(t),\ t \in \mathbb{R},\ |m| \leq N, \tag{5.45}$$

where $g_m(t)$ is a given smooth functions. Then, it is obvious that the additional conditions for $v_m(x,t)$, by (5.41) and (5.45), have the forms

$$v_m(0,t) = g_m(t),\ t > 0,\ |m| \leq N. \tag{5.46}$$

Using these, from the differentiated with respect t equation (5.44) after some transformations, we get

$$\begin{aligned}
&g'_m(t) = \\
&= -\int_0^{\frac{t}{2}} p_m(\xi)\Big[k_0(t-2\xi) + p_m(\xi)\int_\xi^{t-\xi} H_m(t-\tau,\xi)k_0(\tau-\xi)\mathrm{d}\tau\Big]\mathrm{d}\xi \\
&+ \int_0^{\frac{t}{2}}\Big\{\sum_{s=-N}^{N} p_s(\xi)\int_\xi^{t-\xi} k_0(t-\xi-\tau)v_{m-s}(\xi,\tau)\mathrm{d}\tau\mathrm{d}\xi \\
&+ \sum_{s=-N}^{N} p_s(\xi)\int_\xi^{t-\xi} H_m(t-\tau,\xi)\int_0^{\tau-\xi} k_0(\alpha)v_{m-s}(\xi,\tau-\alpha)\mathrm{d}\alpha\mathrm{d}\tau\Big\}\mathrm{d}\xi,\ |m| \leq N.
\end{aligned}$$

By differentiating this equality and denoting $\frac{t}{2} = x$ we obtain the integral equation for $p_m(x)$:

$$\begin{aligned}
&p_m(x) = \\
&= -2g''_m(2x) - 2\int_0^x\Big\{p_m(\xi)\Big[k'_0(2x-2\xi) - \frac{m^2\xi}{2}k_0(2x-2\xi) \\
&+ \int_\xi^{2x-\xi} H'_m(2x-\tau,\xi)k_0(\tau-\xi)\mathrm{d}\tau\Big]\mathrm{d}\xi \\
&- \sum_{s=-N}^{N} p_s(\xi)\Big[v_{m-s}(\xi,2x-\xi) + \int_\xi^{2x-\xi} k'_0(2x-\xi-\tau)v_{m-s}(\xi,\tau)\mathrm{d}\tau\Big]\mathrm{d}\xi
\end{aligned}$$

$$+\frac{m^2\xi}{2}\sum_{s=-N}^{N}p_s(\xi)\int_0^{2x-2\xi}k_0(\tau)v_{m-s}(\xi,2x-\xi-\tau)\mathrm{d}\tau\mathrm{d}\xi$$

$$\left.-\int_{\xi}^{2x-\xi}H_m'(2x-\tau,\xi)\sum_{s=-N}^{N}p_s(\xi)\int_0^{\tau-\xi}k_0(\alpha)v_{m-s}(\xi,\tau-\alpha)\mathrm{d}\alpha\mathrm{d}\tau\right\}\mathrm{d}\xi,\ |m|\leq N,$$

(5.47)

where

$$H_m'(t,x)=\frac{\partial}{\partial t}H_m(t,x)=-m^2\left[\frac{J_1(\zeta)}{\zeta}-m^2t^2\frac{J_2(\zeta)}{\zeta^2}\right]_{\zeta=\sqrt{t^2-x^2}}.$$

It should be noted that integral equations (5.47) determine $v_m(x,t)$ as a function of $p_m(x)$, $m=-N,\ldots,N$. Hence, introducing the operator $U=(U_{-N},\ldots,U_N)$ by the right sides of (5.47), we can rewrite relations (5.47) as the operator equations

$$p_m(x)=U_m\left(p_{-N}(x),\ldots,p_N(x)\right),\ m=0,\pm1,\pm2,\ldots,\pm N,\ t\in[0,\ T/2].\tag{5.48}$$

Let

$$p_m^0(x)=-2g_m''(2x),\ m=0,\pm1,\pm2,\ldots,\pm N.$$

Denote by $\Omega_0\left(N,X,P_0\right)$ the set of functions $p_m(x)$, $-N\leq m\leq N$, satisfying the conditions

$$\left\|p_m(x)-p_m^0(x)\right\|_{C[0,X]}\leq P_0,\ P_0=\max_{-N\leq m\leq N}\left\|p_m^0(x)\right\|_{C[0,X]},$$

$$m=0,\pm1,\pm2,\ldots,\pm N.$$

If functions $p_m(x)\in\Omega_0\left(N,X,P_0\right)$ for $-N\leq m\leq N$ then, obviously, $p_m(x)\in\Omega\left(N,X,2P_0\right)$. For operator equations (5.48) the following theorem holds.

Theorem 5.2 *Let presentation (5.45) be valid. Moreover, let data (5.46) satisfy the following conditions:*

$$g_m(t)\in C^2[0,\ T],\ g_m(0)=g_m'(0)=0,\ m=0,\pm1,\pm2,\ldots,\pm N.$$

and

$$G=\max_{-N\leq m\leq N}\left\|g_m''(t)\right\|_{C[0,T]}.$$

Then, there exists a number $T_0\in(0,\ T)$ such that operator equations (5.48) have a unique solution, belonging to the set $\Omega_0\left(N,T_0/2,2G\right)$.

The **proof** of the theorem can be fulfilled quite identically to the corresponding theorem from the previous section. Therefore, we omit it here.

5.3 Conclusions

This chapter dealt with the inverse problems of recovering a memory kernel in hyperbolic integro-differential equations, having a special dependence with respect to one of the spatial variables. In the first section it is supposed that the unknown kernel is a trigonometric polynomial in the spatial variables with continuous coefficients with respect to the time variable. In the second section it is determined the two-dimensional spatial part of the three-dimensional kernel, and wherein it is supposed that the unknown function is a trigonometric polynomial with respect to one of the spatial variables with continuous coefficients with respect to another. Local solvability theorems and stability estimates for the solution to the inverse problem are obtained.

References

1. Durdiev, D.K., and U.D. Durdiev. 2016. The problem of kernel determination from viscoelasticity system integro-differential equations for homogeneous anisotropic media. *Nanosystems: Physics, Chemistry, Mathematics* 7 (3): 405–409.
2. Durdiev, U.D., and Z.D. Totieva. 2019. A problem of determining a special spatial part of 3D memory kernel in an integro-differential hyperbolic equation. *Mathematical methods in Applied Sciences* 42 (18): 7440–7451.
3. Romanov, V.G. 2016. A problem of recovering a special two-dimention potential in a hyperbolic equation. Eurasian Journal of Mathematical and Computer Applications 7 (6): 573–588.

Chapter 6
Some Inverse Problems of Viscoelasticity System with the Known Kernel

In this chapter, we consider various inverse problems of viscoelasticity system with the known kernel.

The inverse problems for the system of elasticity equations have been considered by many authors. A detailed review of the sources, available in Russian and foreign literature before 1990, is given in the monograph [1], which presents the results of unique solvability, stability estimates, and numerical methods for constructing solutions for many formulations of inverse problems of isotropic and anisotropic elasticity. These studies were continued in the article [2]. It deals with direct and inverse coefficient problems of anisotropic electroelastic medium of cubic structure. Namely, the elastic modulus and piezoelectric modulus are determined from some information about the solution of direct problems. Note also works [3–9], in which the Lame parameters are defined for the equations of isotropic elasticity. The modules of elasticity and piezo module (polarized ceramic) of anisotropic electro-elasticity are also defined there.

One can mention the work [10], related to taking into account the memory effect. Under study is an inverse problem of determining a spatially varying source term in a thermoelastic medium with a memory effect. Based on Carleman's estimate, authors establish a Hölder stability for the inverse source problem only by making a displacement measurement on a given subdomain for sufficiently large times, provided the source is to be known near the boundary. The uniqueness of such an inverse problem is yielded as a direct result.

In the first section of this chapter, we consider a multidimensional inverse problem related to the determination of the function $\rho(x_2, x_3)$, which is the coefficient of the system of integro-differential equations of viscoelasticity (the system of differential Lame equations with memory) in the linear approximation. Some information regarding the solution of the direct linearized problem is given.

© The Author(s), under exclusive license to Springer Nature Singapore Pte Ltd. 2023

D. K. Durdiev and Z. D. Totieva, *Kernel Determination Problems in Hyperbolic Integro-Differential Equations*, Infosys Science Foundation Series in Mathematical Sciences, https://doi.org/10.1007/978-981-99-2260-4_6

The linearization method in inverse problems for hyperbolic equations has been used by some authors. One of the first works was the work of Lavrentiev and Romanov [11], and later, the inverse Lamb problem was solved in the linear approximation [12]. Kabanikhin [13] applied the linearization method to solve two-dimensional inverse problems for the wave equation. In the monograph, Yakhno [1] the linearization principle has found its application in the study of multidimensional inverse problems of determining density for a system of dynamic equations of elasticity. This monograph describes the arguments that allow us to introduce the so-called linearized statements of direct and inverse problems.

First, the one-dimensional inverse problem of determining the density for a system of equations of isotropic elasticity in half-space was studied by Alekseev [14, 15]. Various statements of coefficient inverse problems of elasticity theory are presented in the works of Blagoveshchensky [16–18], Romanov and Volkova [19]. The results of these studies were reflected in the works of Yakhno and Melnikova [20], in particular, the existence, uniqueness, and stability theorems of multidimensional problems of determining the density function for systems of equations of isotropic and anisotropic elasticity. The results obtained are the basis of the theory of inverse problems with pulsed perturbation sources at the boundary of media.

In many articles published over the past ten years, the works [21, 22] can be noted as close to the material of this chapter. In [21] for an integro-differential equation (two-dimensional viscoelasticity problem), the problem of determining the density, elastic modulus, and the spatial part of the kernel is studied, and the unknown functions differ from the given constants only inside the unit circle $D = \{x \in \mathbb{R}^2 \mid |x| < 1\}$. As additional information, one considered a set of solutions to the direct problem, which corresponds to the pulse sources localized on straight lines. Solution traces for a finite time interval are specified on the boundary of the domain D. The problem of determining the Lame coefficients and two kernels of the viscoelasticity system in the three-dimensional case for a given density of the medium is solved in [22].

The second section deals with the problem of determining the thermal-expansion coefficient $\alpha(z),\ z \in [0, Z]$, occurring in the system of integro-differential termo-viscoelasticity equations. The medium density and the Lame parameters are assumed to be functions of one variable. The integrand $h(t),\ t \in [0; T]$ is known. The inverse problem is replaced by the equivalent integral equation for unknown functions. The theorem of unique solvability is proved, and the stability estimate of solving the inverse problem is obtained.

In the third section, we investigate the problem of determining the moduli of elasticity $c_{11}(x_3),\ c_{12}(x_3),\ c_{44}(x_3),\ x_3 > 0$, occurring in the system of integro-differential viscoelasticity equations for homogenous anisotropic medium. The density of medium is constant. The matrix kernel $k(t) = \mathrm{diag}(k_1, k_2, k_3)(t),\ t \in [0, T]$ is known. Additional information is the Fourier transform of the first and third component of the displacements vector for $x_3 = 0$. The results are the theorems of the existence of a unique solution to the inverse problems and the theorems of stability.

Note that the problems discussed in this chapter are primarily interesting for practical purposes (in numerical implementation and analyses of direct problem solution with various kernels).

6.1 The Multidimensional Linearized Problem of Determining the Density Function for a Viscoelasticity System

6.1.1 Statement of Problems and Main Results

The purpose of the research presented in this section is to solve the inverse problem of determining the density of the medium $\rho(x_2, x_3),\ x_2 \in \mathbb{R},\ x_3 > 0$ for the viscoelasticity system (three-dimensional case) from some information about the solution of the direct problem. The Dirac delta function, defined at the boundary of the considered spatial domain, is used as the source of wave propagation. This study is based on the results of monographs [1, 23]. The main results are theorems on necessary and sufficient conditions for the existence of a unique solution to the inverse problem, the stability theorem in a special class of functions, and the convergence theorem. The direct problem is pre-investigated to the extent necessary for the study of the inverse problem. The fundamental solution for the hyperbolic operator with memory is studied.

Let's consider a system of integro-differential equations in the domain $\Omega = \{(x_1, x_2, t) \in \mathbb{R}^3,\ x_3 > 0\}$

$$\rho_0(x_3)\frac{\partial^2 u_i^0}{\partial t^2} = \sum_{j=1}^{3} \frac{\partial T_{ij}^0}{\partial x_j}, \quad i = 1, 2, 3, \tag{6.1}$$

$$u_i^0\big|_{t<0} = 0, \tag{6.2}$$

$$T_{3i}^0\big|_{x_3=+0} = -\delta_{1i}\frac{\delta(t)}{2}, \tag{6.3}$$

where $u^0(x, t) = \big(u_1^0(x, t), u_2^0(x, t), u_3^0(x, t)\big)$ is the displacement vector, $x = (x_1, x_2, x_3)$, $\delta(t)$ is Dirac delta function; δ_{ij} is Kronecker symbol; and T_{ij} is the stress tensor:

$$T_{ij}^0(x, t) = \sigma_{ij}[u^0] + \int_0^t k(t - \tau)\sigma_{ij}[u^0]\,d\tau, \tag{6.4}$$

σ_{ij} are stresses for which, by Hooke's law, the following expression holds

$$\sigma_{ij}[u^0] = \mu_0(x_3)\left(\frac{\partial u_i^0}{\partial x_j} + \frac{\partial u_j^0}{\partial x_i}\right) + \delta_{ij}\lambda_0(x_3)\operatorname{div} u^0. \tag{6.5}$$

$k(t)$ is known and satisfies conditions:

$$k(t) \in C^2[0, T],\ k(0) = 0,\ k'(0) = 0,\ T > 0 \text{ is fixed}. \tag{6.6}$$

We also assume that functions $\rho = \rho(x_3),\ \mu = \mu(x_3),\ \lambda = \lambda(x_3)$ are known and belong to the set Λ:

$$\Lambda := \Big\{\rho(x_3), \lambda(x_3), \mu(x_3) \in C^2([0,\infty))\Big|\rho(x_3) > 0,\ \mu(x_3) > 0,\ \lambda(x_3) > 0;$$

$$\rho'(+0) = \mu'(+0) = \lambda'(+0) = 0\Big\}.$$

The system of Eqs. (6.1)–(6.5) arises in the theory of viscoelastic media with variable density ρ and the Lame coefficients $\lambda,\ \mu$. In applications the integral term responsible for the prehistory of the process enters the system of equations by means of the formula (6.4).

Direct problem for an isotropic viscoelasticity system (the case of vertically inhomogeneous media): determine the vector function $u^0(x,t) = \big(u_1^0(x,t), u_2^0(x,t), u_3^0(x,t)\big)$, satisfying the Eqs. (6.1)–(6.3) in the domain Ω for given $k(t)$ with (6.6) and $\rho_0(x_3)$, $\mu_0(x_3),\ \lambda_0(x_3) \in \Lambda$.

Let $\rho_1(x_2, x_3),\ k(t)$ are known functions, $k(t)$ satisfies conditions (6.6), $u^0(x,t)$ is the solution (6.1)–(6.3).

Direct linearized problem for an isotropic viscoelasticity system: determine the vector function $u^1(x,t) = \big(u_1^1(x,t), u_2^1(x,t), u_3^1(x,t)\big)$, satisfying in the domain Ω the following equations

$$\rho_0(x_3)\frac{\partial^2 u_i^1}{\partial t^2} = \sum_{j=1}^{3}\frac{\partial T_{ij}^1}{\partial x_j} - \rho_1(x_2,x_3)\frac{\partial^2 u_i^0}{\partial t^2},\quad i = 1,2,3, \tag{6.7}$$

$$u_i^1\big|_{t<0} = 0, \tag{6.8}$$

$$T_{3i}^1\big|_{x_3=+0} = 0, \tag{6.9}$$

where $T_{ij}^1(x,t)$ is defined by (6.4), (6.5) (the index "0" is replaced by "1").

The linearized statement of the direct problem is obtained from the original problem by linearization. For clarity, we give the *primary statement* of the problem:

For the given functions $\rho(x_2, x_3),\ k(t)$ determine vector function $u(x,t) = \big(u_1(x,t), u_2(x,t), u_3(x,t)\big)$, satisfying the following equalities in the domain Ω

$$\rho(x_2,x_3)\frac{\partial^2 u_i}{\partial t^2} = \sum_{j=1}^{3}\frac{\partial T_{ij}}{\partial x_j},\quad i = 1,2,3, \tag{6.10}$$

$$u_i\big|_{t<0} = 0, \tag{6.11}$$

$$T_{3i}\big|_{x_3=+0} = -\delta_{1i}\frac{\delta(t)}{2}, \tag{6.12}$$

with $T_{ij}(x,t) = \sigma_{ij}[u] + \int_0^t k(t-\tau)\sigma_{ij}[u]\,d\tau$.

In fact, the principle of linearization is not mathematically substantiated. We are talking about some assumptions under which the problem in the original formulation is simplified. Let we describe the arguments [1], which allows us to take into account linearized assertion. Suppose that

$$\rho(x_2, x_3) = \rho_0(x_3) + \widetilde{\rho}_1(x_2, x_3),$$
$$\mu(x) = \mu_0(x_3) + \widetilde{\mu}_1(x), \quad \lambda(x) = \lambda_0(x_3) + \widetilde{\lambda}_1(x),$$

where $\widetilde{\rho}_1(x_2, x_3), \widetilde{\mu}_1(x), \widetilde{\lambda}_1(x)$ are small by absolute value in comparison with $\rho_0(x_3)$, $\mu_0(x_3)$, $\lambda_0(x_3) \in \Lambda$. Suppose that $\widetilde{\mu}_1(x)$, $\widetilde{\lambda}_1(x)$ are identically equal to zero. We introduce the parameter of smallness ν so that

$$\rho(x_2, x_3) = \rho_0(x_3) + \nu\rho_1(x_2, x_3), \tag{6.13}$$

where $\rho_1(x_2, x_3)$ are comparable in order with $\rho_0(x_3)$.

The solution of the direct problem in the original formulation of $u(x,t)$ is represented as:

$$u(x,t) = \tilde{u}^0(x,t) + \tilde{u}^1(x,t) + \cdots, \tag{6.14}$$

where $\tilde{u}^l(x,t) = \nu^l u^l(x,t)$, $l = 0, 1, 2, \ldots$. Substituting (6.13)–(6.14) into (6.10)–(6.12) and equating the coefficients at ν^l, $l = 0, 1$, we obtain (6.1)–(6.3) and (6.7)–(6.9). The transition form (6.10)–(6.12) to (6.1)–(6.3) and (6.7)–(6.9) is justified if it is known a priori that on the right-hand side of relation (6.14) the contribution of terms with terms for u^l, $l = 2, 3, \ldots$ is negligible.

We introduce classes of functions in which the main results are obtained. For a fixed positive number T, we introduce the set:

$$U(T) = \Big\{ u(x,t) = \big(u_1, u_2, u_3\big)(x,t) \Big| \frac{\partial u}{\partial x_l} \in C\left((0,T);\ W_2^1(\mathbb{R}^2 \times \mathbb{R}_+)\right),$$
$$\frac{\partial}{\partial t}\left(\frac{\partial u}{\partial x_l}\right) \in C\left((0,T);\ W_2^0(\mathbb{R}^2 \times \mathbb{R}_+)\right),$$
$$\frac{\partial u}{\partial x_l}\Big|_{t=+0} = 0,\ \frac{\partial}{\partial t}\left(\frac{\partial u}{\partial x_l}\right)\Big|_{t=+0} = 0,\ l = 1, 2.\Big\}$$

$$\bigcap \big\{ u(x,t) \mid u(x_1, x_2, -x_3, t) = u(x_1, x_2, x_3, t),$$
$$u(x,t) \in C^{0,0}_{x_1,x_2}\left(\mathbb{R}^2;\ L_2(\mathbb{R} \times (0,T)) \cap C^2(\nabla(T))\right) \big\},$$

where

$$\nabla(T) = \{(x_3, t) \mid \psi(x_3) \leqslant t \leqslant T\},\ \psi(x_3) = \int_0^{x_3} \frac{d\xi}{c(\xi)},\ c^2(\xi) = \frac{\mu_0(\xi)}{\rho_0(\xi)},$$

F_{x_2} is Fourier transform operator with respect to the variable x_2 with the parameter ν. Further, by $\psi^{-1}(y)$ we denote the function inverse to $\psi(x_3)$.

Let $s,\ X$ be a fixed positive numbers. We assume that $\bar{\rho}_1(\nu, x_3) = F_{x_2}[\rho_1](\nu, x_3) \in \bar{\Lambda}(s, X)$ if and only if $\bar{\rho}_1(\nu, x_3) \in C(\mathbb{R} \times \mathbb{R}_+)$ and for all fixed $x_3 \in [0, X]$ *supp* $\bar{\rho}_1(\nu, x_3) \subset [-s, s]$. Respectively, $\rho_1(x_2, x_3) \in \Lambda(s, X)$ if and only if $\bar{\rho}_1(\nu, x_3) = F_{x_2}[\rho_1](\nu, x_3) \in \bar{\Lambda}(s, X)$.

For given positive numbers $s,\ T,\ X = \psi^{-1}(T/2)$ we introduce the class of functions $U_1(s, T)$ as follows:

$$U_1(s,T) = \Big\{u(x,t) = (u_1(x,t), u_2(x,t), u_3(x,t)) \mid$$

$$\frac{\partial u_l}{\partial x_1} \in C((0,T);\, W_2^1(\mathbb{R}^2 \times \mathbb{R}_+)),\ \frac{\partial}{\partial t}\left(\frac{\partial u_l}{\partial x_1}\right) \in C((0,T);\, W_2^0(\mathbb{R}^2 \times \mathbb{R}_+)),$$

$$\frac{\partial u_l}{\partial x_1} |_{t=+0}= 0,\ \frac{\partial}{\partial t}\left(\frac{\partial u_l}{\partial x_1}\right) |_{t=+0}= 0,\ l = 1,2,3\Big\}$$

$$\cap \Big\{u(x,t) = (u_1, u_2, u_3) \mid u_l \in C((0,T);\, W_2^1(\mathbb{R}^2 \times \mathbb{R}_+)),$$

$$\frac{\partial u_l}{\partial t} \in C((0,T);\, W_2^0(\mathbb{R}^2 \times \mathbb{R}_+)),\ u_l |_{t=+0}= 0,\ \frac{\partial u_l}{\partial t} |_{t=+0}= 0,\ l = 2,3;$$

$$u_1 = u_1(x_2, x_3, t),\ u_1(x_2, -x_3, t) = u_1(x_2, x_3, t),$$

$$u_1(x_2, x_3, t) = \theta(t - |\phi(x_3)|) F_\nu^{-1}[\bar{u}_1(x_3, t, \nu)],$$

$$\bar{u}_1(x_3, t, \nu) \in C(\mathbb{R} \times \nabla(T))$$

and for fixed $(x_3, t) \in \nabla(T)$ *supp* $\bar{u}_1(x_3, t, \nu) \subset [-s, s]\Big\}$, where F_ν^{-1} is an operator inverse to the Fourier transform operator F_{x_2}, $\theta(t)$ is Heaviside function.

Lemma 6.1 *Let $T > 0$ be a fixed number. Functions $k(t)$ with (6.6), $\lambda_0(x_3)$, $\mu_0(x_3)$, $\rho_0(x_3) \in \Lambda$ are known. Then there is a unique solution $u^0(x, t) \in U(T)$ to the problem (6.1)–(6.3).*

Lemma 6.2 *Let s, T be a fixed positive numbers, $X = \psi^{-1}(T/2)$. Functions $k(t)$ with (6.6), $\lambda_0(x_3)$, $\mu_0(x_3)$, $\rho_0(x_3) \in \Lambda$ are known. Let $u_0(x, t)$ is the solution of the direct problem (6.1)–(6.3), $\rho_1(x_2, x_3) \in \Lambda(s, X)$ is known fixed function. Then there is a unique solution $u^1(x, t) \in U_1(s, T)$ to the problem (6.7)–(6.9).*

Inverse problem: determine function $\rho_1(x_2, x_3) \in \Lambda(s, X)$ in (6.7), if the some information on $u^1(x, t)$ is known:

$$F_{x_2}[u_1^1](x_1, \nu, x_3, t)|_{x_1=0, x_3=+0} = h(\nu, t), \tag{6.15}$$

$(\nu, t) \in \mathbb{R} \times [0, T]$.

Let

$$H(\nu, t) := h(\nu, t) + \int_0^t k(t - \tau) h(\nu, \tau) d\tau.$$

The main results are the following theorems:

Theorem 6.1 *Let s, T be fixed positive numbers, $X = \psi^{-1}(T/2)$. For the existence of a unique solution to the inverse problem $\rho_1(x_2, x_3) \in \Lambda(s, X)$ it is necessary and sufficient that $H(\nu, t) \in C(\mathbb{R} \times [0, T])$ and for any fixed $t \in [0, T]$ supp $H(\nu, t) \subset [-s, s]$.*

Theorem 6.2 *Let s, T be fixed positive numbers, $X = \psi^{-1}(T/2)$; $\rho_1(x_2, x_3)$, $\rho_1^*(x_2, x_3) \in \Lambda(s, X)$ are two solutions of the inverse problem that responds to information $H(\nu, t)$ and $H^*(\nu, t)$, respectively. Then, there is a stability estimate:*

$$\max_{x_3 \in [0, X]} \int_R |\rho_1(x_2, x_3) - \rho_1^*(x_2, x_3)|^2 dx_2 \le \tilde{C} \max_{t \in [0, T]} \int_{-s}^{s} |H(\nu, t) - H^*(\nu, t)|^2 d\nu,$$

where $\tilde{C}$ is constant depending on the values of s, T, $k(t)$, $\mu_0(x_3)$, $\lambda_0(x_3)$, $\rho_0(x_3)$.

Let there be a solution $\rho_1(x_2, x_3) \in L_2(\mathbb{R} \times \mathbb{R}_+)$ and information $H(\nu, t)$. We define the set of functions $H_s(\nu, t)$ by the rule:

$$H_s(\nu, t) := \theta(s - |\nu|) H(\nu, t).$$

We highlight a family of inverse problems: determine a function $\rho_1^s(x_2, x_3) = F_\nu^{-1}[\bar{\rho}^s(\nu, x_3)]$ on data values $H_s(\nu, t)$. We formulate a convergence theorem.

Theorem 6.3 *This family is regularized, that is:*

(1) for every $s > 0$ the inverse problem is correct;
(2) if the data are such that the solution of the original (correct) problem exists, then, at $s \to \infty$ a sequence of solutions the problem of the family with data $H_s(\nu, t)$ tends to the solution of the original (ill-posed) problem.

6.1.2 Direct Problem of Isotropic Viscoelasticity and Properties of Its Solution

To prove Lemma 6.1, we first prove one auxiliary statement.

Lemma 6.3 *Let $T > 0$ be fixed, $V^1(x, t) = (V_1^1(x, t), V_2^1(x, t), V_3^1(x, t))$ is defined in domain $x \in \mathbb{R}^3$, $t \in [0, T]$, belongs to $C\left((0, T);\ W_2^1(\mathbb{R}^3)\right)$, $\frac{\partial V^1}{\partial t}(x, t) \in C\left((0, T);\ W_2^0(\mathbb{R}^3)\right)$ and satisfy the equalities:*

$$\rho_0(x_3) \frac{\partial^2 V_i^1}{\partial t^2} = \sum_{j=1}^{3} \frac{\partial T_{ij}^*}{\partial x_j}, \quad i = 1, 2, 3, \tag{6.16}$$

$$V_i^1\big|_{t<0} = 0, \tag{6.17}$$

where

$$T_{ij}^*(x,t) = \sigma_{ij}[V^1] + \int_0^t k(t-\tau)\sigma_{ij}[V^1]\,d\tau,$$

$$\sigma_{ij}[V^1] = \mu_0(x_3)\left(\frac{\partial V_i^1}{\partial x_j} + \frac{\partial V_j^1}{\partial x_i}\right) + \delta_{ij}\lambda_0(x_3)\,\mathrm{div}\,V^1,$$

$\lambda_0(x_3)$, $\mu_0(x_3)$, $\rho_0(x_3) \in \Lambda$ *are given and* $k(t)$ *satisfy (6.6).*

Then $V^1(x,t) \equiv 0$.

Proof Let $v_i^1(x,t) = V_i^1(x,t) + k * V_i^1(x,t)$ (the symbol "$*$" means a convolution operation on a variable t), then,

$$V_i^1(x,t) = v_i^1(x,t) + r * v_i^1(x,t),$$

where $r(t) = -k(t) - \int_0^t k(t-\tau)r(\tau)\,d\tau$, moreover, $r(t)$ has the same properties as $k(t)$. Therefore, there exists a constant K, such that

$$\max_{t\in[0,T]} |r'(t)| \leqslant K.$$

In terms in $v_i^1(x,t)$ we rewrite (6.16)–(6.17) :

$$\rho_0(x_3)\frac{\partial^2 v_i^1}{\partial t^2} = \sum_{j=1}^{3} \frac{\partial \sigma_{ij}[v^1]}{\partial x_j} - \int_0^t r'(\tau)\frac{\partial v_i^1}{\partial t}(x,t-\tau)\,d\tau, \quad i = 1,2,3, \tag{6.18}$$

$$v_i^1\big|_{t<0} = 0. \tag{6.19}$$

We reduce (6.18)–(6.19) to the system differential equations for $p_i := (v_i^1)_t$ and $\sigma_{11}, \sigma_{12}, \sigma_{13}, \sigma_{22}, \sigma_{23}, \sigma_{33}$, $i = 1,2,3$.

Denote $-g_i(x,t) = \int_0^t r'(t-\tau)p_i(x,\tau)\,d\tau$, $p = (p_1, p_2, p_3)$. The resulting system has the following form:

$$\begin{aligned} \rho_0(x_3)\frac{\partial p_i}{\partial t} &= \sum_{j=1}^{3} \frac{\partial \sigma_{ij}}{\partial x_j} + g_i(x,t), \\ \frac{\partial \sigma_{ij}}{\partial t} &= \mu_0(x_3)\left(\frac{\partial p_i}{\partial x_j} + \frac{\partial p_j}{\partial x_i}\right) + \delta_{ij}\lambda_0(x_3)\,\mathrm{div}\,p. \end{aligned} \tag{6.20}$$

Note that the following equality holds:

$$v_i^1(x,t) = v_i^1(x,0) + \int_0^t \frac{\partial v_i^1}{\partial \tau}(x,\tau)\,d\tau. \tag{6.21}$$

Denote $Y = (p_1, p_2, p_3, \sigma_{11}, \sigma_{12}, \sigma_{13}, \sigma_{22}, \sigma_{23}, \sigma_{33}) = (Y_i)_{i=1}^9$.
For system (6.20), the energy inequality holds [24]:

$$\|Y\|(t) \leqslant C \int_0^t e^{\gamma(t-\tau)} \|g\|(\tau)\,d\tau,$$

where C, γ are constant depends on T, and $\|Y\|(t) = \left(\sum_{j=1}^9 \int_{\mathbb{R}^3} Y_j^2(x,t)\,\mathrm{d}x\right)^{1/2}$, $\|g\|(t) = \left(\sum_{j=1}^3 \int_{\mathbb{R}^3} g_j^2(x,t)\,\mathrm{d}x\right)^{1/2}$, $\|p\|(t) = \left(\sum_{j=1}^3 \int_{\mathbb{R}^3} p_j^2(x,t)\,\mathrm{d}x\right)^{1/2}$. So as $\|Y\|(t) \geq \|p\|(t)$, we get

$$\begin{aligned}
\|p\|(t) &\leqslant C \int_0^t e^{\gamma(t-\tau)} \left(\sum_{j=1}^3 \int_{\mathbb{R}^3} g_j^2(x,t)\,\mathrm{d}x\right)^{1/2} d\tau \\
&\leqslant CM \int_0^t \left(\sum_{j=1}^3 \int_{\mathbb{R}^3} \left(\int_0^\tau r'(\tau-\eta) p_j(x,\eta)\,d\eta\right)^2 \mathrm{d}x\right)^{1/2} d\tau \\
&\leqslant CMKT \int_0^t \|p\|(\tau)\,d\tau,
\end{aligned}$$

where M is a constant value that restricts from above the values of the exponential function to the domain $[0, T]$.

By Gronwall's lemma, we conclude the following: $\|p\|(t) = 0 \Rightarrow (v_i^1)_t(x,t) \equiv 0$. From the equality (6.21) it follows that $v_i^1(x,t) \equiv v_i^1(x,0) \equiv 0$. So as $V_i^1(x,t) = v_i^1(x,t) + r * v_i^1(x,t)$ we get $V_i^1(x,t) \equiv 0$ in $\mathbb{R}^3 \times [0,T]$. Thus, Lemma 6.3 proven.

Proof Now we prove Lemma 6.1. Let $u^0(x,t) \in U(T)$. Consider the system (6.1) with the conditions (6.2), (6.3). We differentiate the right and left sides of the system with respect to the variable x_1 and introduce the notation $w = \frac{\partial u^0}{\partial x_1}$. For the function w, extended into the domain $x_3 < 0$ in an even way, we get (6.16), (6.17), where $V^1(x,t) \equiv w(x,t)$. Using the result of Lemma 6.3, we see that $w \equiv 0$, $(x,t) \in \mathbb{R}^2 \times \mathbb{R}_+ \times [0,T]$. Similarly, $\frac{\partial u^0}{\partial x_2} = 0$ if $(x,t) \in \mathbb{R}^2 \times \mathbb{R}_+ \times [0,T]$. Thus, $u^0(x,t) \in U(T)$ does not depend on x_1 and x_2, i. e. $u^0 = u^0(x_3,t)$. Therefore, the problem of defining $u^0(x,t) \in U(T)$ can be rewritten as a vector function definition problem $u^0(x_3,t)$, for which the following relations are valid for $x_3 > 0$:

$$\rho_0(x_3)\frac{\partial^2 u_{1,2}^0}{\partial t^2} = \frac{\partial}{\partial x_3}\left(\mu_0(x_3)\frac{\partial u_{1,2}^0}{\partial x_3}\right) + \int_0^t k(t-\tau)\frac{\partial}{\partial x_3}\left(\mu_0(x_3)\frac{\partial u_{1,2}^0}{\partial x_3}\right) d\tau,$$

$$\rho_0(x_3)\frac{\partial^2 u_3^0}{\partial t^2} = \frac{\partial}{\partial x_3}\left((\lambda_0(x_3) + 2\mu_0(x_3))\frac{\partial u_3^0}{\partial x_3}\right)$$
$$+ \int_0^t k(t-\tau)\frac{\partial}{\partial x_3}\left((\lambda_0(x_3) + 2\mu_0(x_3))\frac{\partial u_3^0}{\partial x_3}\right) d\tau,$$

$$u_i^0\big|_{t<0} = 0, \quad i = 1, 2, 3,$$

$$\left[\mu_0(x_3)\frac{\partial u_1^0}{\partial x_3} + \int_0^t k(t-\tau)\mu_0(x_3)\frac{\partial u_1^0}{\partial x_3}\, d\tau\right]_{x_3=+0} = -\frac{\delta(t)}{2},$$

$$\left[\mu_0(x_3)\frac{\partial u_2^0}{\partial x_3} + \int_0^t k(t-\tau)\mu_0(x_3)\frac{\partial u_2^0}{\partial x_3}\, d\tau\right]_{x_3=+0} = 0,$$

$$\left[(\lambda(x_3)+2\mu_0(x_3))\frac{\partial u_3^0}{\partial x_3} + \int_0^t k(t-\tau)(\lambda(x_3)+2\mu_0(x_3))\frac{\partial u_3^0}{\partial x_3}\, d\tau\right]_{x_3=+0} = 0.$$

From these equalities it follows that the solution of the direct problem falls into three independently solved by definition components (u_1^0, u_2^0, u_3^0). For $u_2^0(x_3, t)$, $u_3^0(x_3, t)$ we get a problem with zero data. Continuing in an even manner the data functions to the domain $x_3 < 0$ and using the result of Lemma 6.3 (assuming $V^1(x, t) \equiv (u_2^0(x_3, t), u_3^0(x_3, t)))$, we find $u_2^0(x_3, t) \equiv u_3^0(x_3, t) \equiv 0$. For $u_1^0(x_3, t)$ we have:

$$\rho_0(x_3)\frac{\partial^2 u_1^0}{\partial t^2} = \frac{\partial}{\partial x_3}\left(\mu_0(x_3)\frac{\partial u_1^0}{\partial x_3}\right) + \int_0^t k(t-\tau)\frac{\partial}{\partial x_3}\left(\mu_0(x_3)\frac{\partial u_1^0}{\partial x_3}\right) d\tau, \quad (6.22)$$

$$u_1^0\big|_{t<0} = 0, \quad (6.23)$$

$$\left[\mu_0(x_3)\left(\frac{\partial u_1^0}{\partial x_3} + \int_0^t k(t-\tau)\frac{\partial u_1^0}{\partial x_3}\, d\tau\right)\right]_{x_3=+0} = -\frac{\delta(t)}{2}. \quad (6.24)$$

Let

$$\frac{u_1^0(\psi^{-1}(y), t)}{S(y)} = V^0(y, t), \quad S(y) = \sqrt{\frac{c(+0)\rho_0(+0)}{c(\psi^{-1}(y))\rho_0(\psi^{-1}(y))}}.$$

Then, equality (6.22)–(6.24) will take the form:

$$\frac{\partial^2 V^0}{\partial t^2} = \frac{\partial^2 V^0}{\partial y^2} + q(y)V^0 + \int_0^t k(t-\tau)\left(\frac{\partial^2 V^0}{\partial y^2} + q(y)V^0\right) d\tau, \tag{6.25}$$

$$V^0\big|_{t<0} = 0, \tag{6.26}$$

$$\left[\frac{\partial V^0}{\partial y} + \int_0^t k(t-\tau)\frac{\partial V^0}{\partial y}\, d\tau\right]_{y=+0} = -\frac{a_0}{2}\,\delta(t), \tag{6.27}$$

$$q(y) = \frac{S''(y)}{S(y)} - 2\left(\frac{S'(y)}{S(y)}\right)^2, \quad a_0 = \frac{1}{\sqrt{\mu_0(+0)\rho_0(+0)}}.$$

Let $V^0(y,t) + k * V^0(y,t) = Y^0(y,t)$, $V^0(y,t) = Y^0(y,t) + r * Y^0(y,t)$. We rewrite (6.25)–(6.27) in terms of $Y^0(y,t)$:

$$\frac{\partial^2 Y^0}{\partial t^2} = \frac{\partial^2 Y^0}{\partial y^2} + q(y)Y^0 - \frac{\partial^2}{\partial t^2}[r * Y^0], \tag{6.28}$$

$$Y^0\big|_{t<0} = 0, \tag{6.29}$$

$$\frac{\partial Y^0}{\partial y}\bigg|_{y=+0} = -\frac{a_0}{2}\,\delta(t). \tag{6.30}$$

The solution (6.28)–(6.30) has the structure

$$Y^0(y,t) = \frac{a_0}{2}\cdot\theta(t-|y|) + \widetilde{Y}^0(y,t),$$

where $\widetilde{Y}^0(y,t) \in C^1\big((0,T);\ L_2(\mathbb{R})\big) \cap C^0\big((0,T);\ W_2^1(\mathbb{R})\big)$. For the new function $\widetilde{Y}^0(y,t)$ we get:

$$\begin{aligned}\frac{\partial^2 \widetilde{Y}^0}{\partial t^2} = \frac{\partial^2 \widetilde{Y}^0}{\partial y^2} &+ q(y)\frac{a_0}{2}\theta(t-|y|) + q(y)\widetilde{Y}^0\\ &- \frac{\partial^2}{\partial t^2}\left[\frac{a_0}{2}\theta(t-|y|)\int_{|y|}^t r(t-\tau)\, d\tau\right] - \frac{\partial^2}{\partial t^2}\left[\int_0^t r(t-\tau)\widetilde{Y}^0(y,\tau)\, d\tau\right],\end{aligned} \tag{6.31}$$

$$\widetilde{Y}^0\big|_{t=+0} = 0, \quad \frac{\partial \widetilde{Y}^0}{\partial t}\bigg|_{t=+0} = 0, \tag{6.32}$$

$$\left.\frac{\partial \widetilde{Y}^0}{\partial y}\right|_{y=+0} = 0. \tag{6.33}$$

We extend $q(y)$, $\widetilde{Y}^0$ in an even way to the domain $y < 0$. The problem (6.31)–(6.33) is equivalent to the problem of determining even function by y $\widetilde{Y}^0(y,t)$, satisfying (6.31), (6.32). For a fixed $T > 0$, there exists a unique even by y solution to the problem (6.31), (6.32).

We prove the uniqueness of the solution. Let's say, that there are two solutions $\widetilde{Y}_1^0(y,t)$ and $\widetilde{Y}_2^0(y,t)$. Denote $\overline{Y}^0(y,t) = \widetilde{Y}_1^0(y,t) - \widetilde{Y}_2^0(y,t)$. Then, for the function $\overline{Y}^0(y,t)$ we get the problem:

$$\frac{\partial^2 \overline{Y}^0}{\partial t^2} = \frac{\partial^2 \overline{Y}^0}{\partial y^2} + q(y)\overline{Y}^0 - \frac{\partial^2}{\partial t^2}\left[\int_0^t r(t-\tau)\overline{Y}^0(y,\tau)\,d\tau\right],$$

$$\left.\overline{Y}^0\right|_{t=+0} = 0, \quad \left.\frac{\partial \overline{Y}^0}{\partial t}\right|_{t=+0} = 0.$$

Using the energy estimates [25], we deduce:

$$\begin{aligned}
\left\|\overline{Y}^0\right\|_{L_2(\mathbb{R})}(t) &= \left(\int_{\mathbb{R}} \left(\overline{Y}^0\right)^2(y,t)\,dy\right)^{1/2} \\
&\leqslant C\int_0^t \left(\int_{\mathbb{R}} \left[q(y)\overline{Y}^0(y,\tau) - \int_0^\tau r_{\tau\tau}(\tau-\eta)\overline{Y}^0(y,\eta)\,d\eta\right]^2 dy\right)^{1/2} d\tau \\
&\leqslant C\int_0^t \left(\int_{\mathbb{R}} \left[q\overline{Y}^0\right]^2 dy + \int_{\mathbb{R}} \left[\int_0^\tau r_{\tau\tau}\overline{Y}^0(y,\eta)\,d\eta\right]^2 dy \right. \\
&\quad \left. + 2\int_{\mathbb{R}} q\overline{Y}^0 \int_0^\tau r_{\tau\tau}(\tau-\eta)\overline{Y}^0(y,\eta)\,d\eta\,dy\right)^{1/2} d\tau \\
&\leqslant \widetilde{C}\int_0^t \|\overline{Y}^0\|_{L_2(\mathbb{R})} d\tau + (C+2\widetilde{C})K\sqrt{T}\int_0^t \left(\int_{\mathbb{R}}\int_0^\tau |\overline{Y}^0(y,\eta)|^2 d\eta\,dy\right)^{1/2} d\tau.
\end{aligned}$$

By Gronwall's lemma $\|\overline{Y}^0\|_{L_2(\mathbb{R})}(t) = 0$, thus, $\overline{Y}^0(y,t) = 0$ in the domain $\mathbb{R} \times [0,T]$. Hence the equality of solutions $\widetilde{Y}_1^0(y,t)$ and $\widetilde{Y}_2^0(y,t)$ in the domain under consideration. Uniqueness is proven.

Note, that $\widetilde{Y}^0(y,t) = 0$ if $t < |y|$. Hence,

$$Y^0(y,t) = \theta(t-|y|)\left[\frac{a_0}{2} + \widetilde{Y}^0(y,t)\right].$$

For the function $\widetilde{Y}^0(y, t)$, the integral equation is valid:

$$\widetilde{Y}^0(y,t) = \theta(t-|y|)\Bigg[\int\limits_{\frac{y-t}{2}}^{\frac{y+t}{2}}\int\limits_{|\xi|}^{t-|\xi-y|}\big(q(\xi) - r'(t-|y-\xi|-\tau)\big) \times \left(\frac{a_0}{2} + \widetilde{Y}^0(\xi,\tau)\right) d\tau\, d\xi\Bigg]. \tag{6.34}$$

The integral Eq. (6.34) is an integral equation Volterra of the second kind with a continuous core and a continuous free term. It follows from the theory of integral equations that it has only one solution in $\nabla(T)$.

Using the properties of differentiation by parameter, it is easy to prove that $\widetilde{Y}^0(y,t) \in C^2(t \geqslant |y|)$. It follows from the integral Eq. (6.34) that $\widetilde{Y}^0(y,t)\Big|_{t=|y|} = 0$, thus, the function $V^0(y,t) = Y^0(y,t) + r * Y^0(y,t)$ belongs to $C^2(t \geqslant |y|)$ and $V^0(y,t) = \theta(t-|y|)\left[\frac{a_0}{2} + \widetilde{V}^0(y,t)\right]$, $\widetilde{V}^0(y,t) \in C^2(t \geqslant |y|)$, $t \in [0,T]$.

6.1.3 The Fundamental Solution of the Cauchy Problem for a Hyperbolic Operator with Memory

Let us present some operators:

$$\tilde{L}V = \frac{\partial^2 V}{\partial t^2} - \frac{\partial^2 V}{\partial y^2} - q(y,\nu)V + \frac{\partial^2}{\partial t^2}[r * V], \quad \nu \in \mathbb{R}\text{—parameter},$$

$$\tilde{L}_0 V = \frac{\partial^2 V}{\partial t^2} - \frac{\partial^2 V}{\partial y^2} - q(y,\nu)V.$$

In this subsection, we define the fundamental solution of the Cauchy problem for the operator $\tilde{L}$ and study its differential properties in the volume in which they will be needed to study the inverse problem. We denote the required solution by G.

Function G satisfies the following equalities

$$\tilde{L}G = \delta(t-t_0, y-y_0), \tag{6.35}$$

$$G\Big|_{t<0} = 0. \tag{6.36}$$

We have $\tilde{L}V = \tilde{L}_0 V + \frac{\partial^2}{\partial t^2}[r * V]$. Hence, if $G_0(y, y_0, t, t_0, \nu)$ is the fundamental solution of the Cauchy problem of the operator $\tilde{L}_0$:

$$\tilde{L}_0 G_0 = \delta(t-t_0, y-y_0),$$

$$G_0\Big|_{t<0} = 0,$$

then $G = G_0 + \widetilde{G}$, where $\widetilde{G}$ satisfies the following relations:

$$\frac{\partial^2 \widetilde{G}}{\partial t^2} - \frac{\partial^2 \widetilde{G}}{\partial y^2} - q(y,\nu)\widetilde{G} = -\frac{\partial^2}{\partial t^2}[r * \widetilde{G}] - \frac{\partial^2}{\partial t^2}[r * \widetilde{G}], \tag{6.37}$$

$$\widetilde{G}\Big|_{t<0} = 0. \tag{6.38}$$

Lemma 6.4 *Let $T > 0$ be fixed and G satisfies the relations (6.35)–(6.36). There exists the only solution of Cauchy problem (6.35)–(6.36), where $G(y, y_0, t, t_0, \nu)$ has the structure of $G = G_0 + \widetilde{G}$, moreover*

$$\widetilde{G}(y, y_0, t, t_0, \nu)\big|_{t-t_0 \leqslant |y-y_0|} = 0, \quad \widetilde{G} \in C\big(t - t_0 \geqslant |y - y_0|\big),$$
$$\frac{\partial \widetilde{G}}{\partial t} \in C\big(t - t_0 \geqslant |y - y_0|\big), \ t \in [0, T].$$

Proof Notice that the structure of G_0 is well studied [26]:

$$G_0(y, y_0, t, t_0, \nu) = \theta\big(t - t_0 - |y - y_0|\big)\left[\frac{1}{2} + \overline{G}_0(y, y_0, t, t_0, \nu)\right],$$

$$\overline{G}_0(y, y_0, t, t_0, \nu) \in C^2\big(t - t_0 \geqslant |y - y_0|\big),$$

$$\overline{G}_0(y, y_0, t, t_0, \nu)\big|_{t-t_0=|y-y_0|} = 0.$$

From (6.37) and (6.38) follows that

$$\widetilde{G} = G_0 * \left[-\frac{\partial^2}{\partial t^2}(r * G_0)\right] + G_0 * \left[-\frac{\partial^2}{\partial t^2}(r * \widetilde{G})\right]. \tag{6.39}$$

For clarity, we write out the first term of the equality (6.39), denoting it as $F(y, t, y_0, t_0)$:

$$F(y, y_0, t, t_0, \nu) = -\theta(\Gamma)\Bigg[\iint\limits_{\Diamond(y,t)} \frac{1}{2}\left(\frac{1}{2} + \overline{G}_0(\xi, y_0, \tau, t_0, \nu)\right)$$
$$\times r'\big(t - |y - \xi| - \tau\big)\, d\tau d\xi$$
$$+ \iint\limits_{\Diamond(y,t)} \frac{\partial \overline{G}_0}{\partial t}(\xi, y_0, t - \tau, t_0, \nu) \int\limits_{t_0+|\xi-y_0|}^{\tau} \left(\frac{1}{2} + \overline{G}_0(\xi, y_0, s, t_0, \nu)\right)$$
$$\times r'(\tau - s)\, ds\, d\tau d\xi\Bigg].$$

From now and so on

$$\Gamma = t - t_0 - |y - y_0|, \quad \iint\limits_{\Diamond(y,t)} \quad \text{means} \quad \int\limits_{\frac{y+y_0-(t-t_0)}{2}}^{\frac{y+y_0+(t-t_0)}{2}} \int\limits_{t_0+|y_0-\xi|}^{t-|y-\xi|} .$$

Notice that $F(y, y_0, t, t_0, \nu)$ in $\Delta(y, t) = \{(\xi, \tau)|\ 0 \le \tau \le t - |\xi - y|\}$ is continuous.

Let $D' = \{(y, t) : t - t_0 \leqslant |y - y_0|,\ t \in [0, T]\}$. Let us prove that $\widetilde{G}(y, y_0, t, t_0, \nu) \equiv 0$ in D'.

Let us take an arbitrary fixed point $(y_1, t_1) \in D'$ and draw two characteristics through it. Then for all points $(y, t) \in \Delta(y_1, t_1)$, given that

$$F(y, y_0, t, t_0, \nu)|_{t-t_o<|y-y_0|} = 0$$

and

$$\begin{aligned}\widetilde{G}(y, y_0, t, t_0, \nu) = &-\frac{1}{2}\iint\limits_{\Delta(y,t)} r'\big(t - |y - \xi| - \tau\big)\widetilde{G}(\xi, y_0, \tau, t_0, \nu)\, d\tau\, d\xi \\ &- \iint\limits_{\Delta(y,t)} \frac{\partial^2 \overline{G}_0}{\partial t^2}(\xi, y_0, t - \tau, t_0, \nu) \\ &\times \int\limits_0^{\tau} r\big(\tau - s\big)\widetilde{G}(\xi, y_0, s, t_0, \nu)\, ds\, d\tau\, d\xi.\end{aligned}$$

Suppose $\widetilde{G}(y, t, y_0, t_0)$ is continuous in D'.

Let

$$\max_{y_1-(t_1-t)\leqslant y\leqslant y_1+(t_1-t)} \left|\widetilde{G}(y, y_0, t, t_0, \nu)\right| = g(y_0, t, t_0, \nu),$$

$$\max_{(y,t)\in\Delta(y_1,t_1)} \left|r'(t)\right| = M_0, \qquad \max_{(y,t)\in\Delta(y_1,t_1)} \left|r(t)\right| = M_1,$$

$$\max_{(y,t)\in\Delta(y_1,t_1)} \left|\frac{\partial^2 \overline{G}_0}{\partial t^2}(y_0, y, t_0, t, \nu)\right| = M_2, \quad 0 \leqslant t \leqslant t_1 \leqslant T.$$

Next, estimate is fair:

$$\begin{aligned}g(y_0, t, t_0, \nu) &\leqslant \frac{1}{2} M_0 T \int\limits_0^t g(y_0, \tau, t_0, \nu)\, d\tau + M_2 T^2 M_1 \int\limits_0^t g(y_0, \tau, t_0, \nu)\, d\tau \\ &= \left(\frac{1}{2} M_0 T + M_2 T^2 M_1\right) \int\limits_0^t g(y_0, \tau, t_0, \nu)\, d\tau.\end{aligned} \tag{6.40}$$

By Gronwall's lemma $g(y_0, t, t_0, \nu) \equiv 0$ if $t \in [0, t_1]$, thus, $\widetilde{G}(y, y_0, t, t_0, \nu) \equiv 0$ in $\Delta(y_1, t_1)$. As the point was chosen arbitrarily, $\widetilde{G}(y, y_0, t, t_0, \nu) \equiv 0$ in D'.

Using a proven property for $\widetilde{G}(y, y_0, t, t_0, \nu)$, we get integral equation:

$$\widetilde{G}(y, y_0, t, t_0, \nu) = -\theta(\Gamma)\Big[\frac{1}{2}\iint\limits_{\Diamond(y,t)} r'(t - |y - \xi| - s)\,\widetilde{G}(\xi, y_0, s, t_0, \nu)\,ds\,d\xi$$
$$+\iint\limits_{\Diamond(y,t)} \frac{\partial \overline{G}_0}{\partial t}(\xi, y_0, t-\tau, t_0, \nu)\int\limits_{t_0+|y_0-\xi|}^{\tau} r'(\tau - s)\widetilde{G}(\xi, y_0, s, t_0, \nu)ds\,d\tau\,d\xi\Big]$$
$$+ F(y, y_0, t, t_0, \nu). \tag{6.41}$$

From (6.41) it follows that $\widetilde{G}(y, y_0, t, t_0, \nu)|_{t-t_0\le|y-y_0|} = 0$.

The solution (6.41) can be represented in the form

$$\widetilde{G}(y, y_0, t, t_0, \nu) = \sum_{n=0}^{\infty} \widetilde{G}_n(y, y_0, t, t_0, \nu), \tag{6.42}$$

where

$$\widetilde{G}_0 = F(y, y_0, t, t_0, \nu),$$

$$\widetilde{G}_n(y, y_0, t, t_0, \nu) = -\theta(\Gamma)$$
$$\times\Big[\frac{1}{2}\iint\limits_{\Diamond(y,t)} r'(t - |y - \xi| - \tau)\widetilde{G}_{n-1}(\xi, y_0, \tau, t_0, \nu)\,d\tau\,d\xi$$
$$+\iint\limits_{\Diamond(y,t)} \frac{\partial \overline{G}_0}{\partial t}(\xi, y_0, t - \tau, t_0, \nu) \tag{6.43}$$
$$\times \int\limits_{t_0+|y_0-\xi|}^{\tau} r'(\tau - s)\widetilde{G}_{n-1}(\xi, y_0, s, t_0, \nu)\,ds\,d\tau\,d\xi\Big].$$

Let

$$\max_{(\xi,\tau)\in\Diamond(y,t)} \left|r'(\tau)\right| = M_3,$$

$$\max_{(\xi,\tau)\in\Diamond(y,t)} \left|\frac{\partial \overline{G}_0}{\partial t}(\xi, y_0, t - \tau, t_0, \nu)\right| = M_4,$$

$$\max_{(\xi,\tau)\in\Diamond(y,t)} \left|F(\xi, y_0, \tau, t_0, \nu)\right| = M_5.$$

It follows from (6.43) the estimate :

$$\left|\widetilde{G}_n(y, y_0, t, t_0, \nu)\right| \leqslant \left[\frac{1}{2} + M_4(T - t_0)\right]^n M_3^n M_5 \frac{(T - t_0)^{2n}}{2 \cdot n!}.$$

Series (6.42) is majored by convergent numerical series, which means (6.42) converges uniformly, consequently, $\widetilde{G}(y, t, y_0, t_0)$ is continuous in $\{(y, t) : |y - y_0| \leqslant t - t_0 \leqslant T - |y - y_0|\}$.

Let us prove the uniqueness of the solution to the problem (6.35)–(6.36). Suppose, there exist two solutions G_1 and G_2, $G_1 = G_0 + \widetilde{G}_1$ and $G_2 = G_0 + \widetilde{G}_2$, since the fundamental solution of operator is unique. Let $\widehat{G} = G_1 - G_2$ or $\widehat{G} = \widetilde{G}_1 - \widetilde{G}_2$, which are the same. Then, we get a problem with a function $\widehat{G} = \widehat{G}(y, t, y_0, t_0)$:

$$\frac{\partial^2 \widehat{G}}{\partial t^2} - \frac{\partial^2 \widehat{G}}{\partial y^2} - q(y, \nu)\widehat{G} = -\frac{\partial^2}{\partial t^2}\left[r * \widehat{G}\right], \tag{6.44}$$

$$\widehat{G}\Big|_{t<0} = 0. \tag{6.45}$$

As a result of solving problem (6.44)–(6.45) we get integral equation, which is an analogue of (6.41), with $F(y, y_0, t, t_0, \nu) \equiv 0$. Using the estimate of (6.40) kind and Gronwall's inequality, we now deduce $\widehat{G}(y, y_0, t, t_0, \nu) \equiv 0$, which proves the uniqueness of problem solution.

Differentiating (6.41) with respect to t, we get:

$$\begin{aligned}
\frac{\partial \widetilde{G}}{\partial t}(y, y_0, t, t_0, \nu) = -\theta(\Gamma)\Bigg[&\frac{1}{2} \iint\limits_{\Diamond(y,t)} r''(t - |y - \xi| - \tau) \times \widetilde{G}(\xi, y_0, \tau, t_0, \nu)\, d\tau\, d\xi \\
&+ \iint\limits_{\Diamond(y,t)} \frac{\partial \overline{G}_0}{\partial t}(\xi, y_0, t - \tau, t_0, \nu) \int\limits_{t_0+|y_0-\xi|}^{\tau} r''(\tau - s) \\
&\times \widetilde{G}(\xi, y_0, s, t_0, \nu)\, ds\, d\tau\, d\xi \\
&+ \iint\limits_{\Diamond(y,t)} \frac{1}{2}\left(\frac{1}{2} + \overline{G}_0(\xi, y_0, \tau, t_0, \nu)\right) r''(t - |y - \xi| - \tau)\, d\tau\, d\xi \\
&+ \iint\limits_{\Diamond(y,t)} \frac{\partial \overline{G}_0}{\partial t}(\xi, y_0, t - \tau, t_0, \nu) \int\limits_{t_0+|y_0-\xi|}^{\tau} \left(\frac{1}{2} + \overline{G}_0(\xi, y_0, s, t_0, \nu)\right) \\
&\times r''(t - |y - \xi| - s)\, ds\, d\tau\, d\xi \Bigg].
\end{aligned}$$

Using features of the integrand, we conclude that from continuousness of the right part of the last equality follows continuousness of the left parts in domain $t - t_0 \geqslant |y - y_0|$. Lemma 6.4 is proved.

6.1.4 Direct and Inverse Linearized Problem

In this section, we present the solution of the direct problem (6.7)–(6.9). All properties are studied to the extent necessary to solve the inverse linearized problem. First, we prove Lemma 6.2.

Proof Differentiate the left and right parts of the Eqs. (6.7)–(6.9) by the variable x_1 and introduce the notation $w = \frac{\partial u^1}{\partial x_1}$. Then, given the independence of ρ_1, u^0 from x_1 we get a problem of the form (6.7)–(6.9) for w with zero data. According to the proved Lemma 6.3 in the class $U_1(s, T)$ has only zero solution. Hence, $u^1 = u^1(x_2, x_3, t)$.

For brevity, we define the linear integral operator L by the formula

$$L\,[k(t), u(x_2, x_3, t)] = u(x_2, x_3, t) + \int_0^t k(t-\tau)u(x_2, x_3, \tau)d\tau$$

(here u—scalar function).

In this material sometimes we will not specify the dependence of functions on variables in L, implying dependence of the first function on variables t, and the second on x_2, x_3, t.

The problem of determining u^1 can be rewritten as the problem of determining $u^1 = u^1(x_2, x_3, t)$, which at $x_3 > 0$, $(x_2, t) \in \mathbb{R}^2$, satisfies the following equalities:

$$\rho_0(x_3)\frac{\partial^2 u_1^1}{\partial t^2} = L\left[k, \frac{\partial}{\partial x_3}\left(\mu_0(x_3)\frac{\partial u_1^1}{\partial x_3}\right) + \mu_0(x_3)\frac{\partial^2 u_1^1}{\partial x_2^2}\right] - \rho_1(x_2, x_3)\frac{\partial^2 u_i^0}{\partial t^2}, \tag{6.46}$$

$$\begin{aligned}\rho_0(x_3)\frac{\partial^2 u_2^1}{\partial t^2} = L\Big[k, &\frac{\partial}{\partial x_3}\left(\mu_0(x_3)\frac{\partial u_2^1}{\partial x_3}\right) + (\lambda_0(x_3) + 2\mu_0(x_3))\frac{\partial^2 u_2^1}{\partial x_2^2}\\ &+ \lambda_0(x_3)\frac{\partial^2 u_3^1}{\partial x_3 \partial x_2} + \frac{\partial}{\partial x_3}\left(\mu_0(x_3)\frac{\partial u_3^1}{\partial x_2}\right)\Big],\end{aligned} \tag{6.47}$$

$$\begin{aligned}\rho_0(x_3)\frac{\partial^2 u_3^1}{\partial t^2} = L\Big[k, &\frac{\partial}{\partial x_3}\left((\lambda_0 a(x_3) + 2\mu_0(x_3))\frac{\partial u_3^1}{\partial x_3}\right)\\ &+ (\lambda_0(x_3) + \mu_0(x_3))\frac{\partial^2 u_2^1}{\partial x_2 \partial x_2}\Big],\end{aligned} \tag{6.48}$$

$$u_1^1 \,|_{t<0} \equiv 0, \tag{6.49}$$

$$u_i^1 \,|_{t=+0} = 0, \quad \frac{\partial u_i^1}{\partial t} \,|_{t=+0} = 0, \ i = 2, 3, \tag{6.50}$$

$$L\left[k, \mu_0(x_3)\frac{\partial u_1^1}{\partial x_3}\right] |_{x_3=+0} = 0, \tag{6.51}$$

$$L\left[k, \mu_0(x_3)\left(\frac{\partial u_2^1}{\partial x_3} + \frac{\partial u_3^1}{\partial x_2}\right)\right]|_{x_3=+0} = 0, \tag{6.52}$$

$$L\left[k, (\lambda_0(x_3) + 2\mu_0(x_3))\frac{\partial u_3^1}{\partial x_3} + \lambda_0\frac{\partial u_2^1}{\partial x_2}\right]|_{x_3=+0} = 0. \tag{6.53}$$

The solution splits into two independent solvable problems (6.46), (6.49), (6.51) and (6.47), (6.48), (6.50), (6.52), (6.53) related to the definition of functions $u_1^1 = u_1^1(x_2, x_3, t)$, $u_2^1 = u_2^1(x_2, x_3, t)$, $u_3^1 = u_3^1(x_2, x_3, t)$, respectively. At the same time the task (6.47), (6.48), (6.50), (6.52), (6.53) is a problem for a homogeneous system of equations with homogeneous boundary conditions and zero initial data. Let us continue in an even way to the domain $x_3 < 0$ of the function u_2 and u_3. Using the result of Lemma 6.3, if $V^1(x_2, x_3, t) = \left(u_2^1(x_2, x_3, t), u_3^1(x_2, x_3, t)\right)$, we get that $u_2^1 \equiv u_3^1 \equiv 0$ when $(x_2, x_3, t) \in \mathbb{R} \times \mathbb{R} \times [0, T]$.

We rewrite the Eqs. (6.46), (6.49), (6.51) in an equivalent way in terms of the variable y and functions

$$V(x, y, t) := \frac{F_{x_2}[u_1^1]\left(\nu, \psi^{-1}(y), t\right)}{s(y)}, \quad V^0 := \frac{u^0\left(\psi^{-1}(y), t\right)}{s(y)},$$

$$s(y) := \sqrt{\frac{v(+0)\rho_0(+0)}{v\left(\psi^{-1}(y)\right)\rho_0\left(\psi^{-1}(y)\right)}}.$$

Let $\bar{V}(\nu, y, t) := L\left[k, V(\nu, y, t)\right]$, then $V(\nu, y, t) = L\left[r, \bar{V}(\nu, y, t)\right]$, where $r(t) = -L[k, r(t)]$.

For $\bar{V}$, $t \in \mathbb{R}$, $y > 0$ we get:

$$\frac{\partial^2 \bar{V}}{\partial t^2} = \frac{\partial^2 \bar{V}}{\partial y^2} + q(\nu, y)\bar{V} - \frac{\partial^2}{\partial t^2}\left[r * \bar{V}\right] - \frac{\bar{\rho}_1(\nu, \psi^{-1}(y))}{\rho_0(\psi^{-1}(y))}\frac{\partial^2 V^0}{\partial t^2}, \tag{6.54}$$

$$\bar{V}|_{t<0} \equiv 0, \tag{6.55}$$

$$\frac{\partial \bar{V}}{\partial y}|_{y=+0} = 0, \tag{6.56}$$

where

$$q(\nu, y) := \frac{s''(y)}{s(y)} - 2\left[\frac{s'(y)}{s(y)}\right]^2 - \nu^2\frac{\mu_0(\psi^{-1}(y))}{\rho_0(\psi^{-1}(y))},$$

Note that in by virtue of (6.46) the function $q(\nu, y)$ defined above will be continuous on the variables (ν, y).

If we continue $q(\nu, y)$, $\bar{\rho}_1(\nu, \psi^{-1}(y))$, V^0 in an even manner to the domain $y < 0$, then the problem of determining the function $\bar{V}(\nu, y, t)$ (6.54)–(6.56) for y

is equivalent to the problem of determining the even function $\bar{V}(\nu, y, t)$ satisfying for $y \in \mathbb{R},\ t \in \mathbb{R}$ the equalities (6.54), (6.55).

The previous section shows that there is a unique fundamental solution to the operator $\tilde{L}$:

$$\tilde{L}G = \frac{\partial^2 G}{\partial t^2} - \frac{\partial^2 G}{\partial y^2} - q(\nu, y)G - \frac{\partial^2}{\partial t^2}[r * G]. \tag{6.57}$$

Using the properties of the fundamental solution $G(y, y_0, t, t_0, \nu)$, the solution of the problem (6.54), (6.55) at $y \in \mathbb{R},\ t \in \mathbb{R}$ is written by the formula

$$\begin{aligned}
\bar{V}(\nu, y, t) &= -\iint\limits_{\mathbb{R}^2} G(y, \xi, t, \tau, \nu)\frac{\bar{\rho}_1(\nu, \psi^{-1}(\xi))}{\rho_0(\psi^{-1}(\xi))}\frac{\partial^2 V^0}{\partial \tau^2}\mathrm{d}\tau\mathrm{d}\xi \\
&= -\frac{\partial^2}{\partial t^2}\iint\limits_{\mathbb{R}^2} \theta(t-\tau-|\xi-y|)\theta(\tau-|\xi|)\left(\frac{1}{2}+\overline{G}_0+\widetilde{G}\right)(y, \xi, t-\tau, \nu) \\
&\times \frac{\bar{\rho}_1(\nu, \psi^{-1}(\xi))}{\rho_0(\psi^{-1}(\xi))}\left(\frac{a_0}{2}+\widetilde{V}^0(\xi, \tau)\right)\mathrm{d}\tau\mathrm{d}\xi \\
&= \theta(t-|y|)\Bigg\{\frac{a_0}{8}\left[\frac{\bar{\rho}_1(\nu, \psi^{-1}(\frac{y+t}{2}))}{\rho_0(\psi^{-1}(\frac{y+t}{2}))}+\frac{\bar{\rho}_1(\nu, \psi^{-1}(\frac{y-t}{2}))}{\rho_0(\psi^{-1}(\frac{y-t}{2}))}\right] \\
&+ \int\limits_{\frac{y-t}{2}}^{\frac{y+t}{2}}\Bigg[\frac{a_0}{2}\left(\frac{\partial \overline{G}_0}{\partial t}+\frac{\partial \widetilde{G}}{\partial t}\right)(y, \xi, |\xi-y|, \nu)(y, \xi, |\xi-y|, \nu) \\
&+ \frac{1}{2}\frac{\partial \widetilde{V}}{\partial t}(\xi, t-|\xi-y|) \\
&+ \left(\frac{\partial \overline{G}_0}{\partial t}+\frac{\partial \widetilde{G}}{\partial t}\right)(y, \xi, |\xi-y|, \nu)\frac{\partial \widetilde{V}}{\partial t}(\xi, t-|\xi-y|) \\
&+ \int\limits_{\xi}^{t-|\xi-y|}\left(\frac{\partial^2 \overline{G}_0}{\partial t^2}+\frac{\partial^2 \widetilde{G}}{\partial t^2}\right)(y, \xi, |\xi-y|, \nu)\left(\frac{a_0}{2}+\widetilde{V}^0(\xi, \tau)\right)\mathrm{d}\tau\Bigg]\frac{\bar{\rho}_1(\nu, \psi^{-1}(\xi))}{\rho_0(\psi^{-1}(\xi))}\mathrm{d}\xi\Bigg\}.
\end{aligned}$$

From equality (6.57) follows the validity of the statement of Lemma 6.2, since uniqueness is proved completely analogous to how it was done for the problem (6.35), (6.36).

Let

$$H(\nu, t) := L[k, h(\nu, t)]. \tag{6.58}$$

Proof The proof of Theorem 6.1 is analogous to the proof of a similar theorem from the monograph [1]. Indeed, given the properties and structure of the functions V_0 and G (Sects. 6.1.2 and ch6:sec6.4), it follows from (6.57):

$$\bar{V}(\nu, 0, t) = -\frac{a}{4}\frac{\bar{\rho}_1(\nu, \psi^{-1}(t/2))}{\rho_0(\psi^{-1}(t/2))} - \int_{-t/2}^{t/2} \tilde{K}_1(\nu, \xi, t)\frac{\bar{\rho}_1(\nu, \psi^{-1}(\xi))}{\rho_0(\psi^{-1}(\xi))}\mathrm{d}\xi, \quad t > 0, \tag{6.59}$$

where $a := [\mu_0(+0)\rho_0(+0)]^{-\frac{1}{2}}$, $\tilde{K}_1(\nu, \xi, t)$ is known continuous even by ξ function at $\xi \in [-t/2, t/2],\ t \in [0, T],\ \nu \in \mathbb{R}$.

Since $\bar{V}(\nu, 0, t) = \bar{L}\Big[k, F_{x_2}[u_1^1](\nu, x_3, t)|_{x_3=+0}\Big]$, then from (6.58)–(6.59) we obtain the Volterra integral equation of the second kind with respect to $\bar{\rho}_1$:

$$\frac{\bar{\rho}_1(\nu, \psi^{-1}(t/2))}{\rho_0(\psi^{-1}(t/2))} + \frac{8}{a}\int_{-t/2}^{t/2} K_1(\nu, \xi, t)\frac{\bar{\rho}_1(\nu, \psi^{-1}(\xi))}{\rho_0(\psi^{-1}(\xi))}\mathrm{d}\xi = -\frac{4}{a}H(\nu, t). \tag{6.60}$$

It follows from the theory of integral equations that the equation (6.60) has a unique continuous at $t \in [0, T],\ \nu \in \mathbb{R}$ solution $\bar{\rho}_1(\nu, \psi^{-1}(y)) = F_{x_2}[\rho_1](\nu, x_3)|_{x_3=\psi^{-1}(y)}$, hence the validity of Theorem 6.1.

The validity of Theorem 6.2 follows from the estimates for Eq. (6.60) and the isometric property of the Fourier transform operator.

Now let us consider the question of convergence of the sequence of solutions of some family of problems to the desired solution. Let there be a solution $\rho_1(x_2, x_3) \in L(\mathbb{R} \times \mathbb{R}_+)$ corresponding to the information $H(\nu, t)$.

Theorems 6.1 and 6.2 show the well posedness of the inverse problem in a special class of functions (well-posedness class). Now, we show that

$$\lim_{s\to\infty}\sup_{x_3\in[0,X]}\int_{\mathbb{R}} |\rho_1^s(x_2, x_3) - \rho_1(x_2, x_3)|^2\mathrm{d}x_2 = 0.$$

Indeed,

$$\begin{aligned}
&\lim_{s\to\infty}\sup_{x_3\in[0,X]}\int_{\mathbb{R}} |\rho_1^s(x_2, x_3) - \rho_1(x_2, x_3)|^2\mathrm{d}x_2 \\
&= \lim_{s\to\infty}\sup_{x_3\in[0,X]}\Big(\int_{-s}^{s} |\rho_1^s(\nu, x_3) - \rho_1(\nu, x_3)|^2\mathrm{d}\nu \\
&+ \int_{-\infty}^{-s} |\rho_1(\nu, x_3)|^2\mathrm{d}\nu + \int_{s}^{+\infty} |\rho_1(\nu, x_3)|^2\mathrm{d}\nu\Big) \\
&= \lim_{s\to\infty}\sup_{x_3\in[0,X]}\left(\int_{-\infty}^{-s} |\rho_1(\nu, x_3)|^2\mathrm{d}\nu + \int_{s}^{+\infty} |\rho_1(\nu, x_3)|^2\mathrm{d}\nu\right) = 0.
\end{aligned}$$

Thus, Theorem 6.3 is proved.

6.2 The Problem of Determining the Coefficient of Thermal Enlargement of Thermoviscoelasticity Equation

Let us consider a system of integro-differential equations of dynamic thermoviscoelasticity for $x = (x_1, \ x_2, \ x_3)$, $(x_1, x_2) \in \mathbb{R}^2, \ t > 0, \ x_3 > 0$

$$\rho \frac{\partial^2 u_i}{\partial t^2} = \sum_{j=1}^{3} \frac{\partial T_{ij}}{\partial x_j}, \ i = 1, 2, 3, \tag{6.61}$$

$$\frac{\partial H}{\partial t} = k \Delta H, \tag{6.62}$$

$$u_i \mid_{t=+0} = 0, \ \frac{\partial u_i}{\partial t} \mid_{t=+0} = 0, \ i = 1, 2, 3, \tag{6.63}$$

$$H \mid_{t=+0} = 0, \tag{6.64}$$

$$T_{3j} \vert_{x_3=+0} = 0, \ j = 1, 2, 3, \tag{6.65}$$

$$\left(\frac{\partial H}{\partial x_3} - \gamma H \right) \vert_{x_3=+0} = -\gamma (\tilde{T}_1 - \tilde{T}_0), \tag{6.66}$$

where $u(x, t) = (u_1(x, t), u_2(x, t), u_3(x, t))$ is the displacement vector, $H(x, t)$ is the temperature increment, Δ is the Laplace operator in the variable $x_1, x_2, x_3, k > 0$ is constant thermal diffusivity coefficient; $\tilde{T}_1, \ \tilde{T}_0, \ \gamma$ are constant, $\tilde{T}_1 > \tilde{T}_0, \gamma > 0$, T_{ij} is the stress tensor:

$$T_{ij}(x, t) = \sigma_{ij}[u](x, t) + \int_0^t h(t - \tau) \sigma_{ij}[u](x, \tau) d\tau, \tag{6.67}$$

σ_{ij} are stresses (1.87).

The problem of determining the displacement vector $u(x, t)$ and the temperature increment $H(x, t)$ satisfying (in the generalized sense) equalities (6.61)–(6.66), for the given functions $\alpha(z), \ h(t), \ \rho(x_3), \ \mu(x_3), \ \lambda(x_3)$ and the given constants $k, \ \gamma, \ \tilde{T}_0, \ \tilde{T}_1$ called *the direct problem.*

In what follows, we shall see that the solution of the initial– boundary value problem (6.62), (6.64), (6.66) will depend only on the variables x_3, t. Thus, all the coefficients appearing in the system of Eqs. (6.61), (6.67), and (1.87) depend only on these variables. Hence, in view of the assumptions on the density of the medium and the Lame coefficients, taking conditions (6.63), (6.65) into account, we conclude that: $u_3(x,t) \equiv u_3(x_3,t) \not\equiv 0$, $u_1 \equiv u_2 \equiv 0$.

The inverse problem consists in finding the coefficient of thermal enlargement $\alpha(z)$, $z \in [0, Z]$, $Z = H(0, T)$, appearing in (6.61), by using formula (1.87) if the following additional information about the solution of the problem (6.61)–(6.66) is known:

$$u_3(x_3, t)|_{x_3=+0} = g(t), \ t > 0, \tag{6.68}$$

$g(t)$ is a given function.

For the function $H(x_3, t)$ (see Sect. 1.3) equalities (6.61), (6.63), (6.65) are equivalent to the relations with operator L (Sect. 6.1.4)

$$\begin{aligned} \rho(x_3)\frac{\partial^2 u_3}{\partial t^2} = L\Big[h, \frac{\partial}{\partial x_3}\Big[(\lambda(x_3) + 2\mu(x_3))\frac{\partial u_3}{\partial x_3} \\ - (3\lambda(x_3) + 2\mu(x_3))R(H(x_3, t))\Big]\Big], \end{aligned} \tag{6.69}$$

$$u_3 \,|_{t=+0} = 0, \ \frac{\partial u_3}{\partial t}\,|_{t=+0} = 0, \tag{6.70}$$

$$L\left[h, \frac{\partial u_3}{\partial x_3}\right]_{x_3=+0} = \frac{3\lambda(0) + 2\mu(0)}{\lambda(0) + 2\mu(0)} L\left[h, R(H(0, t))\right], \tag{6.71}$$

where $R(s) = \int_0^s \alpha(z)\mathrm{d}z$.

Let

$$\begin{aligned} \rho(x_3),\ \mu(x_3),\ \lambda(x_3) \in \Lambda_1 = \{\rho(x_3),\ \mu(x_3),\ \lambda(x_3) \in C^2[0, +\infty)| \\ \rho(x_3) > 0,\ \mu(x_3) > 0,\ \lambda(x_3) + 2\mu(x_3) > 0, \\ \rho'(+0) = \mu'(+0) = \lambda'(+0) = 0\}, \end{aligned}$$

$$h(t) \in \Lambda_2 = \{h(t) \in C^2[0, T] \,\big|\, h(+0) = 0,\ h'(+0) = 0\}.$$

Consider the new variable y introduced by the formula

$$y = \psi(x_3) := \int_0^{x_3} \frac{\mathrm{d}\xi}{\nu(\xi)}, \ \nu(x_3) := \sqrt{\frac{\lambda(x_3) + 2\mu(x_3)}{\rho(x_3)}}.$$

By $\psi^{-1}(y)$ we denote the function inverse to $\psi(x_3)$.

Let

$$v(y, t) := \frac{u_3\left(\psi^{-1}(y), t\right)}{s(y)}, \ s(y) := \sqrt{\frac{a(y)}{a(0)}},$$

$$a(y) := \frac{1}{\sqrt{[\lambda(\psi^{-1}(y)) + 2\mu(\psi^{-1}(y))]\rho(\psi^{-1}(y))}}.$$

Then, in terms of the functions introduced above, the variable y the inverse problem (6.69)–(6.71), (6.68) takes the form

$$\frac{\partial^2 v}{\partial t^2} = L\left[h, \frac{\partial^2 v}{\partial y^2} + q(y)v\right] - \sqrt{a(y)a(0)}\frac{\partial}{\partial y}Q[h](y,t),\ y > 0,\ t > 0, \quad (6.72)$$

$$v|_{t=+0} = 0,\ \frac{\partial v}{\partial t}\Big|_{t=+0} = 0 \quad (6.73)$$

$$L[h, \frac{\partial v(y,t)}{\partial y}]\Big|_{y=+0} = a(0)Q[h](0,t), \quad (6.74)$$

$$v(y,t)|_{y=+0} = g(t), \quad t > 0, \quad (6.75)$$

where

$$q(y) := \frac{s''(y)}{s(y)} - 2\left[\frac{s'(y)}{s(y)}\right]^2,\ \tilde{\lambda}(y) := \lambda(\psi^{-1}(y)),$$

$$\tilde{\mu}(y) := \mu(\psi^{-1}(y)),\ \tilde{H}(y,t) := H(\psi^{-1}(y), t),$$

$$Q[h](y,t) := (3\tilde{\lambda}(y) + 2\tilde{\mu}(y))\Big[R(\tilde{H}(y,t)) + \int_0^t h(\tau)R(\tilde{H}(y, t-\tau))d\tau\Big].$$

Let $L[h, v(y,t)] = w(y,t)$. Then $v(y,t) = L[r, w(y,t)]$, where

$$r(t) = -h(t) - \int_0^t h(t-\tau)r(\tau)d\tau. \quad (6.76)$$

With respect to the functions $w(y,t)$ and $r(t)$ Eqs. (6.72)–(6.75) take the form

$$\begin{aligned} \frac{\partial^2 w}{\partial t^2} &= \frac{\partial^2 w}{\partial y^2} + q(y)w - \sqrt{a(y)a(0)}\frac{\partial}{\partial y}Q[h](y,t) \\ &\quad - \int_0^t r''(t-\tau)w(y,\tau)d\tau,\ y > 0,\ t > 0, \end{aligned} \quad (6.77)$$

$$w|_{t=+0} = 0,\ \frac{\partial w}{\partial t}\Big|_{t=+0} = 0, \quad (6.78)$$

$$\left.\frac{\partial w}{\partial y}\right|_{y=+0} = a(0)Q[h](0,t), \tag{6.79}$$

$$w|_{y=+0} = g(t) + \int_0^t h(t-\tau)g(\tau)\mathrm{d}\tau. \tag{6.80}$$

We rewrite Eqs. (6.77)–(6.80) in terms of the function $V(y,t) = \frac{\partial w}{\partial t}(y,t)$:

$$\frac{\partial^2 V}{\partial t^2} = \frac{\partial^2 V}{\partial y^2} + q(y)V - \sqrt{a(y)a(0)}\frac{\partial}{\partial y}Q_t[h](y,t) - \int_0^t r''(\tau)V(y,t-\tau)\mathrm{d}\tau,\ y>0,\ t>0, \tag{6.81}$$

$$V|_{t=+0} = 0,\ \left.\frac{\partial V}{\partial t}\right|_{t=+0} = 0, \tag{6.82}$$

$$\left.\frac{\partial V}{\partial y}\right|_{y=+0} = a(0)Q_t[h](0,t), \tag{6.83}$$

$$V|_{y=+0} = g'(t) + \int_0^t h(\tau)g'(t-\tau)\mathrm{d}\tau, \tag{6.84}$$

where

$$Q_t[h](y,t) = (3\tilde{\lambda}(y) + 2\tilde{\mu}(y)) \times \left[\alpha(\tilde{H}(y,t))\frac{\partial \tilde{H}}{\partial t} + \int_0^t h(\tau)\alpha(\tilde{H}(y,t-\tau))\frac{\partial \tilde{H}}{\partial t}(y,t-\tau)\mathrm{d}\tau\right].$$

In (6.79) we used the equality $h(0) = -r(0)$, which follows from (6.76).

Let

$$\Delta(y,t) = \{(\xi,\tau)|\ 0 \le \tau \le t-y,\ 0 \le \xi \le t+y-\tau\} \cup \{(\xi,\tau)|\ t-y \le \tau \le t,\ t-y+\tau \le \xi \le t+y-\tau\},\ 0 \le y \le t$$

and

$$\Delta(y,t) = \{(\xi,\tau)|\ 0 \le \tau \le t,\ t-y+\tau \le \xi \le t+y-\tau\},\ 0 \le t < y.$$

Applying Green's formula in the domain $\triangle(y,t)$ and the condition (6.82)–(6.84), we get

$$V(y,t)=\frac{1}{2}\Bigg\{g'(t-y)+\int\limits_0^{t-y}h(\tau)g'(t-y-\tau)\mathrm{d}\tau$$
$$+\iint\limits_{\triangle(y,t)}\Big[q(\xi)V(\xi,\tau)-\int\limits_0^{\tau}r''(\eta)V(\xi,\tau-\eta)\mathrm{d}\eta\Big]\mathrm{d}\xi\mathrm{d}\tau$$
$$-\int\limits_0^{t}\sqrt{a(\xi)}b(\xi)L\Big[h,\alpha(\tilde{H}(\xi,\tau))\frac{\partial\tilde{H}}{\partial\tau}(\xi,\tau)\Big]\Big|_{\xi=t+y-\tau}\mathrm{d}\tau$$
$$+\iint\limits_{\triangle(y,t)}\frac{\partial}{\partial\xi}(\sqrt{a(\xi)})b(\xi)L\Big[h,\alpha(\tilde{H}(\xi,\tau))\frac{\partial\tilde{H}}{\partial\tau}(\xi,\tau)\Big]\mathrm{d}\xi\mathrm{d}\tau$$
$$+\int\limits_{t-y}^{t}\sqrt{a(\xi)}b(\xi)L\Big[h,\alpha(\tilde{H}(\xi,\tau))\frac{\partial\tilde{H}}{\partial\tau}(\xi,\tau)\Big]\Big|_{\xi=t-y+\tau}\mathrm{d}\tau\Bigg\},\quad 0\le y\le t,\tag{6.85}$$

$$V(y,t)=\frac{1}{2}\Bigg\{\iint\limits_{\triangle(y,t)}\Big[q(\xi)V(\xi,\tau)-\int\limits_0^{\tau}r''(\eta)V(\xi,\tau-\eta)\mathrm{d}\eta\Big]\mathrm{d}\xi\mathrm{d}\tau$$
$$-\int\limits_0^{t}\sqrt{a(\xi)}b(\xi)L\Big[h,\alpha(\tilde{H}(\xi,\tau))\frac{\partial\tilde{H}}{\partial\tau}(\xi,\tau)\Big]\Big|_{\xi=t+y-\tau}\mathrm{d}\tau$$
$$+\int\limits_0^{t}\sqrt{a(\xi)}b(\xi)L\Big[h,\alpha(\tilde{H}(\xi,\tau))\frac{\partial\tilde{H}}{\partial\tau}(\xi,\tau)\Big]\Big|_{\xi=y-t+\tau}\mathrm{d}\tau$$
$$+\iint\limits_{\triangle(y,t)}\frac{\partial}{\partial\xi}(\sqrt{a(\xi)})b(\xi)L\Big[h,\alpha(\tilde{H}(\xi,\tau))\frac{\partial\tilde{H}}{\partial\tau}(\xi,\tau)\Big]\mathrm{d}\xi\mathrm{d}\tau\Bigg\},\quad 0\le t<y,\tag{6.86}$$

where $b(\xi):=\sqrt{a(0)}(3\tilde{\lambda}(\xi)+2\tilde{\mu}(\xi))$.

Notice that in the Eq. (6.85)

$$\iint\limits_{\triangle(y,t)}\mathrm{d}\xi\mathrm{d}\tau=\int\limits_0^{t-y}\int_0^{t+y-\tau}\mathrm{d}\xi\mathrm{d}\tau+\int\limits_{t-y}^{t}\int_{t-y+\tau}^{t+y-\tau}\mathrm{d}\xi\mathrm{d}\tau,$$

and in the Eq. (6.86)

$$\iint\limits_{\Delta(y,t)} \mathrm{d}\xi\,\mathrm{d}\tau = \int\limits_0^t \int_{t-y+\tau}^{t+y-\tau} \mathrm{d}\xi\,\mathrm{d}\tau.$$

From (6.85) we obtain the equation at $y = 0$:

$$\begin{aligned} V(0,t) = \frac{1}{2}\Bigg\{ & g'(t) + \int\limits_0^t h(\tau)g'(t-\tau)\mathrm{d}\tau \\ & + \int\limits_0^t \int\limits_0^{t-\tau} \Big[q(\xi)V(\xi,\tau) - \int\limits_0^\tau r''(\eta)V(\xi,\tau-\eta)\mathrm{d}\eta \Big] \mathrm{d}\xi\,\mathrm{d}\tau \\ & - \int\limits_0^t \sqrt{a(t-\tau)}b(t-\tau)L\Big[h, \alpha(\tilde{H}(\xi,\tau))\frac{\partial \tilde{H}}{\partial \tau}(\xi,\tau)\Big]\Big|_{\xi=t-\tau} \mathrm{d}\tau \\ & + \int\limits_0^t \int\limits_0^{t-\tau} \frac{\partial}{\partial \xi}(\sqrt{a(\xi)})b(\xi)L\Big[h, \alpha(\tilde{H}(\xi,\tau))\frac{\partial \tilde{H}}{\partial \tau}(\xi,\tau)\Big]\mathrm{d}\xi\,\mathrm{d}\tau \Bigg\}. \end{aligned} \tag{6.87}$$

We rewrite the fourth integral of equality (6.87) by replacing the integration variable τ by s by the rule $s = G(\tau,t) := \tilde{H}(t-\tau,\tau)$. Then

$$\begin{aligned} & \int\limits_0^t \sqrt{a(t-\tau)}b(t-\tau)\Big[\alpha(\tilde{H}(\xi,\tau))\frac{\partial \tilde{H}}{\partial \tau}(\xi,\tau) \\ & \quad + \int\limits_0^\tau h(\tau-\eta)\alpha(\tilde{H}(\xi,\eta))\frac{\partial \tilde{H}}{\partial \eta}(\xi,\eta)\mathrm{d}\eta\Big]\Big|_{\xi=t-\tau} \mathrm{d}\tau \\ & = \int\limits_0^{H(0,t)} A(s,t)\Big[\alpha(s) + \int\limits_0^\tau h(\tau-\eta)\alpha(\tilde{H}(\xi,\eta))\frac{\partial \tilde{H}}{\partial \eta}(\xi,\eta)\mathrm{d}\eta\Big]_{\xi=t-\tau,\tau=G^{-1}(s,t)} \mathrm{d}s, \end{aligned} \tag{6.88}$$

where

$$A(s,t) = \sqrt{a(t-\tau)}b(t-\tau)\left[\frac{\partial \tilde{H}}{\partial \tau}(\xi,\tau)\frac{1}{\frac{\partial}{\partial \tau}G(\tau,t)}\right]_{\xi=t-\tau,\tau=G^{-1}(s,t)},$$

$G^{-1}(s,t)$ is the function inverse to $s = G(\tau,t)$ at a fixed t.

Using the results of Lemma 4.1 [1], we note that $\frac{\partial \tilde{H}}{\partial \tau}(\xi,\tau) > 0,\ \frac{\partial}{\partial \tau}G(\tau,t) > 0,\ \xi > 0,\ \tau > 0$.

Replacing the variable $p = \tilde{H}(t-\tau, \tau)\|_{\tau=G^{-1}(s,t)} := \phi(s, \eta, t)$ in the convolution integral of (6.88), we obtain

$$\int_0^{H(0,t)} A(s,t)\Big[\alpha(s) + \int_0^s B(s,p,t)\alpha(p)\mathrm{d}p\Big]\mathrm{d}s,$$

where

$$B(s,p,t) = h(\tau-\eta)\Big|_{\tau=G^{-1}(s,t),\eta=\phi^{-1}(s,p,t)}.$$

For $V(0,t)$ taking into account (6.84) and (6.88) we have

$$\begin{aligned}V(0,t) =& \int_0^t\int_0^{t-\tau}\Big[q(\xi)V(\xi,\tau) - \int_0^\tau r''(\eta)V(\xi,\tau-\eta)\mathrm{d}\eta\Big]\mathrm{d}\xi\mathrm{d}\tau\\ &- \int_0^{H(0,t)} A(s,t)\Big[\alpha(s) + \int_0^s B(s,p,t)\alpha(p)\mathrm{d}p\Big]\mathrm{d}s\\ &+ \int_0^t\int_0^{t-\tau}\frac{\partial}{\partial\xi}(\sqrt{a(\xi)})b(\xi)L\Big[h,\alpha(\tilde{H}(\xi,\tau))\frac{\partial\tilde{H}}{\partial\tau}(\xi,\tau)\Big]\mathrm{d}\xi\mathrm{d}\tau.\end{aligned} \tag{6.89}$$

Differentiating (6.88) by the variable t, and then performing a similar replacement of the variable (as it was done in equality (6.88)) in the fifth integral, we deduce:

$$\begin{aligned}V'(0,t) =& \int_0^t\Big[q(t-\tau)V(t-\tau,\tau) - \int_0^\tau r''(\eta)V(t-\tau,\tau-\eta)\mathrm{d}\eta\Big]\mathrm{d}\tau\\ &- A(H(0,t),t)\alpha(H(0,t)) - \int_0^{H(0,t)} B(H(0,t),s,t)\alpha(s)\mathrm{d}s\\ &- \int_0^{H(0,t)}\frac{\partial A}{\partial t}(s,t)\Big[\alpha(s) + \int_0^s\frac{\partial B}{\partial t}(s,p,t)\alpha(p)\mathrm{d}p\Big]\mathrm{d}s\\ &+ \int_0^{H(0,t)} A_1(s,t)\Big[\alpha(s) + \int_0^s B(s,p,t)\alpha(p)\mathrm{d}p\Big]\mathrm{d}s,\end{aligned}$$

where

$$A_1(s,t) = \Big[\frac{\partial}{\partial\xi}(\sqrt{a(\xi)})b(\xi)\frac{\partial\tilde{H}}{\partial\tau}(\xi,\tau)\frac{1}{\frac{\partial}{\partial\tau}G(\tau,t)}\Big]_{\xi=t-\tau,\tau=G^{-1}(s,t)},$$

$$A(H(0,t),t) = a(0)(3\lambda(0) + 2\mu(0))\frac{\partial \tilde{H}}{\partial \tau}(0,\tau)\frac{1}{\frac{\partial}{\partial \tau}G(\tau,t)|_{\tau-t}} > 0.$$

Hence we obtain the equation for α:

$$\alpha(H(0,t)) = \tilde{\alpha}(t) + \int_0^t \Big[\tilde{q}(t-\tau)V(t-\tau,\tau) - \int_0^\tau \tilde{r}''(\eta)V(t-\tau,\tau-\eta)\mathrm{d}\eta\Big]\mathrm{d}\tau + \int_0^{H(0,t)} \Big[K_1(H(0,t),s,t)\alpha(s) + \int_0^s K_2(H(0,t),s,p,t)\alpha(p)\mathrm{d}p\Big]\mathrm{d}s, \tag{6.90}$$

where

$$\tilde{\alpha}(t) = -\frac{g''(t) + \int_0^t h(\tau)g''(t-\tau)\mathrm{d}\tau}{A(H(0,t),t)},$$

$$\tilde{q}(t-\tau) = \frac{q(t-\tau)}{A(H(0,t),t)}, \quad \tilde{r}''(\eta) = \frac{r''(\eta)}{A(H(0,t),t)},$$

$$K_1(H(0,t),s,t) = -\frac{B(H(0,t),s,t) + \frac{\partial A}{\partial t}(s,t) - A_1(s,t)}{A(H(0,t),t)},$$

$$K_2(H(0,t),s,p,t) = -\frac{\frac{\partial A}{\partial t}(s,t)\frac{\partial B}{\partial t}(s,p,t) - A_1(s,t)B(s,p,t)}{A(H(0,t),t)}.$$

From the above reasoning follows the validity of the following lemma.

Lemma 6.5 *Let* $g(t) \in C^2[0,T]$, $g(+0) = 0$, $g'(+0) = 0$, $T > 0$. *Functions* $(\rho,\lambda,\mu) \in \Lambda_1$, $h(t) \in \Lambda_2$. *Then the solution of the inverse problem (6.81)–(6.84) in the class of functions* $\alpha(z) \in C[0,Z]$, $Z = H(0,T)$ *is equivalent to the solution of the Eq. (6.90).*

The main results of this section are the following theorems of unique solvability and stability of the solution of the inverse problem.

Theorem 6.4 *Suppose that the conditions of the lemma hold and* T, $\tilde{T}_0$, $\tilde{T}_1$, k, γ *are fixed positive numbers with* $\tilde{T}_1 > \tilde{T}_0$. *Then there is a unique solution to the inverse problem (6.61)–(6.66)* $\alpha(z) \in C[0,Z]$, $Z = H(0,T)$ *for any fixed* $T > 0$.

Theorem 6.5 *Let* $\alpha(z)$ *and* $\alpha^*(z)$ *be solutions of the inverse problem (6.61)–(6.66) with the data*

$$\{h(t),\ g(t)\},\ \{h^*(t),\ g^*(t)\},$$

respectively. Then there exists a positive number

$$C = C(\alpha_{00},\ k,\ \gamma,\ \tilde{T}_0,\ \tilde{T}_1,\ T),$$

$$\alpha_{00} = \max\Big\{\|\rho(y)\|_{C^2[0,\ \psi^{-1}(T)]},\ \|\lambda(y)\|_{C^2[0,\ \psi^{-1}(T)]},\ \|\mu(y)\|_{C^2[0,\ \psi^{-1}(T)]},$$

$\|h(t)\|_{C^2[0,T]}, \|h^*(t)\|_{C^2[0,T]}\Big\}$, *such that the following stability estimate holds:*

$$\|\alpha - \alpha^*\|_{C[0,H(0,T)]} \le C\Big[\|h - h^*\|_{C[0,T]} + \|g - g^*\|_{C^2[0,T]}\Big]. \tag{6.91}$$

First, we prove Theorem 6.4.

Proof Let $z = H(0,t)$, then $t = f(z)$ is function inverse to z.
We define the sequence of functions $\alpha_n(z)$:

$$\alpha_n(z) = \tilde{\alpha}(f(z))) + \int_0^{f(z)} \Big[\tilde{q}(f(z) - \tau)V_{n-1}((f(z)) - \tau, \tau)$$
$$- \int_0^{\tau} \tilde{r}''(\eta)V_{n-1}(f(z) - \tau, \tau - \eta)\mathrm{d}\xi\Big]\mathrm{d}\tau$$
$$+ \int_0^{z}\Big[K_1(z, s, f(z))\alpha_{n-1}(s) + \int_0^{s} K_2(z, s, p, f(z))\alpha_{n-1}(p)\mathrm{d}p\Big]\mathrm{d}s,\ n \ge 1, \tag{6.92}$$

$\alpha_0(z) = \tilde{\alpha}(f(z))$, V_{n-1} is defined by integral Eqs. (6.85)–(6.86), in which the function $\alpha(\tilde{H}(\xi,\tau))$ is replaced by $\alpha_{n-1}(\tilde{h}(\xi,\tau))$.
Construct a functional series

$$\sum_{n=1}^{\infty}[\alpha_{n+1}(z) - \alpha_n(z)]. \tag{6.93}$$

We prove the uniform convergence of the series (6.93). To do this, we note that from Eqs. (6.85)–(6.86) for the difference $V_n(y,t) - V_{n-1}(y,t)$ the estimate based on the use of the Gronwall's lemma is valid:

$$|V_n(y,t) - V_{n-1}(y,t)| \le C_1 \int_0^{H(y,t)} |\alpha_n(s) - \alpha_{n-1}(s)|\,\mathrm{d}s,\ n \ge 1,$$

$C_1 = C_1(\alpha_{00}, k, \gamma, \tilde{T}_0, \tilde{T}_1, T)$, then

$$|\alpha_{n+1}(z) - \alpha_n(z)| \le C_2 \int_0^z \|\alpha_n(s) - \alpha_{n-1}(s)\| \, ds,$$
$$C_2 = C_2(\alpha_{00}, k, \gamma, \tilde{T}_0, \tilde{T}_1, T). \tag{6.94}$$

In (6.94) we used the inequality $0 \le \tilde{H}(\xi, \tau) \le H(0, t)$ for $\xi \ge 0, \tau \ge 0$, which follows from Lemma 4.1 [1].

From (6.94) follows

$$\|\alpha_{n+1} - \alpha_n\| \le \frac{(C_2 z)^n}{n!} \le \frac{(C_2 Z)^n}{n!}.$$

The series (6.93) is majored by a convergent number series, then (6.93) converges uniformly on the interval [0, Z]. It follows that the sequence $\alpha_n(z)$ uniformly converges to the continuous function $\alpha(z)$. The uniqueness follows from the existence of the trivial solution of the corresponding homogeneous Eq. (6.90) (proof by using Gronwall's lemma). We can also say that the uniqueness of the solution is a consequence of Theorem 6.5 on stability.

Now, we prove Theorem 6.5.

Proof Let $\alpha(z)$ and $\alpha^*(z)$ be solutions of the inverse problem (6.61)–(6.66) with the data

$$\{h(t), \; g(t)\}, \; \{h^*(t), \; g^*(t)\},$$

respectively. For $\alpha(z)$ the Eq. (6.90) holds and for $\alpha^*(z)$ we have the equation

$$\alpha^*(H(0,t)) = \tilde{\alpha}^*(t) + \int_0^t \left[\tilde{q}(t-\tau)V^*(t-\tau,\tau) - \int_0^\tau \tilde{r}''^*(\eta)V^*(t-\tau,\tau-\eta)d\xi\right]d\tau$$
$$+ \int_0^{H(0,t)} \left[K_1^*(H(0,t),s,t)\alpha^*(s) + \int_0^s K_2^*(H(0,t),s,p,t)\alpha^*(p)dp\right]ds,$$

where

$$K_1^*(H(0,t),s,t) = -\frac{B^*(H(0,t),s,t) + \frac{\partial A}{\partial t}(s,t) - A_1(s,t)}{A(H(0,t),t)},$$

$$K_2^*(H(0,t),s,p,t) = -\frac{\frac{\partial A}{\partial t}(s,t)\frac{\partial B^*}{\partial t}(s,p,t) - A_1(s,t)B^*(s,p,t)}{A(H(0,t),t)},$$

$$V^*(y,t) = \frac{1}{2}\Bigg\{g'^*(t-y) + \int\limits_0^{t-y} h^*(\tau)g'^*(t-y-\tau)\mathrm{d}\tau$$

$$+ \iint\limits_{\Delta(y,t)} \Big[q(\xi)V^*(\xi,\tau) - \int\limits_0^{\tau} r''^*(\eta)V^*(\xi,\tau-\eta)\mathrm{d}\eta\Big]\mathrm{d}\xi\mathrm{d}\tau$$

$$- \int\limits_0^{t} \sqrt{a(\xi)}b(\xi)L\Big[h^*, \alpha^*(\tilde{H}(\xi,\tau))\frac{\partial \tilde{H}}{\partial \tau}(\xi,\tau)\Big]\Big|_{\xi=t+y-\tau}\mathrm{d}\tau$$

$$+ \iint\limits_{\Delta(y,t)} \frac{\partial}{\partial \xi}(\sqrt{a(\xi)})b(\xi)L\Big[h^*, \alpha^*(\tilde{H}(\xi,\tau))\frac{\partial \tilde{H}}{\partial \tau}(\xi,\tau)\Big]\mathrm{d}\xi\mathrm{d}\tau$$

$$+ \int\limits_{t-y}^{t} \sqrt{a(\xi)}b(\xi)L\Big[h^*, \alpha^*(\tilde{H}(\xi,\tau))\frac{\partial \tilde{H}}{\partial \tau}(\xi,\tau)\Big]\Big|_{\xi=t-y+\tau}\mathrm{d}\tau\Bigg\}, \quad 0 \le y \le t,$$

the corresponding function replacements are also made for the Eq. (6.86).

For $V(y,t) - V^*(y,t)$ the following estimate is valid:

$$|V(y,t) - V^*(y,t)| \le C_3\Big[\|h-h^*\|_{C[0,T]} + \|g-g^*\|_{C^2[0,T]} + \int_0^{H(y,t)} \|\alpha-\alpha^*\|_{C[0,s]}\mathrm{d}s\Big],$$

where $C_3 = C_3(\alpha_{00}, k, \gamma, \tilde{T}_0, \tilde{T}_1, T)$.

Then, composing the difference $\alpha(z) - \alpha(z)^*$ and considering that

$$\|r(t)\|_{C^2[0,T]} \le \|h(t)\|_{C^2[0,T]} \exp\left(T\|h(t)\|_{C^2[0,T]}\right),$$

we get

$$\|\alpha-\alpha^*\|_{C[0,Z]} \le C_4\Big[\|h-h^*\|_{C[0,T]} + \|g-g^*\|_{C^2[0,T]} + \int\limits_0^{Z} \|\alpha-\alpha^*\|_{C[0,s]}\mathrm{d}s\Big], \tag{6.95}$$

where the constant C_4 depends on the same parameters as C in Theorem 6.5.

From inequality (6.95) and Gronwall's lemma follows an estimate (6.91), where $C = C_4\exp(C_4 H(0,T))$.

6.3 One-Dimensional Inverse Coefficient Problem of Anisotropic Viscoelasticity Equation

For $x = (x_1, x_2, x_3)$, $(x_1, x_2) \in \mathbb{R}^2$, $t \in \mathbb{R}$, $x_3 > 0$ let us consider the following system of integro-differential equations of dynamical viscoelasticity

$$\rho \frac{\partial^2 u_i}{\partial t^2} = \sum_{j=1}^{3} \frac{\partial T_{ij}}{\partial x_j}, \quad i = 1, 2, 3, \tag{6.96}$$

under the following initial and boundary conditions:

$$u_j \mid_{t<0} \equiv 0, \tag{6.97}$$

$$T_{j3}|_{x_3=+0} = f_j(x_1, x_2, t), \ j = 1, 2, 3, \tag{6.98}$$

where $\rho > 0$ is the constant density of the medium.

The stress tensor (in viscoelastic anisotropic media) is represented by

$$T_{ij} = \sum_{m,l=1}^{3} c_{ijkl} \left[S_{ml} + \int_0^t k_i(t-\tau) S_{ml}(x,\tau) d\tau \right], \ i = 1, 2, 3, \quad j = 1, 2, 3, \tag{6.99}$$

where

$$S_{ml} = \frac{1}{2} \left(\frac{\partial u_m}{\partial x_l} + \frac{\partial u_l}{\partial x_m} \right), \ m = 1, 2, 3, \ l = 1, 2, 3,$$

$c_{ijml} = c_{ijml}(x)$ are the elastic moduli; $k(t) = \text{diag}(k_1, k_2, k_3)(t)$ is the medium relaxation function.

Assume that $c_{ijml} = c_{jiml} = c_{ijlm} = c_{mlij}$.

The symmetry of the stress tensor diminishes the number of independent elastic moduli from 81 to 21. Writing $c_{\alpha\beta} = c_{ijml}$, where $\alpha = (ij)$ and $\beta = (ml)$, in accordance with $(11) \to 1$, $(22) \to 2$, $(33) \to 3$, $(23) = (32) \to 4$, $(13) = (31) \to 5$ and $(12) = (21) \to 6$ we can represent the matrix of independent elastic moduli as a symmetric 6×6 matrix since the order in all pairs of indices (i, j) is immaterial and only six distinct pairs exist. We consider anisotropic media of cubical structure, i.e., the matrix of characteristics of the medium has the symmetric block form:

$$c_{\alpha\beta} = \begin{pmatrix} \begin{matrix} c_{11} & c_{12} & c_{12} \\ c_{12} & c_{11} & c_{12} \\ c_{12} & c_{12} & c_{11} \end{matrix} & O_{(3\times3)} \\ O_{(3\times3)} & \begin{matrix} c_{44} & 0 & 0 \\ 0 & c_{44} & 0 \\ 0 & 0 & c_{44} \end{matrix} \end{pmatrix}. \tag{6.100}$$

Let the elastic moduli c_{11}, c_{12}, c_{44} depends only on x_3, and the vector function (c_{11}, c_{12}, c_{44}) belongs to

$$\Lambda_3 = \Big\{(c_{11}(x_3), c_{12}(x_3), c_{44}(x_3)) : c_{11} > 0,\ c_{44} > 0,\ c_{11} > c_{12},\ c_{11} + 2c_{12} > 0,$$
$$c_{11},\ c_{44} \in C^2(\mathbb{R}_+),\ c_{12} \in C(\mathbb{R}_+)\Big\},\ \mathbb{R}_+ = [0, \infty),$$

$k_i(t) \in C^2(\mathbb{R}_+),\ i = 1, 2, 3$. Without limiting the generality, assume that $k_i(0) = 0,\ k_i{}'(0) = 0,\ i = 1, 2, 3$. These conditions further greatly simplify cumbersome calculations, but do not affect the method of study.

For anisotropic media of cubical structure (6.100) we can rewrite (6.96)–(6.98) as

$$\rho \frac{\partial^2 u_1}{\partial t^2} = L\Big[k_1, \frac{\partial}{\partial x_3}\left(c_{44}\frac{\partial u_1}{\partial x_3}\right) + c_{12}\frac{\partial^2 u_3}{\partial x_1 \partial x_3} + \frac{\partial}{\partial x_3}\left(c_{44}\frac{\partial u_3}{\partial x_1}\right)$$
$$+ c_{44}\frac{\partial^2 u_1}{\partial x_2^2} + c_{11}\frac{\partial^2 u_1}{\partial x_1^2} + (c_{12} + c_{44})\frac{\partial^2 u_2}{\partial x_1 \partial x_2}\Big], \tag{6.101}$$

$$\rho \frac{\partial^2 u_2}{\partial t^2} = L\Big[k_2, \frac{\partial}{\partial x_3}\left(c_{44}\frac{\partial u_2}{\partial x_3}\right) + c_{12}\frac{\partial^2 u_3}{\partial x_2 \partial x_3} + \frac{\partial}{\partial x_3}\left(c_{44}\frac{\partial u_3}{\partial x_2}\right)$$
$$+ c_{44}\frac{\partial^2 u_2}{\partial x_1^2} + c_{11}\frac{\partial^2 u_2}{\partial x_2^2} + (c_{12} + c_{44})\frac{\partial^2 u_1}{\partial x_1 \partial x_2}\Big], \tag{6.102}$$

$$\rho \frac{\partial^2 u_3}{\partial t^2} = L\Big[k_3, \frac{\partial}{\partial x_3}\left(c_{11}\frac{\partial u_3}{\partial x_3}\right) + c_{44}\frac{\partial^2 u_1}{\partial x_1 \partial x_3} + \frac{\partial}{\partial x_3}\left(c_{12}\frac{\partial u_1}{\partial x_1}\right)$$
$$+ c_{44}\frac{\partial^2 u_2}{\partial x_2 \partial x_3} + \frac{\partial}{\partial x_3}\left(c_{12}\frac{\partial u_2}{\partial x_2}\right) + c_{44}\frac{\partial^2 u_3}{\partial x_1^2} + c_{44}\frac{\partial^2 u_3}{\partial x_2^2}\Big], \tag{6.103}$$

$$L\left[k_1, c_{44}\frac{\partial u_3}{\partial x_1} + c_{44}\frac{\partial u_1}{\partial x_3}\right]\Big|_{x_3=+0} = f_1(x_1, t), \tag{6.104}$$

$$L\left[k_2, c_{44}\frac{\partial u_3}{\partial x_2} + c_{44}\frac{\partial u_2}{\partial x_3}\right]\Big|_{x_3=+0} = f_2(x_1, t), \tag{6.105}$$

$$\left(L\left[k_3, c_{12}\frac{\partial u_1}{\partial x_1} + c_{12}\frac{\partial u_2}{\partial x_2} + c_{11}\frac{\partial u_3}{\partial x_3}\right]\right)\Big|_{x_3=+0} = f_3(x_1, t), \tag{6.106}$$

$$u_i\ |_{t<0} \equiv 0,\quad i = 1, 2, 3. \tag{6.107}$$

Refer as the direct problem to the problem of finding the displacement vector $u(x, t)$ satisfying (6.96)–(6.98) in the sense of distributions for given $c_{11}(x_3)$, $c_{12}(x_3)$, $c_{44}(x_3)$, $k_j(t)$, $f_j(x_1, t)$, $j = 1, 2, 3$.

We consider that

$$f_j \equiv -\frac{1}{2}\delta(t)\delta(x_1),\ j = 1, 2, 3,$$

in Eqs. (6.104)–(6.106), where $\delta(\cdot)$ is Dirac delta function.

Since the coefficients of the equations and boundary conditions in (6.101)–(6.107) are independent of x_2, so is the solution u, φ to the direct problem (Sect. 6.1).

Let us express (6.101)–(6.107) in terms of the Fourier transform:

$$\rho\frac{\partial^2 U_1}{\partial t^2} = L\Big[k_1, \frac{\partial}{\partial x_3}\left(c_{44}\frac{\partial U_1}{\partial x_3}\right) + i\nu c_{12}\frac{\partial U_3}{\partial x_3} + i\nu\frac{\partial}{\partial x_3}(c_{44}U_3) - \nu^2 c_{11}U_1\Big], \tag{6.108}$$

$$\rho\frac{\partial^2 U_2}{\partial t^2} = L\Big[k_2, \frac{\partial}{\partial x_3}\left(c_{44}\frac{\partial U_2}{\partial x_3}\right) - \nu^2 c_{44}U_2\Big], \tag{6.109}$$

$$\rho\frac{\partial^2 U_3}{\partial t^2} = L\Big[k_3, \frac{\partial}{\partial x_3}\left(c_{11}\frac{\partial U_3}{\partial x_3}\right) + i\nu c_{44}\frac{\partial U_1}{\partial x_3} + i\nu\frac{\partial}{\partial x_3}(c_{12}U_1) - \nu^2 c_{44}U_3\Big], \tag{6.110}$$

$$L\left[k_1, ic_{44}\nu U_3 + c_{44}\frac{\partial U_1}{\partial x_3}\right]|_{x_3=+0} = F_1(\nu, t), \tag{6.111}$$

$$L\left[k_2, c_{44}\frac{\partial U_2}{\partial x_3}\right]|_{x_3=+0} = F_2(\nu, t), \tag{6.112}$$

$$L\left[k_3, i\nu c_{12}U_1 + c_{11}\frac{\partial U_3}{\partial x_3}\right]|_{x_3=+0} = F_3(\nu, t), \tag{6.113}$$

$$U_j\ |_{t<0} \equiv 0, \quad j = 1, 2, 3; \tag{6.114}$$

where

$$U_j(x_3, t, \nu) = \int_{-\infty}^{\infty} u(x_1, x_3, t)\exp(i\nu x_1)\mathrm{d}x_1,\ j = 1, 2, 3,$$

$$F_j(t, \nu) = \int_{-\infty}^{\infty} f_j(x_1, t)\exp(i\nu x_1)\mathrm{d}x_1,\ j = 1, 2, 3,$$

ν is a parameter and $U_j \in C^1(\mathbb{R}; C(\tilde{D}))$, $\tilde{D} = \{(x_3, t) : x_3 \geq 0,\ 0 \leq t \leq T\}$, $T > 0$.

We consider (6.108)–(6.114) as the union of the two subsystems. The first, comprising (6.108), (6.110), (6.111), (6.113), and (6.114), determines U_1 and U_3. The second, comprising (6.109), (6.112), and (6.114), determines U_2.

The subject of the current research is the *inverse problem* of determining the elastic moduli $c_{44}(x_3)$, $c_{11}(x_3)$, $c_{12}(x_3)$, $x_3 > 0$, that is included in the equality (6.96) using

(6.99), when the additional information about the components $U_j,\ j = 1, 3$ of the solution to the direct problem is available:

$$U_1(x_3, t, \nu)\mid_{x_3=+0,\nu=+0} = g_1(t), \tag{6.115}$$

$$U_3(x_3, t, \nu)\mid_{x_3=+0,\nu=+0} = g_3(t), \tag{6.116}$$

$$\frac{\partial U_1}{\partial \nu}(x_3, t, \nu)\mid_{x_3=+0,\nu=+0} = g_2(t), \tag{6.117}$$

where $g_j(t),\ t > 0,\ j = 1, 2, 3$, are given functions.

The solution of the inverse problem can be reduced to the sequential solution of some auxiliary inverse problems for scalar hyperbolic equations.

From the equalities (6.108), (6.111), (6.114), and (6.115) for the function

$$U_1^0(x_3, t) = U_1(x_3, t, \nu)\mid_{\nu=+0}$$

we obtain

$$\rho \frac{\partial^2 U_1^0}{\partial t^2} = L\Big[k_1, \frac{\partial}{\partial x_3}\Big(c_{44}\frac{\partial U_1^0}{\partial x_3}\Big)\Big],\ x_3 > 0, \tag{6.118}$$

$$U_1^0\mid_{t<0} \equiv 0, \tag{6.119}$$

$$L\left[k_1, c_{44}\frac{\partial U_1^0}{\partial x_3}\right]\mid_{x_3=+0} = -\frac{1}{2}\delta(t), \tag{6.120}$$

$$U_1^0\mid_{x_3=0} = g_1(t). \tag{6.121}$$

These equalities allow us to consider the following inverse problems.

Inverse problem 1: determine a function $c_{44}(x_3) > 0,\ c_{44}(x_3) \in C^2(\mathbb{R}_+)$, that is included into the differential equality (6.118), so the problem's solution (6.118)–(6.118) satisfies the equation (6.121), where $g_1(t),\ k_1(t),\ t \in [0, T]$ are known functions, ρ is a known positive constant.

From (6.110), (6.113), (6.114), (6.116) for the function

$$U_3^0(x_3, t) = U_3(x_3, t, \nu)\mid_{\nu=+0}$$

we obtain

$$\rho \frac{\partial^2 U_3^0}{\partial t^2} = L\Big[k_3, \frac{\partial}{\partial x_3}\Big(c_{11}\frac{\partial U_3^0}{\partial x_3}\Big)\Big],\ x_3 > 0, \tag{6.122}$$

$$U_3^0\mid_{t<0} \equiv 0, \tag{6.123}$$

$$L\left[k_3, c_{11}\frac{\partial U_3^0}{\partial x_3}\right] |_{x_3=+0}= -\frac{1}{2}\delta(t), \tag{6.124}$$

$$U_3^0 |_{x_3=0}= g_3(t). \tag{6.125}$$

Inverse problem 2: determine a function $c_{11}(x_3) > 0,\ c_{11}(x_3) \in C^2(\mathbb{R}_+)$, that is included into the differential equality (6.122), so the problem's solution (6.122)–(6.124) satisfies the equation (6.125), where $g_3(t),\ k_3(t),\ t \in [0, T]$ are known functions, ρ is known positive constant.

From (6.109), (6.112), (6.114), (6.117) for the function

$$U_1^1(x_3, t) = \frac{\partial}{\partial \nu}U_1(x_3, t, \nu) |_{\nu=+0}$$

we obtain

$$\rho\frac{\partial^2 U_1^1}{\partial t^2} = L\Big[k_1, \frac{\partial}{\partial x_3}\left(c_{44}\frac{\partial U_1^1}{\partial x_3}\right) + ic_{12}\frac{\partial U_3^0}{\partial x_3} + i\frac{\partial}{\partial x_3}(c_{44}U_3^0)\Big],\ x_3 > 0, \tag{6.126}$$

$$U_1^1 |_{t<0}\equiv 0, \tag{6.127}$$

$$L\left[k_1, c_{44}\frac{\partial U_1^1}{\partial x_3} + ic_{44}U_3^0\right] |_{x_3=+0}= 0, \tag{6.128}$$

$$U_1^1 |_{x_3=0}= g_2(t). \tag{6.129}$$

Inverse problem 3: determine a function $c_{12}(x_3) > 0,\ c_{12}(x_3) \in C^2(\mathbb{R}_+)$, that is included into the differential equality (6.126), so the problem's solution (6.126)–(6.128) satisfies the equation (6.129), where $g_2(t),\ k_1(t),\ t \in [0, T]$ are known functions, ρ is known positive constant, $c_{44},\ c_{11}$ are functions, that are defined as solutions of *inverse problems 1, 2*, and $U_3^0(x_3, t)$ is the direct problem's solution (6.122)–(6.124).

Introduce the new variables y, z as follows:

$$y = \psi_1(x_3) := \int_0^{x_3}\sqrt{\frac{\rho}{c_{44}(\xi)}}\mathrm{d}\xi, \quad z = \psi_2(x_3) := \int_0^{x_3}\sqrt{\frac{\rho}{c_{11}(\xi)}}\mathrm{d}\xi.$$

Let h_0, h_1, T, X be fixed positive numbers, $h_0 < h_1, X = \frac{T}{2}\sqrt{\frac{h_1}{\rho}}$. Let $\Lambda_1(h_0, h_1, X) = \{c(x_3) \in C^2[0, X] : \|c\|_{C[0,X]} \leq h_1,\ c(x_3) \geq h_0\}$.

Theorem 6.6 *Let T be a fixed positive number. The function $g_1(t) \in C^2[0, T]$, $g_1(+0) > 0$, $k_1(t) \in C^2[0, T]$. Then, there is only one solution of the inverse problem 1 $c_{44}(x_3) \in C^2[0, X]$, $X \leq \psi_1^{-1}(T/2)$.*

Theorem 6.7 *Let T, h_0, h_1 be fixed positive numbers, $h_0 \leq h_1$, $X_0 = \frac{T}{2}\sqrt{\frac{h_0}{\rho}}$ and the functions $c_{44}^{(1)}(x_3)$, $c_{44}^{(2)}(x_3)$, which belongs to $\Lambda_1(h_0, h_1, X)$, are the solutions of the inverse problem 1 with data $\{g_1^{(1)}(t), k_1^{(1)}(t)\}$, $\{g_1^{(2)}(t), k_1^{(2)}(t)\}$, respectively. Then there is a positive number $C = C(h_0, h_1, M_1, \rho, T)$, $M_1 = \max\{\|g_1^{(1)}(t)\|_{C^2[0,T]}, \|k_1^{(1)}(t)\|_{C^2[0,T]}, \|g_1^{(2)}(t)\|_{C^2[0,T]}, \|k_1^{(2)}(t)\|_{C^2[0,T]}\}$, so the estimate of stability is fair:*

$$\|c_{44}^{(1)} - c_{44}^{(2)}\|_{C[0,X_0]}) \leq C\left(\|g_1^{(1)} - g_1^{(2)}\|_{C^2[0,T]} + \|k_1^{(1)} - k_1^{(2)}\|_{C^2[0,T]}\right). \quad (6.130)$$

Denote $\|\cdot\|(T) = \|\cdot\|_{C[0,T]}$, $\|\cdot\|_l(T) = \|\cdot\|_{C^l[0,T]}$, $l = 1, 2, \ldots$.

Theorem 6.8 *Let T be a fixed positive number. The function $g_3(t) \in C^2[0, T]$, $g_3(+0) > 0$, $k_3(t) \in C^2[0, T]$. Then there is only one solution of the inverse problem 2 $c_{11}(x_3) \in C^2[0, X]$, $X \leq \psi_2^{-1}(T/2)$.*

Theorem 6.9 *Let T, h_0, h_1 be fixed positive numbers, $h_0 \leq h_1$, $X_0 = \frac{T}{2}\sqrt{\frac{h_0}{\rho}}$ and $c_{11}^{(1)}(x_3)$, $c_{11}^{(2)}(x_3)$, which belongs to $\Lambda_1(h_0, h_1, X)$, are the solutions of the inverse problem 2 with data $\{g_3^{(1)}(t), k_3^{(2)}(t)\}$, $\{g_3^{(1)}(t), k_3^{(2)}(t)\}$, respectively. Then, there is a positive number $C^* = C^*(h_0, h_1, M_2, \rho, T)$, $M_2 = \max\{\|g_3^{(1)}(t)\|_2(T), \|k_3^{(1)}(t)\|_2(T), \|g_3^{(2)}(t)\|_2(T), \|k_3^{(2)}(t)\|_2(T)\}$, so the estimate of stability is fair:*

$$\|c_{11}^{(1)} - c_{11}^{(2)}\|(X_0) \leq C^*\left(\|g_3^{(1)} - g_3^{(2)}\|_2(T) + \|k_3^{(1)} - k_3^{(2)}\|_2(T)\right). \quad (6.131)$$

Theorem 6.10 *Let T be a fixed positive number, $X = \frac{T}{2}\sqrt{\frac{h_1}{\rho}}$. The functions $g_2(t) \in C^1[0, T]$, $k_1(t) \in C^2[0, T]$, $g_2(+0) = 0$, $k_3(t) \equiv 0$. In addition, the following inequality is fair:*

$$\max\left[-g_2'(+0), 2g_2'(+0)\right] < \frac{g_1(+0)}{2\rho[g_1(+0) + g_3(+0)]}.$$

Then there is only one solution of the inverse problem 3 $c_{12}(x_3) \in C[0, X]$.

Theorem 6.11 *Let T, h_0, h_1 be fixed positive numbers, $h_0 \leq h_1$, $X_0 = \frac{T}{2}\sqrt{\frac{h_0}{\rho}}$, $k_3(t) \equiv 0$ and $c_{12}^{(1)}(x_3)$, $c_{12}^{(2)}(x_3)$, which belongs to $\Lambda_2(h_1, X) = \{c(x_3) \in C[0, X] : \|c\| \leq h_1\}$, are the solutions of the inverse problem 3 with data $\{c_{11}^{(1)}(x_3), c_{44}^{(1)}(x_3), g_3^{(1)}(t), k_1^{(1)}(t)\}$, $\{c_{11}^{(2)}(x_3), c_{44}^{(2)}(x_3), g_3^{(2)}(t), k_1^{(2)}(t)\}$, respectively. Herewith the functions $c_{44}^{(1)}(x_3)$, $c_{44}^{(2)}(x_3)$ from $\Lambda_1(h_0, h_1, X)$ are the solutions of the inverse problem 1 with data $\{g_1^{(1)}(t), k_1^{(1)}(t)\}$, $\{g_1^{(2)}(t), k_1^{(2)}(t)\}$; and the functions $c_{11}^{(1)}(x_3)$, $c_{11}^{(2)}(x_3)$ from $\Lambda_1(h_0, h_1, X)$ are the solutions of the inverse problem 2 with data $\{g_3^{(1)}(t), k_3^{(1)}(t)\}$, $\{g_3^{(2)}(t), k_3^{(2)}(t)\}$, respectively, for $t \in [0, T]$. Then there is a positive number $C^{**} = C^{**}(h_0, h_1, M_3, \rho, T)$,*

$$M_3 = \max\{\|g_3^{(1)}(t)\|_2(T),\ \|g_2^{(1)}(t)\|_1(T),\ \|k_1^{(1)}(t)\|_2(T),$$
$$\|g_3^{(2)}(t)\|_2(T),\ \|g_2^{(2)}(t)\|_1(T),\ \|k_1^{(2)}(t)\|_2(T)\},$$

so the estimate of stability is fair:

$$\begin{aligned}\|c_{12}^{(1)} - c_{12}^{(2)}\|(X_0) \le C^{**}\Big(&\|c_{44}^1 - c_{44}^2\|_2(X_0) + \|c_{11}^1 - c_{11}^2\|_2(X_0)\\ &+ \|g_1^{(1)} - g_1^{(2)}\|_2(T) + \|g_2^{(1)} - g_2^{(2)}\|_1(T)\\ &+ \|g_3^{(1)} - g_3^{(2)}\|_2(T) + \|k_1^{(1)} - k_1^{(2)}\|_2(T)\Big).\end{aligned} \tag{6.132}$$

Proofs of the theorems will be presented below. Let us study *inverse problem 1*. Let

$$v(y,t) := \frac{U_1^0\left(\psi_1^{-1}(y), t\right)}{s(y)}, \quad s(y) := \left(\frac{c_{44}(+0)}{c_{44}\left(\psi_1^{-1}(y)\right)}\right)^{1/4}.$$

Then (6.124)–(6.127) in terms of the new functions and variable $y > 0$ takes the form

$$\frac{\partial^2 v}{\partial t^2} = L\Big[k_1, \frac{\partial^2 v}{\partial y^2} + q(y)v\Big], \tag{6.133}$$

$$v|_{t<0} \equiv 0, \tag{6.134}$$

$$L\left[k_1, \frac{\partial v(y,t)}{\partial y} + \frac{1}{2}\frac{a'(0)}{a(0)}v(y,t)\right]_{y=+0} = -\frac{a(0)\delta(t)}{2}, \tag{6.135}$$

$$v(y,t)|_{y=+0} = g_1(t), \quad t > 0, \tag{6.136}$$

where $q(y) = \frac{s''(y)}{s(y)} - 2\left[\frac{s'(y)}{s(y)}\right]^2$, $u(y) = \frac{1}{\sqrt{\rho \cdot c_{44}(\psi_1^{-1}(y))}}$.

Let $L[k_1, v(y,t)] = w(y,t)$. Then $v(y,t) = L[r_1, w(y,t)]$, where

$$r_1(t) = -k_1(t) - \int_0^t k_1(t-\tau)r_1(\tau)d\tau. \tag{6.137}$$

For the function $w(y,t)$ we find:

$$\frac{\partial^2 w}{\partial t^2} = \frac{\partial^2 w}{\partial y^2} + q(y)w - \int_0^t r_1''(t-\tau)w(y,\tau)d\tau, \ y > 0,\ t > 0, \tag{6.138}$$

$$w|_{t<0} \equiv 0, \tag{6.139}$$

$$\left[\frac{\partial w(y,t)}{\partial y}+\frac{1}{2}\frac{a'(0)}{a(0)}w(y,t)\right]_{y=+0}=-\frac{a(0)\delta(t)}{2}, \tag{6.140}$$

$$w(y,t)|_{y=+0}=L\left[k_1,g_1\right],\ t>0. \tag{6.141}$$

The above constructions lead to the following *inverse problem 1**: to determine the function $q(y)\in C(0,T/2)$ included in the differential Eq. (6.138) if the solution $w(y,t)$ of the problem (6.138)–(6.140) satisfies the equality (6.141).

Direct problem solution (6.138)–(6.141) has the form:

$$w(y,t)=\theta(t-y)\left[\frac{a(0)}{2}+\tilde{w}(y,t)\right],$$

where $\theta(t-y)$ is the Heaviside function, the unknown function $\tilde{w}$ satisfies the following equations:

$$\begin{aligned}\frac{\partial^2\tilde{w}}{\partial t^2}&=\frac{\partial^2\tilde{w}}{\partial y^2}+q(y)\theta(t-y)\left[\frac{a(0)}{2}+\tilde{w}(y,t)\right]\\&-\theta(t-y)\int\limits_y^t r_1''(t-\tau)\left[\frac{a(0)}{2}+\tilde{w}(y,\tau)\right]d\tau,\quad y>0,\ t>0,\end{aligned} \tag{6.142}$$

$$\tilde{w}|_{t<0}\equiv 0, \tag{6.143}$$

$$\left[\frac{\partial \tilde{w}(y,t)}{\partial y}+\frac{a'(0)}{4}+\frac{1}{2}\frac{a'(0)}{a(0)}\tilde{w}(y,t)\right]_{y=+0}=0, \tag{6.144}$$

$$\tilde{w}(y,t)|_{y=+0}=L\left[k_1,g_1\right]-\frac{a(0)}{2},\ t>0. \tag{6.145}$$

Given (6.145) condition (6.144) can be rewritten as

$$\left.\frac{\partial\tilde{w}(y,t)}{\partial y}\right|_{y=+0}=-\frac{1}{2}\frac{a'(0)}{a(0)}L[k_1,g_1]\equiv f(t). \tag{6.146}$$

The solution of the problem (6.142), (6.143), (6.146) $\tilde{w}(y,t)$ can be represented as

$$\tilde{w}(y,t)=w_1(y,t)+w_2(y,t), \tag{6.147}$$

where $w_1(y,t)$ is the solution of the problem:

$$\begin{aligned}\frac{\partial^2 w_1}{\partial t^2}&=\frac{\partial^2 w_1}{\partial y^2}+\theta(t-y)q(y)w_1\\&-\theta(t-y)\int\limits_y^t r_1''(t-\tau)w_1(y,\tau)d\tau,\end{aligned} \tag{6.148}$$

$$w_1|_{t<0} \equiv 0, \tag{6.149}$$

$$\left.\frac{\partial w_1(y,t)}{\partial y}\right|_{y=+0} = f(t) \tag{6.150}$$

and $w_2(y,t)$ is the solution of the problem:

$$\begin{aligned} \frac{\partial^2 w_2}{\partial t^2} &= \frac{\partial^2 w_2}{\partial y^2} + \theta(t-y)\Big[\frac{a(0)}{2}q(y) + w_2(y,t)\Big] \\ &- \theta(t-y)\int\limits_y^t r_1''(t-\tau)\Big[\frac{a(0)}{2} + w_2(y,\tau)\Big]d\tau, \end{aligned} \tag{6.151}$$

$$w_2|_{t<0} \equiv 0, \tag{6.152}$$

$$\left.\frac{\partial w_2(y,t)}{\partial y}\right|_{y=+0} = 0, \tag{6.153}$$

Let us study the problem (6.148)–(6.150). Its solution is presented in the form of:

$$w_1(y,t) = \theta(t-y)\Big[w_1^*(y,t) - F(t-y)\Big], \text{ where } F(t-y) := \int\limits_0^{t-y} f(s)ds, \tag{6.154}$$

then for the function $w_1^*(y,t)$ we find:

$$\begin{aligned} \frac{\partial^2 w_1^*}{\partial t^2} &= \frac{\partial^2 w_1^*}{\partial y^2} + q(y)\theta(t-y)\Big[w_1^*(y,t) - F(t-y)\Big] \\ &- \theta(t-y)\int\limits_y^t r_1''(t-\tau)\big[w_1^*(y,\tau) - F(\tau-y)\big]d\tau, \end{aligned} \tag{6.155}$$

$$w_1^*(y,t)|_{t<0} \equiv 0, \tag{6.156}$$

$$\left.\frac{\partial w_1^*(y,t)}{\partial y}\right|_{y=+0} = 0, \tag{6.157}$$

Let us continue functions $q(y),\ \theta(t-y),\ F(t-y)$ for $y<0$ as even functions. The problem solution (6.155)–(6.157) is equivalent to the problem of determining an even by y function $w_1^*(y,t)$ satisfying the Eqs. (6.155), (6.156).

Then the solution of the problem (6.155), (6.156) is presented as:

$$w_1^*(y,t) = \frac{1}{2}\int\limits_{\frac{y-t}{2}}^{\frac{y+t}{2}}\int\limits_{|\xi|}^{t-|y-\xi|}\int\limits_{|\xi|}^{\tau} r_1''(\tau-\eta)F(\eta-|\xi|)\mathrm{d}\eta\mathrm{d}\tau\mathrm{d}\xi$$

$$+\frac{1}{2}\int\limits_{\frac{y-t}{2}}^{\frac{y+t}{2}}\int\limits_{|\xi|}^{t-|y-\xi|}\Big\{q(\xi)\Big[w_1^*(\xi,\tau)-F(\tau-|\xi|)\Big] \tag{6.158}$$

$$-\int\limits_{|\xi|}^{\tau} r_1''(\tau-\eta)w_1^*(\xi,\eta)\mathrm{d}\eta\Big\}\mathrm{d}\tau\mathrm{d}\xi.$$

From (6.158) it follows, that $w_1^*(y,|y|+0)=0,\ w_1^*(y,+0)=0$.

Similar arguments can be used to show that the solution of the problem (6.151)–(6.153) with an even continuation of the functions $q(y),\ \theta(t-y)$ in the domain $y<0$ is equivalent to the solution of the integral equation:

$$w_2(y,t) = \frac{a(0)}{4}\int\limits_{\frac{y-t}{2}}^{\frac{y+t}{2}} r_1(t-|y-\xi|-|\xi|)\mathrm{d}\xi$$

$$+\frac{1}{2}\int\limits_{\frac{y-t}{2}}^{\frac{y+t}{2}}\int\limits_{|\xi|}^{t-|y-\xi|}\Big\{q(\xi)\Big[\frac{a(0)}{2}+w_2(\xi,\tau)\Big] \tag{6.159}$$

$$-\int\limits_{|\xi|}^{\tau} r_1''(\tau-\eta)w_2(\xi,\tau)\mathrm{d}\eta\Big\}\mathrm{d}\tau\mathrm{d}\xi.$$

It follows from (6.159) that $w_2(y,|y|+0)=0,\ w_2(y,+0)=0$.

Theorem 6.12 *(On the local unique solvability of the inverse problem 1*). Let T be a fixed positive number, $g_1(t)\in C^2[0,T],\ g_1(+0)>0,\ g_1'(+0)>0,\ k_1(t)\in C^2[0,T]$. Then, for sufficiently small T, there exists a unique solution of the inverse problem 1*, which belongs to $C[-T/2,T/2]$.*

We define the set $\tilde{Q}(q_0,T)=\{q(y):\|q\|_{C[-T/2,T/2]}\le q_0$.

Theorem 6.13 *(On the stability of the inverse problem 1*). Let $q^{(1)}(y),\ q^{(2)}(y)\in\tilde{Q}(q_0,T)$ two solutions of the inverse problem 1* with data $\{g_1^{(1)}(t),\ k_1^{(1)}(t)\}$, $\{g_1^{(2)}(t),\ k_1^{(2)}(t)\}$, respectively. Then there is a positive number $C^*=C^*(q_0,\ a(0),\ a'(0),\ M_1,\rho,\ T)$, $M_1=\max\{\|g_1^{(1)}(t)\|_2(T),\ \|k_1^{(1)}(t)\|_2(T),\|g_1^{(2)}(t)\|_2(T),\ \|k_1^{(2)}(t)\|_2(T)\}$, which is a valid stability estimate*

$$\|q^{(1)} - q^{(2)}\|_{C[-T/2,T/2]} \leq C^* \left(g_1^{(1)} - g_1^{(2)}\|_2(T) + \|k_1^{(1)} - k_1^{(2)}\|_2((T)\right). \quad (6.160)$$

Now we prove Theorem 6.12.

Proof Differentiate by t the equality (6.158) and (6.159):

$$\begin{aligned} \frac{\partial w_1^*}{\partial t}(y,t) = \frac{1}{2}\int\limits_{\frac{y-t}{2}}^{\frac{y+t}{2}} \Bigg\{ & q(\xi)\Big[w_1^*(\xi, t-|y-\xi|) - F(t-|y-\xi|-|\xi|)\Big] \\ & - \int\limits_0^{t-|y-\xi|-|\xi|} r_1''(\eta)\Big[w_1^*(\xi, t-|y-\xi|-\eta) \\ & - F(t-|y-\xi|-|\xi|-\eta)\Big]\mathrm{d}\eta\Bigg\}\mathrm{d}\xi, \end{aligned} \quad (6.161)$$

$$\begin{aligned} \frac{\partial w_2}{\partial t}(y,t) = \frac{1}{2}\int\limits_{\frac{y-t}{2}}^{\frac{y+t}{2}} \Bigg\{ & q(\xi)\Big[\frac{a(0)}{2} + w_2(\xi, t-|y-\xi|) \\ & - \int\limits_0^{t-|y-\xi|-|\xi|} r_1''(\eta)\Big[\frac{a(0)}{2} + w_2(\xi, t-|y-\xi|-\eta)\Big]\mathrm{d}\eta\Bigg\}\mathrm{d}\xi, \end{aligned} \quad (6.162)$$

$$\begin{aligned} \frac{\partial^2 w_1^*}{\partial t^2}(y,t) = \frac{1}{2}\int\limits_{\frac{y-t}{2}}^{\frac{y+t}{2}} \Bigg\{ & q(\xi)\Big[\frac{\partial w_1^*}{\partial t}(\xi, t-|y-\xi|) - f(t-|y-\xi|-|\xi|)\Big] \\ & - \int\limits_0^{t-|y-\xi|-|\xi|} r_1''(\eta) \times \Big[\frac{\partial w_1^*}{\partial t}(\xi, t-|y-\xi|-\eta) \\ & - f(t-|y-\xi|-\eta-|\xi|)\Big]\mathrm{d}\eta\Bigg\}\mathrm{d}\xi, \end{aligned} \quad (6.163)$$

$$\begin{aligned} \frac{\partial^2 w_2}{\partial t^2}(y,t) = {} & \frac{a(0)}{8}\left[q\left(\frac{y+t}{2}\right) + q\left(\frac{y-t}{2}\right)\right] \\ & + \frac{1}{2}\int\limits_{\frac{y-t}{2}}^{\frac{y+t}{2}} \Bigg\{q(\xi)\frac{\partial w_2}{\partial t}(\xi, t-|y-\xi|) - \frac{a(0)}{2} r_1''(t-|y-|\xi|) \\ & - \int\limits_0^{t-|y-\xi|-|\xi|} r_1''(\eta)\frac{\partial w_2}{\partial t}(\xi, t-|y-\xi|-\eta)\mathrm{d}\eta\Bigg\}\mathrm{d}\xi, \end{aligned} \quad (6.164)$$

Note that from the presentation (6.147) it follows that

$$\tilde{w}'(0,t) = (w_1^*)'(0,t) + w_2{}'(0,t) + \frac{a'(0)}{2a(0)} L[k_1, g_1(t)].$$

On the other hand, from (6.145) we get $\tilde{w}(0,t) = L[k_1, g_1(t)] - \frac{a(0)}{2}$, therefore,

$$g_1'(t) + \int_0^t k_1'(t-\tau) g_1(\tau) d\tau = (w_1^*)'(0,t) + w_2{}'(0,t) + \frac{a'(0)}{2a(0)} L[k_1, g_1(t)], \tag{6.165}$$

$$(w_1^*)''(0,t) + w_2{}''(0,t) = g_1''(t) + \int_0^t k_1''(t-\tau) g_1(\tau) d\tau - \frac{a'(0)}{2a(0)} \left[g_1'(t) + \int_0^t k_1'(t-\tau) g_1(\tau) d\tau \right]. \tag{6.166}$$

From (6.165) for $t = 0$ we find the value $a'(0) = 4g_1'(0)$, because from (6.145) we get $a(0) = 2g_1(0)$. Thus, the values $a(0),\ a'(0)$ are known.

From (6.163), (6.164) for $y = 0$ and (6.166) we obtain an integral equation for $q(y)$:

$$\begin{aligned} q(y) = &\frac{2a'(0)}{a^2(0)} \left[g_1'(2|y|) + \operatorname{sign}(y) \int_0^{2y} k_1'(2|y| - \tau) g_1(\tau) d\tau \right] \\ &- \frac{4}{a(0)} \left[g_1''(2|y|) + \operatorname{sign}(y) \int_0^{2y} k_1''(2|y| - \tau) g_1(\tau) d\tau \right] \\ &- \frac{4}{a(0)} \operatorname{sign}(y) \int_0^y \Bigg\{ q(\xi) \Big[\frac{\partial w_1^*}{\partial t}(\xi, 2|y| - \xi) - f(2|y| - 2\xi) \Big] \\ &- \int_0^{2|y|-2\xi} r_1''(\eta) \Big[\frac{\partial w_1^*}{\partial t}(\xi, 2|y| - \xi - \eta) - f(2|y| - 2\xi - \eta) \Big] d\eta \Bigg\} d\xi \\ &- \frac{4}{a(0)} \operatorname{sign}(y) \int_0^y \Bigg\{ q(\xi) \frac{\partial w_2}{\partial t}(\xi, 2|y| - \xi) - \frac{a(0)}{2} r_1''(2|y| - 2\xi) \\ &- \int_0^{2|y|-2\xi} r_1''(\eta) \frac{\partial w_2}{\partial t}(\xi, 2|y| - \xi - \eta) d\eta \Bigg\} d\xi. \end{aligned} \tag{6.167}$$

Let $\Diamond(y,t) = \{(\xi,\tau) : |\xi| \le \tau \le t - |y-\xi|\}$ and consider the equalities (6.158), (6.159), (6.161), (6.161), (6.167) in the domain $\Diamond(0,T)$. These equations are a closed system of nonlinear integral equations of the second kind with respect to functions $w_1^*(y,t)$, $w_2(y,t)$, $\left(\frac{\partial w_1^*}{\partial t}\right)(y,t)$, $\left(\frac{\partial w_2}{\partial t}\right)(y,t)$, $q(y)$. Express (6.158), (6.159), (6.161), (6.161), (6.167) as the operator equation

$$\varphi = A\varphi, \tag{6.168}$$

where

$$\varphi := \left[w_1^*(y,t), w_2(y,t), \frac{\partial w_1^*}{\partial t}(y,t), \frac{\partial w_2}{\partial t}(y,t), q(y)\right]$$

is a vector function with components φ_i $(i = 1, 2, 3, 4, 5)$, while the operator A defined on the set of functions $\varphi \in C(\Diamond(0,T))$ in accordance with (6.158), (6.159), (6.161), (6.161), (6.167) is of the form $A = (A_1, A_2, A_3, A_4, A_5)$:

$$A_1\varphi = \varphi_{01} + \frac{1}{2}\int_{\frac{y-t}{2}}^{\frac{y+t}{2}}\int_{|\xi|}^{t-|y-\xi|}\left\{\varphi_5(\xi)\Big[\varphi_1(\xi,\tau) - F(\tau - |\xi|)\Big]\right.$$
$$\left. - \int_{|\xi|}^{\tau} r_1''(\tau-\eta)\varphi_1(\xi,\eta)\mathrm{d}\eta\right\}\mathrm{d}\tau\mathrm{d}\xi,$$

$$A_2\varphi = \varphi_{02} + \frac{1}{2}\int_{\frac{y-t}{2}}^{\frac{y+t}{2}}\int_{|\xi|}^{t-|y-\xi|}\left\{\varphi_5(\xi)\Big[\frac{a(0)}{2} + \varphi_2(\xi,\tau)\Big]\right.$$
$$\left. - \int_{|\xi|}^{\tau} r_1''(\tau-\eta)\varphi_2(\xi,\tau)\mathrm{d}\eta\right\}\mathrm{d}\tau\mathrm{d}\xi,$$

$$A_3\varphi = \varphi_{03} + \frac{1}{2}\int_{\frac{y-t}{2}}^{\frac{y+t}{2}}\left\{\varphi_5(\xi)\Big[\varphi_1(\xi, t-|y-\xi|) - F(t-|y-\xi|-|\xi|)\Big]\right.$$
$$\left. - \int_0^{t-|y-\xi|-|\xi|} r_1''(\eta)\varphi_1(\xi, t-|y-\xi|-\eta)\mathrm{d}\eta\right\}\mathrm{d}\xi,$$

$$A_4\varphi = \varphi_{04} + \frac{1}{2}\int_{\frac{y-t}{2}}^{\frac{y+t}{2}} \Big\{ \varphi_5(\xi)\Big[\frac{a(0)}{2} + \varphi_2(\xi, t - |y - \xi|)\Big]$$

$$- \int_0^{t-|y-\xi|-|\xi|} r_1''(\eta)\varphi_2(\xi, t - |y - \xi| - \eta)\mathrm{d}\eta \Big\}\mathrm{d}\xi,$$

$$A_5\varphi = \varphi_{05} - \frac{4}{a(0)}\mathrm{sign}(y) \int_0^y \Big\{ \varphi_5(\xi)\Big[\varphi_2(\xi, 2|y| - \xi) - f(2|y| - 2\xi)\Big]$$

$$- \int_0^{2|y|-2\xi} r_1''(\eta)\varphi_2(\xi, 2|y| - \xi - \eta)\mathrm{d}\eta \Big\} - \frac{4}{a(0)}\mathrm{sign}(y)\mathrm{d}\xi$$

$$\times \int_0^y \Big\{ \varphi_1(\xi)\varphi_4(\xi, 2|y| - \xi)$$

$$- \int_0^{2|y|-2\xi} r_1''(\eta)\varphi_4(\xi, 2|y| - \xi - \eta)\mathrm{d}\eta \Big\}\mathrm{d}\xi,$$

where

$$\varphi_{01} = \frac{1}{2}\int_{\frac{y-t}{2}}^{\frac{y+t}{2}} \int_{|\xi|}^{t-|y-\xi|} \int_{|\xi|}^{\tau} r_1''(\tau - \eta)F(\eta - |\xi|)\mathrm{d}\eta\mathrm{d}\tau\mathrm{d}\xi;$$

$$\varphi_{02} = \frac{a(0)}{4}\int_{\frac{y-t}{2}}^{\frac{y+t}{2}} r_1(t - |y - \xi| - |\xi|)\mathrm{d}\xi;$$

$$\varphi_{03} = \frac{1}{2}\int_{\frac{y-t}{2}}^{\frac{y+t}{2}} \int_0^{t-|y-\xi|-|\xi|} r_1''(\eta)F(t - |y - \xi| - |\xi| - \eta)\mathrm{d}\eta\mathrm{d}\xi;$$

$$\varphi_{04} = -\frac{a(0)}{4}\int_{\frac{y-t}{2}}^{\frac{y+t}{2}} r_1'(t - |y - \xi| - |\xi|)\mathrm{d}\xi;$$

$$\varphi_{05} = \frac{2a'(0)}{a^2(0)}\left[g_1'(2|y|) + \text{sign}(y)\int_0^{2y} k_1'(2|y|-\tau)g_1(\tau)d\tau\right]$$
$$-\frac{4}{a(0)}\left[g_1''(2|y|) + \text{sign}(y)\int_0^{2y} k_1''(2|y|-\tau)g_1(\tau)d\tau\right]$$
$$-\frac{4}{a(0)}\text{sign}(y)\int_0^{y}\int_0^{2|y|-2\xi} r_1''(\eta)f(2|y|-2\xi-\eta)d\eta d\xi$$
$$+\,2\text{sign}(y)\int_0^{y} r_1''(2|y|-2\tau)d\tau.$$

Let

$$\varphi_0 = [\varphi_{01}, \varphi_{02}, \varphi_{03}, \varphi_{04}, \varphi_{05}],$$

$$\|\varphi\|(T) = \max\left\{\max_{(y,t)\in\Diamond(0,T)}|\varphi_i|,\ i=1,2,3,4;\ \max_{y\in[-T/2,T/2]}|\varphi_5|\right\}.$$

The system (6.168) is a closed system of Volterra integral equations of the second kind with continuous free terms and kernels with respect to unknown functions. As a small parameter, the system (6.168) contains an integration interval that does not exceed the number T. Therefore, for small T the Banach principle is applicable to it, which ensures the existence of a unique system's solution.

Define the set of functions $Q = \{\varphi|\ \|\varphi - \varphi_0\| \le \|\varphi_0\}$ in the space $C(\Diamond(0,T))$. There is an estimate for $\varphi \in Q$: $\|\varphi\| \le 2\|\varphi_0\|$.

We show that for sufficiently small T, the operator A maps the set Q to itself. Evaluating the integrals included in the formulas of the system (6.168), we find:

$$\|A_1\varphi - \varphi_{01}\| \le T^2\|\varphi_0\|\Big[2\|\varphi_0\| + \|F\| + T\|r''\|\Big],$$

$$\|A_2\varphi - \varphi_{02}\| \le T^2\|\varphi_0\|\Big[\frac{a(0)}{2} + 2\|\varphi_0\| + T\|r_1''\|\Big],$$

$$\|A_3\varphi - \varphi_{03}\| \le T\|\varphi_0\|\Big[2\|\varphi_0\| + \|F\| + T\|r_1''\|\Big],$$

$$\|A_4\varphi - \varphi_{04}\| \le T\|\varphi_0\|\Big[\frac{a(0)}{2} + 2\|\varphi_0\| + T\|r_1''\|\Big],$$

$$\|A_5\varphi - \varphi_{05}\| \le \frac{4}{a(0)}T\|\varphi_0\|\Big[2\|\varphi_0\| + \|f\| + T\|r_1''\|\Big].$$

Hence

$$\|A\varphi - \varphi_0\|(T) \le \alpha\|\varphi_0\|,$$

where $\alpha = \mu \cdot \left[\frac{a(0)}{2} + 2\|\varphi_0\| + T\|r_1''\| + \|f\| \max\{1, T\}\right]$, $\mu = \max\left\{T, T^2, \frac{4T}{a(0)}\right\}$.

Obviously, there is a number T^* such that $\alpha < 1$ and for all $T \in (0, T^*)$ the operator A maps the set Q to itself.

Take two arbitrary elements φ^1 and φ^2 of $Q(\varphi_0, \|\varphi_0\|)$. Using the auxiliary inequalities

$$|\varphi_i^1\varphi_j^1 - \varphi_i^2\varphi_j^2| \le |\varphi_i^1|\,|\varphi_j^1 - \varphi_j^2| + |\varphi_j^2|\,|\varphi_i^1 - \varphi_i^2| \le 4\|\varphi_0\|\,\|\varphi^1 - \varphi^2\|,$$

we arrive

$$\|A\varphi^1 - A\varphi^2\| \le \tilde{\alpha}\,\|\varphi^1 - \varphi^2\|,$$

where

$$\tilde{\alpha} = \frac{1}{2}\mu \cdot \left[\frac{a(0)}{2} + 4\|\varphi_0\| + T\|r_1''\| + \|f\| \max\{1, T\}\right].$$

Note that $\alpha < 1$ implies $\tilde{\alpha} < 1$. Therefore, we can conclude that the operator A is a contraction mapping on $Q(\varphi_0, \|\varphi_0\|)$. Then according to the Banach principle the Eq. (6.168) has a unique solution in $Q(\varphi_0, \|\varphi_0\|)$ for all $T \in (0, T^*)$.

The system of Eq. (6.168) is solved by the method of successive approximations. Hence, we can find a unique vector function φ in domain $\Diamond(0, T)$ for all $T \in (0, T^*)$ and thus determine the function $q(y) \in C[-T/2, T/2]$. Theorem 6.12 is proved.

Now we prove Theorem 6.13.

Proof It is known that the domain $\Diamond(0, T)$ has an equivalent description

$$\left\{(y, t) |\, 0 \le t \le T,\ 0 \le |y| \le \frac{T}{2} - |t - \frac{T}{2}|\right\}.$$

Let $\varphi^{(j)}$ be vector functions that are solutions of (6.168) with the data $\{k_1^{(j)}(t), g_1^{(j)}(t)\}$, respectively, i.e., $\varphi^{(j)} = A\varphi^{(j)}$ for $j = 1, 2$.

Let

$$\tilde{\varphi}(t) = \max\Big\{ \max_{0 \le |\xi| \le \frac{t}{2} - |\tau - \frac{t}{2}|} |\varphi_i^{(1)}(\xi, \tau) - \varphi_i^{(2)}(\xi, \tau)|,\ i = \overline{1, 4};$$
$$\max_{\xi \in [-\frac{t}{2}, \frac{t}{2}]} |\varphi_5^{(1)}(\xi) - \varphi_5^{(2)}(\xi)|\Big\}$$

Note that the estimate

$$\|r_1(t)\|_2(T) \le \|k(t)\|_2(T) \exp\left(T\|k(t)\|_2(T)\right)$$

follows from (6.137). The known functions $k_1^{(j)}$, $g_1^{(j)}$ appear in the free terms of integral equations (6.168) in the functions $r_1^{\prime\prime(j)}(t),\ F^{(j)}(t),\ f^{(j)}(t),\ j=1,2$. Passing in these expressions to the differences $k_1^{(1)}-k_1^{(2)}$, $g_1^{(1)}-g_1^{(2)}$ as in [1], we find the estimate

$$\tilde{\varphi}(t)\le C_1\left(\|g^{(1)}-g^{(2)}\|_{C^2[0,T]}+\|k^{(1)}-k^{(2)}\|_{C^2[0,T]}\right)+C_2\int\limits_0^t\tilde{\varphi}(\tau)\mathrm{d}\tau,$$

where $C_1=C_1(a(0),\ a'(0),\ M_1,\ \rho,\ T)$, $C_2=C_2(q_0,\ a(0),\ a'(0),\ M_1,\ \rho,\ T)$.
Using Gronwall's inequality, we find

$$\tilde{\varphi}(t)\le C_1\exp\left(C_2T\right)\left(\|g_1^{(1)}-g_1^{(2)}\|_2(T)+\|k_1^{(1)}-k_1^{(2)}\|_2(T)\right). \tag{6.169}$$

From (6.169) it follows the estimate (6.160). Theorem 6.13 is proved.

The stability Theorem 6.13 implies the validity of the uniqueness theorem for any fixed T.

Theorem 6.14 *Suppose that the conditions of the Theorem 6.13 hold. If the solution of the inverse problem* exists and belongs to* $C[-T/2,T/2]$, *then it is unique.*

Further, we consider $q(y)$ to be a known function.

Let us prove Theorems 6.6–6.9.

We define $K(y)=\frac{\mathrm{d}}{\mathrm{d}y}\ln a(y)$ as the solution of the Cauchy problem for the Riccati equation:

$$K'(y)+\frac{1}{2}K^2(y)=2q(y),\quad K(0)=\frac{a'(0)}{a(0)}. \tag{6.170}$$

The existence of a unique solution to the Cauchy problem (6.170) under the condition $T\left(\left|\frac{a'(0)}{a(0)}\right|+Tq_0\right)\le 1$ for all $T\in(0,\min(T^*,T_1)$ (here T_1 is a positive root of the corresponding square equation with respect to T) follows from the contraction-mapping principle for the integral equation:

$$K(y)=\frac{a'(0)}{a(0)}+2\int\limits_0^y q(\xi)\mathrm{d}\xi+\frac{1}{2}\int\limits_0^y K^2(\xi)\mathrm{d}\xi.$$

We find $c_{44}(\psi_1^{-1}(y))$ by the known function $a(y)$:

$$a(y)=a(0)\exp\left(\int\limits_0^y K(\xi)\mathrm{d}\xi\right)\Rightarrow c_{44}(\psi_1^{-1}(y))=\frac{1}{\rho a^2(y)}.$$

Now we determine $\psi_1^{-1}(y)$. Differentiating with respect to y the equality

$$y = \int_0^{\psi_1^{-1}(y)} \sqrt{\frac{c_{44}(\xi)}{\rho}} d\xi$$

we find

$$1 = \sqrt{\frac{\rho}{c_{44}(\psi_1^{-1}(y))}} \frac{d\psi_1^{-1}(y)}{dy} \Rightarrow \frac{1}{\rho a(y)} = \frac{d\psi_1^{-1}(y)}{dy}.$$

Finally, $\psi_1^{-1}(y) = \frac{1}{\rho} \int_0^y \frac{1}{a(\xi)} d\xi$.

The latter formula defines the function $\psi_1^{-1}(y)$ for all $y \in (0, T/2)$ by known function $a(y) \in C^2[0, T/2]$ with positive values. Theorem 6.6 is proved.

By integrating from 0 to y the Eq. (6.170) for $K^{(1)}(y)$ and $K^{(2)}(y)$, then subtracting one equation from the other, we can get the inequality:

$$|K^{(1)}(y) - K^{(2)}(y)| \le \frac{1}{2} \int_0^y |K^{(1)}(\xi) - K^{(2)}(\xi)||K^{(1)}(\xi) + K^{(2)}(\xi)| d\xi \\ + 2 \int_0^y |q^{(1)}(\xi) - q^{(2)}(\xi)| d\xi. \tag{6.171}$$

Using Gronwall's inequality to (6.171), we find

$$\|K^{(1)} - K^{(2)}\|_{C[-T/2,T/2]} \le C_3(h_0, h_1, \rho, T)\|q^{(1)} - q^{(2)}\|_{C[-T/2,T/2]} \\ \le C_3 C^* \left(\|g_1^{(1)} - g_1^{(2)}\|_2(T) + \|k_1^{(1)} - k_1^{(2)}\|_2(T) \right). \tag{6.172}$$

From the definition of functions $K^{(1)}(y)$ and $K^{(2)}(y)$ it follows that

$$a^{(j)}(y) = a(0) \exp\left(\int_0^{\psi_1^{(j)}(x_3)} K^{(j)}(\xi) d\xi \right), \quad j = 1, 2,$$

hence

$$c_{44}^{(j)}(x_3) = \frac{1}{a^2(0)\rho} \exp\left(-2 \int_0^{\psi_1^{(j)}(x_3)} K^{(j)}(\xi) d\xi \right), \quad j = 1, 2.$$

Making up the difference

$$c_{44}^{(1)}(x_3) - c_{44}^{(2)}(x_3) = \frac{1}{a^2(0)\rho}\left[\exp\left(-2\int_0^{\psi_1^{(1)}(x_3)} K^{(1)}(\xi)\mathrm{d}\xi\right) - \exp\left(-2\int_0^{\psi_1^{(2)}(x_3)} K^{(2)}(\xi)\mathrm{d}\xi\right)\right]$$

and referring to identical calculations [2], we obtain the required stability estimate (6.130).

Thus, we proved Theorem 6.7.

Let

$$\hat{v}(z,t) := \frac{U_3^0\left(\psi_2^{-1}(z),t\right)}{p(z)}, \quad p(z) := \left(\frac{c_{11}(+0)}{c_{11}\left(\psi_2^{-1}(z)\right)}\right)^{1/4}.$$

Then, the inverse problems (6.128)–(6.131) in terms of the new functions and variable z take the form

$$\frac{\partial^2 \hat{v}}{\partial t^2} = L\left[k_3, \frac{\partial^2 \hat{v}}{\partial z^2} + \tilde{q}(z)\hat{v}\right], \quad z > 0, \ t \in \mathbb{R}, \tag{6.173}$$

$$\hat{v}|_{t<0} \equiv 0, \tag{6.174}$$

$$L\left[k_3, \frac{\partial \hat{v}(z,t)}{\partial z} + \frac{1}{2}\frac{b'(0)}{b(0)}\hat{v}(z,t)\right]_{z=+0} = -\frac{b(0)\delta(t)}{2}, \tag{6.175}$$

$$\hat{v}(z,t)|_{z=+0} = g_3(t), \quad t > 0, \tag{6.176}$$

where $\tilde{q}(z) = \frac{p''(z)}{p(z)} - 2\left[\frac{p'(z)}{p(z)}\right]^2$, $b(z) = \frac{1}{\sqrt{\rho \cdot c_{11}(\psi_2^{-1}(z))}}$.

Using in Eqs. (6.173)–(6.176) the same substitutions as in the case of the *inverse problem 1*, we obtain the problem similar to the problem (6.138)–(6.140):

$$\frac{\partial^2 \hat{w}}{\partial t^2} = \frac{\partial^2 \hat{w}}{\partial z^2} + \tilde{q}(z)\hat{w} - \int_0^t r_3''(t-\tau)\hat{w}(z,\tau)\mathrm{d}\tau, \ z > 0, \ t \in \mathbb{R}, \tag{6.177}$$

$$\hat{w}|_{t<0} \equiv 0, \tag{6.178}$$

$$\left[\frac{\partial \hat{w}(z,t)}{\partial z} + \frac{1}{2}\frac{b'(0)}{b(0)}\hat{w}(z,t)\right]_{z=+0} = -\frac{b(0)\delta(t)}{2}, \tag{6.179}$$

$$\hat{w}(z,t)|_{z=+0} = L\left[k_3, g_3\right],\ t > 0, \tag{6.180}$$

where

$$\hat{w}(z,t) = L[k_3, \hat{v}(z,t)],\quad \hat{v}(z,t) = L[r_3, \hat{w}(z,t)],$$

$$r_3(t) = -k_3(t) - \int_0^t k_3(t-\tau)r_3(\tau)d\tau.$$

Note that the *inverse problem 2* (6.177)–(6.180) is similar to the inverse problem (6.138)–(6.141). Therefore, Theorems 6.8 and 6.9 are proved like Theorems 6.6 and 6.7.

For further reasoning, we will need a structure of functions U_3^0, $\frac{\partial U_3^0}{\partial x_3}$. Since the direct problem (6.177)–(6.179) is similar to the problem (6.138)–(6.140) for $x_3 \in R,\ t \in R$ we find

$$U_3^0(x_3,t) = \theta(t - |\psi_2(x_3)|)\tilde{U}(x_3,t),$$

where

$$\tilde{U}(x_3,t) = p(\psi_2(x_3))L\left[r_3, \frac{b(0)}{2} + \tilde{W}(\psi_2(x_3),t)\right] \tag{6.181}$$

is even function to the variable x_3, $\tilde{U}(x_3,t) \in C^1(\nabla(T))$, $\nabla(T) = \{(x_3,t)|\ |\psi_2(x_3)| \le t \le T\}$ (here function $\tilde{W}(\psi_2(x_3),t)$ is an analogue of the regular part of function $\tilde{w}(y,t)$ in *inverse problem 1*).

From (6.181), given the structure of function $\tilde{W}(\psi_2(x_3),t)$, we obtain

$$\frac{\partial U_3^0}{\partial x_3} = D_1(x_3)\delta(t - |\psi_2(x_3)|) + \theta(t - |\psi_2(x_3)|)D_2(x_3,t),$$

$$D_1(x_3) = -\sqrt{\frac{\rho}{c_{11}(x_3)}}\tilde{U}(x_3, |\psi_2(x_3)|)\mathrm{sign}(x_3),\ D_2(x_3,t) = \frac{\partial \tilde{U}(x_3,t)}{\partial x_3}. \tag{6.182}$$

Let us study *inverse problem 3*.

Let

$$v_1(y,t) := \frac{U_1^1\left(\psi_1^{-1}(y),t\right)}{s(y)},\quad s(y) := \left(\frac{c_{44}(+0)}{c_{44}\left(\psi_1^{-1}(y)\right)}\right)^{1/4}.$$

Then (6.124)–(6.127) in terms of the new functions and variable y takes the form for $y > 0,\ t \in \mathbb{R}$:

$$\begin{aligned} \frac{\partial^2 v_1}{\partial t^2} &= L\Big[k_1, \frac{\partial^2 v_1}{\partial y^2} + q(y)v_1 \\ &+ \frac{i}{\rho s(y)}\left(\tilde{c}_{12}(y)\frac{\partial U_3^0}{\partial x_3} + \frac{\partial}{\partial x_3}(c_{44}U_3^0)\right)\Big|_{x_3=\psi_1^{-1}(y)}\Big], \end{aligned} \tag{6.183}$$

$$v_1|_{t<0} \equiv 0, \tag{6.184}$$

$$L\left[k_1, \frac{\partial v_1(y,t)}{\partial y}\right]_{y=+0} = f_1(t), \tag{6.185}$$

$$v_1(y,t)|_{y=+0} = g_2(t), \quad t > 0, \tag{6.186}$$

where

$$\tilde{c}_{12}(y) = c_{12}(\psi_1^{-1}(y)),$$

$$f_1(t) = -i\sqrt{\frac{c_{44}(+0)}{\rho}} L[k_1, g_3(t)] - \frac{1}{2}\frac{a'(0)}{a(0)} L[k_1, g_2(t)].$$

Let $L[k_1, v_1(y,t)] = W(y,t) \Rightarrow v_1(y,t) = L[r_1, W(y,t)]$.

Then

$$\frac{\partial^2 W}{\partial t^2} = \frac{\partial^2 W}{\partial y^2} + q(y)W - \int_0^t r_1''(t-\tau)W(y,\tau)d\tau \\ + \frac{i}{\rho s(y)} L\left[k_1, \tilde{c}_{12}(y)\frac{\partial U_3^0}{\partial x_3} + \frac{\partial}{\partial x_3}(c_{44}U_3^0)\right]\Big|_{x_3=\psi_1^{-1}(y)}, \tag{6.187}$$

$$W|_{t<0} \equiv 0, \tag{6.188}$$

$$\frac{\partial W}{\partial y}\Big|_{y=+0} = f_1(t), \tag{6.189}$$

$$W(y,t)|_{y=+0} = L\left[k_1, g_2\right], \quad t > 0. \tag{6.190}$$

We seek a solution to the direct problem (6.187)–(6.189) in the form:

$$W(y,t) = W_1(y,t) + W_2(y,t),$$

where $W_1(y,t)$ is the solution of the problem

$$\frac{\partial^2 W_1}{\partial t^2} = \frac{\partial^2 W_1}{\partial y^2} + q(y)W_1 - \int_0^t r_1''(t-\tau)W_1(y,\tau)d\tau, \tag{6.191}$$

$$W_1|_{t<0} \equiv 0, \tag{6.192}$$

$$\frac{\partial W_1}{\partial y}\Big|_{y=+0} = f_1(t) \tag{6.193}$$

and $W_2(y,t)$ is the solution of the problem

$$\frac{\partial^2 W_2}{\partial t^2} = \frac{\partial^2 W_2}{\partial y^2} + q(y)W_2 - \int_0^t r_1''(t-\tau)W_2(y,\tau)d\tau \\ + \frac{i}{\rho s(y)} L\left[k_1, \tilde{c}_{12}(y)\frac{\partial U_3^0}{\partial x_3} + \frac{\partial}{\partial x_3}(c_{44}U_3^0)\right]\Big|_{x_3=\psi_1^{-1}(y)}, \tag{6.194}$$

$$W_2|_{t<0} \equiv 0, \tag{6.195}$$

$$\left.\frac{\partial W_2}{\partial y}\right|_{y=+0} = 0. \tag{6.196}$$

The solution of (6.191)–(6.193) is similar to the solution of (6.148)–(6.150) with the well-known function $q(y)$. Therefore, we assume that $W_1(y,t)$ is a known function.

The function $W_2(y,t)$ extend for $y<0$ as even function and $\tilde{c}_{12}(y)$ as an odd function for $y<0$, now (6.195) holds automatically. Here the functions $c_{44}(x_3)$, $c_{11}(x_3)$, $U_3^0(x_3,t)$ are even functions by variable x_3 (see *inverse problems 1, 2*).

For even function $W_2(y,t)$ we obtain:

$$\frac{\partial^2 W_2}{\partial t^2} = \frac{\partial^2 W_2}{\partial y^2} + q(y)W_2 - \int_0^t r_1''(t-\tau)W_2(y,\tau)d\tau \\ + \frac{i}{\rho s(y)} L\left[k_1, \tilde{c}_{12}(y)\frac{\partial U_3^0}{\partial x_3} + \frac{\partial}{\partial x_3}(c_{44}U_3^0)\right]\Big|_{x_3=\psi_1^{-1}(y)}, \quad y \in \mathbb{R}, \tag{6.197}$$

$$W_2|_{t<0} \equiv 0, \tag{6.198}$$

The problem solution (6.197)–(6.198) will take a form of:

$$W_2(y,t) = \iint_{R^2} G(y,t,\xi,\tau)\frac{i}{\rho s(\xi)} L\left[k_1, \tilde{c}_{12}(\xi)\frac{\partial U_3^0}{\partial x_3} + \frac{\partial}{\partial x_3}(c_{44}U_3^0)\right]_{x_3=\psi_1^{-1}(\xi)} d\tau d\xi, \tag{6.199}$$

where $G(y,t,y_0,t_0)$ is a fundamental solution of operator $\tilde{L}$:

$$\tilde{L}W_2 = \frac{\partial^2 W_2}{\partial t^2} - \frac{\partial^2 W_2}{\partial y^2} - q(y)W_2 + \int_0^t r''(t-\tau)W_2(y,\tau)$$

and as it was shown in Sect. 6.1.3, takes a structure:

$$G(y,t,y_0,t_0)=\theta(t-t_0-|y-y_0|)\left[\frac{1}{2}+\tilde{G}(y,t,y_0,t_0)\right],\quad y_0,\ t_0\in\mathbb{R}. \tag{6.200}$$

Let us consider in more detail the expressions in (6.199) :

$$\begin{aligned} L\left[k_1,\tilde{c}_{12}(\xi)\frac{\partial U_3^0}{\partial x_3}\right]_{x_3=\psi_1^{-1}(\xi)} &= \tilde{c}_{12}(\xi)\Big[D_1(x_3)\delta(\tau-|\psi_2(x_3)|) \\ &+\theta(\tau-|\psi_2(x_3)|)\Big(D_2(x_3,\tau) \\ &+k_1(\tau-|\psi_2(x_3)|)D_1(x_3) \\ &+\int\limits_{|\psi_2(x_3)|}^{\tau}k_1(\tau-s)D_2(x_3,s)\mathrm{d}s\Big)\Big]_{x_3=\psi_1^{-1}(\xi)}, \end{aligned} \tag{6.201}$$

$$\begin{aligned} &L\left[k_1,\frac{\partial}{\partial x_3}(c_{44}U_3^0)\right]_{x_3=\psi_1^{-1}(\xi)} \\ &=\Big[c_{44}(x_3)D_1(x_3)\delta(\tau-|\psi_2(x_3)|) \\ &+\theta(\tau-|\psi_2(x_3)|)\Big(\frac{\mathrm{d}c_{44}}{\mathrm{d}x_3}\tilde{U}(x_3,\tau)+c_{44}(x_3) \\ &\times[D_2(x_3,\tau)+D_1(x_3)k_1(\tau-|\psi_2(x_3)|)] \\ &+\int\limits_{|\psi_2(x_3)|}^{\tau}k_1(\tau-s)\Big[\frac{\mathrm{d}c_{44}}{\mathrm{d}x_3}\tilde{U}(x_3,s)+c_{44}(x_3)D_2(x_3,s)\Big]\mathrm{d}s\Big)\Big]_{x_3=\psi_1^{-1}(\xi)}. \end{aligned} \tag{6.202}$$

Let

$$d(y)=\psi_2(\psi_1^{-1}(y))),\quad \Delta(y,t)=\{(\xi,\tau):\ 0\le\tau\le t-|y-\xi|\},$$
$$\nabla=\{(\xi,\tau):|d(\xi)|\le\tau\},\quad \Theta=\{(y,t):\ 0<t\le\psi(y)\}.$$

The function $\psi(y)$ is defined by the values of y and $d(y)$ exactly as in the work [2].

Taking into account (6.200)–(6.202), we rewrite the formula (6.199)

$$
\begin{aligned}
W_2(y,t) = & \iint\limits_{\mathbb{R}^2} \theta(t-\tau-|y-\xi|)\left[\frac{1}{2}+\tilde{G}(y,t,\xi,\tau)\right] \\
& \times \frac{i}{\rho s(\xi)}\tilde{c}_{12}(\xi)\Big[D_1(x_3)\delta(\tau-|\mathrm{d}(\xi)|) \\
& + \theta(\tau-|\mathrm{d}(\xi)|)\Big(D_2(x_3,\tau)+k_1(\tau-|\mathrm{d}(\xi)|)D_1(x_3) \\
& + \int_{|\mathrm{d}(\xi)|}^{\tau} k_1(\tau-s)D_2(x_3,s)\mathrm{d}s\Big)\Big]_{x_3=\psi_1^{-1}(\xi)} \mathrm{d}\tau\mathrm{d}\xi \\
& + \iint\limits_{\mathbb{R}^2} \theta(t-\tau-|y-\xi|) \\
& \times \left[\frac{1}{2}+\tilde{G}(y,t,\xi,\tau)\right]\frac{i}{\rho s(\xi)}\Big[c_{44}(x_3)D_1(x_3)\delta(\tau-|\psi_2(x_3)|) \\
& + \theta(\tau-|\psi_2(x_3)|)\Big(\frac{\mathrm{d}c_{44}}{\mathrm{d}x_3}\tilde{U}(x_3,\tau)+c_{44}(x_3) \\
& \times [D_2(x_3,\tau)+D_1(x_3)k_1(\tau-|\psi_2(x_3)|)] \\
& + \int\limits_{|\psi_2(x_3)|}^{\tau} k_1(\tau-s)\Big[\frac{\mathrm{d}c_{44}}{\mathrm{d}x_3}\tilde{U}(x_3,s)+c_{44}(x_3)D_2(x_3,s)\Big]\mathrm{d}s\Big)\Big]_{x_3=\psi_1^{-1}(\xi)} \mathrm{d}\tau\mathrm{d}\xi.
\end{aligned}
\tag{6.203}
$$

The definition of the functions $\psi(y)$, $\mathrm{d}(y)$ and sets $\Delta(y,t)$, ∇, Θ implies the following formulas:

$$
\begin{aligned}
& \Delta(y,t)\bigcap\nabla=\varnothing,\ (y,t)\in\Theta, \\
& \Delta(y,t)\bigcap\nabla=\bigcup_{i=1}^{n}B_i(y,t),\ (y,t)\in\mathbb{R}^2\setminus\Theta, \\
& \quad B_i(y,t)=\{(\xi,\tau):\ y_{*i}(y,t)\le\xi\le y_i^*(y,t),\ |\mathrm{d}(\xi)|\le\tau\le t-|y-\xi|\},
\end{aligned}
$$

at that $\Delta(0,t)\bigcap\nabla=B_1(0,t)$ the functions $y_{*i}(y,t)$, $y_i^*(y,t)$ are the solutions of equation $t-|y-\xi|=|\mathrm{d}(\xi)|$ with respect to variable ξ and $y_{*i}(y,t)\le y_i^*(y,t)$.

Then

$$
\begin{aligned}
W_2(y,t) = & F(y,t)+\sum_{i=1}^{n}\int\limits_{y_{*i}(y,t)}^{y_i^*(y,t)}\tilde{c}_{12}(\xi)\hat{D}_1(\xi)\left[\frac{1}{2}+\tilde{G}(y,t,\xi,|\mathrm{d}(\xi)|)\right]\mathrm{d}\xi \\
& + \theta(t-|\psi(y)|)\sum_{i=1}^{n}\int\limits_{y_{*i}(y,t)}^{y_i^*(y,t)}\int\limits_{|\mathrm{d}(\xi)|}^{t-|y-\xi|}\tilde{c}_{12}(\xi)\left[\frac{1}{2}+\tilde{G}(y,t,\xi,\tau)\right]
\end{aligned}
\tag{6.204}
$$

$$\times\left(\hat{D}_2(\xi,\tau)+k_1(\tau-|\mathrm{d}(\xi)|)\hat{D}_1(\xi)\right.$$
$$\left.+\int\limits_{|\mathrm{d}(\xi)|}^{\tau} k_1(\tau-s)\hat{D}_2(\xi,s)\mathrm{d}s\right)\mathrm{d}\tau\mathrm{d}\xi,$$

where

$$\hat{D}_1(\xi)=\frac{i}{\rho s(\xi)}D_1(\psi_1^{-1}(\xi)),\ \hat{D}_2(\xi,\tau)=\frac{i}{\rho s(\xi)}D_2(\psi_1^{-1}(\xi),\tau),$$

$$F(y,t)=\sum_{i=1}^{n}\int\limits_{y_{*i}(y,t)}^{y_i^*(y,t)} c_{44}\hat{D}_1(\xi)\left[\frac{1}{2}+\tilde{G}(y,t,\xi,|\mathrm{d}(\xi)|)\right]\mathrm{d}\xi$$
$$+\theta(\tau-|\psi(y)|)\sum_{i=1}^{n}\int\limits_{y_{*i}(y,t)}^{y_i^*(y,t)}\int\limits_{|\mathrm{d}(\xi)|}^{t-|y-\xi|}\left[\frac{1}{2}+\tilde{G}(y,t,\xi,\tau)\right]\frac{i}{\rho s(\xi)}$$
$$\times\left[\frac{\mathrm{d}c_{44}}{\mathrm{d}x_3}\tilde{U}(x_3,\tau)+c_{44}(x_3)\Big(D_2(x_3,\tau)+D_1(x_3)k_1(\tau-|\mathrm{d}(\xi)|)\Big)\right.$$
$$\left.+\int\limits_{|\mathrm{d}(\xi)|}^{\tau} k_1(\tau-s)\Big(\frac{\mathrm{d}c_{44}}{\mathrm{d}x_3}\tilde{U}(x_3,s)+c_{44}(x_3)D_2(x_3,s)\Big)\mathrm{d}s\right]_{x_3=\psi_1^{-1}(\xi)}\mathrm{d}\tau\mathrm{d}\xi. \tag{6.205}$$

Substituting in (6.204) $y=0$, for $t>0$ we obtain

$$W_2(0,t)=F(0,t)+\int\limits_{y_{*1}(0,t)}^{y_1^*(0,t)}\tilde{c}_{12}(\xi)\left\{\hat{D}_1(\xi)\left[\frac{1}{2}+\tilde{G}(0,t,\xi,|\mathrm{d}(\xi)|)\right]\right.$$
$$+\int\limits_{|\mathrm{d}(\xi)|}^{t-|\xi|}\left[\frac{1}{2}+\tilde{G}(0,t,\xi,\tau)\right]\times\left(\hat{D}_2(\xi,\tau)+k_1(\tau-|\mathrm{d}(\xi)|)\hat{D}_1(\xi)\right.$$
$$\left.\left.+\int\limits_{|\mathrm{d}(\xi)|}^{\tau} k_1(\tau-s)\hat{D}_2(\xi,s)\mathrm{d}s\right)\mathrm{d}\tau\right\}\mathrm{d}\xi. \tag{6.206}$$

Differentiating (6.206) with respect to the variable t, taking into account the parity of the integrand and the fact that $W_2(0,t)=L[k_1,g_2(t)]-W_1(0,t)$, we have

$$\tilde{c}_{12}(y_1^*(0,t))\hat{D}_1(y_1^*(0,t))y_1^{*\prime}(0,t)+\int\limits_{0}^{y_1^*(0,t)}\tilde{c}_{12}(\xi)K(\xi,t)\mathrm{d}\xi \tag{6.207}$$
$$=F'(0,t)-L[k_1,g_2'(t)]+W_1'(0,t),$$

where

$$
\begin{aligned}
K(\xi,t) &= 2\hat{D}_1(\xi)\tilde{G}'_t(0,t,\xi,|\mathrm{d}(\xi)|) + \hat{D}_2(\xi,t-|\xi|) \\
&\quad + k_1(t-|\xi|-|\mathrm{d}(\xi)|)\hat{D}_1(\xi) \\
&\quad + \int\limits_{|\mathrm{d}(\xi)|}^{t-|\xi|} \Big\{ k_1(t-|\xi|-\tau)\hat{D}_2(\xi,\tau) + 2\tilde{G}'_t(0,t,\xi,\tau) \\
&\quad \times \Big[\hat{D}_2(\xi,\tau) + k_1(\tau-|\mathrm{d}(\xi)|)\hat{D}_1(\xi) \\
&\quad + \int\limits_{|\mathrm{d}(\xi)|}^{\tau} k_1(\tau-s)\hat{D}_2(\xi,s)\mathrm{d}s \Big]\Big\}\mathrm{d}\tau,
\end{aligned}
\tag{6.208}
$$

here it was taken into account that $W'_2(0,t) = L[k_1, g'_2(t)] - W'_1(0,t)$ by (6.190) and (6.191).

We show that $-\hat{D}_1(y_1^*(0,t)) \cdot y_1^{*\prime}(0,t) > 0$ if we ignore the third memory component, i.e., $k_3(t) \equiv 0 \Rightarrow r_3(t) \equiv 0$. From a physical point of view, this is justified by the fact that in some environments with memory there may be a lack of viscosity in the vertical direction.

Indeed, since $y_1^{*\prime}(0,t)$ satisfies the equality $t - \xi = \mathrm{d}(\xi)$, then

$$
1 - y_1^{*\prime}(0,t) = d'(y_1^*(0,t))y_1^{*\prime}(0,t) \;\Rightarrow\; y_1^{*\prime}(0,t) = \frac{1}{1 + d'(y_1^*(0,t))},
$$

$$
d'(y_1^*(0,t)) = \sqrt{\frac{c_{44}(\psi_1^{-1}(y_1^*(0,t)))}{c_{11}(\psi_1^{-1}(y_1^*(0,t)))}},
$$

hence, $y_1^{*\prime}(0,t) > 0$. So, $-\hat{D}_1(y_1^*(0,t)) > 0$ due to (6.181), (6.182) and the fact that $\tilde{W}(\psi_2(x_3), |\psi_2(x_3)|) \equiv 0$. Then (6.207) can be rewritten as

$$
\tilde{c}_{12}(y_1^*(0,t)) = \int\limits_0^{y_1^*(0,t)} \tilde{c}_{12}(\xi)\tilde{K}(\xi,t)\mathrm{d}\xi + \tilde{F}(t), \tag{6.209}
$$

$$
\tilde{K}(\xi,t) = -\frac{K(\xi,t)}{\hat{D}_1(y_1^*(0,t))y_1^{*\prime}(0,t)}, \quad \tilde{F}(t) = \frac{F'(0,t) - L[k_1, g'_2(t)] + W'_1(0,t)}{\hat{D}_1(y_1^*(0,t))y_1^{*\prime}(0,t)}. \tag{6.210}
$$

Equation (6.209) is the second-kind Volterra integral equation. Here $\tilde{K}(\xi,t) \in C(\Omega(T))$, $\Omega(T) = \{(\xi,t):\ 0 \le \xi \le t \le T/2\}$. This obviously follows from the continuity in this domain of the right side of (6.208). We show that $\tilde{F}(t) \in C[0,T]$.

Let us differentiate (6.205) with respect to the variable t for $y = 0$, taking into account the parity of the integrand functions:

$$F'(0,t) = \int_0^{y_1^*(0,t)} \frac{i}{\rho s(\xi)} \Bigg[2c_{44}(x_3)D_1(x_3)\tilde{G}_t'(0,t,\xi,|\mathrm{d}(\xi)|) + \frac{\mathrm{d}c_{44}}{\mathrm{d}x_3}\tilde{U}(x_3,t-|\xi|)$$
$$+ c_{44}(x_3)\Big(D_2(x_3,t-|\xi|) + D_1(x_3)k_1(t-|\xi|-|\mathrm{d}(\xi)|)\Big)$$
$$+ \int_{|\mathrm{d}(\xi)|}^{t-|\xi|} k_1(t-|\xi|-s)\Big(\frac{\mathrm{d}c_{44}}{\mathrm{d}x_3}\tilde{U}(x_3,s) + c_{44}(x_3)D_2(x_3,s)\Big)\mathrm{d}s\Bigg]_{x_3=\psi_1^{-1}(\xi)} \mathrm{d}\xi$$
$$+2\int_0^{y_1^*(0,t)}\int_{|\mathrm{d}(\xi)|}^{t-|\xi|} \tilde{G}_t'(0,t,\xi,\tau)\frac{i}{\rho s(\xi)}\Bigg[\frac{\mathrm{d}c_{44}}{\mathrm{d}x_3}\tilde{U}(x_3,\tau)$$
$$+ c_{44}(x_3)\Big(D_2(x_3,\tau) + D_1(x_3)k_1(\tau-|\mathrm{d}(\xi)|)\Big)$$
$$+ \int_{|\mathrm{d}(\xi)|}^{\tau} k_1(\tau-s)\Big(\frac{\mathrm{d}c_{44}}{\mathrm{d}x_3}\tilde{U}(x_3,s) + c_{44}(x_3)D_2(x_3,s)\Big)\mathrm{d}s\Bigg]_{x_3=\psi_1^{-1}(\xi)} \mathrm{d}\tau\mathrm{d}\xi. \tag{6.211}$$

The continuity of the right side of (6.211) implies the continuity of the left side in the domain of integration.

Further, $W_1'(0,t) \in C[0,T]$ by virtue of the analogy with (6.154) and the continuity of the right side of (6.161) for $y = 0$. Thus, Eq. (6.209) is a second-kind Volterra integral equation for the function $c_{12}(y_1^*(0,t))$ with a continuous kernel and a continuous free term. As is known, in this case, there is a unique continuous solution.

After determining $\tilde{c}_{12}(y_1^*(0,t)) \in C[0,T/2]$ we find $c_{12}(x_3) = \tilde{c}_{12}(\psi_1(x_3))$ for $x_3 \in [0, \psi_1^{-1}(T/2)]$. Since the functions c_{11}, c_{12}, c_{44} belongs to the class Λ, for the solvability of the *inverse problem*, (to determine the elastic moduli) it is necessary to require the inequality specified in [2] and following from (6.207) for $t = 0$:

$$\tilde{c}_{12}(+0)\hat{D}_1(0))y_1^{*\prime}(0,+0) = -g_2'(+0)$$
$$\Rightarrow -\tilde{c}_{12}(+0)\frac{i}{2\rho c_{11}(+0)}\frac{g_1(+0)}{g_1(+0)+g_3(+0)} = -g_2'(+0)$$
$$\Rightarrow \max\left[-g_2'(+0), 2g_2'(+0)\right] < \frac{g_1(+0)}{2\rho[g_1(+0)+g_3(+0)]}$$

by virtue of $c_{11} > \max\left[-c_{12}, 2c_{12}\right]$.

Theorem 6.10 is proved.

Let us prove Theorem 6.11.

Proof For simplicity, we denote $\tilde{y} = y_1^*(0,t)$.

Now we consider two integral equations of (6.209) type with its sets of given functions $c_{11}^{(1)}(x_3)$, $c_{44}^{(1)}(x_3)$, $k_1^{(1)}(t)$, $g_2^{(1)}(t)$, $g_3^{(1)}(t)$ and $c_{44}^{(2)}(x_3)$, $c_{11}^{(2)}(x_3)$, $k_1^{(2)}(t)$, $g_2^{(2)}(t)$, $g_3^{(2)}(t)$, respectively:

$$\tilde{c}_{12}^{(j)}(\tilde{y}) = \int_0^{\tilde{y}} c_{12}^{(j)}(\xi)\tilde{K}^j(\xi,t)d\xi + \tilde{F}^{(j)}(t), \quad j = 1, 2.$$

We have

$$\begin{aligned} \tilde{c}_{12}^1(\tilde{y}) - \tilde{c}_{12}^2(\tilde{y}) = \int_0^{\tilde{y}} \Big[\left(c_{12}^{(1)}(\xi) - c_{12}^{(2)}(\xi) \right) \tilde{K}^1(\xi,t) \\ + \left(K^{(1)}(\xi,t) - K^{(2)}(\xi,t) \right) c_{12}^{(2)}(\xi) \Big] d\xi \\ + \tilde{F}^{(1)}(t) - \tilde{F}^{(2)}(t). \end{aligned} \tag{6.212}$$

Taking into account $(w_1^*(0,t))' \equiv W_1'(0,t) + f_1(t), \;\; F(t) \equiv \int_0^t f_1(s)ds$, from (6.158) and (6.161) the estimate follows:

$$\begin{aligned} \|W'^{(1)}_1(0,t) - W'^{(2)}_1(0,t)\|(T) \le C_4(h_0, h_1, M_3, \rho, T) \\ \cdot \left(\|c_{44}^{(1)} - c_{44}^{(2)}\|_2(X) + \|g_2^{(1)} - g_2^{(2)}\|(T) \right. \\ \left. + \|g_3^{(1)} - g_3^{(2)}\|(T) + \|k_1^{(1)} - k_1^{(2)}\|_2(T) \right), \end{aligned} \tag{6.213}$$

from equality (6.211) we obtain

$$\begin{aligned} \|F'^{(1)}_1(0,t) - F'^{(2)}_1(0,t)\|(T) \le C_5(h_0, h_1, M_3, \rho, T, \max_{(y,t)\in B_1(0,T)} \tilde{G}_t') \\ \times \left(\|c_{44}^{(1)} - c_{44}^{(2)}\|_1(X) + \|c_{11}^{(1)} - c_{11}^{(2)}\|_2(X) \right. \\ \left. + \|g_3^{(1)} - g_3^{(2)}\|_2(T) + \|k_1^{(1)} - k_1^{(2)}\|(T) \right) \end{aligned} \tag{6.214}$$

by virtue of the following estimates, taking into account $k_3(t) \equiv 0$:

$$\|D_1^{(1)}(\tilde{y}) - D_1^{(2)}(\tilde{y})\|(T/2) \le C_6(h_0, h_1, \rho) \cdot \|c_{11}^{(1)} - c_{11}^{(2)}\|(X),$$

$$\begin{aligned} \|\tilde{U}^{(1)}(x_3,t) - \tilde{U}^{(2)}(x_3,t)\|_{C^1(\nabla(T))} \le C_7(h_0, h_1, \rho, T, M_2)\Big(\|g_3^{(1)} - g_3^{(2)}\|_2(T) \\ + \|c_{11}^{(1)} - c_{11}^{(2)}\|_2(X) \Big). \end{aligned}$$

Consider $\tilde{F}^{(1)}(t) - \tilde{F}^{(2)}(t)$:

$$\begin{aligned} \tilde{F}^{(1)}(t) - \tilde{F}^{(2)}(t) = \frac{F'^{(1)}_1(0,t) - L[k_1^{(1)}, g'^{(1)}_2(t)] + W'^{(1)}_1(0,t)}{\hat{D}_1^{(1)}(\tilde{y})\tilde{y}'^{(1)}} \\ - \frac{F_1'^{(2)}(0,t) - L[k_1^{(2)}, g_2'^{(2)}(t)] + W_1'^{(2)}(0,t)}{\hat{D}_1^{(2)}(\tilde{y})\tilde{y}'^{(2)}}. \end{aligned}$$

Passing in these expressions to the differences as in [1], taking into account (6.213), (6.214) and estimates

$$\|\hat{D}_1^{(1)}(\tilde{y}) - \hat{D}_1^{(2)}(\tilde{y})\|(T/2) \le C_8(h_0, h_1, \rho)\Big(\|c_{44}^{(1)} - c_{44}^{(2)}\|(X) + \|c_{11}^{(1)} - c_{11}^{(2)}\|(X)\Big),$$

$$\|\tilde{y}'^{(1)} - \tilde{y}'^{(2)}\|(T_2) \le C_9(h_0, h_1)\Big(\|c_{44}^{(1)} - c_{44}^{(2)}\|(X) + \|c_{11}^{(1)} - c_{11}^{(2)}\|(X)\Big),$$

using Gronwall's lemma in (6.212) , we arrive at the estimate (6.132). Theorem 6.11 is proved.

6.4 Conclusions

In this chapter, inverse coefficient problems for a system of viscoelasticity equations are studied. The known kernels of integral operators (modeling the memory phenomenon) are assumed to be both scalar and diagonal matrix type. Sufficient conditions for unique solvability of inverse problems in some classes of functions are formulated; stability estimates for the unknown coefficients are obtained. For example, the paragraph shows that inverse coefficient problems are solvable only if the third (vertical) memory component is equal to zero. These problems are of interest for analyzing the effect of the memory function under imposed conditions on the properties of a viscoelastic medium. To this end, it is possible to numerically implement systems of integral equations, that is, to make software packages and conduct a series of computational experiments.

References

1. Yakhno, V.G. 1990. *Inverse problems for differential equations of elasticity*. Novosibirsk: Nauka (Russian).
2. Yakhno, V.G., and I.Z. Merazhov. 2000. Some direct problems and one-dimensional inverse electroelasticity problem for "slow" waves. *Siberian Advances in Mathematics* 10 (1): 87–150.
3. Bogachev, I.V., A.O. Vatulyan, and O.V. Yavruyan. 2012. Identification of properties of inhomogeneous electroelastic medium. *Applied Mathematics and Mechanics* 76 (5): 860–866.
4. Nakamura, G., and G. Uhlmann. 1993. Identification of Lame parameters by boundary observations. *American Journal of Mathematics* 115: 1161–1189.
5. Nakamura, G., and G. Uhlmann. 1994. Global uniqueness for an inverse boundary value problems arising in elasticity. *Inventiones Mathematicae* 118: 457–474.
6. Uhlmann, G. 1992. Inverse boundary value problems and applications. *Asterisque* 207: 153–211.
7. Yakhno, V.G. 1998. Multidimensional inverse problems in ray formulation for hyperbolic equations. *Journal of Inverse and Ill-Posed Problems* 6 (4): 373–386.
8. Yakhno, V.G. 1989. A multidimensional inverse dynamic problem of isotropic elasticity. *Doklady Mathematics* 34 (1): 35–36.

9. Yakhno, V.G. 1989. A two-dimensional inverse problem for a system of dynamical Lame equations. *Doklady Mathematics* 34 (7): 616–617.
10. Wu, B., and J. Liu. 2012. Determination of an unknown source for a thermoelastic system with a memory effect. *Inverse Problems* 28: 095012.
11. Lavrent'ev, M.M., and V.G. Romanov. 1966. On three linearized problems for hyperbolic equations. *Doklady Akademii Nauk SSSR* 171 (6): 1279–1281.
12. Romanov, V.G. 1983. The inverse lamb problem in the linear approximation. In: *The collection "Numerical methods in seismological studies"*, 51–78. Novosibirsk: Nauka (Russian).
13. Kabanikhin, S.I. 1988. *Projection—Difference methods for determining the coefficients of hyperbolic equations*. Novosibirsk: Nauka (Russian).
14. Alekseev, A.S. 1962. Some inverse problems of the theory of wave propagation. *Izvestiya Akademii Nauk, Seriya Geograficheskaya* 11: 1514–1531 (Russian).
15. Alekseev, A.S. 1967. Inverse dynamic problem of seismics. In: *Some methods and applications for interpritation of geophysic data*, 9–84. M.: Nauka (Russian).
16. Blagoveshchenskii, A.S. 1971. The local method of solution of the nonstationary inverse problem for an inhomogeneous string. *Proceedings of the Steklov Institute of Mathematics* 115: 30–41.
17. Blagoveshchenskii, A.S. 1971. The quasi-two-dimensional inverse problem for the wave equation. *Proceedings of the Steklov Institute of Mathematics* 115: 63–76.
18. Blagoveshchenskii, A.S. 2001. *Inverse problems of wave processes*. The Netherlands: VSP.
19. Romanov, V.G., and E.A. Volkova. 1982. Inverse dynamic problems of anisotropic elastic medium. *Docl. USSR Academy of Sciences* 267 (4): 780–783 (Russian).
20. Melnikova, T.V., and V.G. Yakhno. 1985. *One-dimensional inverse dynamic problem of isotropic elasticity for a spherically symmetric model of the earth*, 44. Novosibirsk (Preprint/as USSR, Sib.ed.) (Russian).
21. Romanov, V.G. 2012. A two-dimensional inverse problem for the viscoelasticity equation. *Siberian Mathematical Journal* 53 (6): 1128–1138.
22. Romanov, V.G. 2014. On the determination of the coefficients in the viscoelasticity equations. *Siberian Mathematical Journal* 55 (3): 503–510.
23. Romanov, V.G. 1987. *Inverse problems of mathematical physics*. Utrecht: VNU Science Press.
24. Mizohata, S. 1979. *The theory of partial differential equations*. Cambridge University Press.
25. Courant, R., and D. Hilbert. 1962. *Methods of mathematical physics, II*, 830. New York-London: Interscience Publ.
26. Romanov, V.G. 2011. The problem of determinig the kernel of electrodinamics equations for dispersion media. *Doklady Mathematics* 84 (2): 613–616.

Chapter 7
Kernel Identification Problems in a Viscoelasticity System

In this chapter, we study one and two-dimensional problems of determining the memory kernel from a system of viscoelasticity and thermoviscoelasticity equations. The first two sections are devoted to inverse problems of determining the one and two-dimensional kernels of the system of viscoelastic equations. Wherein, under the assumption that the coefficients of the system depend on one spatial variable and the applied concentrated force is directed along one of the coordinate axes, the system is reduced to second-order equation. Based on the methods of Chaps. 2 and 3, the global and local solvability theorems are proved and estimates of the stability of the solution are established.

In the third section the problem of determining the one-dimensional matrix kernel of the system of viscoelasticity is studied. This problem is reduced to the sequential determination of the diagonal elements of the matrix kernel from second-order hyperbolic equations.

In the fourth section, the problem of recovering the memory kernel of thermoviscoelasticity equation is investigated. The research of this problem is also based on the methods used in Chap. 1.

7.1 The Problem of Determining the One-Dimensional Kernel of Viscoelasticity Equation

For $(x_1,\ x_2,\ t) \in \mathbb{R}^3$, $x_3 > 0$ consider the following system of integro-differential equations:

$$\rho \frac{\partial^2 u_i}{\partial t^2} = \sum_{j=1}^{3} \frac{\partial T_{ij}}{\partial x_j}, \quad i = 1, 2, 3, \tag{7.1}$$

© The Author(s), under exclusive license to Springer Nature Singapore Pte Ltd. 2023

D. K. Durdiev and Z. D. Totieva, *Kernel Determination Problems in Hyperbolic Integro-Differential Equations*, Infosys Science Foundation Series in Mathematical Sciences, https://doi.org/10.1007/978-981-99-2260-4_7

at the initial and boundary conditions:

$$u_i \mid_{t<0} \equiv 0, \tag{7.2}$$

$$T_{3j}|_{x_3=+0} = -\delta_{1j}\delta'(t)/2, \ \ j = 1, 2, 3, \tag{7.3}$$

where $u(x, t) = (u_1(x, t), u_2(x, t), u_3(x, t))$ is the displacement vector, $x = (x_1, x_2, x_3)$; T_{ij} is the stress tensor (6.4), (6.5) (the index "0" is omitted).

We assume that $\rho = \rho(x_3), \ \mu = \mu(x_3), \ \lambda = \lambda(x_3) \in \Lambda$. Under this assumption from equalities (7.1)–(7.3) it follows that $u_1(x, t) \equiv u_1(x_3, t) \not\equiv 0, \ u_2 \equiv u_3 \equiv 0$ and the function $\lambda(x_3)$ will not enter the remaining equations (Sect. 6.1).

The inverse problem: determine the kernel $k(t), \ t > 0$, included in the integral of equation (7.1) by (6.4), if additional information about the solution of problem (7.1)–(7.3)

$$u_1(x_3, t)|_{x_3=+0} = g(t), \ t > 0 \tag{7.4}$$

is given.

Let

$$y = \psi(x_3) := \int_0^{x_3} \frac{d\xi}{\nu(\xi)}, \ \ \nu(x_3) := \sqrt{\frac{\mu(x_3)}{\rho(x_3)}}.$$

The main results of this section are the following theorems of global unique solvability and stability.

Theorem 7.1 *Let*

$$g(t) = \frac{a}{2}\delta(t) + \theta(t)g_0(t), \ a = [\mu(+0)\rho(+0)]^{-\frac{1}{2}} \tag{7.5}$$

and $g_0 \in C^2\,[0, \ T]$, $\theta(t)$ *is the Heaviside function. In addition, we also assume* $(\rho, \mu) \in C^3\left[0, \psi^{-1}(T/2)\right)$. *Then, there exists a unique solution of the inverse problem (7.1)–(7.4)* $k(t) \in C^2\,[0, \ T]$ *for any fixed* $T > 0$.

Let $K(h_0)$ is the set of the functions $k(t) \in C^2[0, T]$, satisfying for $t \in [0, T]$ the inequality $\|k(t)\|_{C^2[0,T]} \le h_0$ with a fixed positive constant h_0. This constant is defined in (7.37).

Theorem 7.2 *Let* $k^1(t) \in K(h_0)$, $k^2(t) \in K(h_0)$ *are two solutions of the inverse problem (7.1)–(7.4) with data sets*

$$\left\{\rho^1(\psi^{-1}(y)), \ \mu^1(\psi^{-1}(y)), \ g_0^1(t)\right\},$$

$$\left\{\rho^2(\psi^{-1}(y)), \ \mu^2(\psi^{-1}(y)), \ g_0^2(t)\right\},$$

respectively. Then, there is a positive number $C = C(h_0, \ h_{00}, \ T)$,

$$h_{00} = \max\Big\{ \|\rho^1(y)\|_{C^3[0,\ \psi^{-1}(T/2)]},\ \|\mu^1(y)\ \|_{C^3[0,\ psi^{-1}(T/2)]},\ \|g_0^1(t)\|_{C^2[0,\ T]},$$

$$\|\rho^2(y)\|_{C^3[0,\psi^{-1}(T/2)]},\ \|\mu^2(y)\ \|_{C^3[0,\psi^{-1}(T/2)]},\ \|g_0^2(t)\|_{C^2[0,\ T]}\Big\}$$

such that the following stability estimate is valid:

$$\|k^1(t) - k^2(t)\|_{C^2[0,T]} \le C\Big[\|\rho^1 - \rho^2\|_{C^3[0,\psi^{-1}(T/2)]}$$
$$+\|\mu^1 - \mu^2\|_{C^3[0,\psi^{-1}(T/2)]} + \|g_0^1 - g_0^2\|_{C^2[0,T]}\Big]. \tag{7.6}$$

These theorems will be proved below.

We rewrite (7.1)–(7.3) with respect to the nonzero component of the displacement vector $u_1(x_3, t)$:

$$\rho(x_3)\frac{\partial^2 u_1}{\partial t^2} = L\left[k, \frac{\partial}{\partial x_3}\left(\mu(x_3)\frac{\partial u_1}{\partial x_3}\right)\right], \tag{7.7}$$

$$u_1|_{t<0} \equiv 0, \tag{7.8}$$

$$L\left[k, \mu(x_3)\frac{\partial u_1}{\partial x_3}\right]_{x_3=+0} = -\frac{\delta'(t)}{2}. \tag{7.9}$$

Let

$$v(y,t) := \frac{u_1\left(\psi^{-1}(y), t\right)}{s(y)}, \quad s(y) := \sqrt{\frac{\nu(+0)\rho(+0)}{\nu\left(\psi^{-1}(y)\right)\rho\left(\psi^{-1}(y)\right)}}.$$

Then the inverse problem (7.7)–(7.9), (7.4) in terms of the introduced functions and the variable $y > 0$ takes the form

$$\frac{\partial^2 v}{\partial t^2} = L\left[k, \frac{\partial^2 v}{\partial y^2} + q(y)v\right], \tag{7.10}$$

$$v|_{t<0} \equiv 0, \tag{7.11}$$

$$L\left[k, \frac{\partial v(y,t)}{\partial y}\right]_{y=+0} = -\frac{a\delta'(t)}{2}, \tag{7.12}$$

$$v(y,t)|_{y=+0} = g(t), \quad t > 0, \tag{7.13}$$

where $q(y) = \frac{s''(y)}{s(y)} - 2\left[\frac{s'(y)}{s(y)}\right]^2$.

Let $\left[v(y,t) + \int_0^t k(t-\tau)v(y,\tau)d\tau\right]\exp(-k(0)t/2) = w(y,t)$, then

$$v(y,t) = \exp\left(k(0)t/2\right) w(y,t) + \int_0^t r(t-\tau) \exp\left(k(0)\tau/2\right) w(y,\tau) d\tau,$$

where

$$r(t) = -k(t) - \int_0^t k(t-\tau) r(\tau) d\tau. \tag{7.14}$$

In terms of the new functions $w(y,t)$ and $r(t)$ (7.10)– (7.13) take the form

$$\frac{\partial^2 w}{\partial t^2} = \frac{\partial^2 w}{\partial y^2} + H(y)w - \int_0^t h(t-\tau) w(y,\tau) d\tau, \tag{7.15}$$

$$w|_{t<0} \equiv 0, \tag{7.16}$$

$$\left.\frac{\partial w}{\partial y}\right|_{y=+0} = -\frac{a\delta'(t)}{2} + \frac{a}{4} r(0)\delta(t), \tag{7.17}$$

$$w|_{y=+0} = \tilde{g}(t) + \int_0^t k_0(t-\tau)\tilde{g}(\tau) d\tau, \tag{7.18}$$

where

$$H(y) := q(y) + \frac{r^2(0)}{4} - r'(0), \;\; h(t) := r''(t) \exp\left(r(0)t/2\right),$$

$$\tilde{g}(t) := \frac{a\delta(t)}{2} + \tilde{g}_0(t)\theta(t), \;\; \tilde{g}_0(t) := g_0(t) \exp\left(r(0)t/2\right),$$

$$k_0(t) := k(t) \exp\left(r(0)t/2\right).$$

Here we used the equality $k(0) = -r(0)$, which is a consequence of (7.14).

The theory of hyperbolic equations implies that, as a solution to the direct problem (7.15)–(7.17), the function $w(y,t)$ satisfies $w \equiv 0$ for $t < y$ and $y > 0$ and near the characteristic line $t = y$ has the structure

$$w(y,t) = \frac{a}{2}\delta(t-y) + \theta(t-y)\tilde{w}(y,t), \tag{7.19}$$

where $\tilde{w}(y,t)$ is a regular function.

Substituting the function (7.19) into equations (7.15)–(7.18) and using the method of singularity extraction, we find that the function $\tilde{w}(y,t)$ in the domain $t > y > 0$ satisfies the equations ($w = \tilde{w}$ for $t > y$):

$$\frac{\partial^2 w}{\partial t^2} = \frac{\partial^2 w}{\partial y^2} + H(y)w - \frac{a}{2}h(t-y) - \int_y^t h(t-\tau)w(y,\tau)\mathrm{d}\tau, \tag{7.20}$$

$$w|_{t=y+0} = -\frac{ar(0)}{4} + \frac{a}{4}\int_0^y H(\xi)\mathrm{d}\xi := \beta(y), \tag{7.21}$$

$$\left.\frac{\partial w}{\partial y}\right|_{y=+0} = 0, \tag{7.22}$$

$$w|_{y=+0} = \tilde{g}_0(t) + \int_0^t k_0(t-\tau)\tilde{g}_0(\tau)\mathrm{d}\tau + \frac{a}{2}k_0(t). \tag{7.23}$$

Provided that for $y = t = 0$ $w(y,t)$, $\left(\frac{\partial w}{\partial y}\right)(y,t)$ are continuous functions, we can express $r(0)$ and $r'(0)$ from (7.21)–(7.23) in terms of the known numbers as

$$r(0) = \frac{4}{a}g_0(0),\ r'(0) = -q(0) - \frac{4}{a^2}g_0^2(0) + \frac{4}{a}\tilde{g}_0'(0). \tag{7.24}$$

To derive these equalities, we used the relations

$$k'(t) = -r'(t) - r(0)k(t) - \int_0^t r'(t-\tau)k(\tau)\mathrm{d}\tau,\ \ k'(0) = -r'(0) + r^2(0).$$

Further suppose that in $H(y)$, the values of $r(0)$ and $r'(0)$ are replaced with (7.24).

Lemma 7.1 *Under the conditions of Theorem 7.1 the problem (7.20)–(7.23) for* $(y,t) \in D_T$, $D_T = ((y,t)|0 \le y \le t \le T - y)$ *is equivalent to the problem of determining the functions* $w(y,t)$, $\left(\frac{\partial w}{\partial t}\right)(y,t)$, $h(t)$, $k_0(t)$, $k_0'(t)$, $k_0''(t)$ *from the following system of integral equations:*

$$w(y,t) = \beta(y) + \int_y^t \frac{\partial w}{\partial \tau}(y,\tau)\mathrm{d}\tau, \tag{7.25}$$

$$\begin{aligned}\frac{\partial w}{\partial t}(y,t) &= \frac{1}{2}(\tilde{g}_0'(t-y) - r(0)\tilde{g}_0(t-y)) - \frac{a}{8}H\left(\frac{y+t}{2}\right)\\ &- \frac{a}{4}h(t-y)y + \frac{a}{4}k_0'(t-y) + \frac{1}{2}\int_0^{t-y} k_0'(t-y-\tau)\tilde{g}_0(\tau)\mathrm{d}\tau\\ &+ \frac{1}{2}\int_0^{y}\left[H(\xi)w(\xi,t-y+\xi) - \int_0^{t-y} h(\tau)w(\xi,t-y+\xi-\tau)\mathrm{d}\tau\right]\mathrm{d}\xi\\ &+ \frac{1}{2}\int_y^{\frac{t+y}{2}}\Bigg[H(\xi)w(\xi,t+y-\xi) - \frac{a}{2}h(t+y-2\xi)\\ &- \int_0^{t+y-2\xi} h(\tau)w(\xi,t+y-\xi-\tau)\mathrm{d}\tau\Bigg]\mathrm{d}\xi,\end{aligned} \tag{7.26}$$

$$\begin{aligned}h(t) &= \frac{1}{2}H'\left(\frac{t}{2}\right)\\ &- \frac{4}{a}\left[\tilde{g}_0''(t) - r(0)\tilde{g}_0'(t) + \left(\frac{r^2(0)}{2} - r'(0)\right)\tilde{g}_0(t) - \frac{1}{2}H\left(\frac{t}{2}\right)\beta\left(\frac{t}{2}\right)\right]\\ &- 2k_0''(t) - \frac{4}{a}\int_0^t k_0''(t-\tau)\tilde{g}_0(\tau)\mathrm{d}\tau - \frac{2}{a}\int_0^t h(\tau)\beta\left(\frac{t-\tau}{2}\right)\mathrm{d}\tau\\ &+ \frac{4}{a}\int_0^{\frac{t}{2}}\left[H(\xi)\frac{\partial w}{\partial t}(\xi,t-\xi) - \int_0^{t-2\xi} h(\tau)\frac{\partial w}{\partial t}(\xi,t-\xi-\tau)\mathrm{d}\tau\right]\mathrm{d}\xi,\end{aligned} \tag{7.27}$$

$$k_0(t) = -r(0) + \left(\frac{r^2(0)}{2} - r'(0)\right)t + \int_0^t (t-\tau)k_0''(\tau)\mathrm{d}\tau, \tag{7.28}$$

$$k_0'(t) = \frac{r^2(0)}{2} - r'(0) + \int_0^t k_0''(\tau)\mathrm{d}\tau, \tag{7.29}$$

$$k_0''(t) = -h(t) + \left(\frac{r^2(0)}{4} - r'(0)\right)k_0(t) - \int_0^t h(t-\tau)k_0(\tau)\mathrm{d}\tau. \tag{7.30}$$

Proof To prove the Lemma 7.1, we use the equalities (3.63). Taking this into account, we integrate (7.20) along the corresponding characteristics of the differential operators of the first order for $(y, t) \in D_T$. First, we integrate along the characteristic of operator $\frac{\partial}{\partial t} - \frac{\partial}{\partial y}$ from point (y, t) to the point $((y+t)/2,\ (y+t)/2)$ in the plane of the variables $(\xi,\ \tau)$. Using the equality $\left(\frac{\partial}{\partial t} + \frac{\partial}{\partial y}\right) w((y+t)/2,\ (y+t)/2) = (a/4)H((y+t)/2)$, which follows from (7.21) after differentiation with respect to y, we obtain

$$\left(\frac{\partial}{\partial t} + \frac{\partial}{\partial y}\right) w(y,t) = \frac{a}{4} H\left(\frac{y+t}{2}\right) + \int\limits_y^{(y+t)/2} \Bigg[H(\xi) w(\xi, t+y-\xi)$$

$$- \frac{a}{2} h(t+y-2\xi) - \int\limits_0^{t+y-2\xi} h(\tau) w(\xi, t+y-\xi-\tau) d\tau \Bigg] d\xi. \tag{7.31}$$

Now we integrate along the characteristic of operator $\frac{\partial}{\partial t} + \frac{\partial}{\partial y}$ from point $(0,\ t-y)$ to the point (y, t). Using the equalities (7.22) and (7.23), we find

$$\left(\frac{\partial}{\partial t} - \frac{\partial}{\partial y}\right) w(y,t)$$

$$= \tilde{g}_0'(t-y) - r(0)\tilde{g}_0(t-y)$$

$$- \frac{a}{2} h(t-y) y + \frac{a}{2} k_0'(t-y) + \int\limits_0^{t-y} k_0'(t-y-\tau)\tilde{g}_0(\tau) d\tau \tag{7.32}$$

$$\int\limits_0^y \Bigg[H(\xi) w(\xi, t-y+\xi) - \int\limits_0^{t-y} h(\tau) w(\xi, t-y+\xi-\tau) d\tau \Bigg] d\xi.$$

From (7.31) and (7.32) it is easy to obtain equation (7.26). In (7.31), setting $y = 0$ and using conditions (7.22) and (7.23), we find

$$\tilde{g}_0'(t) - r(0)\tilde{g}_0(t) + \frac{a}{2} k_0'(t) + \int\limits_0^t k_0'(t-\tau)\tilde{g}_0(\tau) d\tau = \frac{a}{4} H\left(\frac{t}{2}\right)$$

$$+ \int\limits_0^{t/2} \Bigg[H(\xi) w(\xi, t-\xi) - \frac{a}{2} h(t-2\xi) - \int\limits_0^{t-2\xi} h(\tau) w(\xi, t-\xi-\tau) d\tau \Bigg] d\xi.$$

Differentiating this equation, after simple computations, we arrive at equation (7.27).

Equations (7.28)–(7.30) are used to close the system of integral equations (7.25)–(7.27).

Under the conditions of Theorem 7.1, the equivalence of the system of integral equations (7.25)–(7.30) and the inverse problem (7.20)–(7.23) is established in the usual way. □

Let us prove Theorem 7.1. The proof of the theorem is quite analogous to the proof of the Theorem 3.1 from Sect. 3.1. Express (7.25)–(7.30) as the operator equation

$$\varphi = A\varphi, \tag{7.33}$$

where

$$\begin{aligned}\varphi &= \Big[\varphi_1(y,t), \varphi_2(y,t), \varphi_3(t), \varphi_4(t), \varphi_5(t), \varphi_6(t)\Big]\\ &= \Big[w(y,t),\ \frac{\partial w}{\partial t}(y,t) + \frac{a}{4}h(t-y)y - \frac{a}{4}k_0'(t-y),\ h(t)\\ &\quad - 2k_0''(t),\ k_0(t),\ k_0'(t),\ k_0''(t) + h(t) - r_{00}k_0(t)\Big]\end{aligned}$$

is the vector function with components φ_i $(i = 1, 2, \ldots, 6)$ and $r_{00} := \frac{r^2(0)}{4} - r'(0)$, while the operator A defined on the set of functions $\varphi \in C[D_T]$ in accordance with (7.25)–(7.30), is the form

$$A = (A_1,\ A_2,\ A_3,\ A_4,\ A_5,\ A_6):$$

$$\begin{aligned}A_1\varphi &= \varphi_{01} + \int_y^t \Big[\varphi_2(y,\tau)\\ &\quad - \frac{a}{12}y\left(\varphi_3(\tau-y) + 2\varphi_6(\tau-y) + 2r_{00}\varphi_4(\tau-y)\right) + \frac{a}{4}\varphi_5(\tau-y)\Big]\mathrm{d}\tau,\\ A_2\varphi &= \varphi_{02} + \frac{1}{2}\int_0^{t-y} \varphi_5(t-y-\tau)\tilde{g}_0(\tau)\mathrm{d}\tau + \frac{1}{2}\int_0^y \Big[H(\xi)\varphi_1(\xi, t-y+\xi)\\ &\quad - \frac{1}{3}\int_0^{t-y} ((\varphi_3(\tau) + 2\varphi_6(\tau) + 2r_{00}\varphi_4(\tau))\varphi_1(\xi, t-y+\xi-\tau)\mathrm{d}\tau\Big]\mathrm{d}\xi\\ &\quad + \frac{1}{2}\int_y^{\frac{t+y}{2}} \Big[H(\xi)\varphi_1(\xi, t+y-\xi)\\ &\quad - \frac{a}{6}(\varphi_3(t+y-2\xi) + 2\varphi_6(t+y-2\xi) + 2r_{00}\varphi_4(t+y-2\xi))\\ &\quad - \frac{1}{3}\int_0^{t+y-2\xi} (\varphi_3(\tau) + 2\varphi_6(\tau) + 2r_{00}\varphi_4(\tau))\varphi_1(\xi, t+y-\xi-\tau)\mathrm{d}\tau\Big]\mathrm{d}\xi,\end{aligned} \tag{7.34}$$

$$A_3\varphi = \varphi_{03} + \frac{4}{3a}\int_0^t (\varphi_6(t-\tau) - \varphi_3(t-\tau) + r_{00}\varphi_4(t-\tau))\,\tilde{g}_0(\tau)d\tau$$

$$+\frac{2}{3a}\int_0^t (\varphi_3(\tau) + 2\varphi_6(\tau) + 2r_{00}\varphi_4(\tau))\,\beta\left(\frac{t-\tau}{2}\right)d\tau$$

$$\frac{4}{a}\int_0^{t/2}\Bigg[H(\xi)\Bigg(\varphi_2(\xi, t-\xi) - \frac{a\xi}{12}\Big(\varphi_3(t-2\xi) + 2\varphi_6(t-2\xi) - 2r_{00}\varphi_4(t-2\xi)\Big)$$

$$+\frac{a}{4}\varphi_5(t-2\xi)\Bigg) - \frac{1}{3}\int_0^{t-2\xi} (\varphi_3(\tau) + 2\varphi_6(\tau) + 2r_{00}\varphi_4(\tau))$$

$$\times\Bigg(\varphi_2(\xi, t-\xi-\tau) - \frac{a\xi}{12}(\varphi_3(t-2\xi-\tau) + 2\varphi_6(t-2\xi-\tau)$$

$$+2r_{00}\varphi_4(t-2\xi-\tau)) + \frac{a}{4}\varphi_5(t-2\xi-\tau)\Bigg)d\tau\Bigg]d\xi,$$

$$A_4\varphi = \varphi_{04} + \frac{1}{3}\int_0^t (t-\tau)\,(\varphi_6(\tau) - \varphi_3(\tau) + r_{00}\varphi_4(\tau))\,d\tau,$$

$$A_5\varphi = \varphi_{05} + \frac{1}{3}\int_0^t (\varphi_6(\tau) - \varphi_3(\tau) + r_{00}\varphi_4(\tau))\,d\tau,$$

$$A_6\varphi = \varphi_{06} - \frac{1}{3}\int_0^t (\varphi_3(t-\tau) + 2\varphi_6(t-\tau) + 2r_{00}\varphi_4(t-\tau))\,\varphi_4(\tau)d\tau,$$

where

$$\varphi_0(y,t)$$
$$= (\varphi_{01}, \varphi_{02}, \varphi_{03}, \varphi_{04}, \varphi_{05}, \varphi_{06})$$
$$:= \Bigg[\beta(y), \frac{1}{2}\left(\tilde{g}_0'(t-y) - r(0)\tilde{g}_0(t-y)\right) - \frac{a}{8}H\left(\frac{y+t}{2}\right), \frac{1}{2}H'\left(\frac{t}{2}\right)$$
$$-\frac{4}{a}\left(\tilde{g}_0''(t) - r(0)\tilde{g}_0'(t) + r_{00}\tilde{g}_0(t) + \frac{1}{2}H\left(\frac{t}{2}\right)\beta\left(\frac{t}{2}\right)\right), -r(0) + r_{00}t,\ r_{00},\ 0\Bigg]. \tag{7.35}$$

Denote by C_σ the Banach space of continuous functions with the family of weighted norms

$$\|\varphi\|_\sigma = \max\left\{ \sup_{(y,t)\in D_T} \left|\varphi_{1,2}(y,t)e^{-\sigma t}\right|, \sup_{t\in[0,T]} \left|\varphi_{4,5,6}(t)e^{-\sigma t}\right| \right\}, \quad \sigma \geq 0.$$

It is obvious that for $\sigma = 0$ this is the space of continuous functions with the usual norm. We denote this norm by $\|\varphi\|$. The norms $\|\varphi\|_\sigma$ and $\|\varphi\|$ are equivalent for every fixed $T \in (0, \infty)$ due to (3.17) $|\varphi\|_\sigma$. We choose the number σ below. Take the ball $Q_\sigma(\varphi_0, \|\varphi_0\|) =: \{\varphi | \|\varphi - \varphi_0\|_\sigma \leq \|\varphi_0\|\}$ of radius $\|\varphi_0\|$ centered at the point φ_0 of some weighted space C_σ with $\sigma \geq 0$, where the function φ_0 is defined using (7.35) and

$$\|\varphi_0\| = \max_{1\leq j\leq 6} \left\{\|\varphi_{0j}\|\right\}.$$

Observe that $\varphi \in Q_\sigma(\varphi_0, \|\varphi_0\|)$ satisfies the estimate $\|\varphi\|_\sigma \leq \|\varphi_0\|_\sigma + \|\varphi_0\| \leq 2\|\varphi_0\|$.

Let $\varphi(y,t) \in Q_\sigma(\varphi_0, \|\varphi_0\|)$ and verify that for a suitable choice of $\sigma > 0$ the operator A maps balls into balls, i.e., $A\varphi \in Q_\sigma(\varphi_0, \|\varphi_0\|)$. Indeed, calculating the norm of the differences $||A_1\varphi - \varphi_{01}||_\sigma$ like a proof of the Theorem 3.1, in view of (7.34), for $(y,t) \in D_T$ we have

$$||A_1\varphi - \varphi_{01}||_\sigma \leq 2\|\varphi_0\|\mu_1(a, T, |r_{00}|)\frac{1}{\sigma},$$

$$||A_2\varphi - \varphi_{02}||_\sigma \leq 2\|\varphi_0\|\mu_2(a, T, |r_{00}|, G_0, H_0, \|\varphi_0\|)\frac{1}{\sigma},$$

$$||A_3\varphi - \varphi_{03}||_\sigma \leq 2\|\varphi_0\|\mu_3(a, T, |r_{00}|, G_0, H_0, \|\varphi_0\|, \beta_0)\frac{1}{\sigma},$$

$$||A_4\varphi - \varphi_{04}||_\sigma \leq 2\|\varphi_0\|\mu_4(T, |r_{00}|)\frac{1}{\sigma},$$

$$||A_5\varphi - \varphi_{05}||_\sigma \leq 2\|\varphi_0\|\mu_5(|r_{00}|)\frac{1}{\sigma},$$

$$||A_6\varphi - \varphi_{06}||_\sigma \leq 2\|\varphi_0\|\mu_6(T, |r_{00}|, |r(0)|, \|\varphi_0\|)\frac{1}{\sigma},$$

where $H_0 := \max_{y\in[0,T/2]} |H(y)|$, $\beta_0 := \max_{y\in[0,T/2]} |\beta(y)|$, $G_0 := \max_{t\in[0,T]} |\tilde{g}_0(t)|$. We obtained the last inequality using the fourth equation of (7.34). Choosing

$$\sigma \geq \sigma_0 := 2 \max_{1\leq j\leq 6} \left\{\mu_j\right\},$$

we infer that A maps the ball $Q_\sigma(\varphi_0, \|\varphi_0\|)$ into the ball $Q_\sigma(\varphi_0, \|\varphi_0\|)$.

Take two arbitrary elements φ^1 and φ^2. Using the auxiliary inequalities

$$|\varphi_i^1\varphi_j^1 - \varphi_i^2\varphi_j^2| e^{-\sigma t} \le |\varphi_i^1| |\varphi_j^1 - \varphi_j^2| e^{-\sigma t} + |\varphi_j^2| |\varphi_i^1 - \varphi_i^2| e^{-\sigma t}$$

$$\le 4\|\varphi_0\| \left\|\varphi^1 - \varphi^2\right\|_\sigma, \ (y, t) \in D_T,$$

we find

$$\left\|(A\varphi^1 - A\varphi^2)\right\|_\sigma \le \left\|\varphi^1 - \varphi^2\right\|_\sigma \le \frac{\sigma_{00}}{\sigma} \left\|\varphi^1 - \varphi^2\right\|_\sigma,$$

where

$$\sigma_{00} = \max\Big\{\mu_1(a, T, |r_{00}|),\ \mu_2(a, T, |r_{00}|, G_0, H_0, 2\|\varphi_0\|),$$
$$\mu_3(a, T, |r_{00}|, G_0, H_0, 2\|\varphi_0\|, \beta_0),\ \mu_4(T, |r_{00}|),$$
$$\mu_5(|r_{00}|), \mu_6(T, |r_{00}|, |r(0)|, 2\|\varphi_0\|)\Big\}$$

The resulting estimates imply that if σ satisfies $\sigma > \sigma^* := \max\{\sigma_0, \sigma_{00}\}$, then, A is a contraction mapping on $Q_\sigma(\varphi_0, \|\varphi_0\|)$. Then, according to the Banach principle, (7.33) has a unique solution in $Q_\sigma(\varphi_0, \|\varphi_0\|)$ for each fixed $T > 0$.

Since $k_0(t) := \exp(r(0)t/2)k(t)$, we find

$$k(t) = \exp[-r(0)t/2]k_0(t). \tag{7.36}$$

The proof of Theorem 7.1 is complete.

Now let us prove the Theorem 7.2. Since the Theorem 7.1 holds, the solution (7.33) belongs to the set $Q_\sigma(\varphi_0, \|\varphi_0\|)$ and $\|\varphi_i\|_\sigma \le 2\|\varphi_0\|,\ i = 1, 2, \cdots, 6$. Thus,

$$\max_{t\in[0,T]} |k(t)| \le 2\|\varphi_0\| \exp(|r(0)|T) := h_0. \tag{7.37}$$

Take the vector φ^j, for $j = 1, 2$ with the data $\left\{\rho^j(\psi^{-1}(y)), \mu^j(\psi^{-1}(y)), g_0^j(t)\right\}$ respectively; i.e., $\varphi^j = A\varphi^j$. The known functions $\rho^j\left[\psi^{-1}(y)\right],\ \mu^j\left[\psi^{-1}(y)\right]$, appear in the free terms of these integral equations in the compositions $H^j(y), q^j(y), s^j(y),\ j = 1, 2$. Passing in these expressions to the differences $\rho^1 - \rho^2$, $\mu^1 - \mu^2$, as in [1], by the arguments in the proof of Theorem 7.1 for $\sigma \ge \sigma^*$ we arrive at the estimate

$$\left\|\varphi^1 - \varphi^2\right\|_\sigma \le C_0\gamma + \frac{\sigma^*}{\sigma} \left\|\varphi^1 - \varphi^2\right\|_\sigma, \tag{7.38}$$

where

$$\gamma := \|\rho^1 - \rho^2\|_{C^3[0,\psi^{-1}(T/2)]} + \|\mu^1 - \mu^2\|_{C^3[0,\psi^{-1}(T/2)]} + \|g_0^1 - g_0^2\|_{C^2[0,T]}$$

and the constant C_0 depends on the same parameter as C in Theorem 7.2. Now (3.17) and (7.38) imply the estimate

$$\left\|k_0^1 - k_0^2\right\| \le C_1 \gamma,$$

with $C_1 = \sigma C_0/(\sigma - \sigma^*)$.

Considering (7.36) for $\{k^1, k_0^1\}$ and $\{k^2, k_0^2\}$ and using (7.38), we obtain (7.6).

7.2 The Problem of Determining the Multidimensional Kernel of Viscoelasticity Equation

Consider the system (7.1)–(7.3) with

$$T_{ij}(x,t) = \sigma_{ij}[u](x,t) + \int_0^t k(x_2, t-\tau)\sigma_{ij}[u](x,\tau)\mathrm{d}\tau, \tag{7.39}$$

σ_{ij} are stresses (6.5).

We assume that $\rho = \rho(x_3),\ \mu = \mu(x_3),\ \lambda = \lambda(x_3)$ are known and belong to the set Λ.

Under this assumptions from equalities (7.1)–(7.3),(7.39) it follows that $u_1(x,t) \equiv u_1(x_3,t) \not\equiv 0,\ u_2 \equiv u_3 \equiv 0$ and the function $\lambda(x_3)$ will not enter the remaining equations.

The inverse problem: determine the kernel $k(x_2,t),\ t > 0$, included in the integral of equation (7.1) by (7.39), if additional information about the solution of problem (7.1)–(7.3), (7.39) is known

$$u_1(x_2, x_3, t)|_{x_3=+0} = g(x_2, t),\ t > 0. \tag{7.40}$$

We rewrite (7.1)–(7.3) with (7.39) with respect to the nonzero component of the displacement vector $u_1(x_2, x_3, t)$:

$$\rho(x_3)\frac{\partial^2 u_1}{\partial t^2} = \frac{\partial}{\partial x_3}\left(\mu(x_3)L\left[k, \frac{\partial u_1}{\partial x_3}\right]\right) + \mu(x_3)\frac{\partial}{\partial x_2}\left(L\left[k, \frac{\partial u_1}{\partial x_2}\right]\right), \tag{7.41}$$

$$u_1|_{t<0} \equiv 0, \tag{7.42}$$

$$L\left[k, \frac{\partial u_1}{\partial x_3}\right]_{x_3=0} = -\frac{\delta'(t)}{2\mu(0)}. \tag{7.43}$$

Further, the index “2” of the variable x will be omitted.

Let

$$y = \phi(x_3) := \int_0^{x_3} \frac{d\xi}{\nu(\xi)}, \quad \nu(x_3) := \sqrt{\frac{\mu(x_3)}{\rho(x_3)}}$$

and

$$v(x, y, t) := \frac{u_1\left(x, \phi^{-1}(y), t\right)}{s(y)}, \quad s(y) := \sqrt{\frac{\nu(+0)\rho(+0)}{\nu\left(\phi^{-1}(y)\right)\rho\left(\phi^{-1}(y)\right)}}.$$

Then, in terms of the functions introduced above and the variable y, the inverse problem (7.41)–(7.43) takes the form

$$\frac{\partial^2 v}{\partial t^2} = L\left[k, \frac{\partial^2 v}{\partial y^2} + q(y)v\right] + \nu^2 \frac{\partial}{\partial x} L\left[k, \frac{\partial v}{\partial x}\right], \quad y > 0, \ (x, t) \in \mathbb{R}^2, \tag{7.44}$$

$$v(x, y, t)|_{t<0} \equiv 0, \tag{7.45}$$

$$\frac{\partial}{\partial y} L\,[k, v]_{y=+0} = -\frac{a\delta'(t)}{2}, \tag{7.46}$$

$$v(x, y, t)|_{y=+0} = g(x, t), \quad t > 0, x \in \mathbb{R}, \tag{7.47}$$

where $q(y) := \frac{s''(y)}{s(y)} - 2\left[\frac{s'(y)}{s(y)}\right]^2$, $a = [\mu(+0)\rho(+0)]^{-\frac{1}{2}}$.

In the equation (7.44) the function $\nu\left(\phi^{-1}(y)\right)$ is denoted by ν.

Consider the operator equation for the function $v(x, y, t)$

$$L\,[k, v] = \exp\left(k(x, 0)t/2\right) w(x, y, t).$$

It is easy to see that v is expressed by the formula

$$v(x, y, t) = L\left[r, \exp\left(k(x, 0)t/2\right) w\right], \tag{7.48}$$

where

$$r(x, t) = -k(x, t) - \int_0^t k(x, t - \tau) r(x, \tau) d\tau. \tag{7.49}$$

With respect to the functions $w(x, y, t)$, $v(x, y, t)$ and $r(x, t)$ for $y > 0$, $(x, t) \in \mathbb{R}^2$, the equations (7.44)–(7.47) take the form

$$\frac{\partial^2 w}{\partial t^2} = \frac{\partial^2 w}{\partial y^2} + H(x, y)w - \int_0^t h(x, t - \tau)w(x, y, \tau)d\tau$$
$$+\nu^2 \exp\left(r(x, 0)t/2\right) \frac{\partial}{\partial x} L\left[k, \frac{\partial v}{\partial x}\right], \tag{7.50}$$

$$w(x, y, t)|_{t<0} \equiv 0, \tag{7.51}$$

$$\left.\frac{\partial w}{\partial y}\right|_{y=+0} = -\frac{a\delta'(t)}{2} + \frac{a}{4} r(x, 0)\delta(t), \tag{7.52}$$

$$w|_{y=+0} = L\left[k_0, \ \tilde{g}(x, t)\right], \tag{7.53}$$

where

$$H(x, y) := q(y) + \frac{r^2(x, 0)}{4} - \frac{\partial r}{\partial t}(x, 0), \ h(x, t) := \frac{\partial^2 r}{\partial t^2}(x, t) \exp\left(r(x, 0)t/2\right),$$

$$\tilde{g}(x, t) := \frac{a\delta(t)}{2} + \tilde{g}_0(x, t)\theta(t), \ \tilde{g}_0(x, t) := g_0(x, t) \exp\left(r(x, 0)t/2\right),$$

$$k_0(x, t) := k(x, t) \exp\left(r(x, 0)t/2\right), r_0(x, t) := r(x, t) \exp\left(r(x, 0)t/2\right).$$

Here we used the equality $k(x, 0) = -r(x, 0)$, which follows from (7.49). Further, the equations (7.50)–(7.53) will be considered when performing (7.48).

The solution of the direct problem (7.48), (7.50)–(7.52) is presented in the form

$$w(x, y, t) = \frac{a}{2}\delta(t - y) + \theta(t - y)\tilde{w}(x, y, t), \tag{7.54}$$

where $\tilde{w}(x, y, t)$ is regular function. Then, for the function v we find from equality (7.48):

$$v(x, y, t) = \frac{a}{2} \exp\left(k(x, 0)y/2\right) \delta(t - y)$$
$$+ \frac{a}{2} r(x, t - y) \exp\left(k(x, 0)y/2\right) \theta(t - y) + \theta(t - y)\tilde{v}(x, y, t), \tag{7.55}$$

where

$$\tilde{v}(x, y, t) = L_0\left[r, \exp\left(k(x, 0)t/2\right) \tilde{w}\right],$$

the operator L_0 differs from the operator L only in that the lower limit in the integral of L is y.

The direct calculations show that the last term in (7.50) takes the form

$$\frac{\nu^2 a}{2}\left(\frac{\partial^2 k}{\partial x^2}(x,0)\frac{y}{2} + \left(\frac{\partial k}{\partial x}\right)^2 (x,0)\frac{y^2}{4}\right)\delta(t-y)$$
$$+\frac{\nu^2 a}{2}\Bigg[\left(\frac{\partial^2 k}{\partial x^2}(x,0)(x,0)\frac{y}{2} + \left(\frac{\partial k}{\partial x}\right)^2 (x,0)\frac{y^2}{4}\right)k(x,t-y)$$
$$+\frac{\partial k}{\partial x}(x,0)\frac{\partial k}{\partial x}(x,t-y)\Bigg]\theta(t-y) + \nu^2\exp\left(r(x,0)t/2\right)$$
$$\times\frac{\partial}{\partial x}L_0\left[k,\ \frac{\partial}{\partial x}\left(r(x,t-y)\exp\left(k(x,0)y/2\right)\right) + \frac{\partial\tilde{v}}{\partial x}\right]\theta(t-y).$$

Note that $w = \tilde{w}$ at $t > y > 0$. Let $\beta(x,y) := w(x,y,y)$. For a fixed $x \in \mathbb{R}$ we obtain a differential equation for $\beta(x,y)$

$$\frac{\partial}{\partial y}\beta(x,y) = \frac{a}{4}H(x,y) + \frac{a\nu^2}{4}\left(\frac{\partial^2 k}{\partial x^2}(x,0)\frac{y}{2} + \left(\frac{\partial k}{\partial x}\right)^2 (x,0)\frac{y^2}{4}\right)$$

with initial condition

$$\beta(x,0) = -\frac{a}{4}r(x,0).$$

we find

$$\beta(x,y) = -\frac{a}{4}r(x,0) + \frac{a\nu^2}{4}\left(\frac{\partial^2 k}{\partial x^2}(x,0)\frac{y^2}{4} + \left(\frac{\partial k}{\partial x}\right)^2 (x,0)\frac{y^3}{12}\right) + \frac{a}{4}\int_0^y H(x,\xi)d\xi. \tag{7.56}$$

Hence, the functions $w(x,y,t)$, $k(x,t)$ in domain $t > y > 0$ satisfy the equations

$$\frac{\partial^2 w}{\partial t^2} = \frac{\partial^2 w}{\partial y^2} + H(x,y)w - \frac{a}{2}h(x,t-y)$$
$$+ k_{00}(x,y,t) - \int_y^t h(x,t-\tau)w(x,y,\tau)d\tau \tag{7.57}$$
$$+ \nu^2\exp\left(r(x,0)t/2\right)\frac{\partial}{\partial x}L_0\left[k, r_{00} + \frac{\partial\tilde{v}}{\partial x}\right],$$

$$w(x,y,t)|_{t=y} = \beta(x,y), \tag{7.58}$$

$$\left.\frac{\partial w}{\partial y}\right|_{y=+0} = 0, \tag{7.59}$$

$$w|_{y=+0} = \frac{a}{2}k_0(x,t) + L\left[k_0,\ \tilde{g}_0\right], \tag{7.60}$$

where

$$k_{00}(x, y, t) :=$$
$$\frac{\nu^2 a}{2}\left[\left(\frac{\partial^2 k}{\partial x^2}(x,0)\frac{y}{2} + \left(\frac{\partial k}{\partial x}\right)^2(x,0)\frac{y^2}{4}\right)k(x,t-y) + \frac{\partial k}{\partial x}(x,0)\frac{\partial k}{\partial x}(x,t-y)\right], \tag{7.61}$$

$$r_{00}(x, y, t) := \frac{\partial}{\partial x}\Big(r(x, t-y)\exp\left(k(x,0)y/2\right)\Big). \tag{7.62}$$

Provided that for $y = t = 0$ $w(x, y, t)$, $\left(\frac{\partial w}{\partial y}\right)(x, y, t)$ are continuous functions, we can express $r(x, 0)$ and $\frac{\partial r}{\partial t}(x, 0)$ from (7.58)–(7.60) in terms of the available values of functions as

$$r(x,0) = \frac{4}{a}g_0(x,0), \quad \frac{\partial r}{\partial t}(x,0) = -q(0) - \frac{4}{a^2}g_0^2(x,0) + \frac{4}{a}\frac{\partial \tilde{g}_0}{\partial t}(x,0). \tag{7.63}$$

To derive this, we use the relations

$$\frac{\partial k}{\partial t}(x,t) = -\frac{\partial r}{\partial t}(x,t) - r(x,0)k(x,t) - \int_0^t \frac{\partial r}{\partial t}(x, t-\tau)k(x,\tau)d\tau,$$

$$\frac{\partial k}{\partial t}(x,0) = -\frac{\partial r}{\partial t}(x,0) + r^2(x,0).$$

Assume henceforth that in the relations for $H(x, y)$ we replace $r(x, 0)$ and $\frac{\partial r}{\partial t}(x, 0)$ with their values (7.58).

Further, we assume that $r(x, 0)$, $\frac{\partial r}{\partial t}(x, 0)$ are known functions, their values are calculated by formulas (7.63). It follows that the functions $k(x, 0)$, $\frac{\partial k}{\partial x}(x, 0)$, $(x, 0)$, $\frac{\partial^2 k}{\partial x^2}(x, 0)$, $\frac{\partial k}{\partial t}(x, 0)$, and hence $k_{00}(x, y, t)$, $r_{00}(x, y, t)$ at $t = y$ are also known.

We construct a system of integro-differential equations for functions w, $\tilde{v}$, h, k_0. Using Dalamber's formula we find from (7.57), (7.59), (7.60)

$$\begin{aligned} w(x,y,t) &= w_0(x,y,t) \\ &+ \frac{1}{2}\int_0^y \int_{t-y+\xi}^{t+y-\xi} \Big\{ H(x,\xi)w(x,\xi,\tau) - \frac{a}{2}h(x,\tau-\xi) + k_{00}(x,\xi,\tau) \\ &- \int_0^{\tau-\xi} h(x,\gamma)w(x,\xi,\tau-\gamma)d\gamma \\ &+ \nu^2 \exp\left(r(x,0)\tau/2\right)\frac{\partial}{\partial x}L_0\left[k, r_{00} + \frac{\partial \tilde{v}}{\partial x}\right](x,\xi,\tau)\Big\} d\tau d\xi, \end{aligned} \tag{7.64}$$

$x \in \mathbb{R},\ 0 < y < t,$

where

$$w_0(x,y,t) = \frac{1}{2}\Big\{\frac{a}{2}(k_0(x,t+y) + k_0(x,t-y)) + \tilde{g}_0(x,t+y) + \int_0^{t+y} k_0(x,\tau)\tilde{g}_0(x,t+y-\tau)d\tau + \tilde{g}_0(x,t-y) + \int_0^{t-y} k_0(x,\tau)\tilde{g}_0(x,t-y-\tau)d\tau\Big\}.$$

Passing to the limit in the formula (7.64) at $t \to y+0$ taking into account the conditions (7.59), (7.60) we find

$$2\beta(x,y) - w_0(x,y,y+0) = \int_0^y \int_\xi^{2y-\xi} \Big\{H(x,\xi)w(x,\xi,\tau) - \frac{a}{2}h(x,\tau-\xi) + k_{00}(x,\xi,\tau) - \int_0^{\tau-\xi} h(x,\gamma)w(x,\xi,\tau-\gamma)d\gamma + \nu^2 \exp(r(x,0)\tau/2)\frac{\partial}{\partial x}L_0\Big[k, r_{00} + \frac{\partial\tilde{v}}{\partial x}\Big](x,\xi,\tau)\Big\}d\tau d\xi.$$

Differentiating by y the last equality, we obtain:

$$\begin{aligned}\frac{\partial\beta}{\partial y}(x,y) - \frac{1}{2}\frac{dw_0}{dy}(x,y,y+0) - \int_0^y \Big\{&H(x,\xi)w(x,\xi,2y-\xi) - \frac{a}{2}h(x,2(y-\xi)) + k_{00}(x,\xi,2y-\xi) - \int_0^{2(y-\xi)} h(x,\gamma)w(x,\xi,2y-\xi-\gamma)d\gamma \\ &+ \nu^2\exp(r(x,0)(2y-\xi)/2)\frac{\partial}{\partial x}L_0\Big[k, r_{00} + \frac{\partial\tilde{v}}{\partial x}\Big](x,\xi,2y-\xi)\Big\}d\xi.\end{aligned} \tag{7.65}$$

The equation for h is obtained by differentiating equality (7.65) by y after replacing the variable in the second integral $2(y-\xi)$ by ξ':

$$
\begin{aligned}
h(x,y) &= h_0(x,y) + \frac{2}{a}\int_0^y k_0(x,\xi)\frac{\partial^2 \tilde{g}_0}{\partial t^2}(x,y-\xi)\mathrm{d}\xi \\
&+ \frac{1}{a}\int_0^{y/2}\Big\{H(x,\xi)\frac{\partial w}{\partial t}(x,\xi,y-\xi) + \frac{\partial k_{00}}{\partial t}(x,\xi,y-\xi) \\
&- \frac{1}{2}h(x,y-2\xi)\beta(x,\xi) \\
&- \int_0^{y-2\xi} h(x,\tau)\frac{\partial w}{\partial t}(x,\xi,y-\xi-\tau)\mathrm{d}\tau + \nu^2\exp\left(r(x,0)(y-\xi)/2\right) \\
&\times\Big\{\frac{\partial}{\partial x}L_0\Big[k(x,y-\xi),\Big(\frac{\partial r_{00}}{\partial t} + \frac{\partial}{\partial x}\Big(\frac{\partial \tilde{v}}{\partial t}\Big)\Big)(x,\xi,y-\xi)\Big] \\
&+ \frac{r(x,0)}{2}\frac{\partial}{\partial x}L_0\Big[k(x,y-\xi),(r_{00} + \frac{\partial \tilde{v}}{\partial x})(x,\xi,y-\xi)\Big] \\
&+ \frac{1}{2}\frac{\partial}{\partial x}\Big[k(x,y-2\xi)(r_{00} + \frac{\partial \tilde{v}}{\partial x})(x,\xi,\xi)\Big]\Big\}\mathrm{d}\xi,
\end{aligned}
\tag{7.66}
$$

where

$$
\begin{aligned}
h_0(x,y) &= \frac{\partial^2 k_0}{\partial x^2}(x,y) + \frac{2}{a}\Big[g_0(x,0)\frac{\partial k_0}{\partial t}(x,y) + k_0(x,y)\frac{\partial \tilde{g}_0}{\partial t}(x,0) + \frac{\partial^2 \tilde{g}_0}{\partial t^2}(x,y)\Big] \\
&- \frac{2}{a}\frac{\partial^2}{\partial y^2}\beta(x,y/2) + \frac{2}{a}H(x,y/2)\beta(x,y/2) + \frac{2}{a}k_{00}(x,y/2,y/2) \\
&+ \frac{2}{a}\nu^2\exp(r(x,0)y/4)\left(\frac{\partial r_{00}}{\partial x}(x,y/2,y/2) + \frac{\partial^2 \beta}{\partial x^2}(x,y/2)\right).
\end{aligned}
$$

To calculate the function $\frac{\partial w}{\partial t}$, use the equality (7.64). Applying an equivalent description of the integration domain in the form of $\{(\xi,\tau)|t-y\le\tau\le t+y, 0\le\xi\le y-|t-\tau|\}$, we differentiate the equation by t. Replacing the integration variable $\tau - t = \xi'$, we find

$$\frac{\partial w}{\partial t}(x,y,t)=\frac{\partial w_0}{\partial t}(x,y,t)$$
$$+\frac{1}{2}\int_{-y}^{y} sgn(\xi)\Big\{H(x,y-|\xi|)w(x,y-|\xi|,\xi+\tau)-\frac{a}{2}h(x,\xi+t-y+|\xi|)$$
$$+k_{00}(x,y-|\xi|,t+|\xi|)-\int_0^{t-y+\xi+|\xi|} h(x,\gamma)w(x,y-|\xi|,t+\xi-\gamma)d\gamma$$
$$+v^2\exp\left(r(x,0)\frac{\xi+t}{2}\right)\frac{\partial}{\partial x}L_0\left[k(x,\xi+t),(r_{00}+\frac{\partial\tilde{v}}{\partial x})(x,y-|\xi|,\xi+t)\right]\Big\}d\xi, \tag{7.67}$$

where

$$\frac{\partial w_0}{\partial t}(x,y,t)=\frac{1}{2}\Big\{\frac{a}{2}(\frac{\partial k_0}{\partial t}(x,y,t)(x,t+y)+\frac{\partial k_0}{\partial t}(x,y,t)(x,t-y))$$
$$+g_0(x,0)[k_0(x,t+y)+k_0(x,t-y)]$$
$$+\frac{\partial\tilde{g}_0}{\partial t}(x,y,t)(x,t+y)+\frac{\partial\tilde{g}_0}{\partial t}(x,t-y)$$
$$+\int_0^{t+y} k_0(x,\tau)\frac{\partial\tilde{g}_0}{\partial t}(x,t+y-\tau)d\tau+\int_0^{t-y} k_0(x,\tau)\frac{\partial\tilde{g}_0}{\partial t}(x,t-y-\tau)d\tau\Big\}.$$

To make the system closed (7.64), (7.66), (7.67) we use the following obvious equality:

$$r_0(x,t)=r(x,0)+\frac{\partial r}{\partial t}(x,0)t$$
$$+\int_0^t (y-\xi)(h(x,\xi)+r(x,0)\frac{\partial r_0}{\partial t}(x,\xi)-\frac{r^2(x,0)}{4}r_0(x,\xi))d\xi, \tag{7.68}$$

$$\frac{\partial r_0}{\partial t}(x,t)=\frac{\partial r}{\partial t}(x,0)+\int_0^t\left(h(x,\xi)+r(x,0)\frac{\partial r_0}{\partial t}(x,\xi)-\frac{r^2(x,0)}{4}r_0(x,\xi)\right)d\xi, \tag{7.69}$$

$$k_0(x,t)=-r(x,0)+\Big(\frac{r^2(x,0)}{2}-\frac{\partial r}{\partial t}(x,0)\Big)t+\int_0^t(t-\xi)\frac{\partial^2 k_0}{\partial t^2}(x,\xi)d\xi, \tag{7.70}$$

$$\frac{\partial k_0}{\partial t}(x,t)=\frac{r^2(x,0)}{2}-\frac{\partial r}{\partial t}(x,0)+\int_0^t\frac{\partial^2 k_0}{\partial t^2}(x,\xi)d\xi, \tag{7.71}$$

$$\frac{\partial^2 k_0}{\partial t^2}(x,t) = -\,h(x,t) + \Big(\frac{r^2(x,0)}{4} - \frac{\partial r}{\partial t}(x,0)\Big)k_0(x,t) - \int_0^t h(x,t-\xi)k_0(x,\xi)\mathrm{d}\xi, \tag{7.72}$$

following from the above introduced notation for k, r, k_0, r_0 and equation (7.55), we find

$$\tilde{v}(x,y,t) = \exp\left(k(x,0)t/2\right) L_0\left[r_0, w\right], \tag{7.73}$$

after differentiating by t:

$$\frac{\partial \tilde{v}}{\partial t}(x,y,t) = \exp\left(k(x,0)t/2\right) \times \left(\frac{k(x,0)}{2} L_0\left[r_0, w\right] + L_0\left[r_0, w_t\right] + r_0(x,t-y)\beta(x,y)\right). \tag{7.74}$$

Let $c_0(x) := \frac{r^2(x,0)}{4} - \frac{\partial r}{\partial t}(x,0)$. In further material, the system (7.57)–(7.60) will be considered in the domain

$$D_T = G_T \times \mathbb{R},\ G_T = \{(y,t) \mid 0 \le y \le t \le T-y\},\ T > 0.$$

Now we introduce a vector function φ with components

$$\varphi_1(x,y,t) = w(x,y,t) - \frac{1}{2}\Big\{\frac{a}{2}(k_0(x,t+y) + k_0(x,t-y)) + \int_0^{t+y} k_0(x,\tau)\tilde{g}_0(x,t+y-\tau)\mathrm{d}\tau + \int_0^{t-y} k_0(x,\tau)\tilde{g}_0(x,t-y-\tau)\mathrm{d}\tau\Big\},$$

$$\varphi_2(x,y,t) = \tilde{v}(x,y,t),$$

$$\varphi_3(x,y,t) = \frac{\partial w}{\partial t}(x,y,t) - \frac{1}{2}\Big\{\frac{a}{2}\Big(\frac{\partial k_0}{\partial t}(x,t+y) + \frac{\partial k_0}{\partial t}(x,t-y)\Big) + g_0(x,0)[k_0(x,t+y) + k_0(x,t-y)] + \int_0^{t+y} k_0(x,\tau)\frac{\partial \tilde{g}_0}{\partial t}(x,t+y-\tau)\mathrm{d}\tau + \int_0^{t-y} k_0(x,\tau)\frac{\partial \tilde{g}_0}{\partial t}(x,t-y-\tau)\mathrm{d}\tau\Big\},$$

$$\varphi_4(x,y,t) = \frac{\partial \tilde{v}}{\partial t}(x,y,t),$$

$$\varphi_5(x,y) = 2h(x,y) - c_0(x)k_0(x,y) - \frac{2}{a}\left[g_0(x,0)\frac{\partial k_0}{\partial t}(x,y) + k_0(x,y)\frac{\partial \tilde{g}_0}{\partial t}(x,0)\right],$$

$$\varphi_6(x,y) = r_0(x,y), \quad \varphi_7(x,y) = \frac{\partial r_0}{\partial t}(x,y),$$

$$\varphi_8(x,y) = k_0(x,y), \quad \varphi_9(x,y) = \frac{\partial k_0}{\partial t}(x,y)$$

and the vector function φ^0 with components

$$\varphi_1^0(x,y,t) = \frac{1}{2}\Big[\tilde{g}_0(x,t+y) + \tilde{g}_0(x,t-y)\Big],$$

$$\varphi_3^0(x,y,t) = \frac{1}{2}\Big[\frac{\partial \tilde{g}_0}{\partial t}(x,t+y) + \frac{\partial \tilde{g}_0}{\partial t}(x,t-y)\Big],$$

$$\varphi_5^0(x,y) = \frac{2}{a}\Big[\frac{\partial^2 \tilde{g}_0}{\partial t^2}(x,y) - \frac{\partial^2}{\partial y^2}\beta(x,y/2) + H(x,y/2)\beta(x,y/2) + k_{00}(x,y/2,y/2)$$

$$+\nu^2 \exp(r(x,0)y/4)\Big(\frac{\partial r_{00}}{\partial x}(x,y/2,y/2) + \frac{\partial^2 \beta}{\partial x^2}(x,y/2)\Big)\Big],$$

$$\varphi_6^0(x,y) = r(x,0) + \frac{\partial r}{\partial t}(x,0)y, \quad \varphi_7^0(x) = r_t(x,0),$$

$$\varphi_8^0(x,y) = -r(x,0) + \Big(\frac{r^2(x,0)}{2} - \frac{\partial r}{\partial t}(x,0)\Big)y,$$

$$\varphi_9^0(x) = \frac{r^2(x,0)}{2} - \frac{\partial r}{\partial t}(x,0).$$

Excluding from the equation (7.66) the unknown function $k_{0tt}(x,y)$ (this function is included in the h_0) using (7.72), we rewrite the system of equations (7.64), (7.66)–(7.74) in terms of vector functions φ and φ^0

$$\varphi = \varphi^0 + \tilde{A}\varphi, \tag{7.75}$$

with nonlinear operator $\tilde{A}$, which is constructed similarly to the operator A (Sect. 7.1). Note that

$$w(x, y, t) = \varphi_1(x, y, t) + \frac{1}{2}\Big\{\frac{a}{2}(\varphi_8(x, t + y) + \varphi_8(x, t - y))$$
$$+ \int_0^{t+y} \varphi_8(x, \tau)\tilde{g}_0(x, t + y - \tau)d\tau + \int_0^{t-y} \varphi_8(x, \tau)\tilde{g}_0(x, t - y - \tau)d\tau\Big\},$$
$$w_t(x, y, t) = \varphi_3(x, y, t) + \frac{1}{2}\Big\{\frac{a}{2}(\varphi_9(x, t + y) + \varphi_9(x, t - y))$$
$$+ g_0(x, 0)[\varphi_8(x, t + y) + \varphi_8(x, t - y)]$$
$$+ \int_0^{t+y} \varphi_8(x, \tau)\frac{\partial \tilde{g}_0}{\partial t}(x, t + y - \tau)d\tau + \int_0^{t-y} \varphi_8(x, \tau)\frac{\partial \tilde{g}_0}{\partial t}(x, t - y - \tau)d\tau\Big\},$$
$$h(x, y) = \frac{1}{2}[\varphi_5(x, y) + c_0(x)\varphi_8(x, y)]$$
$$+ \frac{1}{a}\left[g_0(x, 0)\varphi_9(x, y) + \varphi_8(x, y)\frac{\partial \tilde{g}_0}{\partial t}(x, 0)\right]. \tag{7.76}$$

As in Sect. 2.1 let us consider the Banach function space $A_s(r)$, $s > 0$, analytical in the vicinity of the origin point. If $s' \in (0, s)$ and the following inequality holds

$$\left\| \frac{\partial^\alpha}{\partial x^\alpha}\varphi(x)\right\|_{s'} \leq \alpha^\alpha \frac{\|\varphi\|_s(r)}{(s - s')^\alpha} \tag{7.77}$$

for all α.

The main result of this work is a theorem of local unique solvability in the class of analytical functions with the variable x.

Theorem 7.3 *Let* $(\rho, \mu, \lambda) \in \Lambda$, $\left(\tilde{g}_0(x, 0), \frac{\partial \tilde{g}_0}{\partial t}(x, 0)\right) \in A_{s0}$,

$$\left(\tilde{g}_0(x, t), \frac{\partial \tilde{g}_0}{\partial t}(x, t), \frac{\partial^2 \tilde{g}_0}{\partial t^2}(x, t)\right) \in C_t\left([0, T];\ A_{s0}\right),$$

$$\max\left\{\|\tilde{g}_0\|_{s0}(t), \left\|\frac{\partial \tilde{g}_0}{\partial t}\right\|_{s0}(t), \left\|\frac{\partial^2 \tilde{g}_0}{\partial t^2}\right\|_{s0}(t)\right\} \leq R,$$

$$\max\left\{\|\beta\|_{s0}(y),\ \|H\|_{s0}(y),\ \max(1, \nu_0^2)\|\exp(\pm r(x, 0)t/2)\|_{s0}(t)\right\} \leq R,$$

$$\max\left\{\|\varphi_i^0\|_{s0}(t)\right\} \leq R,\ i = 5, 6, 7, 8, 9,$$

$$\nu_0^2 = \max_{0 \leq y \leq \frac{T}{2}} \nu^2(\phi^{-1}(y)),\ t \in [0, T],\ y \in [0, T/2],\ R > 0.$$

Then there is $b \in (0, T/2)$, *that for any* $s \in (0, s_0)$ *in the domain* $\Gamma_{sT} = D_T \bigcap \{(x, y, t)|\, 0 \leq y \leq b(s_0 - s)\}$ *there is a unique solution of the system of equations (7.75), for which*

$$\varphi_1(x, y, t),\ \varphi_3(x, y, t) \in C_{(y,t)}\left(P_{sT};\ A_{s0}\right),$$

$$\varphi_i(x,t) \in C_t\left([0,\ b(s_0-s)];\ A_{s0}\right),\ i=5,6,7,8,9,$$

$$P_{sT} = G_T \bigcap \{(y,t)|\, 0 \le y \le b(s_0-s)\},$$

with

$$\|\varphi_i - \varphi_i^0\|_s \le R,\ i=1,6,8,\ \|\varphi_i - \varphi_i^0\|_s \le \frac{R}{s_0-s},\ i=3,7,9,$$

$$\|\varphi_5 - \varphi_5^0\|_s \le \frac{R}{(s_0-s)^2},\ (y,t) \in P_{sT}.$$

Under the conditions of the Theorem 7.3

$$\varphi_i^0 \in C_{(y,t)}(G_T;\ A_{s0}),\ \|\varphi_i^0\|_s(y,t) \le R,\ i=1,3,\ (y,t) \in G_T,\ 0 < s < s_0.$$

Let b_n be members of a monotonically decreasing sequence (2.17) and the number b is defined by equation (2.18). The number $b_0 \in (0, T/2)$ will be chosen appropriately.

We construct a process of successive approximations with $n = 0, 1, 2, ...$ by the scheme

$$\varphi^{n+1} = \varphi^0 + \widetilde{A}\varphi^n.$$

We define the function $s'(y)$ by the formula

$$s'(y) = \frac{s+\gamma^n(y)}{2},\quad \gamma^n(y) = s_0 - \frac{y}{b_n}. \tag{7.78}$$

Let $\psi^n = \varphi^{n+1} - \varphi^n$. Then the following relations are true:

$$\psi^0 = \widetilde{A}\varphi^0,$$

$$\psi_1^{n+1}(x,y,t)$$
$$= \frac{1}{2}\int_0^y \int_{t-y+\xi}^{t+y-\xi} \Bigg\{ H(x,\xi)\hat{w}^n(x,\xi,\tau) - \frac{a}{2}\hat{h}^n(x,\tau-\xi) + \hat{k}_{00}^n(x,\xi,\tau)$$
$$- \int_0^{\tau-\xi} \left(\hat{h}^n(x,\gamma)w^{n+1}(x,\xi,\tau-\gamma) - h^n(x,\gamma)\hat{w}^n(x,\xi,\tau-\gamma)\right) d\gamma$$
$$+ \nu^2 \exp(r(x,0)\tau/2)\,\frac{\partial}{\partial x}\Big\{ L_0[\varphi_8^{n+1}(x,\tau)\exp(-r(x,0)(\tau)/2),$$
$$\left(\hat{r}_{00}^n + \frac{\partial \psi_2^n}{\partial x}\right)(x,\xi,\tau)]$$
$$- \int_0^{\tau-\xi} \psi_8^n(x,\gamma)\exp(-r(x,0)\gamma/2)\left(r_{00}^n + \frac{\partial \varphi_2^n}{\partial x}\right)(x,\xi,\tau-\gamma)d\gamma \Big\}\Bigg\} d\tau d\xi,$$

$$
\begin{aligned}
&\psi_3^{n+1}(x, y, t) \\
&= \frac{1}{2}\int_{-y}^{y} \operatorname{sgn}(\xi)\Big\{H(x, y-|\xi|)\hat{w}^n(x, y-|\xi|, \xi+\tau) \\
&\quad - \frac{a}{2}\hat{h}^n(x, \xi+t-y+|\xi|) \\
&\quad + \hat{k}_{00}^n(x, y-|\xi|, t+|\xi|) - \int_0^{t-y+\xi+|\xi|} \Big(\hat{h}^n(x,\gamma)w^{n+1}(x, y-|\xi|, t+\xi-\gamma) \\
&\quad - h^n(x,\gamma)\hat{w}^n(x, y-|\xi|, t+\xi-\gamma)\Big)\mathrm{d}\gamma + \nu^2 \exp\left(r(x,0)(\xi+t)/2\right) \\
&\quad \times \frac{\partial}{\partial x}\Big\{L_0[\varphi_8^{n+1}(x, \xi+t)\exp\left(-r(x,0)(\xi+t)/2\right), \left(\hat{r}_{00}^n + \frac{\partial \psi_2^n}{\partial x}\right)(x, y-|\xi|, \xi+t)] \\
&\quad - \int_0^{t-y+\xi+|\xi|} \psi_8^n(x,\gamma)\exp(-r(x,0)\gamma/2)\left(r_{00}^n + \frac{\partial \varphi_2^n}{\partial x}\right)(x, y-\xi, \xi+t-\gamma)\mathrm{d}\gamma\Big\}\Big\}\mathrm{d}\xi, \\
&\psi_5^{n+1}(x, y) \\
&= \int_0^y \Big(\frac{2}{a}\psi_8^n(x,\xi)\tilde{g}_{0tt}(x, y-\xi) - \hat{h}^n(x, y-\xi))\varphi_8^{n+1}(x,\xi) \\
&\quad + h^n(x, y-\xi))\psi_8^n(x,\xi)\Big)\mathrm{d}\xi + \frac{1}{a}\int_0^{y/2}\Big\{H(x,\xi)\hat{w}_t^n(x,\xi,y-\xi) + \hat{k}_{00t}^n(x,\xi,y-\xi) \\
&\quad - \frac{1}{2}\hat{h}^n(x, y-2\xi)\beta(x,\xi) - \int_0^{y-2\xi}\Big(\hat{h}^n(x,\tau)w_t^{n+1}(x,\xi,y-\xi-\tau) \\
&\quad - h^n(x,\tau)\hat{w}_t^n(x,\xi,y-\xi-\tau)\Big)\mathrm{d}\tau + \nu^2\exp\left(r(x,0)(y-\xi)/2\right) \\
&\quad \times \Big\{\frac{\partial}{\partial x}\Big\{L_0[\varphi_8^{n+1}(x, y-\xi)\exp\left(-r(x,0)(y-\xi)/2\right), \left(\hat{r}_{00t}^n + \frac{\partial \psi_4^n}{\partial x}\right)(x,\xi,y-\xi)] \\
&\quad - \int_0^{y-2\xi}\psi_8^n(x,\tau)\exp\left(-r(x,0)\tau/2\right)\left(r_{00t}^n + \frac{\partial \varphi_4^n}{\partial x}\right)(x,\xi,y-\xi-\tau)\mathrm{d}\tau\Big\} \\
&\quad + \frac{r(x,0)}{2}\frac{\partial}{\partial x}\Big\{L_0[\varphi_8^{n+1}(x,\tau)\exp\left(-r(x,0)(y-\xi)/2\right), \left(\hat{r}_{00}^n + \frac{\partial \psi_2^n}{\partial x}\right)(x,\xi,y-\xi)] \\
&\quad - \int_0^{y-2\xi}\psi_8^n(x,\gamma)\exp(-r(x,0)\tau/2)\left(r_{00}^n + \frac{\partial \varphi_2^n}{\partial x}\right)(x,\xi,y-\xi-\tau)\mathrm{d}\gamma\Big\} \\
&\quad + \frac{1}{2}\frac{\partial}{\partial x}\Big[\exp\left(-r(x,0)(y-2\xi)/2\right)\Big(\varphi_8^{n+1}(x, y-2\xi)\left(\hat{r}_{00}^n + \frac{\partial \psi_2^n}{\partial x}\right)(x,\xi,\xi) \\
&\quad - \psi_8^n(x, y-2\xi)\left(r_{00}^n + \frac{\partial \varphi_2^n}{\partial x}\right)(x,\xi,\xi)\Big)\Big]\Big\}\mathrm{d}\xi,
\end{aligned}
$$

$$\psi_6^{n+1}(x,y) = \int_0^y (y-\xi)\left(\hat{h}^n(x,\xi) + r(x,0)\psi_7^n(x,\xi) - \frac{r^2(x,0)}{4}\psi_6^n(x,\xi)\right)\mathrm{d}\xi,$$

$$\psi_7^{n+1}(x,y) = \int_0^y \left(\hat{h}^n(x,\xi) + r(x,0)\psi_7^n(x,\xi) - \frac{r^2(x,0)}{4}\psi_6^n(x,\xi)\right)\mathrm{d}\xi,$$

$$\psi_8^{n+1}(x,y) = \int_0^y (y-\xi)$$

$$\times \left(c_0(x)\psi_8^n(x,\xi) - L[\varphi_8^n, \hat{h}^n] - \int_0^\xi \psi_8^n(x,\xi-\tau)h^{n+1}(x,\tau)\mathrm{d}\tau\right)\mathrm{d}\xi,$$

$$\psi_9^{n+1}(x,y) = \int_0^y (c_0(x)\psi_8^n(x,\xi) - L[\varphi_8^n, \hat{h}^n] - \int_0^\xi \psi_8^n(x,\xi-\tau)h^{n+1}(x,\tau)\mathrm{d}\tau)\mathrm{d}\xi,$$

where $\hat{k}_{00}^n(x,y,t)$, $\hat{r}_{00}^n(x,y,t)$, $\hat{w}^n(x,y,t)$, $\hat{w}_t^n(x,y,t)$, $\hat{h}^n(x,y)$ have the form according to (7.61), (7.62), (7.76) in which the corresponding components of the vector function ψ are supplied.

We show that one can choose $b_0 \in (0, T/2)$ so that for $\forall n = 0, 1, 2, \ldots$ inequalities are satisfied

$$\begin{aligned}
\lambda_n = \max\Bigg\{ &\sup_{(y,t,s)\in F_n}\left[\|\psi_1^n\|_s(y,t)\frac{\gamma^n(y)-s}{y}\right],\\
&\sup_{(y,t,s)\in F_n}\left[\|\psi_3^n\|_s(y,t)\frac{(\gamma^n(y)-s)^2}{y}\right],\\
&\sup_{(y,t,s)\in F_n}\left[\|\psi_5^n\|_s(y)\frac{(\gamma^n(y)-s)^3}{y}\right], \quad \sup_{(y,t,s)\in F_n}\left[\|\psi_{6,8}^n\|_s(y)\frac{\gamma^n(y)-s}{y}\right],\\
&\sup_{(y,t,s)\in F_n}\left[\|\psi_{7,9}^n\|_s(y)\frac{(\gamma^n(y)-s)^2}{y}\right]\Bigg\} < \infty,
\end{aligned} \tag{7.79}$$

$$\begin{aligned}
&\|\varphi_1^{n+1} - \varphi_1^0\|_s(y,t) \le R, \ \|\varphi_i^{n+1} - \varphi_i^0\|_s(y) \le R, \ i = 6, 8,\\
&\|\varphi_3^{n+1} - \varphi_3^0\|_s(y,t) \le \frac{R}{s_0 - s}, \ \|\varphi_i^{n+1} - \varphi_i^0\|_s(y) \le \frac{R}{s_0 - s}, \ i = 7, 9,\\
&\|\varphi_5^{n+1} - \varphi_5^0\|_s(y) \le \frac{R}{(s_0 - s)^2}, \ (y,t,s) \in F_{n+1},
\end{aligned} \tag{7.80}$$

where

$$F_n = \{(y,t,s) | (y,t) \in G_T, 0 \le y \le b_n(s_0 - s), 0 < s < s_0\}.$$

Indeed, using the relations for ψ_i^n at $n = 0$ we obtain:

$$\|\psi_1^0\|_s(y) \le \int_0^y (y-\xi)\Big\{R\|w^0\|_s + \frac{a}{2}\|h^0\|_s + \|k_{00}^0\|_s + T\|h^0\|_s\|w^0\|_s$$

$$+R\|\frac{\partial}{\partial x}L_0\left[\varphi_8^0(x,\tau)\exp(-r(x,0)\tau/2), \left(r_{00}^0 + \frac{\partial\varphi_2^0}{\partial x}\right)\right]\|_s(\xi,\tau)\Big\}\mathrm{d}\xi,$$

where

$$\|w^0\|_s \le R\left(1 + \frac{a}{2} + 2TR\right), \ \|h^0\|_s \le R\left(\frac{1}{2} + R\left(1 + \frac{4}{a}\right)\right),$$

$$\|\varphi_2^0\|_s(y,t) \le R(1+RT)\|w^0\|_s \le R(1+RT)R\left(\frac{1}{2} + R\left(1 + \frac{4}{a}\right)\right),$$

$$\|k_{00}^0\|_s(y,t) \le \frac{a}{2}R^2\left(R\left(1 + \frac{T}{2}\right) + \|\frac{\partial\varphi_8^0}{\partial x}\|_s\right), \quad s \in (0, s_0), \ (y,t) \in G_T.$$

Let us use inequality (2.35) to estimate $\|\frac{\partial\varphi_8^0}{\partial x}\|_s$, $\|r_{00}^0\|_s$, $\|\frac{\partial\varphi_2^0}{\partial x}\|_s$, $\|\frac{\partial}{\partial x}L_0\|_s$:

$$\|\frac{\partial\varphi_8^0}{\partial x}\|_s(y) \le \frac{R}{s'(y)-s}, \quad \|r_{00}^0\|_s(y,t) \le \frac{R^2}{s'(y)-s},$$

$$\|\frac{\partial}{\partial x}L_0\left[\varphi_8^0\exp(-r(x,0)(t \pm y \mp \xi)/2), r_{00}^0 + \frac{\partial\varphi_2^0}{\partial x}\right]\|_s(y,t)$$

$$\le \frac{1}{s'(y)-s}\left(\|r_{00}^0\|_{s'} + \frac{\|\varphi_2^0\|_{s'}}{s'(y)-s} + 2TR^2\left(\|r_{00}^0\|_{s'} + \frac{\|\varphi_2^0\|_{s'}}{s'(y)-s}\right)\right)$$

$$\le \frac{R^2 + \|\varphi_2^0\|_{s'}}{(s'(y)-s)^2}(1+2TR^2).$$

Taking the function $s'(\xi)$ from (7.78) for $n = 0$, taking into account the above inequalities we have:

$$\|\psi_1^0\|_s(y,t) \le \mu_0(a,T,R,s_0)\int_0^y \frac{y-\xi}{(\gamma^0(y)-s)^2}\mathrm{d}\xi \le b_0\mu_0(a,T,R,s_0)\frac{y}{\gamma^0(y)-s}.$$

Similarly,

$$\|\psi_3^0\|_s(y,t) \le \frac{1}{2}\int_{-y}^{y} \frac{\mu_1(a,T,R,s_0)}{(\gamma^0(y)-s)^2}\mathrm{d}\xi \le \mu_1(a,T,R,s_0)\frac{y}{(\gamma^0(y)-s)^2},$$

$$\|\psi_5^0\|_s(y) \le \frac{a}{2}\int_0^y R\left(\frac{2}{a}R + \|h^0\|_s\right)d\xi$$

$$+\frac{1}{a}\int_0^{y/2}\Big\{R\|w_t^0\|_s + \frac{1}{2}R\|h^0\|_s + \|k_{00t}^0\|_s + T\|h^0\|_s\|w_t^0\|_s$$

$$+R\Big\{\|\frac{\partial}{\partial x}L_0\left[\varphi_8^0(x, y-\xi)\exp(-r(x,0)(y-\xi)/2), r_{00t}^0 + \frac{\partial\varphi_4^0}{\partial x}\right]\|_s$$

$$+R\|\frac{\partial}{\partial x}L_0[\varphi_8^0(x, y-\xi)\exp(-r(x,0)(y-\xi)/2), r_{00}^0 + \frac{\partial\varphi_2^0}{\partial x}]\|_s$$

$$+\frac{1}{2}\|\frac{\partial}{\partial x}\left(\varphi_8^0(x, y-\xi)\exp(-r(x,0)(y-\xi)/2)\left(r_{00}^0 + \frac{\partial\varphi_2^0}{\partial x}\right)\right)\|_s\Big\}\Big\}d\xi,$$

where

$$\|w_t^0\|_s \le R\left(1+\frac{a}{2}+R(1+T)\right), \quad \|\varphi_4^0\|_s \le R(1+TR)(R\|w^0\|_s + \|w_t^0\|_s) + R^3,$$

$$\|k_{00t}^0\|_s(y,t) \le R\|k_{00t}^0\|_s + \frac{a}{2}R^2\left(R(1+TR) + R^2 + \|\frac{\partial\varphi_9^0}{\partial x}\|_s(y)\right),$$

$$\|r_{00t}^0\|_s(y,t) \le \frac{R^2(1+R)}{s'(y)-s}, \quad \|\frac{\partial\varphi_9^0}{\partial x}\|_s(y) \le \frac{R}{s'(y)-s}, \quad s\in(0,s_0),\ (y,t)\in G_T.$$

It follows from the above estimates that

$$\|\psi_5^0\|_s(y) \le \mu_2(a,T,R,s_0)\frac{y}{(\gamma^0(y)-s)^2} \le s_0\mu_2(a,T,R,s_0)\frac{y}{(\gamma^0(y)-s)^3}.$$

Doing the same, we have

$$\|\psi_6^0\|_s(y) \le \int_0^y (y-\xi)(\|h^0\|_s + 2R^2)d\xi$$

$$\le s_0RT\left(1/2 + 2R + R\left(1+\frac{4}{a}\right)\right)\frac{y}{\gamma^0(y)-s},$$

$$\|\psi_7^0\|_s(y) \le \int_0^y (\|h^0\|_s + 2R^2)d\xi \le s_0^2R\left(1/2+2R+R\left(1+\frac{4}{a}\right)\right)\frac{y}{(\gamma^0(y)-s)^2},$$

$$\|\psi_8^0\|_s(y) \le \int_0^y (y-\xi)(R^2 + \|h^0\|_s(1+TR))\mathrm{d}\xi$$

$$\le s_0 RT\left(R + \left(1/2 + R\left(1+\frac{4}{a}\right)\right)(1+TR)\right)\frac{y}{\gamma^0(y)-s},$$

$$\|\psi_9^0\|_s(y) \le \int_0^y (R^2 + R\left(1/2 + R\left(1+\frac{4}{a}\right)\right)(1+TR))\mathrm{d}\xi$$

$$\le s_0^2 R(R + (1/2 + R(1+\frac{4}{a}))(1+TR))\frac{y}{\gamma^0(y)-s}.$$

These estimates imply the fulfillment of the inequalities (7.79) with $n = 0$. In addition, for $(y, t, s) \in F_1$ we find

$$\|\varphi_1^1 - \varphi_1^0\|_s(y,t) = \|\psi_1^0\|_s(y,t) \le \frac{\lambda_0 y}{\gamma^0(y)-s} \le \frac{\lambda_0 b_1}{1-b_1/b_0} = \lambda_0 b_0,$$

$$\|\varphi_3^1 - \varphi_3^0\|_s(y,t) = \|\psi_3^0\|_s(y,t) \le \frac{\lambda_0 y}{(\gamma^0(y)-s)^2} \le \frac{2\lambda_0 b_0}{s_0-s},$$

$$\|\varphi_5^1 - \varphi_5^0\|_s(y) = \|\psi_5^0\|_s(y,t) \le \frac{\lambda_0 y}{(\gamma^0(y)-s)^3} \le \frac{4\lambda_0 b_0}{(s_0-s)^3},$$

$$\|\varphi_6^1 - \varphi_6^0\|_s(y) = \|\psi_6^0\|_s(y,t) \le \frac{\lambda_0 y}{\gamma^0(y)-s} \le \frac{\lambda_0 b_1}{1-b_1/b_0} = \lambda_0 b_0,$$

$$\|\varphi_7^1 - \varphi_7^0\|_s(y) = \|\psi_7^0\|_s(y,t) \le \frac{\lambda_0 y}{(\gamma^0(y)-s)^2} \le \frac{2\lambda_0 b_0}{s_0-s},$$

$$\|\varphi_8^1 - \varphi_8^0\|_s(y) = \|\psi_8^0\|_s(y,t) \le \frac{\lambda_0 y}{\gamma^0(y)-s} \le \frac{\lambda_0 b_1}{1-b_1/b_0} = \lambda_0 b_0,$$

$$\|\varphi_9^1 - \varphi_9^0\|_s(y) = \|\psi_9^0\|_s(y,t) \le \frac{\lambda_0 y}{(\gamma^0(y)-s)^2} \le \frac{2\lambda_0 b_0}{s_0-s}.$$

If we choose b_0 so that $4\lambda_0 b_0 \le R$, the inequalities (7.80) will be satisfied for $n = 0$. We show by induction that inequalities (7.79), (7.80) also occur for other n if b_0 is chosen appropriately. Let inequalities (7.79), (7.80) be true for $n = 0, 1, 2, \ldots, j$. Then, for $(y, t, s) \in F_{j+1}$

$$\|\psi_1^{j+1}\|_s(y,t) \le \int_0^y (y-\xi)\Big\{R\|\hat{w}^j\|_s + \frac{a}{2}\|\hat{h}^j\|_s + \|\hat{k}_{00}^j\|_s$$
$$+T(\|\hat{h}^j\|_s\|w^{j+1}\|_s + \|h^j\|_s\|\hat{w}^j\|_s)$$
$$+R\Big\{\|\frac{\partial}{\partial x}L_0[\varphi_8^{j+1}\exp(-r(x,0)t/2), \hat{r}_{00}^j + \frac{\partial\psi_2^j}{\partial x}]\|_s +$$
$$T\|\frac{\partial}{\partial x}\left[\psi_8^j \exp(-r(x,0)t/2)\left(r_{00}^j + \frac{\partial\varphi_2^j}{\partial x}\right)\right]\|_s\Big\}\Big\}d\xi.$$

It's easy to see that

$$\|\hat{w}^j\|_s(y,t) \le \|\psi_1^j\|_s + \frac{1}{2}(a+TR)\|\psi_8^j\|_s \le (1+(a+TR)/2)\frac{\lambda_j y}{\gamma^j(y)-s},$$

$$\|\hat{h}^j\|_s(y) \le \frac{1}{2}\|\psi_5^j\|_s + R\|\psi_8^j\|_s + \frac{2}{a}R(\|\psi_9^j\|_s + \|\psi_8^j\|_s)$$

$$\le (1/2 + R(1+2/a)s_0 + 2Rs_0)\frac{\lambda_j y}{(\gamma^j(y)-s)^3},$$

$$\|\hat{k}_{00}^j\|_s(y,t) \le \frac{a}{2}R^2\left((1+TR/2)\|\psi_8^j\|_s + \frac{\|\psi_8^j\|_s}{s'(y)-s}\right)$$

$$\le \frac{a}{2}R^2(3+TR/2)\frac{\lambda_j y}{(\gamma^j(y)-s)^2},$$

$$\|\hat{r}_{00}^j\|_s(y,t) \le R\frac{\lambda_j y}{(s'(y)-s)(\gamma^j(y)-s)},$$

$$\|\psi_2^j\|_s(y,t) \le R(\|\hat{w}^j\|_s(1+2RT) + T\|\psi_6^j\|_s\|w^j\|_s) \le$$

$$R((1+2RT)(1+a/2+TR/2) + TR(2+a+2TR))\frac{\lambda_j y}{\gamma^j(y)-s},$$

$$\|\frac{\partial}{\partial x}L_0\left[\varphi_8^{j+1}\exp(-r(x,0)t/2), \hat{r}_{00}^0 + \frac{\partial\psi_2^j}{\partial x}\right]\|_s(y,t)$$

$$\le \frac{1}{s'(y)-s}(1+TR)\left(\|\hat{r}_{00}^j\|_s + \frac{\|\psi_2^j\|_s}{s'(y)-s}\right) \le \mu_3(a,T,R)\frac{\lambda_j y}{(\gamma^j(y)-s)^3},$$

where

$$\|w^j\|_s(y,t) \le \|\varphi_1^j\|_s + \frac{a}{2}\|\varphi_8^j\|_s + TR\|\varphi_8^j\|_s \le R(2+a+2TR).$$

Here and in further intermediate calculations the function s' is determined by the equality (7.78) at $n = j$ and the following inequalities are used

$$\|\varphi_i^j\|_s \le 2R,\ i = 1, 6, 8,\ \|\varphi_5^j\|_s \le R\frac{1+s_0^2}{(s_0-s)^2},\ \|\varphi_i^j\|_s \le R\frac{1+s_0}{s_0-s},\ i = 3, 7, 9,$$

fair under the inductive assumption, as well as obvious inequalities $b_j \le b_0$, $\gamma^{j+1}(y) \le \gamma^j(y)$, $\|\varphi_2^j\|_s(y,t) \le R(1+2RT)\|w^j\|_s \le R^2(1+2RT)(2+a+2TR)$,

$$\|\varphi_4^j\|_s \le R(R\|w^j\|_s(1+2TR) + \|w_t\|_s(1+2RT) + 2R^2) \le \frac{\mu_4(a,T,R,s_0)}{s_0-s},$$

$$\|w_t^j\|_s \le \frac{R(1+s_0)(1+aR/2) + 2R^2 s_0(1+T)}{s_0-s}.$$

Finally, for ψ_1^{j+1} we get

$$|\psi_1^{j+1}\|_s(y,t) \le \int_0^y (y-\xi)\mu_5(a,T,R,s_0)\frac{\lambda_j\xi}{(\gamma^j(\xi)-s)^3}d\xi$$
$$\le \lambda_j b_0^2\mu_5(a,T,R,s_0)\frac{y}{\gamma^{j+1}(y)-s}.$$

Similar arguments for ψ_i^{j+1}, $i = 3, 5, 6, 7, 8, 9$, give us the following inequalities

$$\|\psi_3^{j+1}\|_s(y,t) \le \frac{1}{2}\int_{-y}^{y} \mu_6(a,T,R,s_0)\frac{\lambda_j\xi}{(\gamma^j(\xi)-s)^3}d\xi$$
$$\le \lambda_j b_0\mu_6(a,T,R,s_0)\frac{y}{(\gamma^{j+1}(y)-s)^2},$$

$$\|\psi_5^{j+1}\|_s(y,t) \le \int_0^y \mu_7(a,T,R,s_0)\frac{\lambda_j\xi}{(\gamma^j(\xi)-s)^4}d\xi$$
$$\le \lambda_j b_0\mu_7(a,T,R,s_0)\frac{y}{(\gamma^{j+1}(y)-s)^3},$$

$$\|\psi_6^{j+1}\|_s(y) \le \int_0^y (y-\xi)$$
$$\times\left(\frac{\lambda_j\xi(1/2+R(1+2/a)s_0+2Rs_0)}{(\gamma^j(y)-s)^3} + \frac{\lambda_j\xi R}{(\gamma^j(\xi)-s)^2} + \frac{\lambda_j\xi R}{\gamma^j(\xi)-s}\right)d\xi$$
$$\le \lambda_j\mu_8(b_0,a,R,s_0)\frac{y}{(\gamma^{j+1}(y)-s)},$$

$$\|\psi_7^{j+1}\|_s(y)$$
$$\leq \int_0^y \left(\frac{\lambda_j \xi (1/2 + R(1+2/a)s_0 + 2Rs_0)}{(\gamma^j(y) - s)^3} + \frac{\lambda_j \xi R}{(\gamma^j(\xi) - s)^2} + \frac{\lambda_j \xi R}{\gamma^j(\xi) - s} \right) d\xi$$
$$\leq \lambda_j b_0^{-1} \mu_4 8(b_0, s_0, R) \frac{y}{(\gamma^{j+1}(y) - s)^2},$$

$$\|\psi_8^{j+1}\|_s(y) \leq \lambda_j \mu_9(b_0, s_0, R, T) \frac{y}{\gamma^{j+1}(y) - s},$$

$$\|\psi_9^{j+1}\|_s(y) \leq \lambda_j b_0^{-1} \mu_9(b_0, s_0, R, T) \frac{y}{(\gamma^{j+1}(y) - s)^2},\ (y, t, s) \in F_{i+1}.$$

From the estimates obtained, it follows that

$$\lambda_{i+1} \leq \lambda_i \rho,\ \lambda_{i+1} \leq \infty,\ \rho = \max(\mu_5, \mu_6, \mu_7, \max(1, b_0^{-1})\mu_8, \max(1, b_0^{-1})\mu_9)).$$

However, for $(y, t, s) \in F_{i+2}$ we have

$$\|\varphi_i^{j+2} - \varphi_i^0\|_s \leq \sum_{n=0}^{j+1} \|\psi_i^n\|_s \leq \sum_{n=0}^{j+1} \frac{\lambda_n y}{\gamma^n(y) - s}$$
$$\leq \sum_{n=0}^{j+1} \frac{\lambda_n b_{j+2}}{1 - b_{i+2}/b_n} \leq \sum_{n=0}^{j+1} \lambda_n b_n (n+1)^2 \leq \lambda_0 b_0 \sum_{n=0}^{j+1} \rho^n (n+1)^2,\ i = 1, 6, 8,$$

$$\|\varphi_5^{j+2} - \varphi_5^0\|_s(y) \leq \sum_{n=0}^{j+1} \|\psi_5^n\|_s(y)$$
$$\leq \sum_{n=0}^{j+1} \frac{\lambda_n y}{(\gamma^n(y) - s)^3} \leq \frac{1}{(s_0 - s)^2} \sum_{n=0}^{j+1} \frac{\lambda_n b_{j+2}}{(1 - b_{i+2}/b_n)^3} \leq \frac{\lambda_0 b_0}{(s_0 - s)^2} \sum_{n=0}^{j+1} \rho^n (n+1)^6,$$

$$\|\varphi_i^{j+2} - \varphi_i^0\|_s \leq \sum_{n=0}^{j+1} \|\psi_i^n\|_s \leq \sum_{n=0}^{j+1} \frac{\lambda_n y}{(\gamma^n(y) - s)^2}$$
$$\leq \frac{1}{s_0 - s} \sum_{n=0}^{j+1} \frac{\lambda_n b_{j+2}}{(1 - b_{i+2}/b_n)^2} \leq \frac{\lambda_0 b_0}{s_0 - s} \sum_{n=0}^{j+1} \rho^n (n+1)^4,\ i = 3, 7, 9.$$

Choose now $b_0 \in (0, T/(2s_0))$ such that

$$\rho \le 1,\ \lambda_0 b_0 \sum_{n=0}^{\infty} \rho^n (n+1)^6 \le R.$$

Then

$$\|\varphi_1^{j+2} - \varphi_1^0\|_s(y,t) \le R,\ \|\varphi_{6,8}^{j+2} - \varphi_{6,8}^0\|_s(y) \le R,$$

$$\|\varphi_5^{j+2} - \varphi_5^0\|_s(y) \le \frac{R}{(s_0 - s)^2},$$

$$\|\varphi_3^{j+2} - \varphi_3^0\|_s(y,t) \le \frac{R}{s_0 - s},\ \|\varphi_{7,9}^{j+2} - \varphi_{7,9}^0\|_s(y) \le \frac{R}{s_0 - s},\ (y,t,s) \in F_{j+2}.$$

Since the choice of b_0 does not depend on the number of approximations, successive approximation φ_i^n, $i = 1, 3, 5, 6, 7, 8, 9$ belong

$$C_{(y,t)}(F;\ A_s),\ F = \bigcap_{n=0}^{\infty} F_n$$

and for them we have the inequality

$$\|\varphi_1^n - \varphi_1^0\|_s(y,t) \le R,\ \|\varphi_{6,8}^n - \varphi_{6,8}^0\|_s(y) \le R,$$

$$\|\varphi_5^n - \varphi_5^0\|_s(y) \le \frac{R}{(s_0 - s)^2},$$

$$\|\varphi_3^n - \varphi_3^0\|_s(y,t) \le \frac{R}{s_0 - s},\ \|\varphi_{7,9}^n - \varphi_{7,9}^0\|_s(y) \le \frac{R}{s_0 - s},\ (y,t,s) \in F.$$

If $s \in (0, s_0)$, the series

$$\sum_{n=0}^{\infty} (\varphi_i^n - \varphi_i^{n-1})$$

converge uniformly in norm of space

$$C_{(y,t)}(P_{sT};\ A_s),\ P_{sT} = G_T \bigcap \{(y,t) |\, 0 \le y \le b(s_0 - s)\},$$

so $\varphi_i^n \to \varphi_i$. The limit function φ_i are elements of $C_{(y,t)}(P_{sT};\ A_s)$ and satisfy the equations (7.75).

Now we prove that the existing solution is unique. Let φ_i and $\bar{\varphi}_i$ be any two solutions satisfying inequalities

$$\|\varphi_1 - \varphi_1^0\|_s(y,t) \le R,\ \|\varphi_i - \varphi_i^0\|_s(y) \le R,\ i = 6, 8,\ \|\varphi_5 - \varphi_5^0\|_s(y) \le \frac{R}{(s_0 - s)^2},$$

$$\|\varphi_3 - \varphi_3^0\|_s(y,t) \le \frac{R}{s_0 - s},\ \|\varphi_i - \varphi_i^0\|_s(y) \le \frac{R}{s_0 - s},\ i = 7, 9,\ (y,t,s) \in F.$$

Let $\tilde{\psi}_i = \varphi_i - \bar{\varphi}_i$ and

$$\lambda := \max\Bigg\{ \sup_{(y,t,s)\in F}\left[\|\tilde{\psi}_1\|_s(y,t)\frac{\gamma(y)-s}{y}\right], \sup_{(y,t,s)\in F}\left[\|\tilde{\psi}_3\|_s(y,t)\frac{(\gamma(y)-s)^2}{y}\right],$$

$$\sup_{(y,t,s)\in F}\left[\|\tilde{\psi}_{6,8}\|_s(y)\frac{\gamma(y)-s}{y}\right], \sup_{(y,t,s)\in F}\left[\|\tilde{\psi}_{7,9}\|_s(y)\frac{(\gamma(y)-s)^2}{y}\right],$$

$$\sup_{(y,t,s)\in F}\left[\|\tilde{\psi}_5\|_s(y)\frac{(\gamma(y)-s)^3}{y}\right], \Bigg\} < \infty,$$

where $\gamma(y) = s_0 - y/b, b = b_0 \prod\limits_{n=0}^{\infty}\left(1 + 1/(n+1)^2\right)^{-1}$. Then for functions $\tilde{\psi}_i$ we can get the relations

$$\tilde{\psi}_1(x,y,t) = \frac{1}{2}\int\limits_0^y \int\limits_{t-y+\xi}^{t+y-\xi} \Bigg\{ H(x,\xi)\tilde{w}(x,\xi,\tau) - \frac{a}{2}\tilde{h}(x,\tau-\xi) + \tilde{k}_{00}(x,\xi,\tau)$$

$$- \int\limits_0^{\tau-\xi} \left(\tilde{h}(x,\gamma)\bar{w}(x,\xi,\tau-\gamma) - h(x,\gamma)\tilde{w}(x,\xi,\tau-\gamma)\right) d\gamma$$

$$+ \nu^2 \exp\left(r(x,0)\tau/2\right)$$

$$\times \frac{\partial}{\partial x}\Bigg\{ L_0\left[\bar{\varphi}_8(x,\tau)\exp\left(-r(x,0)(\tau)/2\right), \left(\tilde{r}_{00} + \frac{\partial\tilde{\psi}_2}{\partial x}\right)(x,\xi,\tau)\right]$$

$$- \int\limits_0^{\tau-\xi} \tilde{\psi}_8(x,\gamma)\exp(-r(x,0)\gamma/2)\left(r_{00} + \frac{\partial\varphi_2}{\partial x}\right)(x,\xi,\tau-\gamma)d\gamma\Bigg\}\Bigg\} d\tau d\xi,$$

$$\tilde{\psi}_3(x,y,t) = \frac{1}{2}\int\limits_{-y}^{y} \operatorname{sgn}(\xi)\Bigg\{ H(x,y-|\xi|)\tilde{w}^n(x,y-|\xi|,\xi+\tau)$$

$$-\frac{a}{2}(\tilde{h}^n(x,\xi+t-y+|\xi|) + \tilde{k}^n_{00}(x,y-|\xi|,t+|\xi|)$$

$$- \int\limits_0^{t-y+\xi+|\xi|} \left(\tilde{h}(x,\gamma)\bar{w}(x,y-|\xi|,t+\xi-\gamma) - h(x,\gamma)\tilde{w}^n(x,y-|\xi|,t+\xi-\gamma)\right)d\gamma$$

$$+\nu^2 \exp\left(r(x,0)(\xi+t)/2\right)$$

$$\times\frac{\partial}{\partial x}\Bigg\{ L_0\left[\bar{\varphi}_8(x,\xi+t)\exp\left(-r(x,0)(\xi+t)/2\right), \left(\tilde{r}^n_{00} + \frac{\partial\tilde{\psi}_2}{\partial x}\right)(x,y-|\xi|,\xi+t)\right]$$

$$- \int_0^{t-y+\xi+|\xi|} \tilde{\psi}_8(x,\gamma)\exp(-r(x,0)\gamma/2)\left(r_{00} + \frac{\partial \varphi_2}{\partial x}\right)(x, y-\xi, \xi+t-\gamma)\mathrm{d}\gamma\Big\}\Big\}\mathrm{d}\xi,$$

$$\tilde{\psi}_5(x,y)$$

$$= \int_0^y \Big(\frac{2}{a}\tilde{\psi}_8(x,\xi)\tilde{g}_{0tt}(x,y-\xi) - \tilde{h}(x,y-\xi))\bar{\varphi}_8(x,\xi)$$

$$+ h(x,y-\xi)\tilde{\psi}_8(x,\xi)\Big)\mathrm{d}\xi + \frac{1}{a}\int_0^{y/2}\Big\{H(x,\xi)\tilde{w}_t(x,\xi,y-\xi) + \tilde{k}_{00t}(x,\xi,y-\xi)$$

$$- \frac{1}{2}\tilde{h}(x,y-2\xi)\beta(x,\xi) - \int_0^{y-2\xi}\Big(\tilde{h}(x,\tau)\bar{w}_t(x,\xi,y-\xi-\tau)$$

$$- h(x,\tau)\tilde{w}_t(x,\xi,y-\xi-\tau)\Big)\mathrm{d}\tau + \nu^2\exp\left(r(x,0)(y-\xi)/2\right)$$

$$\times\Big\{\frac{\partial}{\partial x}\Big\{L_0\left[\bar{\varphi}_8(x,y-\xi)\exp\left(-r(x,0)(y-\xi)/2\right), \left(\tilde{r}^n_{00t} + \frac{\partial\tilde{\psi}_4}{\partial x}\right)(x,\xi,y-\xi)\right]$$

$$- \int_0^{y-2\xi}\tilde{\psi}_8(x,\tau)\exp\left(-r(x,0)\tau/2\right)\left(r_{00t} + \frac{\partial\varphi_4}{\partial x}\right)(x,\xi,y-\xi-\tau)\mathrm{d}\tau\Big\}$$

$$+ \frac{r(x,0)}{2}\frac{\partial}{\partial x}\Big\{L_0\left[\bar{\varphi}_8(x,\tau)\exp\left(-r(x,0)(y-\xi)/2\right), \left(\tilde{r}_{00} + \frac{\partial\tilde{\psi}_2}{\partial x}\right)(x,\xi,y-\xi)\right]$$

$$- \int_0^{y-2\xi}\tilde{\psi}_8(x,\gamma)\exp(-r(x,0)\tau/2)\left(r_{00} + \frac{\partial\varphi_2}{\partial x}\right)(x,\xi,y-\xi-\tau)\mathrm{d}\gamma\Big\}$$

$$+ \frac{1}{2}\frac{\partial}{\partial x}\Big[\exp\left(-r(x,0)(y-2\xi)/2\right)\Big(\bar{\varphi}_8(x,y-2\xi)\left(\tilde{r}_{00} + \frac{\partial\tilde{\psi}_2}{\partial x}\right)(x,\xi,\xi)$$

$$- \tilde{\psi}_8(x,y-2\xi)\left(r_{00} + \frac{\partial\varphi_2}{\partial x}\right)(x,\xi,\xi)\Big)\Big]\Big\}\mathrm{d}\xi,$$

$$\tilde{\psi}_6(x,y) = \int_0^y (y-\xi)(\tilde{h}(x,\xi) + r(x,0)\tilde{\psi}_7(x,\xi) - \frac{r^2(x,0)}{4}\tilde{\psi}_6(x,\xi))\mathrm{d}\xi,$$

$$\tilde{\psi}_7(x,y) = \int_0^y (\tilde{h}(x,\xi) + r(x,0)\tilde{\psi}_7(x,\xi) - \frac{r^2(x,0)}{4}\tilde{\psi}_6(x,\xi))\mathrm{d}\xi,$$

$$\tilde{\psi}_8(x,y) = \int_0^y (y-\xi)(c_0(x)\tilde{\psi}_8(x,\xi) - L[\varphi_8,\tilde{h}] - \int_0^\xi \tilde{\psi}_8(x,\xi-\tau)\tilde{h}(x,\tau))\mathrm{d}\xi,$$

$$\tilde{\psi}_9(x, y) = \int_0^y (c_0(\tilde{\psi}_8(x, \xi) - L[\varphi_8, \tilde{h}] - \int_0^\xi \tilde{\psi}_8(x, \xi - \tau)\bar{h}(x, \tau))d\xi.$$

Applying to them the above estimates, we find the inequality

$$\lambda \leq \lambda\rho', \ \rho' = \max(\mu_5', \ \mu_6', \ \mu_7', \max(1, b^{-1})\mu_8', \ \max(1, b^{-1})\mu_9')) < \rho < 1,$$

here $\mu_{5,6,7}' = \mu_{5,6,7}(b, a, T, R, s_0)$, $\mu_{8,9}' = \mu_{8,9}(b, T, R, s_0)$.
Hence, $\lambda = 0$. Therefore $\varphi_i = \bar{\varphi}_i$. The Theorem 7.3 is proved.

7.3 The Problem of Determining the One-Dimensional Matrix Kernel of the System of Viscoelasticity

Applied geophysics takes an interest in one-dimensional inverse problem of viscoelasticity with a source which can be described as $\nabla\delta(x - x_0)\delta(t)$ (a source of explosive type). This source is interpretated as an instantaneous center of compression. We notice that there are explosions which have some direction.

A distinctive feature of this paper is the use, first, of the explosive-type source as the right side of the considered system of viscoelasticity, and secondly, the after-effect kernel in the form of a diagonal matrix. Among the papers using this type of source, one can note [2], in which one-dimensional inverse coefficient elasticity problems are investigated, existence, uniqueness, and stability theorems for the solution of problems are proved.

For $x = (x_1, x_2, x_3)$, $(x_1, x_2) \in \mathbb{R}^2$, $x_3 > 0$, $t \in \mathbb{R}$, let us consider the following system of integro-differential equations of dynamical viscoelasticity:

$$\rho(x_3)\frac{\partial^2 u_i}{\partial t^2} = \sum_{j=1}^{3} \frac{\partial T_{ij}}{\partial x_j} + F_i(x, t), \ i = 1, 2, 3; \tag{7.81}$$

under the following initial and boundary conditions:

$$u_i \mid_{t<0} \equiv 0, \tag{7.82}$$

$$T_{3j}|_{x_3=+0} = 0, \ j = 1, 2, 3, \tag{7.83}$$

where $u(x, t) = (u_1(x, t), u_2(x, t), u_3(x, t))$ is the displacement vector; T_{ij} is the stress tensor 6.4–6.5 and

$$\begin{pmatrix} F_1(x, t) \\ F_2(x, t) \\ F_3(x, t) \end{pmatrix} = \begin{pmatrix} A_{11} & A_{12} & A_{13} \\ A_{21} & A_{22} & A_{23} \\ A_{31} & A_{32} & A_{33} \end{pmatrix} \nabla\delta(x)\delta(t). \tag{7.84}$$

Here $\delta(\cdot)$ is the Dirac delta function.

The right side of the system (7.81), represented (7.84), simulates a directional explosion. We have assumed that

$$A_{11} \neq 0 \text{ (or } A_{22} \neq 0), \ A_{33} \neq 0, \ A_{12} \neq 0, \ A_{21} \neq 0.$$

We assume that $\rho = \rho(x_3), \ \mu = \mu(x_3), \ \lambda = \lambda(x_3) \in \Lambda$.

Refer as the direct problem to the problem of finding the displacement vector $u(x,t)$ satisfying (7.81)–(7.83) in the sense of distributions for prescribed functions $\rho(x_3), \ \mu(x_3), \ \lambda(x_3), \ k_i(t), \ F_i(x,t), \ i = 1, 2, 3$.

Let us express (7.81)–(7.83) in terms of the Fourier transform

$$\begin{aligned}\rho(x_3)\frac{\partial^2 \tilde{u}_1}{\partial t^2} &= L\Big[k_1, \frac{\partial}{\partial x_3}\Big(\mu(x_3)\frac{\partial \tilde{u}_1}{\partial x_3}\Big) + i\nu_1\frac{\partial}{\partial x_3}(\mu(x_3)\tilde{u}_3) + i\nu_1\lambda(x_3)\frac{\partial \tilde{u}_3}{\partial x_3} \\ &\quad - [(2\mu(x_3)+\lambda(x_3))\nu_1^2 + \mu(x_3)\nu_2^2]\tilde{u}_1 - (\mu(x_3)+\lambda(x_3))\nu_1\nu_2\tilde{u}_2\Big] \\ &\quad + i(\nu_1 A_{11} + \nu_2 A_{12})\delta(x_3)\delta(t) + A_{13}\delta'(x_3)\delta(t),\end{aligned} \tag{7.85}$$

$$\begin{aligned}\rho(x_3)\frac{\partial^2 \tilde{u}_2}{\partial t^2} &= L\Big[k_2, \frac{\partial}{\partial x_3}\Big(\mu(x_3)\frac{\partial \tilde{u}_2}{\partial x_3}\Big) + i\nu_2\frac{\partial}{\partial x_3}(\mu(x_3)\tilde{u}_3) + i\nu_2\lambda(x_3)\frac{\partial \tilde{u}_3}{\partial x_3} \\ &\quad - [(2\mu(x_3)+\lambda(x_3))\nu_2^2 + \mu(x_3)\nu_1^2]\tilde{u}_2 - (\mu(x_3)+\lambda(x_3))\nu_1\nu_2\tilde{u}_1\Big] \\ &\quad + i(\nu_1 A_{21} + \nu_2 A_{22})\delta(x_3)\delta(t) + A_{23}\delta'(x_3)\delta(t),\end{aligned} \tag{7.86}$$

$$\begin{aligned}&\rho(x_3)\frac{\partial^2 \tilde{u}_3}{\partial t^2} \\ &= L\Big[k_3, \frac{\partial}{\partial x_3}\Big((2\mu(x_3)+\lambda(x_3))\frac{\partial \tilde{u}_3}{\partial x_3}\Big) + i\nu_1\mu(x_3)\frac{\partial \tilde{u}_1}{\partial x_3} \\ &\quad + i\nu_1\frac{\partial}{\partial x_3}(\lambda(x_3)\tilde{u}_1) + \nu_2\mu(x_3)\frac{\partial \tilde{u}_2}{\partial x_3} + i\nu_2\frac{\partial}{\partial x_3}(\lambda(x_3)\tilde{u}_2) - \mu(x_3)(\nu_1^2+\nu_2^2)\tilde{u}_3\Big] \\ &\quad + i(\nu_1 A_{31} + \nu_2 A_{32})\delta(x_3)\delta(t) + A_{33}\delta'(x_3)\delta(t),\end{aligned} \tag{7.87}$$

$$\tilde{u}_j|_{t<0} \equiv 0, \ j = 1, 2, 3, \tag{7.88}$$

$$L\Big[k_l, \mu(x_3)\frac{\partial \tilde{u}_l}{\partial x_3} + i\nu_l\mu(x_3)\tilde{u}_3\Big]_{x_3=+0} = 0, \ l = 1, 2, \tag{7.89}$$

$$L\Big[k_3, (2\mu(x_3)+\lambda(x_3))\frac{\partial \tilde{u}_3}{\partial x_3} + i\nu_1\lambda(x_3)\tilde{u}_2\Big]_{x_3=+0} = 0, \tag{7.90}$$

where

$$\tilde{u}_j(\nu_1, \nu_2, x_3, t, \nu) = \iint_{R^2} u_j(x_1, x_2, x_3, t)\exp(i(\nu_1 x_1 + \nu_2 x_2))dx_1\,dx_2.$$

Let

$$\tilde{v}_1(x_3,t) = \frac{\partial \tilde{u}_1}{\partial \nu_2}\Big|_{\nu_1=\nu_2=0},$$

$$\tilde{v}_2(x_3,t) = \frac{\partial \tilde{u}_2}{\partial \nu_1}\Big|_{\nu_1=\nu_2=0}, \quad \tilde{v}_3(x_3,t) = \frac{\partial^2 \tilde{u}_3}{\partial \nu_1 \partial \nu_2}\Big|_{\nu_1=\nu_2=0}.$$

It is possible to obtain from system (7.85)–(7.90) separate, independently solvable problems for imaginary parts of new functions (and $\tilde{v}_j(x_3,t) \equiv \tilde{v}^i_j(x_3,t),\ j=1,2)$:

$$\rho(x_3)\frac{\partial^2 \tilde{v}_j}{\partial t^2} = L\Big[k_j, \frac{\partial}{\partial x_3}\Big(\mu(x_3)\frac{\partial \tilde{v}_j}{\partial x_3}\Big)\Big] + a_j\delta(x_3)\delta(t), \tag{7.91}$$

$$\begin{aligned}\rho(x_3)\frac{\partial^2 \tilde{v}^i_3}{\partial t^2} = L\Big[k_3, \frac{\partial}{\partial x_3}\Big((\lambda(x_3)+2\mu(x_3))\frac{\partial \tilde{v}^i_3}{\partial x_3}\Big) + \mu(x_3)\frac{\partial \tilde{v}_1}{\partial x_3}\\ + \frac{\partial}{\partial x_3}(\lambda(x_3)(\tilde{v}_1+\tilde{v}_2))\Big],\end{aligned} \tag{7.92}$$

$$\tilde{v}_j|_{t<0} \equiv 0,\ j=1,2;\quad \tilde{v}^i_3|_{t<0} \equiv 0, \tag{7.93}$$

$$L\Big[k_j, \mu(x_3)\frac{\partial \tilde{v}_j}{\partial x_3}\Big]_{x_3=+0} = 0,\ j=1,2, \tag{7.94}$$

$$L\Big[k_3, (2\mu(x_3)+\lambda(x_3))\frac{\partial \tilde{v}^i_3}{\partial x_3}\Big]_{x_3=+0} = 0, \tag{7.95}$$

where $(a_1, a_2) := (A_{12}, A_{21})$.

We assume that $k_3(t) \equiv k_2(t)$. The subject of the current research is the inverse problem of the kernel $k(t) = diag(k_1, k_2, k_2)(t),\ t>0$, that is included in the equality (7.81) and, consequently, in (7.91), (7.92), when the additional information about the solution to the direct problem is available:

$$\tilde{v}_j(+0,t) = g_j(t),\ t>0, \tag{7.96}$$

$g_j(t),\ j=1,2$ are given functions.

Let us enter the variables y and z as follows:

$$y = \psi_1(x_3) := \int_0^{x_3} \frac{d\xi}{c_1(\xi)},\quad c_1(x_3) := \sqrt{\frac{\mu(x_3)}{\rho(x_3)}},$$

$$z = \psi_2(x_3) := \int_0^{x_3} \frac{d\xi}{c_2(\xi)},\quad c_2(x_3) := \sqrt{\frac{\lambda(x_3)+2\mu(x_3)}{\rho(x_3)}}.$$

Let

$$v_{1,2}(y,t) := \frac{\tilde{v}_{1,2}\left(\psi_1^{-1}(y), t\right)}{s(y)},\quad s(y) := \sqrt{\frac{c_1(+0)\rho(+0)}{c_1\left(\psi_1^{-1}(y)\right)\rho\left(\psi_1^{-1}(y)\right)}},$$

$$v_3(z,t) := \frac{\tilde{v}_3^i\left(\psi_2^{-1}(z),t\right)}{p(z)},\ p(z) := \sqrt{\frac{c_2(+0)\rho(+0)}{c_2\left(\psi_2^{-1}(z)\right)\rho\left(\psi_2^{-1}(z)\right)}}.$$

Then, in terms of the functions introduced above and the variables $y,\ z$, the inverse problem (7.91)–(7.96) for $y>0,\ z>0,\ t\in\mathbb{R}$ takes the form

$$\frac{\partial^2 v_j}{\partial t^2} = L\left[k_j, \frac{\partial^2 v_j}{\partial y^2} + q(y)v_j\right] + \tilde{a}_j\delta(y)\delta(t),\ j=1,2, \tag{7.97}$$

$$\begin{aligned}\frac{\partial^2 v_3}{\partial t^2} = L\Bigg[k_2, \frac{\partial^2 v_3}{\partial z^2} + \tilde{q}(z)v_3 \\ + \frac{1}{\rho(\psi_2^{-1}(z))p(z)}\left[\mu(x_3)\frac{\partial\tilde{v}_1}{\partial x_3} + \frac{\partial}{\partial x_3}\left(\lambda(x_3)(\tilde{v}_1+\tilde{v}_2)\right)\right]_{x_3=\psi_2^{-1}(z)}\Bigg],\end{aligned} \tag{7.98}$$

$$v_j|_{t<0} \equiv 0,\ j=1,2,3, \tag{7.99}$$

$$L\left[k_j, \frac{\partial v_j(y,t)}{\partial y}\right]_{y=+0} = 0,\ j=1,2;\quad L\left[k_3, \frac{\partial v_3(z,t)}{\partial z}\right]_{z=+0} = 0, \tag{7.100}$$

$$v_j(+0,t) = g_j(t),\ j=1,2;\ t>0, \tag{7.101}$$

where

$$\tilde{a}_j = \frac{a_j}{\rho(+0)\mu(+0)},\ j=1,2;$$

$$q(y) = \frac{s''(y)}{s(y)} - 2\left[\frac{s'(y)}{s(y)}\right]^2,\ \tilde{q}(z) = \frac{p''(z)}{p(z)} - 2\left[\frac{p'(z)}{p(z)}\right]^2.$$

In (7.97), (7.98) the following equations were used, based on the properties of the Delta function:

$$\frac{1}{\rho(x_3)}\delta(x_3) = \frac{1}{\rho(+0)}\delta(x_3),$$

$$\delta(x_3) = \delta(\psi_1^{-1}(y)) = \frac{\delta(y)}{\frac{d\psi_1^{-1}(y)}{dy}\Big|_{y=+0}} = \delta(y)\frac{\psi_1(x_3)}{dx_3}\Big|_{x=+0} = \delta(y)\sqrt{\frac{\rho(+0)}{\mu(+0)}}.$$

Let $L\left[k_j, v_j(\cdot,t)\right]\exp\left(-k_j(0)t/2\right) = w_j(\cdot,t),\ j=1,2,3.$

Then as is easy to see,

$$v_j(\cdot,t) = \exp\left(k_j(0)t/2\right) w_j(\cdot,t) + \int_0^t r_j(t-\tau)\exp\left(k_j(0)\tau/2\right) w_j(\cdot,\tau)d\tau,$$

where $r_j(t)$ is determined by (7.14) for each j.

In terms of the new functions $w_j(\cdot,t)$ and $r_j(t)$, Eqs. (7.97)–(7.101) for $y > 0,\ z > 0,\ t \in \mathbb{R}$ take the form

$$\frac{\partial^2 w_j}{\partial t^2} = \frac{\partial^2 w_j}{\partial y^2} + H_j(y)w_j - \int_0^t h_j(t-\tau)w_j(y,\tau)d\tau + \tilde{a}_j\delta(y)\delta(t),\ j = 1,2, \quad (7.102)$$

$$\frac{\partial^2 w_3}{\partial t^2} = \frac{\partial^2 w_3}{\partial z^2} + \widetilde{H}(z)w_3 - \int_0^t h_2(t-\tau)w_3(z,\tau)d\tau +$$

$$L\left[k_{02}, \frac{\exp\left(r_2(0)t/2\right)}{\rho(\psi_2^{-1}(z))p(z)}\left[\mu(x_3)\frac{\partial\tilde{v}_1}{\partial x_3} + \frac{\partial}{\partial x_3}\left(\lambda(x_3)(\tilde{v}_1+\tilde{v}_2)\right)\right]_{x_3=\psi_2^{-1}(z)}\right], \quad (7.103)$$

$$w_j|_{t<0} \equiv 0,\ j = 1,2,3; \quad (7.104)$$

$$\left.\frac{\partial w_j}{\partial y}\right|_{y=+0} = 0,\ j = 1,2;\ \left.\frac{\partial w_3}{\partial z}\right|_{z=+0} = 0, \quad (7.105)$$

$$w_j(+0,t) = \tilde{g}_j(t) + \int_0^t k_{0j}(t-\tau)\tilde{g}_j(\tau)d\tau,\ j = 1,2; \quad (7.106)$$

where

$$H_j(y) := q(y) + \frac{r_j^2(0)}{4} - r_j'(0),\ j = 1,2;\ H_3(z) := \tilde{q}(z) + \frac{r_2^2(0)}{4} - r_2'(0),$$

$$h_j(t) := r_j''(t)\exp\left(r_j(0)t/2\right),$$

$$\tilde{g}_j(t) := g_j(t)\exp\left(r_j(0)t/2\right),$$

$$k_{0j}(t) := k_j(t)\exp\left(r_j(0)t/2\right),\ j = 1,2.$$

Here we used the equality $k_j(0) = -r_j(0)$, which is a consequence of (7.14) for each j.

Further, the study is divided into two stages. The result of the first stage will be the determining of $k_1(t)$ and $k_2(t)$ of (7.102), (7.104)–(7.106) (*inverse problem*), at the same stage will be also found $w_1(t)$ and $w_2(t)$. The result of the second stage is the determining $w_3(t)$ from (7.103)–(7.105) (*direct problem*).

The *inverse problem* can be considered as a set of two independently (parallel) solvable problems of determining the pair of functions $(w_1(y,t),\ k_1(t))$ and $(w_2(y,t),\ k_2(t))$ from the system (7.102), (7.104)–(7.106). Since these equals are identical for $j=1$ and for $j=2$, in the subsequent calculations of this paragraph the index will be omitted and the study will be valid for $(w_1(y,t),\ k_1(t))$ and for $(w_2(y,t),\ k_2(t))$. To this end, the functions $(w_1(y,t),\ k_1(t))$ and $(w_2(y,t),\ k_2(t))$ can be replaced by $(w(y,t),\ k(t))$, $\tilde{a}_1$ and $\tilde{a}_2$ – by $\tilde{a}$, $g_1(t)$ and $g_2(t)$ – by $g(t)$. But let's not forget that each pair of functions $\left(w_j(y,t),\ k_j(t)\right)$ have own $\tilde{a}_j$ and $g_j(t),\ j=1,2$.

We extend $w(y,t),\ H(y)$ for $y<0$ as even functions. Now, (7.105) holds automatically. Then problem (7.102), (7.104), (7.105) will be equivalent to the problem of determining $w(y,t)$ even in y from the following equalities:

$$\frac{\partial^2 w}{\partial t^2}=\frac{\partial^2 w}{\partial y^2}+H(y)w-\int_0^t h(t-\tau)w(y,\tau)d\tau+\tilde{a}\delta(y)\delta(t),\ y\in\mathbb{R}, \tag{7.107}$$

$$w|_{t<0}\equiv 0. \tag{7.108}$$

Let's study the properties of the direct problem's solution (7.102), (7.104), (7.105). To this end, we represent $w(y,t)$ in the form

$$\begin{aligned} w(y,t)&=\frac{\tilde{a}}{2}\iint_{\mathbb{R}^2}\delta(\xi)\delta(\tau)\theta(t-\tau-|y-\xi|)d\tau d\xi \\ &+\frac{1}{2}\iint_{\mathbb{R}^2}\Big[H(\xi)w(\xi,\tau)-\int_0^\tau h(\tau-s)w(\xi,s)ds\Big]\theta(t-\tau-|y-\xi|)d\tau d\xi \\ &=\frac{\tilde{a}}{2}\theta(t-|y|)+\frac{1}{2}\int_{y-t}^{y+t}\int_0^{t-|y-\xi|}\Big[H(\xi)w(\xi,\tau)-\int_0^\tau h(\tau-s)w(\xi,s)ds\Big]d\tau d\xi, \end{aligned} \tag{7.109}$$

where $\frac{1}{2}\theta(t-|y|)$ – fundamental solution of the operator $\frac{\partial^2}{\partial t^2}-\frac{\partial^2}{\partial y^2}$, $\theta(\cdot)$ – Heaviside function.

It follows from (7.109) that $w(y,t)|_{t<|y|}\equiv 0$. Therefore, (7.107) can be rewritten for $t>|y|$ in the form

$$w(y,t)=\frac{\tilde{a}}{2}+\frac{1}{2}\int_{\frac{y-t}{2}}^{\frac{y+t}{2}}\int_{|\xi|}^{t-|y-\xi|}\Big[H(\xi)w(\xi,\tau)-\int_{|\xi|}^{\tau}h(\tau-s)w(\xi,s)ds\Big]d\tau d\xi. \tag{7.110}$$

Notice, that

$$w(y,|y|)=\frac{\tilde{a}}{2}. \tag{7.111}$$

We differentiate (7.110) twice with respect to t:

$$\frac{\partial w}{\partial t}(y,t)$$
$$=\frac{1}{2}\int\limits_{\frac{y-t}{2}}^{\frac{y+t}{2}}\Big[H(\xi)w(\xi,t-|y-\xi|)\quad-\int_0^{t-|y-\xi|-|\xi|}h(s)w(\xi,t-|y-\xi|-s)\mathrm{d}s\Big]\mathrm{d}\xi, \tag{7.112}$$

$$\frac{\partial^2 w}{\partial t^2}(y,t)$$
$$=\frac{\tilde{a}}{8}\left[H\left(\frac{y+t}{2}\right)+H\left(\frac{y-t}{2}\right)\right]\quad-\frac{\tilde{a}}{4}yh(t-y)-\frac{\tilde{a}}{8}\left[\int\limits_0^{t-y}h(\tau)\mathrm{d}\tau+\int\limits_0^{t}h(\tau)\mathrm{d}\tau\right]$$
$$+\frac{1}{2}\int\limits_{\frac{y-t}{2}}^{\frac{y+t}{2}}\Big[H(\xi)\frac{\partial w}{\partial t}(\xi,t-|y-\xi|)-\int_0^{t-|y-\xi|-|\xi|}h(s)\frac{\partial w}{\partial t}(\xi,t-|y-\xi|-s)\mathrm{d}s\Big]\mathrm{d}\xi. \tag{7.113}$$

It follows from (7.111)–(7.113) that $w(0,0)=\frac{\tilde{a}}{2}$, $\frac{\partial w}{\partial t}(0,0)=0$, $\frac{\partial^2 w}{\partial t^2}(0,0)=\frac{\tilde{a}}{4}H(0)$. These equalities together with condition (7.106) allow us to find $r(0)$, $r'(0)$.

Indeed,

$$\tilde{g}(0)=\frac{\tilde{a}}{2},$$

$$\tilde{g}'(0)-r(0)\tilde{g}(0)=0\Rightarrow g'(0)+\frac{r(0)}{2}g(0)-r(0)g(0)=0\Rightarrow r(0)=\frac{4g'(0)}{\tilde{a}},$$

$$\frac{\tilde{a}}{4}\left[q(0)+\frac{r^2(0)}{4}-r'(0)\right]=g''(0)+\tilde{a}\frac{r^2(0)}{8}-\frac{\tilde{a}}{2}r'(0)$$

$$\Rightarrow r'(0)=\frac{4}{\tilde{a}}g''(0)+\frac{r^2(0)}{4}-q(0).$$

Substituting the value $y=0$ in (7.113) and taking into account the fact that $w(y,t)$ is even function on y, we find

$$\frac{\partial^2 w}{\partial t^2}(0,t)=\frac{\tilde{a}}{4}H\left(\frac{t}{2}\right)-\frac{\tilde{a}}{4}\int\limits_0^t h(\xi)\mathrm{d}\xi$$
$$+\int\limits_0^{\frac{t}{2}}\Big[H(\xi)\frac{\partial w}{\partial t}(\xi,t-\xi)\quad-\int\limits_0^{t-2\xi}h(s)\frac{\partial w}{\partial t}(\xi,t-\xi|-s)\mathrm{d}s\Big]\mathrm{d}\xi. \tag{7.114}$$

Differentiating (7.114) with respect to t, we obtain an equation for unknown function $h(t)$:

$$h(t) = -\frac{4}{\tilde{a}}\frac{\partial^3 w}{\partial t^3}(0,t) + \frac{1}{2}H'\left(\frac{t}{2}\right) + H\left(\frac{t}{2}\right) + \frac{4}{\tilde{a}}\int_0^{\frac{t}{2}}\left[H(\xi)\frac{\partial^2 w}{\partial t^2}(\xi, t-\xi) - \int_0^{t-2\xi} h(s)\frac{\partial^2 w}{\partial t^2}(\xi, t-\xi-s)ds\right]d\xi, \tag{7.115}$$

where

$$\frac{\partial^3 w}{\partial t^3}(0,t) = \tilde{g}'''(t) + \frac{\tilde{a}}{2}k_0''(t) + k_0'(t)\tilde{g}'(0) + k_0(t)\tilde{g}''(0) + \int_0^t k_0(\tau)\tilde{g}'''(t-\tau)d\tau.$$

Adding to the equations (7.110), (7.112), (7.113), (7.115) the below following relations

$$k_0(t) = -r(0) + \left(\frac{r^2(0)}{2} - r'(0)\right)t + \int_0^t (t-\tau)k_0''(\tau)d\tau, \tag{7.116}$$

$$k_0'(t) = \frac{r^2(0)}{2} - r'(0) + \int_0^t k_0''(\tau)d\tau, \tag{7.117}$$

$$k_0''(t) = -h(t) + \left(\frac{r^2(0)}{4} - r'(0)\right)k_0(t) - \int_0^t h(t-\tau)k_0(\tau)d\tau \tag{7.118}$$

we obtain in the domain $\square(T) = \{(y,t)|\ |y| \le t \le T - |y|\}$ the closed system of integral equations with respect to unknown functions.

From the above arguments we obtain the following lemma:

Lemma 7.2 *Let* $g(t) \in C^3[0,T]$, $(\rho,\mu) \in C^3[0, \psi_1^{-1}(T)]$, $T > 0$ *is fixed. Then problem (7.102), (7.104)– (7.106) for* $(y,t) \in \square(T)$ *is equivalent to the problem of determining the vector functions* $w(y,t)$, $\left(\frac{\partial w}{\partial t}\right)(y,t)$, $\left(\frac{\partial^2 w}{\partial t^2}\right)(y,t)$, $h(t)$, $k_0(t)$, $k_0'(t)$, $k_0''(t)$ *from the following system of integral equations (7.110), (7.112), (7.113), (7.115)–(7.118).*

The following theorems hold.

Theorem 7.4 *Suppose that the conditions of the lemma 7.2 hold. Then for arbitrary fixed* $T > 0$ *there exists a unique inverse problem solution* $\{k_1(t),\ k_2(t)\} \in C^2[0,T]$.

Theorem 7.5 *Suppose that* $k^1(t)$, $k^2(t) \in K(h_0)$ *are solutions to inverse problem with the data*

$$\left\{\rho^1(\psi^{-1}(y)),\ \mu^1(\psi^{-1}(y)),\ g^1(t)\right\},$$

$$\left\{\rho^2(\psi^{-1}(y)),\ \mu^2(\psi^{-1}(y)),\ g^2(t)\right\},$$

respectively. Then there is a positive number $C = C(h_0, h_{00}, T)$, *where*

$$h_{00} = \max\Big\{\|\rho^1(y)\|_{C^3[0,\psi^{-1}(T/2)]}, \|\mu^1(y)\|_{C^3[0,\psi^{-1}(T/2)]}, \|g^1(t)\|_{C^3[0,T]},$$

$$\|\rho^2(y)\|_{C^3[0,\psi^{-1}(T/2)]}, \|\mu^2(y)\|_{C^3[0,\psi^{-1}(T/2)]}, \|g^2(t)\|_{C^3[0,T]}\Big\},$$

such that we have the stability estimate

$$\|k^1(t) - k^2(t)\|_{C^2[0,T]} \leq C\Big[\|\rho^1 - \rho^2\|_{C^3[0,\psi^{-1}(T/2)]} + \|\mu^1 - \mu^2\|_{C^3[0,\psi^{-1}(T/2)]}$$

$$+ \|g^1 - g^2\|_{C^3[0,T]}\Big]. \tag{7.119}$$

The proofs of these theorems are similar to the proofs of Theorems 7.1, 7.2.

Now let's consider the *direct problem* of determining $w_3(y, t)$.

Let

$$\widetilde{L}(z, t) = L\Big[k_{02}, \frac{\exp(r_2(0)t/2)}{\rho(\psi_2^{-1}(z))p(z)}\Big[\mu\frac{\partial \tilde{v}_1}{\partial x_3} + \frac{\partial}{\partial x_3}(\lambda(\tilde{v}_1 + \tilde{v}_2))\Big]_{x_3=\psi_2^{-1}(z)}\Big]$$

$$:= \widetilde{L}_1(z, t) + \widetilde{L}_2(z, t), \tag{7.120}$$

where

$$\widetilde{L}_1(z, t) = \frac{\exp(r_2(0)t/2)}{\rho(\psi_2^{-1}(z))p(z)}\Big[\mu\frac{\partial \tilde{v}_1}{\partial x_3} + \frac{\partial}{\partial x_3}(\lambda(\tilde{v}_1 + \tilde{v}_2))\Big]_{x_3=\psi_2^{-1}(z)},$$

$$\widetilde{L}_2(z, t)$$

$$= \int_0^t k_{02}(t - \tau)\frac{\exp(r_2(0)\tau/2)}{\rho(\psi_2^{-1}(z))p(z)}\Big[\mu\frac{\partial \tilde{v}_1}{\partial x_3} + \frac{\partial}{\partial x_3}(\lambda(\tilde{v}_1 + \tilde{v}_2))\Big]_{x_3=\psi_2^{-1}(z)}d\tau.$$

Here

$$\tilde{v}_j(x_3, t) = s(\psi_1(x_3)) \cdot \Big[\exp(k_j(0)\tau/2)\Big[\frac{\tilde{a}_j}{2}\theta(t - |y|)$$

$$+ \frac{1}{2}\int_{\frac{y-t}{2}}^{\frac{y+t}{2}}\int_{|\xi|}^{t-|y-\xi|} F_j(\xi, \tau)d\tau d\xi\Big] + \int_0^t r_j(t - \tau)\exp(k_j(0)\tau/2) \tag{7.121}$$

$$\times\Big[\frac{\tilde{a}_j}{2}\theta(\tau - |y|) + \frac{1}{2}\int_{\frac{y-\tau}{2}}^{\frac{y+\tau}{2}}\int_{|\xi|}^{\tau-|y-\xi|} F_j(\xi, \eta)d\eta d\xi\Big]d\tau\Big]_{y=\psi_1(x_3)},$$

where

$$F_j(\xi,\tau) = H_j(\xi)w_j(\xi,\tau) - \int\limits_0^{\tau-|\xi|} h_j(s)w_j(\xi,\tau-s)\mathrm{d}s, \quad j = 1,2.$$

Notice, that $\tilde{v}_j(x_3,t)$ has a structure

$$\tilde{v}_j(x_3,t) = \Big[\theta(t-|y|)\widetilde{V}_j(y,t)\Big]_{y=\psi_1(x_3)}$$

for $j = 1, 2$, since for $t \to |y| + 0$ the measure of the set $\square(y,t)$ tends to zero. It follows from Theorem7.4 that $\widetilde{V}_j(y,t) \in C(\square(T))$.

Then we have

$$\frac{\partial \tilde{v}_j}{\partial x_3} = -\sqrt{\frac{\rho(x_3)}{\mu(x_3)}}\Big[\mathrm{sign}(y)\delta(t-|y|)\widetilde{V}_j(y,|y|) + \theta(t-|y|)\frac{\partial \widetilde{V}_j(y,t)}{\partial y}\Big]_{y=\psi_1(x_3)}$$

.

Indeed,

$$\frac{\partial \tilde{v}_j}{\partial x_3} = s'(\psi_1(x_3))\sqrt{\frac{\rho(x_3)}{\mu(x_3)}}$$

$$\times\Big[\exp(k_j(0)t/2)\Big[\frac{\tilde{a}_j}{2}\theta(t-|y|) + \frac{1}{2}\int\limits_{\frac{y-t}{2}}^{\frac{y+t}{2}}\int\limits_{|\xi|}^{t-|y-\xi|} F_j(\xi,\tau)\mathrm{d}\tau\mathrm{d}\xi\Big]$$

$$+\int\limits_0^t r_j(t-\tau)\exp(k_j(0)\tau/2)$$

$$\times\Big[\frac{\tilde{a}_j}{2}\theta(\tau-|y|) + \frac{1}{2}\int\limits_{\frac{y-\tau}{2}}^{\frac{y+\tau}{2}}\int\limits_{|\xi|}^{\tau-|y-\xi|} F_j(\xi,\eta)d\eta\mathrm{d}\xi\Big]\mathrm{d}\tau\Big]_{y=\psi_1(x_3)}$$

$$+ s(\psi_1(x_3))\Big[\exp(k_j(0)t/2)\Big[-\frac{\tilde{a}_j}{2}\sqrt{\frac{\rho(x_3)}{\mu(x_3)}}\mathrm{sign}(y)\delta(t-|y|)$$

$$+\frac{1}{2}\sqrt{\frac{\rho(x_3)}{\mu(x_3)}}\int\limits_{\frac{y-t}{2}}^{\frac{y+t}{2}} F_j(\xi,t-|y-\xi|)\mathrm{sign}(\xi-y)\mathrm{d}\xi\Big]$$

$$+\frac{\tilde{a}_j}{2}\sqrt{\frac{\rho(x_3)}{\mu(x_3)}}\mathrm{sign}(y)r_j(t-|y|)\exp(k_j(0)|y|/2) - \frac{1}{2}\sqrt{\frac{\rho(x_3)}{\mu(x_3)}}$$

$$\times\int\limits_0^t r_j(t-\tau)\exp(k_j(0)\tau/2)\int\limits_{\frac{y-\tau}{2}}^{\frac{y+\tau}{2}} F_j(\xi,t-|y-\xi|)\mathrm{sign}(\xi-y)\mathrm{d}\xi\mathrm{d}\tau\Big]_{y=\psi_1(x_3)}. \tag{7.122}$$

We extend the functions $w_3(z,t)$, $\widetilde{H}(z)$, $\widetilde{L}(z,t)$ as even functions for $z<0$. Then the problem of determining the $w_3(z,t)$ (7.103)–(7.105) is equivalent to the problem of determining the even function $w_3(z,t)$ (respect to z) satisfying the equalities (7.103), (7.104).

It follows from the theory of hyperbolic equations that the solution of the problem (7.103), (7.104) is equivalent to the relation:

$$w_3(z,t)=\frac{1}{2}\iint_{\mathbb{R}^2}\Big[\widetilde{H}(\xi)w_3(\xi,\tau)-\int_0^{\tau}h_2(\tau-s)w_3(s,\tau)\mathrm{d}s\Big]\theta(t-\tau-|z-\xi|)\mathrm{d}\tau\mathrm{d}\xi$$

$$+\frac{1}{2}\iint_{\mathbb{R}^2}\widetilde{L}(\xi,\tau)\theta(t-\tau-|z-\xi|)\mathrm{d}\tau\mathrm{d}\xi. \tag{7.123}$$

Since $\widetilde{L}_1(\xi,\tau)$ contains the term with Delta function, we consider separately the double integrating of this term in equation (7.123):

$$\iint_{\mathbb{R}^2}\theta(t-\tau-|z-\xi|)\delta(\tau-|\mathrm{d}(\xi)|)Q(\xi,\tau)\mathrm{d}\xi\mathrm{d}\tau=\sum_{i=1}^{n}\int_{z_{*i}(z,t)}^{z_i^*(z,t)}Q(\xi,|\mathrm{d}(\xi)|)\mathrm{d}\xi\mathrm{d}\tau, \tag{7.124}$$

where

$$\mathrm{d}(\xi)=\psi_1(\psi_2^{-1}(\xi)),$$

$$Q(\xi,\tau)=-\frac{\exp\left(r_2(0)\tau/2\right)}{\sqrt{\rho(\psi_2^{-1}(\xi))\mu(\psi_2^{-1}(\xi))p(\xi)}}s(\mathrm{d}(\xi))\mathrm{sign}(\mathrm{d}(\xi))$$

$$\times\Big[\frac{a_1}{2}\exp\left(k_1(0)\tau/2\right)\Big(\mu(\psi_2^{-1}(\xi))+\lambda(\psi_2^{-1}(\xi))\Big)$$

$$+\frac{a_2}{2}\exp\left(k_2(0)\tau/2\right)\lambda(\psi_2^{-1}(\xi))\Big],$$

$z_{*i}(z,t)$, $z_i^*(z,t)$ are solutions of the equation $t-|z-\xi|=|\mathrm{d}(\xi)|$ with respect to ξ and $z_{*i}(z,t)\le z_i^*(z,t)$.

As we can see, the right side (7.123) is continuous. The remaining terms on the right side of the equations (7.121), (7.122) will also be continuous by double integrating. It is obvious that the question of continuity $L_2(z,t)$ has already been solved, since this component part contains triple integrating. The only question is what will be the domain of continuity $w_3(z,t)$.

Let's introduce the following symbols:

$$\Delta(z,t)=\{(\xi,\tau):\ 0\le\tau\le t-|z-\xi|\},\ \nabla=\{(\xi,\tau):|\mathrm{d}(\xi)|\le\tau\},$$

$$\Theta=\{(z,t):\ 0<t\le\psi(z)\}.$$

The function $\psi(z)$ is defined by the values z and $\mathrm{d}(z)$ exactly as in work [3] (we omit the definition due to its bulkiness and refer the reader to this work).

The following formulas follow from the definition of functions $\psi(z)$, $\mathrm{d}(z)$ and sets $\Delta(z,t)$, ∇, Θ :

$$\Delta(z,t)\bigcap\nabla=\varnothing,\ (z,t)\in\Theta,$$

$$\Delta(z,t)\bigcap\nabla=\bigcup_{i=1}^{n}D_i(z,t),\ (z,t)\in R^2\setminus\Theta,$$

$$D_i(z,t)=\{(\xi,\tau):\ z_{*i}(z,t)\le\xi\le z_i^*(z,t),\ |\mathrm{d}(\xi)|\le\tau\le t-|z-\xi|\}.$$

Integrating terms with $\theta(t-|d(z)|)$ in $L_1(\xi,\tau)$, in particular, we obtain:

$$\iint\limits_{\mathbb{R}^2}\theta(t-\tau-|z-\xi|)\theta(t-|d(\xi)|)P(\xi,\tau)\mathrm{d}\xi\mathrm{d}\tau=\theta(t-\psi(z))\iint\limits_{\Delta(z,t)\bigcap\nabla}P(\xi,\tau)\mathrm{d}\xi\mathrm{d}\tau$$

$$=\theta(t-\psi(z))\sum_{i=1}^{n}\iint\limits_{D_i(z,t)}P(\xi,\tau)\mathrm{d}\xi\mathrm{d}\tau$$

$$=\theta(t-\psi(z))\sum_{i=1}^{n}\int\limits_{z_{*i}(z,t)}^{z_i^*(z,t)}\int\limits_{|d(\xi)|}^{t-|z-\xi|}P(\xi,\tau)\mathrm{d}\xi\mathrm{d}\tau,$$

$$P(\xi,\tau)=\frac{\exp\left(r_2(0)\tau/2\right)}{\rho(\psi_2^{-1}(\xi))p(\xi)}s(\mathrm{d}(\xi))$$

$$\times\left(\frac{\tilde{a}_1}{2}\exp\left(k_1(0)\tau/2\right)+\frac{\tilde{a}_2}{2}\exp\left(k_2(0)\tau/2\right)\right)\frac{\mathrm{d}\lambda}{\mathrm{d}x_3}\bigg|_{x_3=\psi_2^{-1}(\xi)}.$$

Other terms c $\theta(t-|d(z)|)$ will be presented in the same way.

Thus, taking into account the structure $\tilde{v}_j(x_3,t)$ and $\frac{\partial\tilde{v}_j}{\partial x_3}$, in the domain $t>\psi(z)$ we find

$$w_3(z,t)=\frac{1}{2}\iint\limits_{\Delta(z,t)\bigcap\nabla}\left[\widetilde{H}(\xi)w_3(\xi,\tau)-\int\limits_0^\tau h_2(\tau-s)w_3(s,\tau)\mathrm{d}s\right]\mathrm{d}\tau\mathrm{d}\xi$$
$$+\frac{1}{2}\iint\limits_{\Delta(z,t)\bigcap\nabla}\widetilde{L}(\xi,\tau)\mathrm{d}\tau\mathrm{d}\xi. \quad (7.125)$$

The equation (7.125) is a Volterra integral equation of the second kind with a continuous kernel and a continuous free term in the domain $\Delta(z,t)\bigcap\nabla,\ t\in[0,T]$ for any positive number t. From the theory of integral equations it follows that (7.125) has a unique solution in the domain $\Delta(z, t) \bigcap \nabla$. This solution can be found by the method of successive approximations. Each iteration leads to a continuous function. The limit function is a continuous solution of the equation (7.125).

7.4 The Problem of Determining the One-Dimensional Kernel Thermoviscoelasticity Equation

For $x = (x_1, x_2, x_3), (x_1, x_2) \in \mathbb{R}^2, \ x_3 > 0, \ t \in \mathbb{R}$ let's consider the following system of integro-differential equations of dynamical thermoviscoelasticity:

$$\rho \frac{\partial^2 u_i}{\partial t^2} = \sum_{j=1}^{3} \frac{\partial T_{ij}}{\partial x_j}, \ i = 1, 2, 3, \tag{7.126}$$

$$\frac{\partial H}{\partial t} = k \Delta H, \tag{7.127}$$

under the following initial and boundary conditions:

$$u_i \mid_{t<0} \equiv 0, \ i = 1, 2, 3, \tag{7.128}$$

$$H \mid_{t<0} \equiv 0, \tag{7.129}$$

$$T_{3j}|_{x_3=+0} = \delta_{3j} \delta'(t), \ j = 1, 2, 3, \tag{7.130}$$

$$\left(\frac{\partial H}{\partial x_3} - \gamma H \right) |_{x_3=+0} = -\gamma (\widetilde{T}_1 - \widetilde{T}_0) \theta_1(t), \tag{7.131}$$

where $u(x, t) = (u_1(x, t), u_2(x, t), u_3(x, t))$ is the displacement vector, $H(x, t)$ is the temperature increment, Δ is the Laplace operator in the variables x_1, x_2, x_3; $\delta'(t)$ is the derivative of the Dirac delta function; $\widetilde{T}_1, \ \widetilde{T}_0, \ \gamma$ are constants, $\widetilde{T}_1 > \widetilde{T}_0, \gamma > 0, \theta_1(t) = t\theta(t), \theta(t) = 1$ for $t \geq 0$, $\theta(t) = 0$, for $t < 0$, T_{ij} is the stress tensor (6.67), (1.87). Here $\rho = \rho(x_3), \ \mu = \mu(x_3), \ \lambda = \lambda(x_3) \in \Lambda_1$.

The inverse problem: find the kernel $h(t)$, $t > 0$, appearing in (7.126), by using formula (6.67) if the following additional information about the solution of problem (7.126)–(7.131) is known:

$$u_3(x_3, t)|_{x_3=+0} = g(t), \ t > 0, \tag{7.132}$$

$g(t)$ is given function.

For the function $H(x_3, t)$ (see Sect. 1.3) equalities (7.126), (7.128), (7.130) are equivalent to the relations

$$\rho(x_3) \frac{\partial^2 u_3}{\partial t^2}$$

$$= L\left[h, \frac{\partial}{\partial x_3}\left((\lambda(x_3) + 2\mu(x_3))\frac{\partial u_3}{\partial x_3} - (3\lambda(x_3) + 2\mu(x_3))R(H(x_3, t))\right)\right], \tag{7.133}$$

$$u_3|_{t<0} \equiv 0, \tag{7.134}$$

$$\begin{aligned} L\left[h, \frac{\partial u_3}{\partial x_3}\right]_{x_3=+0} &= \frac{\delta'(t)}{\lambda(0) + 2\mu(0)} + \\ &\frac{3\lambda(0) + 2\mu(0)}{\lambda(0) + 2\mu(0)} L\left[h, R(H(0, t))\right], \end{aligned} \tag{7.135}$$

where $R(s) = \int_0^s \alpha(z)dz$.

Let

$$y = \psi(x_3) := \int_0^{x_3} \frac{\mathrm{d}\xi}{\nu(\xi)}, \quad \nu(x_3) := \sqrt{\frac{\lambda(x_3) + 2\mu(x_3)}{\rho(x_3)}}.$$

and

$$v(y, t) := \frac{u_3\left(\psi^{-1}(y), t\right)}{s(y)}, \quad s(y) := \sqrt{\frac{a(y)}{a(0)}},$$

$$a(y) := \frac{1}{\sqrt{[\lambda(\psi^{-1}(y)) + 2\mu(\psi^{-1}(y))]\rho(\psi^{-1}(y))}}.$$

Then, in terms of the functions introduced above and the variable y, the inverse problem (7.133)–(7.135), (7.132) takes the form

$$\frac{\partial^2 v}{\partial t^2} = L\left[h, \frac{\partial^2 v}{\partial y^2} + q(y)v\right] - \sqrt{a(y)a(0)}\frac{\partial}{\partial y} Q[h](y, t), \quad y > 0, \tag{7.136}$$

$$v|_{t<0} \equiv 0, \tag{7.137}$$

$$L\left[h, \frac{\partial v(y, t)}{\partial y}\right]_{y=+0} = a(0)\delta'(t) + a(0)Q[h](0, t), \tag{7.138}$$

$$v(y, t)|_{y=+0} = g(t), \quad t > 0, \tag{7.139}$$

where

$$q(y) := \frac{s''(y)}{s(y)} - 2\left[\frac{s'(y)}{s(y)}\right]^2, \quad \tilde{\lambda}(y) := \lambda(\psi^{-1}(y)),$$

$$\tilde{\mu}(y) := \mu(\psi^{-1}(y)), \quad \widetilde{H}(y, t) := H(\psi^{-1}(y), t),$$

$$Q[h](y, t) := (3\tilde{\lambda}(y) + 2\tilde{\mu}(y))\left[R(\widetilde{H}(y, t)) + \int_0^t h(t - \tau)R(\widetilde{H}(y, \tau))\mathrm{d}\tau\right].$$

Let $\left[v(y,t) + \int_0^t h(t-\tau)v(y,\tau)d\tau\right] \exp\left(-h(0)t/2\right) = w(y,t)$.
Then as is easy to see

$$v(y,t) = \exp\left(h(0)t/2\right) w(y,t) + \int_0^t r(t-\tau) \exp\left(h(0)\tau/2\right) w(y,\tau)d\tau,$$

where $r(t)$ is determined by 6.76.

In terms of the functions $w(y,t)$ and $r(t)$, equations (7.136)–(7.139) take the form

$$\frac{\partial^2 w}{\partial t^2} = \frac{\partial^2 w}{\partial y^2} + B(y)w - \sqrt{a(y)a(0)}\frac{\partial}{\partial y}Q_0[h](y,t) - \int_0^t \tilde{h}(t-\tau)w(y,\tau)d\tau, \quad (7.140)$$

$$w|_{t<0} \equiv 0, \quad (7.141)$$

$$\left.\frac{\partial w}{\partial y}\right|_{y=+0} = a(0)\delta'(t) - \frac{a(0)}{2}r(0)\delta(t) + a(0)Q_0[h](0,t), \quad (7.142)$$

$$w|_{y=+0} = \tilde{g}(t) + \int_0^t h_0(t-\tau)\tilde{g}(\tau)d\tau, \quad (7.143)$$

where

$$B(y) := q(y) + \frac{r^2(0)}{4} - r'(0), \; \tilde{h}(t) := r''(t) \exp\left(r(0)t/2\right),$$

$$\tilde{g}(t) := -a(0)\delta(t) + \tilde{g}_0(t)\theta(t), \; \tilde{g}_0(t) := g_0(t) \exp\left(r(0)t/2\right),$$

$$h_0(t) := h(t) \exp\left(r(0)t/2\right),$$

$$Q_0[h](y,t) := Q[h](y,t) \exp\left(r(0)t/2\right).$$

It follows from the theory of hyperbolic equations that the function $w(y,t)$ being a solution of the direct problem (7.140)–(7.142) possesses the property $w \equiv 0$, $t < y$, $y > 0$ and, in a neighborhood of the characteristic line $t = y$, has the following structure:

$$w(y,t) = -a(0)\delta(t-y) + \theta(t-y)\tilde{w}(y,t), \quad (7.144)$$

where $\tilde{w}(y,t)$ is a regular function.

The function $\tilde{w}(y,t)$ satisfies the equations ($w = \tilde{w}$ for $t > y$):

$$\frac{\partial^2 w}{\partial t^2}=\frac{\partial^2 w}{\partial y^2}+B(y)w+a(0)\tilde{h}(t-y)-\sqrt{a(y)a(0)}\frac{\partial}{\partial y}Q_0[h](y,t)\\ -\int\limits_y^t \tilde{h}(t-\tau)w(y,\tau)\mathrm{d}\tau, \tag{7.145}$$

$$w|_{t=y+0}=\frac{a(0)r(0)}{2}-\frac{a(0)}{2}\int\limits_0^y B(\xi)\mathrm{d}\xi:=\beta(y), \tag{7.146}$$

$$\left.\frac{\partial w}{\partial y}\right|_{y=+0}=a(0)Q_0[h](0,t), \tag{7.147}$$

$$w|_{y=+0}=\tilde{g}_0(t)+\int\limits_0^t h_0(t-\tau)\tilde{g}_0(\tau)\mathrm{d}\tau-a(0)h_0(t). \tag{7.148}$$

Provided that, for $y=t=0$ the functions $w(y,t)$ and $\left(\frac{\partial w}{\partial y}\right)(y,t)$ are continuous, using equations (7.146)–(7.148), we can (as in Sect. 7.1) easily express $r(0),\ r'(0)$ in terms of the known values:

$$r(0)=-\frac{2}{a(0)}g_0(0),\ r'(0)=-q(0)-\frac{1}{a^2(0)}g_0^2(0)-\frac{2}{a(0)}\tilde{g}_0'(0). \tag{7.149}$$

Lemma 7.3 *Let the function $g(t)$ be expressed as*

$$g(t)=-a(0)\delta(t)+\theta(t)g_0(t),\ a(0)=\{[\lambda(+0)+2\mu(+0)]\rho(+0)\}^{-\frac{1}{2}}$$

and let $g_0(t)\in C^2\,[0,\ T]$, $\theta(t)$ be the Heaviside function. In addition, let

$$(\rho,\lambda,\mu)\in C^3\left[0,\psi^{-1}(T/2\right],\ \alpha(z)\in C[0,\widetilde{H}(T/2,T/2)].$$

Then, for $(y,t)\in D_T$, $D_T:=((y,t)|0\le y\le t\le T-y)$ problem (2.14)–(2.17) is equivalent to the problem of determining the vector functions $w(y,t)$, $\left(\frac{\partial w}{\partial t}\right)(y,t)$, $\tilde{h}(t)$, $h_0(t)$, $h_0'(t)$, $h_0''(t)$ from the following system of integral equations:

$$w(y,t)=\beta(y)+\int\limits_y^t \frac{\partial w}{\partial \tau}(y,\tau)\mathrm{d}\tau, \tag{7.150}$$

$$\begin{aligned}\frac{\partial w}{\partial t}(y,t) &= \frac{1}{2}(\tilde{g}_0'(t-y) - r(0)\tilde{g}_0(t-y)) \\ &- \frac{a(0)}{4}B\left(\frac{y+t}{2}\right) + \frac{a(0)}{2}\tilde{h}(t-y)y - \frac{a(0)}{2}h_0'(t-y) \\ &- \frac{\sqrt{a(\frac{y+t}{2})a(0)}}{2}Q_0[h]\left(\frac{y+t}{2},\frac{y+t}{2}\right) + \frac{1}{2}\int_0^{t-y} h_0'(t-y-\tau)\tilde{g}_0(\tau)\mathrm{d}\tau \\ &+ \frac{1}{2}\int_0^y \Bigg[B(\xi)w(\xi,t-y+\xi) + \sqrt{\frac{a(0)}{4a(\xi)}}a'(\xi)Q_0[h](\xi,t-y+\xi) \\ &- \int_0^{t-y}\tilde{h}(\tau)w(\xi,t-y+\xi-\tau)\mathrm{d}\tau\Bigg]\mathrm{d}\xi \\ &+ \frac{1}{2}\int_y^{\frac{t+y}{2}} \Bigg[B(\xi)w(\xi,t+y-\xi) + \sqrt{\frac{a(0)}{4a(\xi)}}a'(\xi)Q_0[h](\xi,t+y-\xi) \\ &+ a(0)\tilde{h}(t+y-2\xi) - \int_0^{t+y-2\xi}\tilde{h}(\tau)w(\xi,t+y-\xi-\tau)\mathrm{d}\tau\Bigg]\mathrm{d}\xi,\end{aligned} \tag{7.151}$$

$$\begin{aligned}\tilde{h}(t) &= \frac{1}{2}B'\left(\frac{t}{2}\right) - 2h_0''(t) \\ &+ \frac{2}{a(0)}\left[\tilde{g}_0''(t) - r(0)\tilde{g}_0'(t) + \left(\frac{r^2(0)}{2} - r'(0)\right)\tilde{g}_0(t) - \frac{1}{2}B\left(\frac{t}{2}\right)\beta\left(\frac{t}{2}\right)\right] \\ &+ 4\frac{a(\frac{t}{2})}{a^2(0)}\frac{d}{dt}Q_0[h]\left(\frac{t}{2},\frac{t}{2}\right) + \frac{2}{a(0)}\int_0^t h_0''(t-\tau)\tilde{g}_0(\tau)\mathrm{d}\tau \\ &+ \frac{1}{a(0)}\int_0^t \tilde{h}(\tau)\beta\left(\frac{t-\tau}{2}\right)\mathrm{d}\tau \\ &- \frac{2}{a(0)}\int_0^{\frac{t}{2}}\Bigg[B(\xi)\frac{\partial w}{\partial t}(\xi,t-\xi) - \sqrt{\frac{a(0)}{4a(\xi)}}a'(\xi)\frac{\partial}{\partial t}Q_0[h](\xi,t-\xi) \\ &- \int_0^{t-2\xi}\tilde{h}(\tau)\frac{\partial w}{\partial t}(\xi,t-\xi-\tau)\mathrm{d}\tau\Bigg]\mathrm{d}\xi,\end{aligned} \tag{7.152}$$

$$h_0(t) = -r(0) + \left(\frac{r^2(0)}{2} - r'(0)\right)t + \int_0^t (t-\tau)h_0''(\tau)\mathrm{d}\tau, \tag{7.153}$$

$$h_0'(t) = \frac{r^2(0)}{2} - r'(0) + \int_0^t h_0''(\tau)d\tau, \tag{7.154}$$

$$h_0''(t) = -\tilde{h}(t) + \left(\frac{r^2(0)}{4} - r'(0)\right) h_0(t) - \int_0^t \tilde{h}(t-\tau)h_0(\tau)d\tau. \tag{7.155}$$

Proof of the Lemma 7.3 is similar to the proof of the Lemma 7.1.

The main results of this section are the following theorems on global unique solvability and stability of the solution of the inverse problem.

Theorem 7.6 *Let the assumptions of the lemma 7.3 hold, and let $T,\ \widetilde{T}_0,\ \widetilde{T}_1,\ k,\ \gamma$ be fixed positive numbers, where $\widetilde{T}_1 > \widetilde{T}_0$. Then there exists a unique solution of the inverse problem (7.126)–(7.131) $h(t) \in C^2[0,\ T]$ for any fixed $T > 0$.*

Let $K(h_0^*)$ be the set of functions $h(t) \in C^2[0, T]$, satisfying the inequality $\|h(t)\|_{C^2[0,T]} \le h_0^*$, where $t \in [0, T]$ and h_0^* is a fixed positive constant.

Theorem 7.7 *Let $h^1(t),\ h^2(t) \in K(h_0^*)$ be solutions of the inverse problem (7.126)–(7.131) with the data*

$$\left\{\rho^1(\psi^{-1}(y)), \lambda^1(\psi^{-1}(y)), \mu^1(\psi^{-1}(y)), \alpha^1(z), g_0^1(t)\right\},$$

$$\left\{\rho^2(\psi^{-1}(y)), \lambda^2(\psi^{-1}(y)), \mu^2(\psi^{-1}(y)), \alpha^2(z), g_0^2(t)\right\},$$

respectively. Then, there exists a positive number $C = C(h_0^*, h_{00}^*, \gamma,\ \widetilde{T}_0,\ \widetilde{T}_1, T)$ and

$$h_{00}^* = \max\left\{\|\rho^j(y)\|_{C^3[0,\psi^{-1}(T/2)]}, \|\lambda^j(y)\|_{C^3[0,\psi^{-1}(T/2)]},\right.$$

$$\left.\|\mu^j(y)\|_{C^3[0,\psi^{-1}(T/2)]}, \|\alpha^j(z)\|_{C[0,\widetilde{H}(T/2,T/2)]}, \|g_0^j(t)\|_{C^2[0,T]},\ j = 1,2\right\},$$

such that the following stability estimate holds:

$$\|h^1(t) - h^2(t)\|_{C^2[0,T]} \le C\Big[\|\rho^1 - \rho^2\|_{C^3[0,\psi^{-1}(T/2)]} + \|\lambda^1 - \lambda^2\|_{C^3[0,\psi^{-1}(T/2)]}$$
$$+ \|\alpha^1 - \alpha^2\|_{C[0,\widetilde{H}(T/2,T/2)]} + \|\mu^1 - \mu^2\|_{C^3[0,\psi^{-1}(T/2)]} + \|g_0^1 - g_0^2\|_{C^2[0,T]}\Big]. \tag{7.156}$$

Proofs of the Theorem 7.6, 7.7 are similar to the proofs of the Theorems 7.1, 7.2.

7.5 Conclusions

In this chapter, the methods discussed in Chaps. 1, 2 and 3 were developed. It is shown that the resolvent of the memory kernel makes it possible to reduce the viscoelasticity equation to the studied model problems of determining kernel. For one-dimensional problems, global solvability is investigated and stability estimates in the appropriate classes of functions are obtained. In addition, methods have been developed for finding matrix diagonal kernel in the case of a source of explosive type, which is relevant in geophysics. Sufficient conditions for the global unique solvability of the problem of determining the diagonal memory kernel are formulated. For scalar kernels depending on two or more variables, the method of scales of Banach spaces, discussed in Chap. 2, is used. The theorem of local unique solvability of the finding of a two-dimensional kernel for the viscoelasticity equations is formulated.

References

1. Yakhno, V.G. 1990. *Inverse problems for differetial equations of elasticity*. Novosibirsk: Nauka (Russian).
2. Karchevsky, A.L., and V.G. Yakhno. 1999. One-dimensional inverse problems for systems of elasticity with a source of explosive type. *Journal of Investigation Ill - Posed Problems* 7 (4): 347–364.
3. Yakhno, V.G., and I.Z. Merazhov. 2000. Some direct problems and one-dimensional inverse electroelasticity problem for "slow" waves. *Siberian Advance Mathematical* 10 (1): 87–150.

Chapter 8
Dirichlet-to-Neumann Maps Method in Kernel Determining Problems

In this chapter, the Dirichlet-to-Neumann maps method is used to solve multidimensional inverse problems. In the first section, we study the problem of determining the multidimensional coefficient at the lowest derivative of a second-order hyperbolic equation for a half space.

The next two sections investigate the problem of finding the multidimensional kernel of an integral in an integro-differential equation. Considered in these sections, problems resemble that of [1, 2] in their forms; however, in the proof it applies the Fourier transform in the second section and Laplace transform in the third section in time to reduce the problems to a family of stationary problems and employ the technique developed for elliptic equations.

8.1 The Dirichlet-to-Neumann Maps Method for a Half Space

Consider the problem of determining the coefficient $q(x, y)$ of the equation:

$$u_{tt} = u_{xx} + \Delta_y u - q(x, y)u, \;\; y \in \mathbb{R}^n, \;\; x \in \mathbb{R}_+, \;\; t \in [0, T], \tag{8.1}$$

in the case when for the solution of the initial–boundary value problem with the data

$$\begin{gathered} u(x, y, 0) = 0, \quad u_t(x, y, 0) = 0, \\ (x, y) \in \mathbb{R}^{n+1}_+ = \{(x, y) : \; y \in \mathbb{R}^n, \; x > 0\}, \end{gathered} \tag{8.2}$$

$$u_x(0, y, t) = g(y, t), \quad y \in \mathbb{R}^n, \quad t \in [0, T], \tag{8.3}$$

© The Author(s), under exclusive license to Springer Nature Singapore Pte Ltd. 2023
D. K. Durdiev and Z. D. Totieva, *Kernel Determination Problems in Hyperbolic Integro-Differential Equations*, Infosys Science Foundation Series in Mathematical Sciences, https://doi.org/10.1007/978-981-99-2260-4_8

the additional information

$$u(0, y, t) = f_g(y, t), \quad y \in \mathbb{R}^n, \quad t \in [0, T]. \tag{8.4}$$

is given.

The uniqueness of the inverse problem solution in the class $L^\infty(\mathbb{R}_+^{n+1})$ can be proved only in the overdetermined statement. Rakesh [3] extended the Sylvester-Ullman method [4] to the inverse problem (8.1)–(8.4). Following [4] let us briefly describe the scheme of the study.

We define the Dirichlet-to-Neumann map

$$\Lambda_q : L_2(\mathbb{R}^n \times [0, T]) \to H^1(\mathbb{R}^n \times [0, T]),$$

which each function $g(y, t) \in L_2(\mathbb{R}^n \times [0, T])$ maps to the function $f_g(y, t) \in H^1(\mathbb{R}^n \times [0, T])$ which is the trace of the solution of the direct problem (8.1)–(8.3) on the plane $x = 0$. The definitions of the classes of functions $L_2(G)$ and $H^1(G) = W_2^1(G)$, where G is a connected open domain in $\mathbb{R}^n$ the reader can find in the book [5].

We denote by $B(r) \subset R^{n+1}$ an open ball of radius r with the center at zero:

$$B(r) = \left\{ (x, y) : \sqrt{x^2 + \sum_{i=1}^{n} y_i^2} < r \right\}.$$

Theorem 8.1 *Suppose that the functions q_1, q_2 belong to the space $L^\infty(\mathbb{R}_+ \times \mathbb{R}^n)$ and are constant outside the closed ball $\overline{B(r)}$.*

Then, if $(\pi + 1)r < T$, and from the condition

$$\Lambda_{q_1}(g) = \Lambda_{q_2}(g), \quad g \in L_2(\mathbb{R}^n \times [0, T]),$$

it follows $q_1 \equiv q_2$.

Proof Suppose that outside the ball $\overline{B(r)}$ the functions q_1 and q_2 are equal to the same constant k (the general case will follow from Lemma 8.3).

Let $g(y, t)$ be a sufficiently smooth function that has a compact support for every $t \in [0, T]$. We denote $p = q_2 - q_1$, $w = u_1 - u_2$, where u_j is the solution of the direct problem (8.1)–(8.3) with boundary data $g(y, t)$ and a fixed coefficient $q(x, y) = q_j(x, y)$, $j = 1, 2$. Note that since $g(y, t)$ has a compact support for a fixed $t \in [0, T]$, the solutions u_j will also have a compact support for a fixed $t \in [0, T]$.

From the condition $\Lambda_{q_1}(g) = \Lambda_{q_2}(g)$, it follows that

$$w_{tt} = w_{xx} + \Delta_y w - q_1 w + p u_2, \; x > 0, \; y \in \mathbb{R}^n, \; t \in [0, T], \tag{8.5}$$

$$w(x, y, 0) = 0, \quad w_t(x, y, 0) = 0, \; x > 0, \; y \in \mathbb{R}^n, \tag{8.6}$$

$$w(0, y, t) = 0, \quad w_x(0, y, t) = 0, \; y \in \mathbb{R}^n, \; r \in [0, T]. \tag{8.7}$$

Note that $p(x, y) = 0$ outside the ball $\overline{B(r)}$.

Let $\varepsilon = (T - (\pi + 1)r)/6$, $T_* = T - \pi(r + \varepsilon)$, the function $v(x, y, t)$ satisfies the following conditions

$$\begin{gathered} v_{tt} = v_{xx} + \Delta_y v - q_1(x, y)v, \\ (x, y \in H = \{(x, y) \in \overline{B(r)} : x > 0\}, \end{gathered} \tag{8.8}$$

$$v(x, y, T_*) = 0, \quad v_t(x, y, T_*) = 0, \quad (x, y) \in H \tag{8.9}$$

and is quite smooth.

Then, using (8.5) and integrating in parts, we get

$$\int\limits_0^{T_*}\int\limits_H pu_2 v \, \mathrm{d}x\mathrm{d}y\mathrm{d}t = \int\limits_0^{T_*}\int\limits_H (w_{tt} - w_{xx} - \Delta_y w + q_1 w)v\mathrm{d}x\mathrm{d}y\mathrm{d}t$$

$$= \int\limits_0^{T_*}\int\limits_H (v_{tt} - v_{xx} - \Delta_y v + q_1 v)w\mathrm{d}x\mathrm{d}y\mathrm{d}t$$

$$+ \int\limits_0^{T_*}\int\limits_{\partial H} \left(w\frac{\partial v}{\partial \nu} - v\frac{\partial w}{\partial \nu}\right)\mathrm{d}S\mathrm{d}t + \int\limits_H (w_t v - w v_t)\Big|_{t=0}^{t=T_*} \mathrm{d}x\mathrm{d}y,$$

where $\frac{\partial}{\partial \nu}$ is the normal derivative on the surface of the hemisphere H. It can be shown [3] that

$$w(x, y, t) = 0, \quad \frac{\partial w}{\partial \nu}(x, y, t) = 0 \text{ if } (\mathrm{x}, \mathrm{y}) \in \partial \mathrm{H}, \; \mathrm{t} \in [0, \mathrm{T}_*].$$

Hence,

$$\int\limits_0^{T_*}\int\limits_{\partial H} p(x, y)u_2(x, y, t)v(x, y, t)\mathrm{d}x\mathrm{d}y\mathrm{d}t = 0 \tag{8.10}$$

for all $g \in C^\infty(\mathbb{R}^n \times [0, T])$, for all corresponding solutions u_2 of the problem (8.1)–(8.3) and all smooth v satisfying (8.8)–(8.9). It follows from (8.10) that $p(x, y) = 0$. We will highlight the sets of functions u_2 and v that concentrate along some lines lying in space (x, y). The construction of such functions is based on the following Lemma [3].

Lemma 8.1 *Let $z \in \mathbb{R}^{n+1}$ be an arbitrary unit vector, $\sigma > 0$, the function $\alpha(x, y) \in C^\infty(\mathbb{R}^{n+1})$ is finite, and*

$$\text{supp}\{\alpha\} \subset N_\epsilon = \{(x, y) \in B(2\epsilon) : x < -\epsilon\} \tag{8.11}$$

Then, for any locally bounded functions $q(x, y)$, we can construct functions $v(x, y, t)$ and $u_2(x, y, t)$ of the forms

$$\begin{aligned} v(x, y, t) &= \alpha\,((x, y) + tz)\,\mathrm{e}^{-i\sigma(\langle(x,y),z\rangle+t)} + \beta_1(x, y, t), \\ u_2(x, y, t) &= \alpha\,((x, y) + tz)\,\mathrm{e}^{+i\sigma(\langle(x,y),z\rangle+t)} + \beta_2(x, y, t), \end{aligned} \tag{8.12}$$

such that u_2 is the solution of (8.1)–(8.3) with the appropriate choice of $g(y, t)$, v satisfies (8.8)–(8.9), and there are estimates

$$\|\beta_1\|_{L_2(H\times[0,T_*])} \leq \frac{c}{\sigma}, \quad \|\beta_2\|_{L_2(\mathbb{R}^{n+1}_+\times[0,T])} \leq \frac{c}{\sigma},$$

in which the constant c depends only on T, α, and q.

Using Lemma 8.1, we define the functions v and u_2 as (8.12) and substitute them in (8.10):

$$\begin{aligned} 0 &= \int_0^{T_*}\int_{\partial H} p(x, y)u_2(x, y, t)v(x, y, t)\mathrm{d}x\mathrm{d}y\mathrm{d}t \\ &= \int_0^{T_*}\int_{\partial H} p(x, y)\alpha^2((x, y) + tz)\mathrm{d}x\mathrm{d}y\mathrm{d}t + \gamma(z, \sigma). \end{aligned}$$

It can be shown [3] that

$$|\gamma(z, \sigma)| \leq \frac{c}{\sigma},$$

where the constant c is independent on σ. Therefore,

$$\begin{aligned} 0 &= \lim_{\sigma\to\infty}\int_0^{T_*}\int_{\partial H} p(x, y)u_2(x, y, t)v(x, y, t)\mathrm{d}x\mathrm{d}y\mathrm{d}t \\ &= \int_0^{T_*}\int_{\partial H} p(x, y)\alpha^2((x, y) + tz)\mathrm{d}x\mathrm{d}y\mathrm{d}t \\ &= \int_{R^{n+1}}\left[\int_0^{T_*} p((x, y) - tz)\mathrm{d}t\right]\alpha^2(x, y)\mathrm{d}x\mathrm{d}y. \end{aligned}$$

Here we used the fact that $p(x, y) = 0$ outside H.

In view of the completeness of the class of functions $\alpha^2(x, y)$ with the condition (8.11) in $L_2(N_\epsilon)$, it follows that

$$\int_0^{T_*} p(a - tz)dt = 0$$

for all $a \in N_\epsilon$, $z \in \mathbb{R}^{n+1}$, $\|z\| = 1$. Due to the finiteness of $p(x, y)$, we get

$$\int_0^{\infty} p(a - tz)dt = 0$$

for all $a \in N_\epsilon$, $z \in \mathbb{R}^{n+1}$, $\|z\| = 1$.

It remains to apply Lemma 9.1.2 [6] and Lemma 9.1.3 [3] to complete the proof of the Theorem 8.1.

Lemma 8.2 *Let $p(x, y)$ be a function bounded on $\mathbb{R}^{n+1}$ with a compact support whose convex hull is separated from some infinite subset $A \subset R^{n+1}$. Then if for all $a \in A$ and $z \in \mathbb{R}^{n+1}$, $\|z\| = 1$ the equation*

$$\int_0^{\infty} p(a + sz)dt = 0,$$

is fair, then $p(x, y) = 0$.

Lemma 8.3 . *Suppose that k_j $(j = 1, 2)$ are the real numbers and for $j = 1, 2$ the functions $u^j(x, y, t)$ satisfy the relations*

$$u^j_{tt} = u^j_{xx} + \Delta_y u^j - k_j u^j, \quad y \in \mathbb{R}^n, \quad x \geq 0, \ t \in [0, T],$$

$$u^j(x, y, 0) = 0, \quad u^j_t(x, y, 0) = 0, \quad y \in \mathbb{R}^n, \ x \geq 0,$$

$$\frac{\partial u^j}{\partial x}(0, y, t) = g(y, t), \quad y \in \mathbb{R}^n, \quad t \in [0, T].$$

If $u^1(0, 0, t) = u^2(0, 0, t)$ for $t \in [0, T]$ for all $g \in C^2(\mathbb{R}^n \times [0, T])$, such that $\operatorname{supp} g \subset \{y : \|y\| < 2T\} \times [0, T]$, then $k_1 = k_2$.

8.2 Stability of Memory Reconstruction from the Dirichlet-to-Neumann Operator

In this section, we consider the problem of reconstructing the spatially inhomogeneous memory $k(x, t)$ in the integro-differential wave equation over a given

Dirichlet–Neumann operator. An estimate of the conditional stability of the solution to the problem is obtained. The presentation of this section follows [7].

8.2.1 The Problem Statement and Results

Let Ω be a bounded domain in $\mathbb{R}^n$, $n \geq 3$, with a smooth boundary $\partial\Omega$. Given the domain $\Omega \times \mathbb{R}$ considers the following integro-differential equation with memory:

$$u_{tt}(x,t) - \Delta u(x,t) - \int_{-\infty}^{t} u(x,\tau)k(x,t-\tau)\mathrm{d}\tau = 0. \tag{8.13}$$

Here $k(x,t)$ is a sufficiently smooth memory function in $\Omega \times (0,\infty)$. We assume $k(x,t)$ extended by zero to $t < 0$. Suppose that for a certain class of functions $g(x,t)$ given on the boundary $\partial\Omega \times \mathbb{R}$, there exists a sufficiently smooth solution $u(x,t)$ of the Eq. (8.13) with the boundary condition

$$u(x,t)\Big|_{\partial\Omega\times\mathbb{R}} = g(x,y). \tag{8.14}$$

On the indicated class of functions we define the nonstationary Dirichlet-to-Neumann operator H that assigns to $g(x,t)$ the normal derivative $\partial_\nu u$ of a solution to (8.13), (8.14) on the boundary $\partial\Omega \times \mathbb{R}$. We consider the inverse problem consisting in reconstructing an unknown memory $k(x,t)$ given the Dirichlet-to-Neumann operator H. We prove conditional stability of a solution to the problem; i.e., we show that for some class of functions $k(x,t)$ we may guarantee closeness of two different memories, provided that the corresponding Dirichlet-to-Neumann operators are close. We suppose that the function $k(x,t)$ belongs to the well-posedness set

$$W(K) := \{k(x,t) \in C^2(\bar{\Omega} \times [0,\infty)) \mid \|k\|_{C^2(\bar{\Omega}\times[0,\infty))} \leq K\},$$

where K some positive number.

Theorem 8.2 *Let $k_j(x,t)$, $j = 1, 2$, be two (different) functions of the class $W(K)$. There is $\gamma_0 > 0$ depending on Ω and K and such that, for every $\gamma \geq \gamma_0$, the corresponding Dirichlet-to-Neumann operators H_j are well defined as operators from $H^1_\gamma(\partial\Omega \times \mathbb{R})$ into $H^0_\gamma(\partial\Omega \times \mathbb{R})$ (see the definition of the spaces and the corresponding norms in the next section) and the estimate*

$$\|k_1 - k_2\|_{\gamma,0} \leq w(\|H_1 - H_2\|),$$

holds with $w(\varepsilon) \sim C_{\Omega,K,\gamma,n}(C_\Omega/\ln \varepsilon^{-1})^{1/(n+2)}$ as $\varepsilon \to 0$.

Here and in the sequel, the letters C with indices stand for constants whose indices indicate the quantities which the constants depend on.

8.2.2 The Direct Problem

Assume that $\gamma > 0$. Denote by $H^s_\gamma(\Omega \times \mathbb{R})$ the space of functions $u(x, t)$ such that $e^{-\gamma t}u(x, t) \in H^s_\gamma(\Omega \times \mathbb{R})$ and denote the corresponding norm by $\|\cdot\|_{\gamma,s}$. Define $H^s_\gamma(\partial\Omega \times \mathbb{R})$ similarly, with the corresponding norm denoted by $\langle\cdot\rangle|_{\gamma,s}$. Solvability of the direct problem is established in the following theorem

Theorem 8.3 *Assume that $K > 0$. There are positive constants γ_0 and C depending on Ω and K and such that, for arbitrary $\gamma \geq \gamma_0$, $k(x, t) \in C(\bar{\Omega} \times (0, \infty))$, $\|k\|_C \leq K$, and $g(x, t) \in H^1_\gamma(\partial\Omega \times \mathbb{R})$, there is a unique solution $u(x, t) \in H^1_\gamma(\Omega \times \mathbb{R})$ to (8.13) with the boundary condition (8.14); moreover, the solution satisfies the estimate*

$$\gamma \|u\|^2_{\gamma,1} + \langle\partial_\nu u\rangle^2_{\gamma,0} \leq \frac{C}{\gamma}\langle g\rangle^2_{\gamma,1}. \tag{8.15}$$

Thus, the Dirichlet-to-Neumann operator acts correctly from $H^1_\gamma(\partial\Omega \times \mathbb{R})$ into $H^0_\gamma(\partial\Omega \times \mathbb{R})$.

The bulk of the proof is contained in the following assertion which is a consequence of a more general result on solvability of boundary value problems for hyperbolic equations ([8], Theorems 11.1 and 13.5):

Theorem 8.4 *There are $\gamma_0 \geq 0$ and C depending on Ω and such that, for arbitrary $\gamma \geq \gamma_0$, $g(x, t) \in H^1_\gamma(\partial\Omega \times \mathbb{R})$, and $f(x, t) \in H^0_\gamma(\partial\Omega \times \mathbb{R})$, there is a unique solution $u(x, t) \in H^1_\gamma(\Omega \times \mathbb{R})$ to the equation*

$$u_{tt} - \Delta u = f(x, t) \tag{8.16}$$

given condition (8.14); moreover, the following estimate is valid:

$$\gamma \|u\|^2_{\gamma,1} + \langle\partial_\nu u\rangle^2_{\gamma,0} \leq C\left(\frac{1}{\gamma}\|f\|_{\gamma,0} + \langle g\rangle^2_{\gamma,1}\right). \tag{8.17}$$

Proof Proof of Theorem 8.3. We seek a solution to (8.13) and (8.14) in the form $u = u^0 + v$, where u^0 is a solution to the problem

$$u^0_{tt} - \Delta u^0 = 0, \quad u^0|_{\partial\Omega\times\mathbb{R}} = g(x, t),$$

and v satisfies the equation

$$v_{tt} - \Delta v = k * (v + u^0) \tag{8.18}$$

and the homogeneous condition

$$v|_{\partial\Omega\times\mathbb{R}} = 0. \tag{8.19}$$

Here $*$ stands for convolution in time. According to Theorem 8.4, for $\gamma \geq \gamma_0$ there is a well-defined bounded linear operator $E : H^0_\gamma(\Omega \times \mathbb{R}) \to H^1_\gamma(\Omega \times \mathbb{R})$ which assigns to a function $f \in H^0_\gamma(\partial\Omega \times \mathbb{R})$ a solution u to Eq. (8.16) with the homogeneous boundary conditions. Moreover, the norm $\|E\|$ depends only on the domain Ω and γ_0. To solve the problem (8.18), (8.19), it suffices to solve the equation

$$v = E(k * v) + E(k * u^0) \tag{8.20}$$

in the space $H^1_\gamma(\Omega \times \mathbb{R})$. Simple calculations using Young's inequality show that

$$\|k * v\|^2_{\gamma,0} \leq \|(\mathrm{e}^{-\gamma t}k) * (\mathrm{e}^{-\gamma t}v)\|^2_{0,0} = \int_\Omega \int_R |(\mathrm{e}^{-\gamma t}k) * (\mathrm{e}^{-\gamma t}v)|^2 \,\mathrm{d}t\mathrm{d}x$$

$$\leq \int_\Omega \left(\int_0^\infty |\mathrm{e}^{-\gamma t}k|\mathrm{d}t\right)^2 \left(\int_0^\infty |\mathrm{e}^{-\gamma t}v|^2\mathrm{d}t\right) \mathrm{d}x \leq \|k\|^2_C \frac{1}{\gamma^2}\|v\|^2_{\gamma,0}.$$

Increasing γ_0 of Theorem 8.4, if necessary, so as to have $\|E\|K/\gamma_0 \leq q$ for some $q,\ 0 < q < 1$, we obtain

$$\|E(k * v)\|_{\gamma,1} \leq q\|v\|_{\gamma,0}, \quad \gamma \geq \gamma_0,$$

which guarantees unique solvability of Eq. (8.20). It is obvious that a solution to (8.20) is a unique solution to (8.18).

Prove (8.16). Using (8.17), from (8.18) we derive

$$\gamma\|v\|^2_{\gamma,1} + \langle\partial_\nu v\rangle^2_{\gamma,0} \leq \frac{C}{\gamma}\left(\|k * v\|^2_{\gamma,0} + \|k * u^0|^2_{\gamma,0}\right) \leq \frac{C\|k\|^2_C}{\gamma^3}\left(\|v\|^2_{\gamma,0} + \|u^0|^2_{\gamma,0}\right).$$

Increasing γ_0 once again so as to have $\gamma_0 - CK/\gamma_0^3 \geq 1$ and applying the estimate (8.18) for $u^0 : \ \|u^0\|^2_{\gamma,1} \leq C\langle g\rangle^2_{\gamma,1}$ for $\gamma \geq \gamma_0$ we obtain

$$\gamma\|u\|^2_{\gamma,1} + \langle\partial_\nu u\rangle^2_{\gamma,0} \leq \frac{C}{\gamma}\langle g\rangle^2_{\gamma,1}$$

with some new constant C. Theorem 8.3 is proven.

8.2.3 Auxiliary Assertions

In Theorem 8.2 we claim existence of some $\gamma > 0$. First we take γ_0 to be the quantity of Theorem 8.3. During the proof (of auxiliary assertions as well as the theorem) we possibly increase it.

The Fourier-Laplace transform. Given a function $u(x,t)$ such that $ue^{-\gamma t} \in L_2(\Omega \times \mathbb{R})$ (or $ue^{-\gamma t} \in L_2(\partial\Omega \times \mathbb{R}^n)$), introduce the Fourier-Laplace transform in time as follows:

$$(Fu)(x,\theta) = \hat{u}(x,\theta) := \frac{1}{\sqrt{2\pi}} \int e^{-it\theta} u(x,t)\mathrm{d}t,$$

where $\theta = \sigma - i\gamma,\ \sigma \in \mathbb{R}$. Then the Parseval identity holds:

$$\langle u(x,t)\rangle_{\gamma,0} = \|\hat{u}(x,\sigma - i\gamma)\|_{L_2(\partial\Omega\times\mathbb{R})}, \tag{8.21}$$

where $\hat{u}$ is considered as a function of x and σ.

Applying the Fourier-Laplace transform to Eq. (8.13) and the boundary condition (8.14), we obtain

$$\Delta\hat{u}(x.\theta) + (\hat{k}(x,\theta) + \theta^2)\hat{u}(x,\theta) = 0, \tag{8.22}$$

$$\hat{u}(x,\theta)|_{\partial\Omega\times\mathbb{R}} = \hat{g}(x,\theta). \tag{8.23}$$

Thus, $\hat{u}(x,\theta)$ is a solution to the family of the stationary problems (8.22), (8.23) depending on the parameter $\theta = \sigma - i\gamma,\ \sigma \in \mathbb{R}$. It follows from unique solvability of the problem (8.13), (8.14) that the Dirichlet problem (8.22), (8.23) has a unique solution for almost every $\theta = \sigma - i\gamma,\ \sigma \in \mathbb{R}$. Therefore, the inverse Fourier-Laplace transform of a solution to the family of the Dirichlet problems (8.22), (8.23) (with a suitable $\hat{g}(x,\theta)$ is a solution to the nonstationary problem (8.13), (8.14).

The basic identity. Assume that $q_j(x) \in C^2(\bar{\Omega}),\ j = 1,\ 2$. Consider the following equation in Ω:

$$\Delta v_j(x) + q_j v_j(x) = 0. \tag{8.24}$$

Suppose that the Dirichlet problem for (8.24) in Ω has at most one solution. For each $q_j(x)$, define the so-called (stationary) Dirichlet-to-Neumann operator Λ_j as follows:

$$\Lambda_j g := \partial_\nu v_j|_{\partial\Omega},$$

where v_j is a solution to (8.24) given the condition $v|_{\partial\Omega} = g$. The operator Λ_j is a bounded operator from $H^1(\partial\Omega)$ into $H^0(\partial\Omega)$. This fact is a consequence of the general results on solvability of elliptic equations in the scale $H^s(\Omega)$ (see, for instance, [9], Chap. 2, Theorem 7.4) and the trace theorem (see, for instance, [9], Chap. 1, Theorem 9.4).

Lemma 8.4 *The following identity holds for sufficiently smooth solutions v_j to (8.24), $j = 1, 2$:*

$$\int_{\Omega} (q_2 - q_1)v_1 v_2 \mathrm{d}x = \int_{\partial\Omega} v_1(\Lambda_1 - \Lambda_2)v_2 \mathrm{d}S. \tag{8.25}$$

Here and in the sequel, the notation like $\Lambda_1 v_1$ means that we apply the operator Λ_1 to the trace $v_1|_{\partial\Omega}$. In the case of real potentials $q_j(x)$, (14) was proven in [10]. The case of general elliptic operators was considered in [11].

Proof Multiply (8.24) written for v_1 by v_2 and integrate the result over Ω:

$$\int_{\Omega} (\nabla \cdot (\nabla v_1 v_2) - \nabla v_1 \cdot \nabla v_2 + q_1 v_1 v_2)\,\mathrm{d}x = 0.$$

Interchanging v_1 and v_2, we similarly obtain

$$\int_{\Omega} (\nabla \cdot (\nabla v_2 v_1) - \nabla v_2 \cdot \nabla v_1 + q_2 v_2 v_1)\,\mathrm{d}x = 0.$$

Subtracting these equalities from one another, applying the Gauss-Ostrogradskii formula, and recalling that

$$\int_{\Omega} (q_2 - q_1)\, v_1 v_2 \mathrm{d}x = \int_{\partial\Omega} ((\Lambda_1 v_1)v_2 - (\Lambda_2 v_2)v_1)\,\mathrm{d}S.$$

To complete the proof, we have to "transpose" the operator Λ_2 on the right-hand side from v_2 to v_1. Let $\bar{\Lambda}_2$ be the Dirichlet-to-Neumann operator corresponding to the potential $\bar{q}_2$. The bar over q_2 stands for complex conjugation. We prove that the operators Λ_2 and $\bar{\Lambda}_2$ are mutually adjoint with respect to the inner product of $L_2(\partial\Omega)$. Suppose that $v(x)$ is a solution to the equation $\Delta v_* + \bar{q}_2 v_* = 0$, and $v_*(x)$ is a solution to the equation $\Delta v_* + \bar{q}_2 v_* = 0$. Multiplying the equations in v and v_* by $\bar{v}_*$ and $\bar{v}$, respectively, we obtain

$$\nabla \cdot (\nabla v \bar{v}_*) - \nabla v \cdot \nabla \bar{v}_* + q_2 v \bar{v}_* = 0; \quad \nabla \cdot (\nabla v_* \bar{v}) - \nabla v_* \cdot \nabla \bar{v} + \bar{q}_2 v_* \bar{v} = 0$$

Take the complex conjugation of the first of the equalities, subtract them from one another, and integrate the result over Ω on using the Gauss-Ostrogradskii formula:

$$\int_{\partial\Omega} \overline{\frac{\partial v}{\partial \nu}} v_* \mathrm{d}S = \int_{\partial\Omega} \frac{\partial v_*}{\partial \nu} \bar{v} \mathrm{d}S.$$

Hence, by the definition of Λ_j

$$\int_{\partial\Omega} (\overline{\Lambda_2 v}) v_* \mathrm{d}S = \int_{\partial\Omega} (\Lambda_2 v_*) \bar{v} \mathrm{d}S,$$

i.e.,

$$(v_*, \Lambda_2 v) = (\bar{\Lambda}_2 v_*, v),$$

where the parentheses stand for the inner product of $L_2(\partial\Omega)$. For arbitrary data $f,\ g \in H^1(\partial\Omega)$ we can find solutions v and v_* to the corresponding equations such that $v|_{\partial\Omega} = g$ and $v_*|_{\partial\Omega} = f$. Thus,

$$(f, \Lambda_2 g) = (\bar{\Lambda}_2 f, g),$$

i.e., Λ_2 and $\bar{\Lambda}_2$ are mutually adjoint with respect to the inner product of $L_2(\partial\Omega)$. Λ_2 and $\bar{\Lambda}_2$ are mutually adjoint with respect to the inner product of $L_2(\partial\Omega)$.

Since $\bar{v}_2$ satisfies the equation $\Delta\bar{v}_2 + \bar{q}_2\bar{v}_2 = 0$, we have

$$\overline{\Lambda_2 v_2} = \overline{\frac{\partial v_2}{\partial \nu}} = \frac{\partial \bar{v}_2}{\partial \nu} = \bar{\Lambda}_2 \bar{v}_2.$$

Using the fact that the operators Λ_2 and $\bar{\Lambda}_2$ are mutually adjoint, we write down

$$\int_{\partial\Omega} v_1 \Lambda_2 v_2 \mathrm{d}S = \int_{\partial\Omega} v_1 \overline{\overline{\Lambda_2 v_2}} \mathrm{d}S = \int_{\partial\Omega} v_1 \overline{\bar{\Lambda}_2 \bar{v}_2} \mathrm{d}S$$

$$= (v_1, \bar{\Lambda}_2 \bar{v}_2) = (\Lambda_2 v_1, \bar{v}_2) = \int_{\partial\Omega} (\Lambda_2 v_1) v_2 \mathrm{d}S$$

which proves (8.25).

Special solutions. Consider the equation

$$(\Delta + \theta^2) v(x) + q(x) v(x) = 0. \tag{8.26}$$

To prove Theorem 8.1, we need some generalization of one theorem by Sylvester and Uhlmann [4] (see also [12]).

Theorem 8.5 *Suppose that $q(x) \in C^s(\bar{\Omega})$ for some integer $s \geq 0$. Then there are constants C_1 and C_2 such that, for every $\zeta \in \mathbb{C}^n$ satisfying the conditions $\zeta \cdot \zeta + \theta^2 = 0$, $|\Im\zeta| \geq \sqrt{2}$, and $|\zeta| \geq C_1 \|q\|_{C^s(\bar{\Omega})}$, equation (8.26) has a solution of the form*

$$v(x, \zeta) = \mathrm{e}^{\zeta \cdot x}(1 + w(x, \zeta)), \tag{8.27}$$

where $w(x, \zeta)$ satisfies the estimate

$$\|w(\cdot, \zeta)\|_{H^s(\Omega)} \leq \frac{C_2}{|\zeta|} \|q\|_{C^s(\bar{\Omega})}. \tag{8.28}$$

The constants C_1 and C_2 depend only on Ω and s.

In the statement of Theorem 8.5, the dot stands for the "inner product without complex conjugation"; i.e., $x \cdot y = x_1 y_1 + \ldots + x_n y_n$. Theorem 8.5 was proven in [12] for $s = 0$. Our proof essentially repeats that of [12] with slight modifications.

Proof Inserting (8.27) into (8.26) and recalling that $\zeta \cdot \zeta + \theta^2 = 0$, we obtain the following equation in w:

$$\Delta w + 2\zeta \cdot \Delta w = -q(1 + w). \tag{8.29}$$

Consider the operator $P := \Delta + 2\zeta \cdot \nabla$. According to Theorem 10.3.7 of [13], there is a bounded linear operator $E : L_2(\Omega) \to L_2(\Omega)$ such that

$$\mathrm{PE} f = f \quad \forall f \in L_2(\Omega);$$

moreover, for each constant coefficient differential operator $Q(D)$ weaker than $P(D)$ the operator $Q(D)E$ is bounded in $L_2(\Omega)$. However, we need the operator $Q(D)E$ to be bounded in $H^s(\Omega)$. To show this, we repeat the proof of Theorem 10.3.7 [13] with minor changes.

Introduce some "local" notations needed only in this proof (cf. [13]). Given a linear differential operator $P(D)$, $D = -i\partial$, with constant coefficients, define the function $\tilde{P}(\xi)$, $\xi \in \mathbb{R}^n$, by the relation

$$\tilde{P}^2(\xi) = \sum_{|\alpha| \geq 0} |\partial^\alpha P(\xi)|^2.$$

Suppose that $k(\xi), \xi \in \mathbb{R}^n$, is a tempered weight function (see [13], Definition 10.1.1) for a precise definition; we do not present the definition, only pointing out that the functions 1 and $\tilde{P}$ that will serve as $k(\xi)$ satisfy this definition). Define the space $B_{p,k}$ to be the set of tempered distributions u such that the Fourier transform

$$\hat{u} := (2\pi)^{-n/2} \int_{\mathbb{R}^n} \mathrm{e}^{-i\xi x} u(x) \mathrm{d}x$$

is a function and

$$\|u\|_{p,k} = \left(\int_{\mathbb{R}^n} |k(\xi) \hat{u}(\xi)^p| \mathrm{d}\xi \right)^{1/p} < \infty$$

(for $p = \infty$ we put $\|u\|_{\infty,k} := ess\ \sup|\hat{k}(\xi)\hat{u}(\xi)|$). A fundamental solution E of the operator P is called *regular* if E belongs to the least local extension $B^{\mathrm{loc}}_{\infty,\tilde{P}}$ of the space $B_{\infty,\tilde{P}}$; i.e., E belongs to the space containing $B_{\infty,\tilde{P}}$ and possessing the properties

$$u \in B^{\mathrm{loc}}_{\infty,\tilde{P}},\ \varphi \in C_0^\infty(\mathbb{R}^n) \Rightarrow u\varphi \in B^{\mathrm{loc}}_{\infty,\tilde{P}};$$

$$\forall u \in D'\ \forall \varphi \in C_0^\infty(\mathbb{R}^n)\ u\varphi \in B^{\mathrm{loc}}_{\infty,\tilde{P}} \Rightarrow u \in B^{\mathrm{loc}}_{\infty,\tilde{P}}.$$

According to Theorem 10.2.1 of [13], every differential operator with constant coefficients possesses a regular fundamental solution.

Fix s and take $f \in H^s(\Omega)$. By the well-known results of analysis (see, for instance, [5], Chap. 3, Sect. 3.4, Theorem 1) there is a compactly-supported extension of f to an ε-neighborhood Ω_ε of Ω with the class preserved; i.e., there is $f_0 \in H^s(\mathbb{R}^n)$ such that $f = f_0$ on Ω, $f_0 \equiv 0$ beyond Ω_ε, and

$$\|f_0\|_{H^s(\mathbb{R}^n)} \le C_{\Omega,s}\|f\|_{H^s(\mathbb{R}^n)}$$

with some constant $C_{\Omega,s}$ independent of f. Define Ef as the restriction of $E_0 * f_0$ to Ω, where E_0 is a regular fundamental solution of P, and $*$ stands for convolution. Then $PEf = f$ (see [13], Chap. 4, Sect. 4.4). In line with [13], put $F_0 = E_0\psi$, where $\psi \in C_0^\infty(\mathbb{R}^n)$ and $\psi = 1$ in a neighborhood of $\Omega - \Omega = \{x - y|\ x, y \in \Omega\}$. Then $F_0 \in B_{\infty,\tilde{P}}$ and $F_0 * f_0 = E_0 * f_0$ in Ω by Theorem 4.2.4 of [13]. Consequently (see [13], Theorem 2, p. 40),

$$\|\partial^\alpha|\zeta|Ef\|_{L_2(\Omega)} \le \||\zeta|\partial^\alpha(F_0 * f_0)\|_{2,1}$$

$$\le |\zeta|\|F_0\|_{\infty,1}\|\partial^\alpha f_0\|_{L_2(\mathbb{R}^n)} \le C_{\Omega,s}\|F_0\|_{\infty,\tilde{P}} \sup_\xi \frac{|\zeta|}{\tilde{P}(\xi)}\|f\|_{H^s(\Omega)},$$

where $C_{\Omega,s}$ depends only on Ω and s. It was proven in [12] that $\|F_0\|_{\infty,\tilde{P}}$ is bounded by a constant depending only on Ω, n, and the order of P. We have thus obtained an operator $E : H^s(\Omega) \to H^s(\Omega)$ such that

$$\mathrm{PE}f = f$$

and

$$\|Ef\|_{H^s(\Omega)} \le \frac{C_{\Omega,n,s}}{|\zeta|} \sup_\xi \frac{|\zeta|}{\tilde{P}(\xi)}\|f\|_{H^s(\Omega)}.$$

Below we show that

$$\sup_\xi \frac{|\zeta|}{\tilde{P}(\xi)} \le 1; \tag{8.30}$$

therefore, our estimate would take the form

$$\|Ef\|_{H^s(\Omega)} \le \frac{C_{\Omega,n,s}}{|\zeta|}\|f\|_{H^s(\Omega)}. \tag{8.31}$$

Return to (8.29). Each solution to the equation

$$w = -E(q(1+w)) \tag{8.32}$$

is a solution to (8.29). Since

$$\|qw\|_{H^s(\Omega)} \le C_s\|q\|_{C^s}\|w\|_{H^s(\Omega)}$$

from (8.31) we obtain

$$\|Eqw\|_{H^s(\Omega)} \le 1/2\|w\|_{H^s(\Omega)}$$

for $|\zeta| \ge 2C_{\Omega,n,s}C_s\|q\|_{C^s}$. This implies solvability of (8.32) together with the needed estimate.

To complete the proof, it remains to establish (8.30), which amounts to the inequality

$$\tilde{P}^2(\xi) \ge |\zeta|^2, \quad \xi \in \mathbb{R}^n,$$

Suppose that $\zeta = \lambda + i\mu$, with $\lambda, \mu \in \mathbb{R}^n$. Calculate the derivatives $\partial_\xi^\alpha P(\xi)$:

$$P(\xi) = -\xi^2 + 2i\zeta\cdot\xi = -\xi^2 - 2\mu\cdot\xi + 2i\lambda\cdot\xi;$$
$$\partial_{\xi_j}P(\xi) = -2\xi_j + 2i\zeta_j = -2\xi_j - 2\mu_j + 2i\lambda_j;$$
$$\partial_{\xi_j}\partial_{\xi_i}P(\xi) = \begin{cases} -2 & \text{for } i = j; \\ 0 & \text{otherwise.}\end{cases}$$

Hence,

$$\begin{aligned}\tilde{P}^2(\xi) &= (\xi^2 + 2\mu\cdot\xi)^2 + 4(\lambda\cdot\xi)^2 + 4\sum_{j=1}^n((\xi_j+\mu_j)^2 + \lambda_j^2) + 12\\ &\ge (\xi^2 + 2\mu\cdot\xi + \mu^2 - \mu^2)^2 + 4(\xi+\mu)^2 + 4\lambda^2\\ &= ((\xi+\mu)^2 - \mu^2)^2 + 4(\xi+\mu)^2 + 4\lambda^2.\end{aligned}$$

Since

$$\min_{x\ge 0}((x-\mu^2)^2 + 4x) = 4(\mu^2 - 1) \ge \mu^2$$

for $\mu^2 \ge 2$, under the conditions of the theorem we have $\tilde{P}^2(\xi) \ge \mu^2 + 4\lambda^2 \ge \mu^2 + \lambda^2$. Theorem 8.5 is proven.

Corollary 8.1 *Under the conditions of the preceding theorem*

$$\|v(\cdot,\zeta)\|_{H^s(\Omega)} \leq C_{\Omega,s} e^{C_\Omega|\zeta|}.$$

Proof The estimate follows from the chain of the inequalities

$$\begin{aligned}\|v(\cdot,\zeta)\|_{H^s(\Omega)} &\leq C_s\|e^{\zeta x}\|_{C^s}\|1+w\|_{H^s} \leq C_s|\zeta|^2 e^{C_\Omega|\zeta|}(\|1\|_{H^s}+\|w\|_{H^s})\\ &\leq C_s|\zeta|^2 e^{C_\Omega|\zeta|}\left(\|1\|_{H^s}+\frac{C_2}{|\zeta|}\|q\|_{C^s}\right)\\ &\leq C_s|\zeta|^2 e^{C_\Omega|\zeta|}\left(\|1\|_{H^s}+\frac{C_2}{C_1}\right) \leq C_{\Omega,s}e^{C_\Omega|\zeta|}.\end{aligned}$$

8.2.4 *Proof of the Main Result*

To proof Theorem 8.2 which is the main result of this section, we suppose that $k_j(x,t) \in W(K)$, $j=1,2$, are two (different) memories. Consider Eq. (8.22) with $k=k_j$. Denote a solution to (8.22) by $\hat{u}_j(x,t)$ and denote the corresponding stationary Dirichlet-to-Neumann operators by Λ_j^θ.

By Lemma 8.4, we have the identity

$$\int\limits_\Omega (\hat{k}_2(x,\theta)-\hat{k}_1(x,\theta))\hat{u}_1(x,\theta)\hat{u}_2(x,\theta)\mathrm{d}x = \int\limits_{\partial\Omega} \hat{u}_1(x,\theta)(\Lambda_1^\theta-\Lambda_2^\theta)\hat{u}_2(x,\theta)\mathrm{d}S.$$

By Theorem 8.5, Eq. (8.22) with $k=k_j$ possesses a solution

$$\hat{u}_j(x,\zeta_j,\theta) = e^{\zeta_j\cdot x}(1+w_j(x,\zeta_j,\theta)),$$

where $\zeta_j \in \mathbf{C}^n$, $\zeta_j\cdot\zeta_j+\theta^2=0$, $|\zeta_j| \geq C_1\|k_j(\cdot,\theta)\|_{C^2(\Omega)}$, $|\mathrm{Im}\zeta_j| \geq \sqrt{2}$ and w_j satisfies the estimate

$$\|w_j(\cdot,\zeta_j,\theta)\|_{H^s(\Omega)} \leq \frac{C_2}{|\zeta_j|}\|k_j(\cdot,\theta)\|_{C^2(\Omega)}. \tag{8.33}$$

Inserting the special solutions $\hat{u}_j$ into the identity, we obtain

$$\begin{aligned}&\int\limits_\Omega (\hat{k}_2(x,\theta)-\hat{k}_1(x,\theta))e^{(\zeta_1+\zeta_2)\cdot x}\mathrm{d}x\\ &= \int\limits_{\partial\Omega}\hat{u}_1(x,\zeta_1,\theta)(\Lambda_1^\theta-\Lambda_2^\theta)\hat{u}_2(x,\zeta_2,\theta)\mathrm{d}S + \int\limits_\Omega (\hat{k}_2(x,\theta)-\hat{k}_1(x,\theta))e^{(\zeta_1+\zeta_2)\cdot x}\\ &\times(w_1(x,\zeta_1,\theta)w_2(x,\zeta_2,\theta)+w_1(x,\zeta_1,\theta)+w_2(x,\zeta_2,\theta))\mathrm{d}x.\end{aligned} \tag{8.34}$$

Now, turn to choosing ζ_j. Take an arbitrary $\xi \in \mathbb{R}^n$. Choose ζ_j so as to have

$$\begin{aligned} &\zeta_1 + \zeta_2 = -i\xi, \quad \zeta_j \cdot \zeta_j + \theta^2 = 0, \\ &|\zeta_j| \geq C_1 \|\hat{k}_j(\cdot, \theta)\|_{C^2(\Omega)}, \ |\mathrm{Im}\zeta_j| \geq \sqrt{2}. \end{aligned} \tag{8.35}$$

We seek ζ_j in the form $\zeta_j = -i\xi/2 + (-1)^j(i\mu + \lambda)$, where $\mu, \lambda \in \mathbb{R}^n$ and $\xi \cdot \mu = \xi \cdot \lambda = 0$. The first condition of (8.35) is obviously satisfied. The second leads to the equation

$$\begin{aligned} \zeta_j \cdot \zeta_j &= -\frac{\xi^2}{4} - \mu^2 + \lambda^2 + (-1)^j(\xi \cdot \mu) + 2i(\lambda \cdot \mu) - (-1)^j i(\xi \cdot \lambda) \\ &= -\theta^2 = -(\sigma - i\gamma)^2 = -(\sigma^2 - \gamma^2 - 2i\sigma\gamma) \end{aligned}$$

which amounts to the system

$$\frac{\xi^2}{4} + \mu^2 - \lambda^2 = (\sigma^2 - \gamma^2), \quad \lambda \cdot \mu = \sigma\gamma. \tag{8.36}$$

As a solution, take μ and λ in $\mathbb{R}^n$ such that $\mu^2 = \sigma^2 + r^2$ and $\lambda^2 = \frac{\xi^2}{4} + \gamma^2 + r^2$, $r \geq \sqrt{2}$. This is possible, since for such choice of μ^2 and λ^2 the first equation becomes an identity while the second can be fulfilled by choosing the angle between μ and λ as $|\mu| \geq |\sigma|$ and $|\lambda| \geq \gamma$. Consider the last but one condition in (8.35). We have

$$\begin{aligned} |\zeta_j|^2 &= \frac{\xi^2}{4} + \mu^2 + \lambda^2 = \frac{\xi^2}{2} + 2\mu^2 + \gamma^2 - \sigma^2 = \frac{\xi^2}{2} + \sigma^2 + 2r^2 + \gamma^2 \\ &= \frac{\xi^2}{2} + |\theta|^2 = 2r^2. \end{aligned} \tag{8.37}$$

To fulfill the last but one condition in (8.35), take the constant γ_0 in the statement of Theorem 8.2 so that $\gamma_0 \geq C_1 \|\hat{k}(\cdot, \theta)\|_{C^2(\Omega)}$ for every $k \in W(K)$; this is obviously possible. Finally, observe that the last condition in (8.35) is satisfied by virtue of $\xi \cdot \mu = 0$ and $|\mu| \geq \sqrt{2}$ for $r \geq \sqrt{2}$. Introduce the notation

$$\tilde{k}(\xi, \theta) := (2\pi)^{-n/2} \int\limits_{\Omega} (\hat{k}_2(x, \theta) - \hat{k}_1(x, \theta))\mathrm{e}^{-i\xi x}\mathrm{d}x$$

and pass from identity (8.34) to the inequality

$$|\tilde{k}(\xi, \theta)| \leq C_\Omega \mathrm{e}^{C_\Omega|\zeta_1|} \|(\Lambda_1^\theta - \Lambda_2^\theta)\hat{u}_2(x, \zeta_2, \theta)\|_{L_2(\partial\Omega)} + \frac{C_{\Omega,K,\gamma}}{|\zeta_j|}. \tag{8.38}$$

Here we have used the Corollary to Theorem 8.5 and the trace theorem,

$$\|\hat{u}_1\|_{L_2(\partial\Omega)} \le C_\Omega \|\hat{u}_1\|_{H^1(\Omega)} \le C_\Omega \mathrm{e}^{C_\Omega|\zeta_1|},$$

together with the estimate

$$\int\limits_\Omega (\hat{k}_2(x,\theta) - \hat{k}_1(x,\theta))(w_1 w_2 + w_1 + w_2)\mathrm{d}x$$

$$\le C_{\Omega,\gamma} K \left(\frac{C_2 K}{|\zeta_1|} + \frac{C_2 K}{|\zeta_2|} + \frac{C_2^2 K^2}{|\zeta_1||\zeta_2|} \right) \le C_{\Omega,K} \frac{1}{|\zeta_j|},$$

which ensues from (8.33). (Here we suppose that $|\zeta_j| > 1$ which can be fulfilled by choosing $\gamma_0 \ge 1$.)

Introduce the function $\hat{g}(x,\zeta_2,\theta)$, $x \in \partial\Omega$, as follows:

$$\hat{g}(x,\zeta_2,\theta) := \frac{\hat{u}(x,\zeta_2,\theta)}{\mathrm{e}^{C_\Omega|\zeta_2|}(1+\sigma^2)},$$

where the constant C_Ω is taken from the Corollary to Theorem 8.5. From the same Corollary and the trace theorem we obtain

$$\|\hat{g}(\cdot,\zeta_2,\theta)\|_{H^1(\partial\Omega)} \le \frac{C_\Omega}{1+\sigma^2}$$

for $\sigma \in \mathbb{R}$ and every r in the definition of ζ_j. (Observe that ζ_2 depends on θ and r, and so $\hat{g}$ depends actually on x, σ, and r.) Consequently,

$$\int\limits_{\mathbb{R}} \|\hat{g}(\cdot,\zeta_2,\theta)\|_{H^1(\partial\Omega)}(1+\sigma^2)\mathrm{d}\sigma \le \int\limits_{\mathbb{R}} \frac{C_\Omega}{1+\sigma^2}\mathrm{d}\sigma = C_\Omega < \infty,$$

which implies that the inverse Fourier-Laplace transform

$$g(x,t) = \frac{1}{\sqrt{2\pi}} \int\limits_{\infty}^{\infty} \hat{h}(x,\zeta_2,\sigma - i\gamma)\mathrm{e}^{it(\sigma-i\gamma)}\mathrm{d}\sigma$$

belongs to $H^l_\gamma(\partial\Omega \times \mathbb{R})$ and its norm $\langle g\rangle_{\gamma,1}$ does not exceed some constant C_Ω depending only on the domain Ω. (The vector ζ_2 depends on $\theta = \sigma - i\gamma$ and r; therefore, the function $g(x,t)$ depends also on r as on a parameter. However, the norm $\langle g\rangle_{\gamma,1}$ is bounded uniformly in r; therefore, we do not indicate the dependence of g on r.)

Integrating the square of (8.38) over $\{(\sigma, \xi) \in \mathbb{R}^{n+1} | \sigma^2 + \xi^2 \leq \rho^2\}$ with $\rho > 0$, we obtain

$$\int_{\sigma^2+\xi^2\leq\rho^2} |\tilde{k}(\xi, \theta)|^2 \mathrm{d}\xi \mathrm{d}\sigma \leq C_{\Omega,K} \mathrm{e}^{C_\Omega\sqrt{\rho^2+\gamma^2+2r^2}}(1+\rho^2)^2$$
$$\times \int_{\sigma^2+\xi^2\leq\rho^2} \|(\Lambda_1^\theta - \Lambda_2^\theta)\hat{g}(x, \zeta_2, \theta)\|^2_{L_2(\partial\Omega)} \mathrm{d}\sigma \mathrm{d}\xi + \frac{C_{\Omega,K}\rho^{n+1}}{r^2}. \tag{8.39}$$

Continue (8.39) by using the equality $F(H_1 - H_2)g = (\Lambda_1^\theta - \Lambda_2^\theta)\hat{g}$:

$$\int_{\sigma^2+\xi^2\leq\rho^2} |\tilde{k}(\xi, \theta)|^2 \mathrm{d}\xi \mathrm{d}\sigma \leq C_{\Omega,K,\gamma} \mathrm{e}^{C_\Omega\sqrt{\rho^2+2r^2}}(1+\rho^2)^2\rho^n$$
$$\times \int_{\sigma\in R} \|(F(H_1 - H_2)g)(x, \theta)\|^2_{L_2(\partial\Omega)} \mathrm{d}\sigma + \frac{C_{\Omega,K}\rho^{n+1}}{r^2}.$$

Applying the Parseval identity (8.21) and the estimate $\langle g\rangle_{\gamma,1} \leq C_\Omega$, we now find that

$$\int_{\sigma^2+\xi^2\leq\rho^2} |\tilde{k}(\xi, \theta)|^2 \mathrm{d}\xi \mathrm{d}\sigma \leq C_{\Omega,K,\gamma,n} \mathrm{e}^{C_\Omega\sqrt{\rho^2+2r^2}} \|H_1 - H_2\|^2 + \frac{C_{\Omega,K}\rho^{n+1}}{r^2}. \tag{8.40}$$

Since $k(x, t) \in C^2(\Omega \times (0, \infty))$ and $\|k\|_{C^2} \leq K$, the function $\mathrm{e}^{-\gamma t}k(x, t)$ belongs to $H^{l/2}(\Omega \times (0, \infty))$. It follows from the results of ([9], p. 71) that the function $\mathrm{e}^{-\gamma t}k(x, t)$ extended by zero beyond $\Omega \times (0, \infty)$ belongs to the space $H^{1/2}(\mathbb{R}^4)$; moreover, $\|\mathrm{e}^{-\gamma t}k\|_{H^{1/2}(\mathbb{R}^4)} \leq C_\Omega K$. Using the equivalent norm in the space $H^{1/2}(\mathbb{R}^4)$, we infer that

$$\int_{\mathbb{R}^4_{\not\xi}} |\tilde{k}(\xi, \theta)|^2(1 + |\xi|^2 + |\sigma|^2)^{1/2} \mathrm{d}\xi \mathrm{d}\sigma \leq C_\Omega K$$

which implies that

$$\int_{\sigma^2+\xi^2\leq\rho^2} 0|\tilde{k}(\xi, \theta)|^2 \mathrm{d}\xi \mathrm{d}\sigma \leq \frac{C_\Omega K}{(1+\rho^2)^{1/2}} \leq \frac{C_\Omega K}{\rho}. \tag{8.41}$$

Combining (8.40) and (8.41), we obtain

$$\int_{\mathbb{R}^{n+1}} |\tilde{k}(\xi, \theta)|^2 d\xi \mathrm{d}\sigma \leq C_{\Omega,K,\gamma,n} \left(\mathrm{e}^{C_\Omega\sqrt{\rho^2+2r^2}} \|H_1 - H_2\|^2 + \frac{\rho^{n+1}}{r^2} + \frac{1}{\rho} \right).$$

Putting $r = \rho^{(n+2)/2}$, estimate the expression in the parentheses by the quantity

$$e^{C_\Omega\sqrt{\rho^2+2\rho^{(n+2)}}}\|H_1 - H_2\|^2 + \frac{2}{\rho}$$

which behaves as

$$e^{C_\Omega\rho^{n/2+1}}\|H_1 - H_2\|^2 + \frac{2}{\rho}$$

at large ρ. Squaring both sides of our estimate, we obtain

$$\|\tilde{k}\|_{L_2(\mathbb{R}^{n+1})} \leq C_{\Omega,K,\gamma,n}\left(e^{C_\Omega\rho^{n/2+1}}\|H_1 - H_2\|^2 + \frac{1}{\rho^{1/2}}\right)$$

at large ρ; whence, by Corollary 2.1 of [14],

$$\|k_1 - k_2\|_{\gamma,0} \leq \omega(\|H_1 - H_2\|),$$

where $\omega(\varepsilon) \sim C_{\Omega,K,\gamma,n}(C_\Omega/\log\varepsilon^{-1})^{1/(n+2)}$ as $\varepsilon \to 0$. Theorem 8.2 is proven completely.

8.3 Recovering a Time- and Space-Dependent Kernel in a Hyperbolic Integro-Differential Equation from a Restricted Dirichlet-to-Neymann Operator

In this section, it is proved that a space- and time-dependent kernel occurring in a hyperbolic integro differential equation in three space dimensions can be uniquely reconstructed from the restriction of the Dirichlet-to-Neumann operator of the equation into a set of Dirichlet data of the form of products of a fixed time-dependent coefficient times arbitrary space-dependent functions. This problem was considered by Janno [15]. Note that determination of a time- and space-dependent heat flux relaxation function from a parabolic integro-differential equation by means of this technique was investigated in [16].

8.3.1 *Problem Formulation and Results*

First we introduce some notation. Given a Banach space X, $\sigma \in \mathbb{R}$, $k \in \{0, 1, 2, ...\}$ and $p \in [1, \infty)$ we define the following sets of abstract functions with values in X:

$$W_\sigma^{k,p}((0,\infty); X) = \{z : (0,\infty) \to X : e^{-\sigma}z(\cdot) \in W^{k,p}((0,\infty); X)\}$$

$$C_\sigma^k([0,\infty);X) = \{z : [0,\infty) \to X : \mathrm{e}^{-\sigma} z(\cdot) \in C^k([0,\infty);X)\}.$$

Here $W^{k,p}((0,\infty);X)$ is the abstract Sobolev space, and $C^k([0,\infty);X)$ consists of abstract functions bounded and continuous in $[0,\infty)$ together with their derivatives up to the order k. In case $k=0$ we set $L^p_\sigma((0,\infty);X) := W^{0,p}_\sigma((0,\infty);X)$ and $C_\sigma([0,\infty);X) := C^0_\sigma([0,\infty);X)$. Moreover, in case $X=\mathbb{C}$, we use the simplified notation $W^{k,p}_\sigma(0,\infty) := W^{k,p}_\sigma((0,\infty);\mathbb{C})$ and $C^k_\sigma[0,\infty) := C^k_\sigma([0,\infty);\mathbb{C})$. Finally, given $k \in \{0,1,2,...\}$ and $\mu > 0$, we denote by $C^{k,\mu}(\Omega)$ the space of functions which are Hölder-continuous of degree λ in Ω together with their derivatives up to the order k.

Let Ω be a three-dimensional domain with a Lipschitz-boundary Γ. In case Ω is filled by the isotropic nonhomogeneous viscoelastic material, the constitutive relations and the system of equations of motion of the material point of Ω contain time- and space-dependent relaxation functions $a(t,x)$ and $b(t,x)$, which correspond to the shear and bulk moduli, respectively (see [17], p. 122–127). We suppose that these functions are unknown. Unfortunately, inverse problems to determine both $a(t,x)$ and $b(t,x)$ in the viscoelasticity system are very complex. Therefore we essentially simplify the situation taking into consideration a partially theoretical scalar model of such kind. If the displacement field is solenoidal, then the system of equations of motion is reduced to three independent equations of divergence type, which can be summarized as

$$\rho(x)\partial_t^2 v(t,x) = \int_{-\infty}^{t} \mathrm{div}[a(t-\tau,x)\nabla\partial_\tau v(\tau,x)]\mathrm{d}\tau, \quad x\in\Omega,\ t\in\mathbb{R} \tag{8.42}$$

where v is an arbitrary linear combination of coordinates of the displacement and ρ is the density. We will pose and study an inverse problem that consists in determining the kernel $a(t,x)$ in (8.42) for $t \geq 0$ and $x \in \Omega$. Let us assume $v(t,x) = 0$ for $t \leq 0$. Then the Eq. (8.42) for twice differentiable $v(t,\cdot)$ is equivalent to the integrated equation

$$\rho(x)\partial_t v(t,x) = \int_0^t (t-\tau)\mathrm{div}[a(t-\tau,x)\nabla\partial_\tau v(\tau,x)]\mathrm{d}\tau, \quad x\in\Omega,\ t\in\mathbb{R}. \tag{8.43}$$

We interpret the latter equation as a generalized form of (8.42) and complement it with the initial and Dirichlet boundary conditions

$$v(0,x) = 0,\ x\in\Omega,\ v(t,x) = \psi(t,x),\ x\in\Gamma,\ t>0. \tag{8.44}$$

Let us associate with v the secondary variable

$$s(t,x) = \int_{-\infty}^{t} a(t-\tau,x)\nabla\partial_\tau v(\tau,x)dt$$

and denote by $\nu(x)$ the outer normal of the boundary of Ω at the point $x \in \Gamma$. Given ρ, a, and ψ, the solution v of the initial–boundary value problem (8.43), (8.44) determines the normal component of s at the boundary:

$$h(t,x) := -\nu(x)\cdot s(t,x) = \int_0^t a(t-\tau,x)\partial_\nu\partial_t v(\tau,x)d\tau,\ x\in\Gamma,\ t>0. \quad (8.45)$$

The operator D, which depends on ρ and a and maps the Dirichlet data ψ via the solution v to the Neumann data h, is called the Dirichlet-to-Neumann operator of the problem (8.43), (8.44).

We begin by eliminating the derivative in the Eq. (8.43) by the change of variable

$$u(t,x) = \partial_t v(t,x) \Leftrightarrow v(t,x) = \int_0^t u(\tau,x)d\tau. \quad (8.46)$$

Defining $\varphi(t,x) := \partial_t\psi(t,x)$, the problem (8.43), (8.44) is transformed to

$$\rho(x)u(t,x) = \int_0^t (t-\tau)\mathrm{div}[a(t-\tau,x)\nabla u(\tau,x)]d\tau, \quad (8.47)$$

$$u(t,x) := \varphi(t,x),\ x\in\Gamma,\ t>0. \quad (8.48)$$

For the obtained problem have the following theorem which will be be proved in Sect. 8.3.2.

Theorem 8.6 *Let*

$$\rho\in L^\infty(\Omega),\ \rho(x)\ge\rho_0>0,\ x\in\Omega \quad (8.49)$$

and

$$a\in W^{1,1}_{\sigma_0}((0,\infty);W^{1,\infty}(\Omega))\cap W^{2,1}_{\sigma_0}((0,\infty);L^\infty(\Omega)),\ a(0,x)\ge a_0>0,\ x\in\Omega, \quad (8.50)$$

with some $\sigma_0\in\mathbb{R}$. Moreover, let

$$\varphi\in W^{5,1}_{\sigma_1}((0,\infty);H^{\frac{3}{2}}(\Gamma)),\ \varphi(0,x)=\partial_t\varphi(0,x)=\ldots=\partial_t^4\varphi(0,x)\equiv 0 \quad (8.51)$$

with some $\sigma_1 \in \mathbb{R}$.

Then there exists σ_a*, which depends on* a *and satisfies the inequalities* $\sigma_a > 0$, $\sigma_a \geq \sigma_0$*, such that the problem (8.47), (8.48) has a solution* u *in the space* $C_{\sigma_u}([0,\infty); H^2(\Gamma))$*, where* $\sigma_u = \max\{\sigma_1, \sigma_a\}$*. Moreover,* $u(0,x) \equiv 0$*. The solution is unique in the space* $L^1_{\sigma_u}((0,\infty); H^2(\Omega))$.

Due to (8.45) and (8.46) the Neumann data h *has the form*

$$h(t,x) = \int_0^t a(t-\tau,x)\partial_\nu u(\tau,x)d\tau, \; x \in \Gamma, \; t > 0. \tag{8.52}$$

Theorem 8.6 with the trace theorem implies the following result.

Corollary 8.2 *Under the assumptions of Theorem 8.6,* $h \in C^1_{\sigma_u}([0,\infty); L^2(\Gamma))$.

Let $\widetilde{D}$ be the operator which assigns to a function φ the function h via the solution u of the problem (8.47), (8.48). Observing Corollary 8.2 we see that $\widetilde{D}$ transforms the set of functions φ satisfying (8.51) into the space $C^1_{\sigma_u}([0,\infty); L^2(\Gamma))$, provided ρ and a meet the conditions (8.49) and (8.50). Moreover, $D\psi = \widetilde{D}\partial_t\psi$. Therefore, the operator D transforms the space

$$\left\{\varphi \in W^{6,1}_{\sigma_1}((0,\infty); H^{\frac{3}{2}}(\Gamma)), \; \psi(0,x) = \partial_t\psi(0,x) = \ldots = \partial_t^5\psi(0,x) \equiv 0\right\}$$

into the space $C^1_{\sigma_u}([0,\infty); L^2(\Gamma))$, provided ρ and a meet the conditions (8.49) and (8.50).

Let us choose a function f such that

$$f \in W^{5,1}_{\sigma_1}(0,\infty), \; f \not\equiv 0, \; Im\, f = 0, \; f(0) = f'(0) = \ldots = f^{IV}(0) = 0, \tag{8.53}$$

where $\sigma_1 \in \mathbb{R}$. For given f, ρ, and a satisfying (8.53), (8.49) and (8.50), respectively, let us define the operator $\lambda_a : H^{\frac{3}{2}}(\Gamma) \to C^1_{\sigma_u}([0,\infty); L^2(\Gamma))$ by the following relation

$$\lambda_a g := \widetilde{D}(f(x)g(x)) = D\left(\int_0^t f(\tau)d\tau g(x)\right).$$

The main result is the following theorem, which asserts that a is uniquely recovered by λ_a, in other words, by the restriction of the Dirichlet-to-Neumann operator D in the set of Dirichlet data of the form $\psi(t,x) = \int_0^t f(\tau)d\tau g(x)$ with fixed f.

Theorem 8.7 *Let* Γ_ϵ *be some neighborhood of* Γ*. Assume that (8.49) and (8.53) are valid and* $\rho \in C^{0,\mu}(\Gamma_\epsilon)$ *with some* $\mu > 0$*. Furthermore, let* a_1 *and* a_2 *be two functions satisfying (8.50) and the relations*

$$a_j \in L^1_{\sigma_0}((0,\infty); W^{2,\infty}(\Omega) \cap C^{1,\mu}(\Gamma_\epsilon)), \; \mathrm{Im}\, a_j = 0, \; j = 1,2. \tag{8.54}$$

Then the equality $\lambda_{a_1} = \lambda_{a_2}$ *implies* $a_1 = a_2$.

The proof of this theorem will be given in Sect. 8.3.4. It uses a preliminary result concerning the boundary identifiability proved in Sect. 8.3.3.

8.3.2 Auxiliary Results

Let X be a complex Banach space and $z \in L^1_\sigma((0,\infty); X)$ with some $\sigma \in \mathbb{R}$. Then the Laplace transform of z, i.e.,

$$Z(p) = \mathcal{L}_{t\to p}(z) = \int_0^\infty \mathrm{e}^{-pt} z(t)\mathrm{d}t \tag{8.55}$$

exists in the half-plane $\mathrm{Re}\, p > \sigma$ and is holomorphic there (see [17]).

Lemma 8.5 *Let* $z \in L^1_\sigma((0,\infty); X)$ *with some* $\sigma \in \mathbb{R}$. *Then*

$$\|Z(p)\|_X \le C(z, \mathrm{Re}\, p) \text{ for } \mathrm{Re}\, p > \sigma \tag{8.56}$$

where

$$C(z,s) = \int_0^\infty \mathrm{e}^{-(s-\sigma)t} \|\mathrm{e}^{-\sigma t} z(t)\|_X \mathrm{d}t \to 0 \text{ as } s \to \infty. \tag{8.57}$$

If, in addition, $z \in W^{k,1}_\sigma((0,\infty); X)$ *with some* $k \in \{1, 2, \ldots\}$ *then*

$$\begin{aligned} \|p^k Z(p) - p^{k-1} z(0) - p^{k-2} z'(0) - \ldots - z^{(k-1)}(0)\|_X &\le C(z^{(k)}, \mathrm{Re}\, p) \\ \text{for } \mathrm{Re}\, p > \sigma, \end{aligned} \tag{8.58}$$

which in case $z(0) = z'(0) = \ldots = z^{(k-1)}(0) = 0$ *implies*

$$|p|^k \|Z(p)\| \le C(z^{(k)}, \mathrm{Re}\, p) \text{ for } \mathrm{Re}\, p > \sigma. \tag{8.59}$$

The proof of the lemma above can be found in [16].

Lemma 8.6 *Let* $Z : \mathbb{C} \to X$ *be holomorphic for* $\mathrm{Re}\, p > \sigma$ *with some* $\sigma \in \mathbb{R}$ *and*

$$|p|^2 \|Z(p)\|_X \le c_0 \text{ for } \mathrm{Re}\, p > \sigma. \tag{8.60}$$

with a constant c_0. *Then there exists a function* $z \in C_\sigma([0,\infty); X)$ *such that* $Z(p) = \mathcal{L}_{t\to p}(z)$. *Moreover,* $z(0) = 0$.

The assertion of the lemma above follows from (see ([17], Proposition 0.2).

Lemma 8.7 *Let a satisfy (8.50) with some $\sigma_0 \in \mathbb{R}$ and $A(p, x) = \mathcal{L}_{t\to p}(a(\cdot, x))$. Then*

$$|p|\|A(p, \cdot)\|_{W^{1,\infty}(\Omega)} \leq c_1 \text{ for } \operatorname{Re} p > \sigma_0 \tag{8.61}$$

with a constant c_1 and

$$pA(p, \cdot) \to a(0, \cdot) \text{ as } \operatorname{Re} p \to \infty \text{ in } W^{1,\infty}(\Omega) \text{ uniformly with respect to } \operatorname{Im} p. \tag{8.62}$$

Moreover, there exists σ_a, satisfying the inequalities $\sigma_a > 0$, $\sigma_a \geq \sigma_0$, such that

$$|p||A(p, x)| \geq k > 0 \text{ for } \operatorname{Re} p > \sigma_a,\ x \in \Omega, \tag{8.63}$$

$$\operatorname{Re}\ p\, a(0, x) - |p\, a(0, x) - p^2 A(p, x)| \geq k > 0 \text{ for } \operatorname{Re} p > \sigma_a,\ x \in \Omega. \tag{8.64}$$

Proof Using the estimate (8.58) for a we obtain

$$\|pA(p, \cdot) - a(0, \cdot)\|_{W^{1,\infty}(\Omega)} \leq C\,(\partial_t a, \operatorname{Re} p) \text{ for } \operatorname{Re} p > \sigma_0, \tag{8.65}$$

$$\|p^2 A(p, \cdot) - pa(0, \cdot) - \partial_t a(0, \cdot)\|_{L^\infty(\Omega)} \leq C\left(\partial_t^2 a, \operatorname{Re} p\right) \text{ for } \operatorname{Re} p > \sigma_0. \tag{8.66}$$

In view of (8.57) and the assumption $a(0, x) = a_0 > 0$ the relation (8.65) implies (8.61)-(8.63) and (8.66) yields (8.64).

Lemma 8.8 *Let φ satisfy (8.51) with some $\sigma_1 \in \mathbb{R}$ and $\Phi(p, x) = \mathcal{L}_{t\to p}(\varphi(\cdot, x))$. Then*

$$|p|^5 \|\Phi(p, \cdot)\|_{H^{\frac{3}{2}}(\Gamma)} \leq c_2 \text{ for } \operatorname{Re} p > \sigma_1 \tag{8.67}$$

with a constant c_2.

The assertion of this lemma follows from (8.59) and (8.57).

8.3.3 Direct Problem

The direct problem (8.47), (8.48) is formally equivalent to the following elliptic boundary value problem derived by means of the Laplace transform:

$$(L_A U)(p, x) \equiv -\operatorname{div}(A(p, x)\nabla U(p, x)) + p\rho(x)U(p, x) = 0,\ x \in \Omega, \tag{8.68}$$

$$U(p, x) = \Phi(p, x),\ x \in \Gamma. \tag{8.69}$$

Here $p \in \mathbb{C}$, $A(p, x) = \mathcal{L}_{t\to p}(a(\cdot, x))$, $\Phi(p, x) = \mathcal{L}_{t\to p}(\varphi(\cdot, x))$ and $U(p, x) = \mathcal{L}_{t\to p}(u(\cdot, x))$.

Proposition 8.1 *Let ρ satisfy (8.49) and a satisfy (8.50) with some $\sigma_0 \in \mathbb{R}$. Moreover, let $\Phi(p,x)$ and $F(p,x)$ be given functions such that*

$$\Phi(p,\cdot) \in H^{\frac{3}{2}}(\Gamma),\ \ F(p,\cdot) \in L^2(\Omega) \text{ for } \mathrm{Re}p > \sigma_1 \tag{8.70}$$

with some $\sigma_1 \in \mathbb{R}$. Then the problem

$$(L_A U)(p,x) = F(p,x),\ \ x \in \Omega,\ \ U(p,x) = \Phi(p,x),\ \ x \in \Gamma \tag{8.71}$$

has a unique solution $U(p,\cdot) \in H^2(\Omega)$ for every Re $p > \sigma_u$ with $\sigma_u = \max\{\sigma_a, \sigma_1\}$ and σ_a from Lemma 8.7. This solution satisfies the estimate

$$\|U(p,\cdot)\|_{H^2(\Omega)} \le c_3|p|^2 \left(\|F(p,\cdot)\|_{L^2(\Omega)} + |p|\|\Phi(p,\cdot)\|_{H^{\frac{3}{2}}(\Gamma)}\right),\ \ \mathrm{Re}p > \sigma_u, \tag{8.72}$$

where the coefficient c_3 depends on $\rho,\ a,\ \Omega$, but is independent of $p,\ \Phi$ and F.

Proof Due to the assumption $\Phi(p,\cdot) \in H^{\frac{3}{2}}(\Gamma)$ for *Re* $p > \sigma_1$, there exists a function $\widetilde{\Phi}(p,x)$ such that $\widetilde{\Phi}(p,\cdot) \in H^2(\Omega)$ and $\widetilde{\Phi}(p,x)\big|_{x\in\Gamma} = \Phi(p,x)$ for *Re* $p > \sigma_1$. Denoting $W = U - \widetilde{\Phi}$, the problem (8.71) reduces to the following problem with homogeneous boundary condition for W:

$$(L_A W)(p,x) = F(p,x) - \left(L_A\widetilde{\Phi}\right)(p,x),\ x \in \Omega,\ \ W(p,x) = 0,\ \ x \in \Gamma. \tag{8.73}$$

The sesquilinear form, associated with the operator L_A in the space $H_0^1(\Omega)$, reads

$$l_p(W_1, W_2) = \int_\Omega A(p,x)\nabla W_1(x)\nabla\overline{W}_2(x)\mathrm{d}x + p\int_\Omega \rho(x)W_1(x)\overline{W}_2(x)\mathrm{d}x. \tag{8.74}$$

By (8.49) and the assertion (8.61) of Lemma 8.7, l_p is bounded in $\left(H_0^1(\Omega)\right)^2$ if Re $p > \sigma_0$. We are going to prove the coercitivity of l_p. For any $W \in H_0^1(\Omega)$ we have

$$\begin{aligned} |l_p(W,W)| \ge & \left|\frac{1}{p}\int_\Omega a(0,x)|\nabla W(x)|^2\mathrm{d}x + p\int_\Omega \rho(x)|W(x)|^2\mathrm{d}x\right| - \\ & - \frac{1}{|p|^2}\int_\Omega \left|pa(0,x) - p^2A(p,x)\right||\nabla W(x)|^2\mathrm{d}x. \end{aligned} \tag{8.75}$$

In order to estimate the first term on the right-hand side of (8.75) from below we use the relation $|c_1p^{-1} + c_2p| \ge \mathrm{Re}\, p(c_1|p|^{-2} + c_2)$, which, as easily can be verified, holds for any $c_1,\ c_2 = 0$ and Re $p > 0$. Applying this relation with $c_1 = \int_\Omega a(0,x)|\nabla W(x)|^2\mathrm{d}x$ and $c_2 = \int_\Omega \rho(x)|W(x)|^2\mathrm{d}x$ in (8.75) we derive

$$
\left|l_p(W, W)\right| \geq \frac{1}{|p|^2} \int_\Omega [Re\, pa(0,x) - |pa(0,x) - p^2 A(p,x)|]|\nabla W(x)|^2 \mathrm{d}x
$$

$$
+ \operatorname{Re} p \int_\Omega \rho(x)|W(x)|^2 \mathrm{d}x
$$

for Re $p > 0$. Applying the estimate (8.64) of Lemma 8.7 and the assumed inequality $\rho(x) \geq \rho_0 > 0$ we obtain

$$
\left|l_p(W, W)\right| \geq c_4 \left(\frac{1}{|p|^2} \|\nabla W\|^2_{L^2(\Omega)} + \|W\|^2_{L^2(\Omega)} \right), \quad \operatorname{Re} p > \sigma_a \tag{8.76}
$$

with a coefficient c_4 depending on a and ρ. Thus, l_p is coercive in $\left(H_0^1(\Omega)\right)^2$ for Re$p > \sigma_a$. By (8.49), (8.70), (8.61) and the relation $\widetilde{\Phi}(p,\cdot) \in H^2(\Omega)$ for $Re\ p > \sigma_1$ the inclusion $F(p,\cdot) - \left(L_A\widetilde{\Phi}\right)(p,\cdot) \in L^2(\Omega) \subset H^{-1}(\Omega)$ holds for Re $p > \sigma_u$. Consequently, due to the Lax-Milgram lemma the problem (8.73) has a unique solution $W(p,\cdot) \in H_0^1(\Omega)$ for Re $p > \sigma_u$. This yields the existence and uniqueness of the solution $U(p,\cdot)$ of the problem (8.71) in the space $H^1(\Omega)$ for any Re $p > \sigma_u$.

To prove the assertion $U(p,\cdot) \in H^2(\Omega)$ for Re $p > \sigma_u$ we rewrite the problem (8.71) in the form

$$
\Delta U(p,x) = F_U(p,x), x \in \Omega, U(p,x) = \Phi(p,x), x \in \Gamma \tag{8.77}
$$

for Re$p > \sigma_u$, where Δ is the Laplacian and

$$
F_U(p,x) = A(p,x)^{-1}[p\rho(x)U(p,x) - \nabla A(p,x)\cdot\nabla U(p,x) - F(p,x)]. \tag{8.78}
$$

In view of the assumptions of Proposition 8.1 and the assertions of Lemma 8.7 we see that $F_U(p,\cdot) \in L^2(\Omega)$ for Re$p > \sigma_u$. Taking this relation and the assumption $\Phi(p,\cdot) \in H^{2/3}(\Gamma)$ into account, the well-known theory of smoothness of the solution of the Dirichlet problem for the Poisson equation (see, e.g., [18], Theorem 27.2) yields $U(p,\cdot) \in H^2(\Omega)$ for Re $p > \sigma_u$ and the estimate

$$
\|U(p,\cdot)\|_{H^2(\Omega)} \leq c_5(\|F_U(p,\cdot)\|_{L^2(\Omega)} + \|\Phi(p,\cdot)\|_{H^{3/2}(\Gamma)}), \operatorname{Re} p > \sigma_u, \tag{8.79}
$$

where c_5 is a constant depending on.

It remains to derive (8.72). We substitute $W(p,\cdot)$ for W_1 and W_2 in the formula (8.74), where W is the solution of (8.73), and apply the divergence theorem for the first integral in the right-hand side of this formula to get $l_p(W,W) = \int_\Omega [F(p,x) - (L_A\widetilde{\Phi})(p,x)]\overline{W}(p,x)\mathrm{d}x$. Thereupon we estimate this expression from above by means of the Cauchy-Schwartz inequality, use the definition of L_A and combine the obtained result with the estimate from below (8.76). This leads to the relation

$$
\|W(p,\cdot)\|_{H^k(\Omega)} \leq c_6|p|^k \left(\|F(p,\cdot)\|_{L^2(\Omega)} + |p|\|\widetilde{\Phi}(p,\cdot)\|_{H^2(\Omega)}\right), \tag{8.80}
$$

for $\operatorname{Re} p > \sigma_u$, where $k \in \{0; 1\}$ and c_6 depend on $a, \ \rho$. Note that we can choose $\widetilde{\Phi}(p, \cdot) \in H^2(\Omega)$, satisfying the relation $\widetilde{\Phi}(p, x)|_{x \in \Gamma} = \Phi(p, x)$, so that the inequality $\|\widetilde{\Phi}(p, \cdot)\|_{H^2(\Omega)} \le c_7 \|\Phi(p, \cdot)\|_{H^{3/2}}(\Gamma)$ holds for $\operatorname{Re} p > \sigma_u$ with a constant c_7, which depends on Ω but is independent of Φ.Using the latter inequality, the estimate (8.80) and the obvious relation $1 < \sigma_u^{-1}|p|$ for $\operatorname{Re} p > \sigma_u$ in the formula $U = W + \widetilde{\Phi}$ we obtain

$$\|U(p, \cdot)\|_{H^k(\Omega)} \le c_8 |p|^k (\|F(p, \cdot)\|_{L^2(\Omega)} + |p| \|\Phi(p, \cdot)\|_{H^{3/2}(\Gamma)}), \tag{8.81}$$

for $\operatorname{Re} p > \sigma_u$, where $k \in \{0; 1\}$ and c_8 depend on a, ρ, Ω. Applying the assumptions imposed on ρ, F, the assertions (8.61), (8.63) of Lemma 8.7 and the estimate (8.81) in (8.78) we derive

$$\|F_U(p, \cdot)\|_{L^2(\Omega)} \le c_9 |p|^2 (\|F(p, \cdot)\|_{L^2(\Omega)} + |p| \|\Phi(p, \cdot)\|_{H^{3/2}(\Gamma)}), \tag{8.82}$$

for $\operatorname{Re} p > \sigma_u$, where c_9 depends on a, ρ, Ω. Finally, (8.79) with (8.82) implies (8.72).

Proposition 8.2 *Let the assumptions of Proposition 8.1 hold for ρ, a, and Φ. Also let $A(p, \cdot)$ and $\Phi(p, \cdot)$ be holomorphic in* $\operatorname{Re} p > \sigma_u$ *with values in $W^{1,\infty}(\Omega)$ and $H^{3/2}(\Gamma)$, respectively. Then the solution $U(p, \cdot)$ of (8.68), (8.69) is holomorphic in* $\operatorname{Re} p > \sigma_u$ *with values in $H^2(\Omega)$.*

Proof Let p and q be arbitrary numbers such that $\operatorname{Re} p > \sigma_u$, $\operatorname{Re} q > \sigma_u$. Define $A_q(p, x) := A(q, x) - A(p, x)$, $\quad \Phi_q(p, x) := \Phi(q, x) - \Phi(p, x)$, $\quad U_q(p, x) := U(q, x) - U(p, x)$. From (8.68) and (8.69) we obtain the following problem for U_q:

$$\begin{aligned} (L_A U_q)(p, x) &= (\operatorname{div}) \left\{ A_q(p, x) \nabla [U_q(p, x) + U(p, x)] \right\} \\ &\quad - (q - p) \rho(x) [U_q(p, x) + U(p, x)], \\ U_q(p, x)|_{x \in \Gamma} &= \Phi_q(p, x). \end{aligned} \tag{8.83}$$

Applying estimate (8.72) of Proposition 8.1 to this problem we obtain

$$\|U_q(p, \cdot)\|_{H^2(\Omega)} \le c_3 |p|^2 \big\{ \big[\|A_q(p, \cdot)\|_{W^{1,\infty}(\Omega)} + |q - p| \|\rho\|_{L^\infty(\Omega)} \big]$$
$$\times \big[\|U_q(p, \cdot)\|_{H^2(\Omega)} + \|U(p, \cdot)\|_{H^2(\Omega)} \big] + |p| \|\Phi_q(p, \cdot)\|_{H^{3/2}(\Gamma)} \big\}.$$

This estimate due the relations $\|A_q(p, \cdot)\|_{W^{1,\infty}(\Omega)} \to 0$ and $\|\Phi_q(p, \cdot)\|_{H^{3/2}(\Gamma)} \to 0$ as $q \to p$, following from the assumptions of the proposition, yields

$$\|U_q(p, \cdot)\|_{H^2(\Omega)} \to 0 \ \text{as} \ q \to p. \tag{8.84}$$

Further, let $A'(p, x)$ and $\Phi'(p, x)$ be the derivatives of $A(p, x)$ and $\Phi(p, x)$ with respect to p, respectively, and let $\widehat{U}$ be the solution of the problem

$$(L_A\widehat{U})(p,x) = \operatorname{div}(A'(p,x)\nabla U(p,x)), \quad \widehat{U}(p,x)|_{x\in\Gamma} = \Phi'(p,x). \tag{8.85}$$

Denoting $\widehat{U}_q(p,x) := \frac{U_q(p,x)}{q-p} - \widehat{U}(p,x) = \frac{U(q,x)-U(p,x)}{q-p} - \widehat{U}(p,x)$ and subtracting (8.85) from (8.83) divided by $q-p$ we obtain the problem

$$\begin{aligned}(L_A\widehat{U})(p,x) &= \operatorname{div}\left\{\left[\frac{A_q(p,x)}{q-p} - A'(p,x)\right]\nabla U(p,x) + \frac{A_q(p,x)}{q-p}\nabla U_q(p,x)\right\}\\ &\quad - \rho(x)U_q(p,x),\\ \widehat{U}_q(p,x)|_{x\in\Gamma} &= \frac{\Phi_q(p,x)}{q-p} - \Phi'(p,x).\end{aligned}$$

Using the estimate (8.72) for this problem we have

$$\begin{aligned}&\|\widehat{U}_q(p,\cdot)\|_{H^2(\Omega)}\\ &\le c_3|p|^2\left\{\|U(p,\cdot)\|_{H^2(\Omega)}\left\|\frac{A_q(p,\cdot)}{q-p} - A'(p,\cdot)\right\|_{W^{1,\infty}(\Omega)}\right.\\ &\quad + \|U_q(p,\cdot)\|_{H^2(\Omega)} \times \left[\left\|\frac{A_q(p,\cdot)}{q-p}\right\|_{W^{1,\infty}(\Omega)} + \|\rho\|_{L^\infty(\Omega)}\right]\\ &\quad \left. + |p|\left\|\frac{\Phi_q(p,\cdot)}{q-p} - \Phi'(p,\cdot)\right\|_{H^{3/2}(\Gamma)}\right\}.\end{aligned} \tag{8.86}$$

Due to the assumptions of the proposition we have

$$\left\|\frac{A_q(p,\cdot)}{q-p} - A'(p,\cdot)\right\|_{W^{1,\infty}(\Omega)} \to 0$$

and $\left\|\frac{\Phi_q(p,\cdot)}{q-p} - \Phi'(p,\cdot)\right\|_{H^{3/2}(\Gamma)} \to 0$ as $q \to p$.

Using these relations as well as (8.84) in (8.86) we obtain $\|\widehat{U}_q(p,\cdot)\|_{H^2(\Omega)} \to 0$ as $q \to p$, or equivalently, $\frac{U(q,\cdot)-U(p,\cdot)}{q-p} \to \widehat{U}(p,\cdot)$ as $q \to p$ in $H^2(\Omega)$. This yields the differentiability, hence holomorphy of $U(p,\cdot)$ at p.

Proof of Theorem 8.6 In virtue of Proposition 8.1 the problem (8.68), (8.69) has a unique solution $U(p,.) \in H^2(\Omega)$ for any Re $p > \sigma_u$. Applying the estimate (8.72) to this solution and observing Lemma 8.8 we obtain

$$|p|^2\|U(p,\cdot)\|_{H^2(\Omega)} \le c_2c_3 \quad \text{for} \quad \operatorname{Re} p > \sigma_u.$$

Further, since A and Φ are the Laplace transforms of abstract functions a and φ with values in $W^{1,\infty}(\Omega)$ and $H^{\frac{3}{2}}(\Gamma)$, respectively, $A(p,\cdot) \in W^{1,\infty}(\Omega)$ and $\Phi(p,\cdot) \in H^{\frac{3}{2}}(\Gamma)$ are holomorphic in the half-plane Re $p > \sigma_u$. Proposition 8.2 yields the holomorphy of $U(p,\cdot) \in H^2(\Omega)$ for Re $p > \sigma_u$. Summing up, the assumptions of Lemma 8.6 are valid for U. Consequently, there exists a function $u \in C_{\sigma_u}([0,\infty); H^2(\Omega))$

such that $L_{t\to p}(u(\cdot,x)) = U(p,x)$ and $u(0,x) \equiv 0$. Applying the inverse transform to (8.68), (8.69) we see that u satisfies the problem (8.47), (8.48). This proves the existence assertion of Theorem 8.6.

To prove the uniqueness assertion let us suppose that (8.47), (8.48) has two solutions $u^j \in L^1_{\sigma_u}((0,\infty); H^2(\Omega))$, $j = 1, 2$. Denote $U^j(p,x) = L_{t\to p}(u^j(\cdot,x))$, $j = 1, 2$. Then $U^j(p,\cdot) \in H^2(\Omega)$, $j = 1, 2$ solve (8.68), (8.69) for $\operatorname{Re} p > \sigma_u$. Proposition 8.1 implies $U^1(p,\cdot) = U^2(p,\cdot)$ for $\operatorname{Re} p > \sigma_u$. Finally, by the uniqueness of the inverse transform the relation $u^1(t,\cdot) = u^2(t,\cdot)$ for almost any $t \in (0,\infty)$ follows.

8.3.4 Uniqueness on the Boundary

In this section we will prove that the assumption $\lambda_{a_1} = \lambda_{a_2}$ implies the equalities $a_1 = a_2$ and $\partial_\nu a_1 = \partial_\nu a_2$ on the boundary Γ.

We begin by introducing some additional notation. Let $u_{a,g}$ denotes the solution of (8.47), (8.48) corresponding to the kernel a and the boundary condition $\varphi(t,x) = f(t)g(x)$. Define $U_{a,g}(p,x) = L_{t\to}(u_{a,g}(\cdot,x))$ and $F(p) = L_{t\to p}(f)$. Then $U_{a,g}$ solves (8.68), (8.69) with the boundary condition $\Phi(p,x) = F(p)g(x)$. Further, let Λ_a stands for the operator that assigns to every function $g \in H^{3/2}(\Gamma)$ the Laplace transform of $\lambda_a g$, namely

$$
\begin{aligned}
(\Lambda_a g)(p,x) &= L_{t\to p}((\lambda_a g)(\cdot,x)) = L_{t\to p}\left(\int_0^{\cdot} a(\cdot-\tau,x)\partial_\nu u_{a,g}(\tau,x)\mathrm{d}x\right) \\
&= A(p,x)\partial_\nu U_{a,g}(p,x), \quad x \in \Gamma, \quad \operatorname{Re} p > \sigma_u.
\end{aligned}
\tag{8.87}
$$

Finally, for any pair of functions a_1 and a_2 satisfying (2.9) we define $A_j(p,x) = L_{t\to p}(a_j(\cdot,x))$, $j = 1, 2$ and $\sigma_{12} := \max\{\sigma_1, \sigma_{u1}, \sigma_{u2}\}$.

Let us prove some lemmas. First one is an analog of the Alessandrini's equality for the inverse conductivity problem [10].

Lemma 8.9 *Let (8.49), (8.53) hold, a_1, a_2 satisfy (8.50) and $g_1, g_2 \in H^{3/2}(\Gamma)$. Then*

$$
\begin{aligned}
&\int_\Omega [A_1(p,x) - A_2(p,x)]\nabla U_{a_1,g_1}(p,x)\cdot\nabla U_{a_2,g_2}(p,x)\mathrm{d}x \\
&= F(p)\int_\Gamma [(\Lambda_{a_1} - \Lambda_{a_2})g_1](p,x)g_2(x)\mathrm{d}\Gamma_x, \quad \operatorname{Re} p > \sigma_{12},
\end{aligned}
\tag{8.88}
$$

where $d\Gamma_x$ is the Lebesgue surface measure of Γ.

Proof. Let a, b be arbitrary functions satisfying (8.50), $A = L_{t\to p}(a)$, $B = L_{t\to p}(b)$ and g, γ be arbitrary functions in $H^{3/2}(\Gamma)$. Multiplying the Eq. (8.68) for $U_{a,g}$ by

$U_{b,\gamma}$, using the divergence formula and the definition of Λ_a we obtain the equality $E(A, g; B, \gamma) = 0$, where

$$E(A, g; B, \gamma) = \int_{\Omega} [A(p, x)\nabla U_{a,g}(p, x) \cdot \nabla U_{b,\gamma}(p, x) + p\rho(x)U_{a,g}(p, x)U_{b,\gamma}(p, x)]\mathrm{d}x - F(p)\int_{\Gamma} (\Lambda_a g)(p, x)\gamma(x)\mathrm{d}\Gamma_x$$

This equality implies the relation

$$E(A_1, g_1; A_2, g_2) - E(A_2, g_2; A_1, g_1) - E(A_2, g_1; A_2, g_2) + E(A_2, g_2; A_2, g_1),$$

which is identical to (8.88).

Lemma 8.10 *Let $z \in \Gamma$ and B be a neighborhood of z. Let α and β be given functions such that $\alpha \in W^{1,\infty}(\Omega) \cap C^{1,\mu}(B)$, $\beta \in L^{\infty}(\Omega) \cap C^{0,\mu}(B)$ with some $\mu > 0$ and $\alpha(x) \geq \alpha_0 > 0$, $\beta(x) \geq \rho_0 > 0$ for $x \in \Omega \cup B$. Then, for any $y \in B \setminus \overline{\Omega}$, there exists a function $\omega_y : \Omega \cup B \setminus \{y\} \to R$ satisfying the following conditions:*

1. *w_y solves the equation*

$$-\operatorname{div}(\alpha(x)\nabla w_y(x)) + \beta(x)w_y(x) = 0,\ x \in \Omega \cup B \setminus \{y\}; \tag{8.89}$$

2. *$w_y(x)|_{x\in\Omega}$ belongs to $H^2(\Omega)$. Moreover, for any $x \in B \setminus \{y\}$ the relation*

$$w_y(x) = \frac{\alpha(y)}{4\pi|x - y|} + \omega_y(x) \tag{8.90}$$

is valid, where

$$|\omega_y(x)| \leq c_{10}|x - y|^{\delta-1},\ |\partial_{x_i}\omega_y(x)| \leq c_{10}|x - y|^{\delta-2},\ i = 1, 2, 3 \tag{8.91}$$

with a coefficient c_{10} and an exponent $\delta \in (0, 1)$ independent of $x \in B \setminus \{y\}$ and $y \in B \setminus \overline{\Omega}$.

The proof of this lemma is given in [16]. Now we can prove the main result of the section.

Theorem 8.8 *Let the assumptions of Theorem 8.7 be satisfied for ρ, f, and a_1, a_2. Then $\lambda_{a_1} = \lambda_{a_2}$ implies $a_1(t, z) = a_2(t, z)$ and $\partial_\nu a_1(t, z) = \partial_\nu a_2(t, z)$ for $z \in \Gamma$ and a.e. $t \in (0, \infty)$.*

Proof The proof uses in an adapted form the method of singular solutions due to Alessandrini [19]. First, we note that the assumption $\lambda_{a_1} = \lambda_{a_2}$ implies $\Lambda_{a_1} = \Lambda_{a_2}$. This, by Lemma 8.9 yields

$$\int_{\Omega} [A_1 - A_2](p, x) \nabla U_{a_1, g_1}(p, x) \cdot \nabla U_{a_2, g_2}(p, x) \mathrm{d}x = 0 \quad \text{for} \quad \operatorname{Re} p > \sigma_{12}. \tag{8.92}$$

Let us choose some $p_0 \in \mathbb{R}$, $p_0 > \sigma_{12}$, such that $F(p_0) \neq 0$. In view of the assumptions $Im\, a_j = 0$, $a_j(0, x) \leq a_0 > 0$ and the assertions (8.62), (8.63) of Lemma 8.7, the functions A_j satisfy the inequality $A_j(p_0, x) \geq \frac{k}{p_0} > 0$ for $x \in \Omega$. Moreover, due to (8.54) we have $A_j(p_0, .) \in W^{2,\infty}(\Omega) \cap C^{1,\mu}(\Gamma_\varepsilon)$, $j = 1, 2$.

Let z be an arbitrary point on Γ. Observing the properties of $A_j(p_0, x)$ and the assumptions imposed on ρ, we can choose a neighborhood $B_z \subset \Gamma_\varepsilon$ of z such that the assumptions of Lemma 8.10 are fulfilled for $B = B_z$, $\beta = p_0 \rho$ and $\alpha = A_j(p_0, .)$ with $j \in \{1, 2\}$. Let $w_{j,y}$ with $j \in \{1, 2\}$ be a function which fulfills the assertions of Lemma 8.10 for $\alpha = \alpha_j = A_j(p_0, .)$, $\beta = p_0 \rho$. Thus,

$$-\operatorname{div}(A_j(p_0, x) \nabla w_{j,y}(x)) + p_0 \rho(x) w_{j,y}(x) = 0, \quad x \in \Omega \cup B_z \setminus \{y\}, j = 1, 2, \tag{8.93}$$

Observing the assertions of Lemma 8.10 and the inequality $\alpha_j(y) \geq \frac{k}{p_0} > 0$, one can verify that

$$\nabla w_{1,y}(x) \cdot \nabla w_{2,y}(x) \sim \frac{\alpha_1(y) \alpha_2(y)}{|x - y|^4} \quad \text{as} \quad y \to x \tag{8.94}$$

Let us set $g_j(x) := w_{j,y}(x)|_{x \in \Gamma}$ for $j = 1, 2$. Then the function $U_{a_j, g_j}(p, x)$ solves (8.68), (8.69) with $A = A_j$ and $\Phi(p, x) = F(p) g_j(x) = F(p) w_{j,y}(x) | x \in \Gamma$. Since (8.68) in case $p = p_0$ coincides with (8.93), we have $U_{a_j, g_j}(p_0, x) = F(p_0) w_{j,y}(x)$ for $x \in \Omega$. Using this relation in (8.92) and taking the inequality $F(p_0) \neq 0$ into account we obtain

$$I_y := \int_{\Omega} [A_1 - A_2](p_0, x) \nabla w_{1,y}(x) \cdot \nabla w_{2,y}(x) \mathrm{d}x = 0 \tag{8.95}$$

On the other hand, by $A_j(p_0, .) \in C^{1,\mu}(B_z)$ we have

$$(A_1 - A_2)(p_0, x) = (A_1 - A_2)(p_0, z) + \nabla(A_1 - A_2)(z) \cdot (x - z) + O(|x - z|^{1+\mu})$$

for $x \in B_z$. Thus, $I_y = I_y^0 + I_y^1 + I_y^2$, where

$$I_y^0 = (A_1 - A_2)(p_0, z) \int_{\Omega \cup B_z} \nabla w_{1,y}(x) \cdot \nabla w_{2,y}(x) \mathrm{d}x.$$

$$I_y^1 = \nabla(A_1 - A_2)(p_0, z) \int\limits_{\Omega \cup B_z} (x - z)\nabla w_{1,y}(x) \cdot \nabla w_{2,y}(x)\mathrm{d}x.$$

$$I_y^2 = \int\limits_{\Omega \cup B_z} O(|x - z|^{1+\mu})\nabla w_{1,y}(x) \cdot \nabla w_{2,y}(x)\mathrm{d}x$$

$$+ \int\limits_{\Omega / B_z} [A_1 - A_2](p_0, x)\nabla w_{1,y}(x) \cdot \nabla w_{2,y}(x)\mathrm{d}x.$$

Observing that the singularity of $w_{j,y}(x)$ is located at $x = y$, where $w_{j,y}(x)$ satisfies the relation (8.94), and the distance between y and $\Omega \setminus B_z$ is bounded away from 0 as $y \to z$, we see that the term I_y^2 is bounded as $y \to z$. Moreover, supposing $(A_1 - A_2)(p_0, z) \neq 0$, we have $I_y \sim I_y^0$ and $|I_y^0| \to \infty$ as $y \to z$. But this contradicts (8.95). Thus, $(A_1 - A_2)(p_0, z) = 0$. Further, supposing $\nabla(A_1 - A_2)(p_0, z) = 0$, we have $I_y \sim I_y^1$ and $|I_y^1| \to \infty$ as $y \to z$. Again, this contradicts (8.95). Consequently, $\nabla(A_1 - A_2)(p_0, z) = 0$.

Since $z \in \Gamma$ was chosen arbitrarily, we have proven the equalities

$$A_1(p, \cdot) = A_2(p, \cdot), \partial_\nu A_1(p, \cdot) = \partial_\nu A_2(p, \cdot) \tag{8.96}$$

in $C^{1,\lambda}(\Gamma)$ for any $p > \sigma_{12}$ such that $F(p) \neq 0$. The set $p : p > \sigma_{12}, F(p) \neq 0$ has an accumulation point because of the assumption $f \neq 0$. Therefore, by means of the analytic continuation we can extend the equalities (8.96) in $C^{1,\lambda}(\Gamma)$ to all $Rep > \sigma_{12}$. Finally, by uniqueness of the inverse transform we obtain the assertion of the theorem.

In this section, we prove Theorem 8.7. For this end we need the following lemmas.

Lemma 8.11 *Let (8.49), (8.53) hold and a satisfy (8.50), (8.54). Moreover, let* $g \in H^{3/2}(\Gamma)$. *Define*

$$V_{a,g}(p, x) := A^{1/2}(p, x)U_{r,g}(p, x), \tag{8.97}$$

where $A^{1/2}(p, x)$ *is the principal value of the square root of* $A(p, x)$, *namely*

$$A^{1/2}(p, x) = |A(p, x)|^{1/2} \left(\cos \frac{\mathrm{Arg} A(p, x)}{2} + i \sin \frac{\mathrm{Arg} A(p, x)}{2} \right).$$

Then $V_{a,g}(p, \cdot)$ *belongs to* $H^2(\Omega)$ *and solves the problem*

$$-\Delta V_{a,g}(p, x) + K(p, x)V_{a,g}(p, x) = 0, \quad x \in \Omega, \tag{8.98}$$

$$V_{a,g}(p, x) = F(p)A^{1/2}(p, x)g(x), \quad x \in \Gamma \tag{8.99}$$

for $\mathrm{Re}\, p > \sigma_u$, *where* $K(p, \cdot)$ *belongs to* $L^\infty(\Omega)$ *and is given by*

$$K(p,x) = \left[\Delta A^{1/2}(p,x) + p\rho(x)A^{-1/2}(p,x)\right] A^{-1/2}(p,x) \tag{8.100}$$

for $\operatorname{Re} p > \sigma_a$.

Proof In view of the assumption (8.54) we obtain $A(p,\cdot) \in W^{2,\infty}(\Omega)$ for $\operatorname{Re} p > \sigma_0$. This together with the assertion (8.63) of Lemma 8.7 yields

$$A^{1/2}(p,\cdot) \in W^{2,\infty}(\Omega), \quad \left|A^{1/2}(p,x)\right| \geq \frac{\kappa^{1/2}}{|p|^{1/2}} > 0, \quad x \in \Omega \tag{8.101}$$

for $\operatorname{Re} p > \sigma_a$. Applying these relations and (8.49) in (8.100) we deduce $K(p,\cdot) \in L^\infty(\Omega)$ for $\operatorname{Re} p > \sigma_a$. Further, due to Theorem 8.6 $U_{a,g}(p,\cdot) \in H^2(\Omega)$ for $\operatorname{Re} p > \sigma_u$. This in view of (8.97) and (8.101) yields $V_{a,g}(p,\cdot) \in H^2(\Omega)$ for $\operatorname{Re} p > \sigma_u$. Finally, observing that $U_{a,g}(p,x)$ solves the problem (8.68), (8.69) with $\Phi(p,x) = F(p)g(x)$, the change of the unknown by (8.97) in this problem results in (8.98) and (8.99) with K of the form (8.100).

Lemma 8.12 *Let the assumptions of Theorem 8.7 be valid for* f, ρ, a_1, *and* a_2. *Assume that* $g_1, g_2 \in H^{3/2}(\Gamma)$ *and* K_j, $j = 1, 2$, *are defined by (8.100) with* A *replaced by* A_j. *Let* $A_1(p,x) = A_2(p,x)$, $\partial_\nu A_1(p,x) = \partial_\nu A_2(p,x)$ *for any* $x \in \Gamma$ *and* $\operatorname{Re} p > \sigma_{12}$. *Then*

$$\int\limits_\Omega \left[K_1(p,x) - K_2(p,x)\right] V_{a_1,g_1}(p,x) V_{a_2,g_2}(p,x) \mathrm{d}x$$
$$= F(p) \int\limits_\Gamma \left[\left(\Lambda_{a_1} - \Lambda_{a_2}\right) g_1\right](p,x) g_2(x) \mathrm{d}\Gamma_x, \quad \operatorname{Re} p > \sigma_{12}. \tag{8.102}$$

Proof Let a, b be arbitrary functions satisfying (8.50), $A = \mathcal{L}_{t\to p}(a)$, $B = \mathcal{L}_{t\to p}(b)$, K be given by (8.100) and g, γ be arbitrary functions in $H^{\frac{3}{2}}(\Gamma)$. Multiplying the Eq. (8.98) for $V_{a,g}$ by $V_{b,\gamma}$ and using the divergence formula we obtain the equality $\widehat{E}(A, g; B, \gamma)$, where

$$\widehat{E}(A, g; B, \gamma) = \int_\Omega \left[\nabla V_{a,g}(p,x) \nabla V_{b,\gamma}(p,x) + K(p,x) V_{a,g}(p,x) V_{b,\gamma}(p,x)\right] \mathrm{d}x$$
$$- \int\limits_\Gamma \partial_\nu V_{a,g} V_{b,\gamma} \mathrm{d}\Gamma_x.$$

This equality implies the relation

$$\widehat{E}(A_1, g_1; A_2, g_2) - \widehat{E}(A_2, g_2; A_1, g_1) - \widehat{E}(A_2, g_1; A_2, g_2) + \widehat{E}(A_2, g_2; A_2, g_1) = 0,$$

which is identical to

$$\int_{\Omega} [K_1 - K_2]\, V_{a,g} V_{b,\gamma} \mathrm{d}x$$

$$= \int_{\Gamma} \left[\partial_\nu V_{a_1,g_1} V_{a_2,g_2} - \partial_\nu V_{a_2,g_2} V_{a_1,g_1} + \partial_\nu V_{a_2,g_2} V_{a_2,g_1} - \partial_\nu V_{a_2,g_1} V_{a_2,g_2}\right] \mathrm{d}\Gamma_x \tag{8.103}$$

By (8.99) and the proved equalities $A(p,x) = A_1(p,x) = A_2(p,x)$, $\partial_\nu A_1(p,x) = \partial_\nu A_2(p,x)$ for $x \in \Gamma$ we have

$$V_{a_2,g_1}(p,x) = V_{a_1,g_1}(p,x) = F(p) A^{1/2}(p,x) g_1(x),$$

$$V_{a_2,g_2}(p,x) = V_{a_1,g_2}(p,x) = F(p) A^{1/2}(p,x) g_2(x),$$

$$\partial_\nu V_{a_1,g_1}(p,x) - \partial_\nu V_{a_2,g_1}(p,x)$$

$$= \partial_\nu \left[A_1(p,x)^{1/2}\right] U_{a_1,g_1}(p,x) - \partial_\nu \left[A_2(p,x)^{1/2}\right] U_{a_2,g_1}(p,x)$$

$$+ A^{1/2}(p,x) \left[A_1(p,x)\partial_\nu U_{a_1,g_1}(p,x) - A_2(p,x)\partial_\nu U_{a_2,g_1}(p,x)\right]$$

$$= A^{1/2}(p,x) \left[\left(\Lambda_{a_1} - \Lambda_{a_2}\right) g_1\right](p,x)$$

for $x \in \Gamma$. Using these relations in (8.103) we derive (8.102).

Lemma 8.13 *Let $z, \zeta, \eta \in \mathbb{R}^3$ and $\omega > 0$ satisfy the relations*

$$z \cdot \eta = z \cdot \zeta = \eta \cdot \zeta = 0, \quad |\eta| = 1, \quad |\zeta|^2 = \frac{|z|^2}{2} + \omega^2. \tag{8.104}$$

Define

$$\xi_1 = i\left(\frac{z}{2} + \omega\eta\right) + \zeta, \quad \xi_2 = i\left(\frac{z}{2} - \omega\eta\right) - \zeta, \tag{8.105}$$

Furthermore, let $\alpha_j \in L^\infty(\Omega)$, $j = 1, 2$. Then there exists $M \geq 0$ depending on α_1, α_2 such that the equations

$$-\Delta \hat{w}_j(x) + \alpha_j(x) \hat{w}_j(x) = 0, \quad j = 1, 2 \tag{8.106}$$

have solutions $\hat{\omega}_j(x)$, $j = 1, 2$, which belong to $H_2(\Omega)$ and have the form

$$\hat{w}_j(x) = \mathrm{e}^{x\cdot\xi_j}\left[1 + \psi_j(x)\right], \tag{8.107}$$

where ψ_j, $j = 1, 2$ satisfy the estimates

$$\left\|\psi_j\right\|_{L^2(\Omega)} \leq \frac{1}{\left|\xi_j\right|} \|\alpha_j\|_{L^\infty(\Omega)} \text{ for } \left|\xi_j\right| > M, \; j = 1, 2. \tag{8.108}$$

The proof of the above lemma is given in [16].

Proof of Theorem 8.7. To prove this theorem, we will apply in an adapted form, the method due to Sylvester and Uhlmann [4]. Assume that $\lambda_{a_1} = \lambda_{a_2}$. Let us fix some $p_0 \in \mathbb{R}$, $p_0 > \sigma_{12}$, such that $F(p_0) \neq 0$. Further, let us choose $z \in \mathbb{R}^3$ and $\omega > 0$ and let $\zeta, \eta \in \mathbb{R}^3$ satisfy the relations (8.104). Define $\alpha_j(x) := K_j(p_0, x)$ for be the functions satisfying the assertions of Lemma 8.13 with these α_j. By virtue of $j = 1, 2$, where K_j is given by (8.100) with A replaced by A_j. Finally, let $\hat{w}_j$, $j = 1, 2$, be the functions satisfying the assertions of Lemma 8.13 with these α_j . By virtue Theorem 8.8 we have $A_1(p, x) = A_2(p, x) =: A(p, x)$ for $x \in \Gamma$, $Re\, p > \sigma_{12}$. Let us set

$$g_j(x) := \frac{\hat{\mathrm{w}}_j(x)}{F(p_0)A^{1/2}(p_0, x)} \quad \text{for } x \in \Gamma,\ j = 1, 2. \tag{8.109}$$

The functions g_j, $j = 1, 2$ belong to $H^{3/2}(\Gamma)$ because of the relations $\hat{\omega}_j \in H^2(\Omega)$, $F(p_0) \neq 0$ and (8.101). The problem (8.98), (8.99) for V_{a_j,g_j} reads

$$-\Delta V_{a_j,g_j}(p, x) + K_j(p, x)V_{a_j,g_j}(p, x) = 0, \quad x \in \Omega, \tag{8.110}$$

$$V_{a_j,g_j}(p, x) = F(p)A^{1/2}(p, x)g_j(x), \quad x \in \Gamma, \tag{8.111}$$

where Re $p > \sigma_{12}$. Due to the relation $\alpha_j = K_j(p_0, \cdot)$ the Eq. (8.110) for V_{a_j,g_j} coincides with (8.106) for $\hat{\omega}_j$ in case $p = p_0$. Further, in view of (8.109) and (8.111) we have $V_{a_j,g_j}(p_0, x) = \hat{w}_j(x)$ for $x \in \Gamma$ and $j = 1, 2$. Recall that $\hat{w}_j$ solves (8.106). Thus, by the uniqueness of the solution of the boundary value problem (8.110), (8.111), we obtain the equality $V_{a_j,g_j}(p_0, x) = \hat{w}_j(x)$ for $x \in \Omega$ and $j = 1, 2$. By the assumptions of Theorem 8.7, the supposed relation $\lambda_{a_1} = \lambda_{a_2}$ and Theorem 8.8 the assumptions of Lemma 8.12 are satisfied. Moreover, $\lambda_{a_1} = \lambda_{a_2}$ implies $\Lambda_{a_1} = \Lambda_{a_2}$. Hence, the equality (8.102) holds with 0 in the right-hand side. In case $p = p_0$ it reads

$$\int_\Omega [K_1(p_0, x) - K_2(p_0, x)]\, \hat{w}_1(x)\hat{w}_2(x)dx = 0.$$

This, in view of (8.107) and (8.105), implies

$$\int_\Omega [K_1(p_0, x) - K_2(p_0, x)]\, e^{ix\cdot z}\, [1 + \psi_1(x)]\, [1 + \psi_2(x)]\, dx = 0. \tag{8.112}$$

From (8.104) and (8.105) we see that that $|\xi_j| \to \infty$ as $\omega \to \infty$ for fixed z. Hence, by and (8.108) $\|\psi_j\|_{L^2(\Omega)} \to 0$ as $\omega \to \infty$ for fixed z. Thus, passing to the limit $\omega \to \infty$ in (8.112) we deduce

$$\int_{\Omega} [K_1(p_0, x) - K_2(p_0, x)]\, e^{ix\cdot z} dx = 0.$$

Since this equality holds for arbitrary $z \in \mathbb{R}^3$, we obtain $K_1(p_0, x) = K_2(p_0, x)$ for $x \in \Omega$.

Due to this equality, (8.100), and Theorem 8.8, the difference $A_1^{1/2}(p_0, \cdot) - A_2^{1/2}(p_0, \cdot)$ solves the problem

$$-\Delta\left[A_1^{1/2}(p_0, x) - A_2^{1/2}(p_0, x)\right] + \tilde{\alpha}(p_0, x)\left[A_1^{1/2}(p_0, x) - A_2^{1/2}(p_0, x)\right] = 0, \tag{8.113}$$

$$A_1^{1/2}(p_0, x) - A_2^{1/2}(p_0, x) = 0, \quad x \in \Gamma, \tag{8.114}$$

where

$$\tilde{\alpha}(p_0, x) = \frac{1}{A_2^{1/2}(p_0, x)} \times \left[\frac{p_0 \rho(x)}{A_1^{1/2}(p_0, x)} + \frac{1}{A_2^{1/2}(p_0, x)}\left(p_0 \rho(x) + \frac{\Delta A_2(p_0, x)}{2} - \frac{|\nabla A_2(p_0, x)|^2}{4A_2(p_0, x)}\right)\right].$$

Let us study the asymptotic behavior of $\tilde{\alpha}$ as $p_0 \to \infty$. Observe that (8.54) and the assertions (8.56), (8.57) of Lemma 8.5 imply $\|\Delta A_2(p_0, \cdot)\|_{L^\infty(\Omega)} \to 0$ as $p_0 \to \infty$. Moreover, by the assertions (8.61) and (8.63) of Lemma 8.7

$$\frac{|\nabla A_2(p_0, x)|^2}{|A_2(p_0, x)|} = \frac{1}{p_0} \frac{|p_0 \nabla A_2(p_0, x)|^2}{p_0 |A_2(p_0, x)|} \le \frac{c_1}{\kappa p_0} \to 0 \text{ as } p_0 \to \infty.$$

Consequently, in view of (8.49)

$$\tilde{\alpha}(p_0, x) \sim \frac{p_0 \rho(x)}{A_2^{1/2}(p_0, x)}\left[\frac{1}{A_1^{1/2}(p_0, x)} + \frac{1}{A_2^{1/2}(p_0, x)}\right] \text{ as } p_0 \to \infty$$

uniformly in $x \in \Omega$. The assumptions $\mathrm{Im} a_j = 0$, $j = 1, 2$ and $a_j(0, x) \ge a_0 > 0$ imply that $A_j(p_0, x) > 0$ for sufficiently large p_0. Hence, $\tilde{\alpha}(p_0, x) \ge 0$ for $p_0 > \tilde{\sigma}$ with some sufficiently large $\tilde{\sigma}$. This implies that the solution of the problem (8.113), (8.114) is unique if $p_0 > \tilde{\sigma}$. Consequently, $A_1(p_0, x) = A_2(p_0, x)$ for $x \in \Omega$ and $p_0 > \tilde{\sigma}$ such that $F(p_0) \ne 0$.

The set $\{p \mid p > \tilde{\sigma}, F(p) \ne 0\}$ has an accumulation point in view of the assumption $f \ne 0$. Hence, by means of analytic continuation we can extend the equality $A_1(p, \cdot) = A_2(p, \cdot)$ for all $\mathrm{Re}\, p > \sigma_{12}$. Finally, by the uniqueness of the inverse transform we derive $a_1 = a_2$.

8.4 Conclusions

In this chapter we have considered some inverse problems for hyperbolic equation using the Dirichlet–Neumann operator which maps Dirichlet data into Neumann data. The uniqueness of solutions is proved, and estimates of the conditional stability of multidimensional inverse problems are obtained. This method makes it possible to study the problems of recovering the coefficients at the lowest derivative and kernels of convolution-type integral operators that model the spatial inhomogeneities of a medium.

References

1. Nachman A. 1988. Reconstructions from boundary measurements. *Annals of Mathematic* (2), 128(3): 531–576 .
2. Rakesh, and Symes, W.W. 1988. Uniqueness for an inverse problem for the wave equation. *Communication of Partial Differential Equations* 13(1): 87–96.
3. Rakesh. 1993. An inverse problem for the wave equation in the half plane. *Inverse Problems* 9(3): 433–441.
4. Sylvester, J., and Uhlmann, G. 1987. A global uniqueness theorem for an inverse boundary value problem. Annals of Mathematic (2), 1(251): 153–169.
5. Mikhailov, V.P. 1976. *Partial Differential Equations*. Moscow: Nauka [in Russian].
6. Hamaker, C., Smith, K.T., Solmin, D.C, Wagner, S.C. 1980. The divergent beam x-ray transform. *Rocky Mountain Journal of Mathematics, Numerical Mathematic* 253–283.
7. Bukhgeim, A.L., G.V. Dyatlov, and V.M. Isakov. 1997. Stability of memory reconstruction from the Dirichlet-Neumann operator. *Siberian Mathematic Journal* 38 (4): 636–646.
8. Volevich, L.R., and S.G. Gindikin. 1980. The method of energy estimates in the mixed problem. *Russian Mathematics Surveys* 35 (5): 57–137.
9. Lions, J.L., and E. Magenes. 1972. *Non-Homogeneous Boundary Value Problems and Applications*, vol. 1. Berlin: Springer-Verlag, Berlin Heidelberg.
10. Alessandrini, G. 1988. Stable determination of conductivity by boundary measurements. *Application of Analysis* 1–3 (27): 153–172.
11. Isakov, V. 1995. *Inverse Problems in PDE*. Berlin: Springer-Verlag.
12. Isakov, V. 1991. Completeness of products of solutions and some inverse problems for PDE. *Journal Differential Equations* 92 (2): 305–316.
13. Hermander, L. 1986. *The Analysis of Linear Partial Differential Operators [Russian translation]*. Moscow: Mir.
14. Bukhgeim, A.L. 1983. *Volterra Equations and Inverse Problems*. Novosibirsk: Nauka [in Russian].
15. Janno, J. 2004. Recovering a time- and space-dependent kernel in a hyperbolic integro-differential equation from a restricted Dirichlet-to-Neymann operator. *Electronic Journal of Differential Equations* 2004 (67): 1–16.
16. Janno, J. 2004. Determination of a time- and space-dependent heat flux relaxation function by means of a restricted Dirichlet-to-Neumann operator. *Mathematical Methods of Application Science* 27 (11): 1241–1260.
17. Prüss, J. 1993. *Evolutionary Integral Equations and Applications*. Basel: Birkh-auser.
18. Treves, F. 1975. *Basic Linear Partial Differential Equations*. New York: Academic Press.
19. Alessandrini, G. 1990. Singular solutions of elliptic equations and the determination of conductivity by boundary measurements. *Journal of Differential Equation* 84: 252–273.

Chapter 9
Open Problems

In this chapter we will list some outstanding research problems.

Problem 1 Many applied problems (e.g., in viscoelasticity, electrodynamics in dispersed media) lead to the consideration of equations of the forms

$$u_{tt} - \tilde{L}u = \int_0^t k(\tau)\tilde{L}u(x, t-\tau)\mathrm{d}\tau, \quad x \in \mathbb{R}^3,\ t > 0 \tag{9.1}$$

or

$$u_{tt} - \tilde{L}u = \int_0^t r(\tau)u_{\tau\tau}(x, t-\tau)\mathrm{d}\tau, \quad x \in \mathbb{R}^3,\ t > 0, \tag{9.2}$$

in which $\tilde{L}$ is an uniformly elliptic operator defined by (0.18) (i.e., a_{ij} satisfy the condition (0.15)), $k(t)$ is a relaxation kernel. In Eq. (9.2) $r(t)$ is the resolvent of the kernel $k(t)$.

Equations (9.1) and (9.2) are equivalent. In fact, considering (9.1) as a Volterra integral equation of the second kind with respect to $\tilde{L}u$ and solving it, we get (9.2). In this case, $r(t)$ and $k(t)$ are related by the formula

$$r(t) = -k(t) - \int_0^t k(t-\tau)r(\tau)\mathrm{d}\tau.$$

The study of an inverse problem of determining the relaxation kernel in the formulation of Sect. 1.2 for a more general elliptic operator $\tilde{L}$ than the operator L (see (1.43)) is an open problem.

© The Author(s), under exclusive license to Springer Nature Singapore Pte Ltd. 2023

D. K. Durdiev and Z. D. Totieva, *Kernel Determination Problems in Hyperbolic Integro-Differential Equations*, Infosys Science Foundation Series in Mathematical Sciences, https://doi.org/10.1007/978-981-99-2260-4_9

Problem 2 In the paper [1] the questions of uniqueness of finding the multidimensional kernel of the integral term of the equation

$$u_{tt} - \Delta u = \int_{-\infty}^{t} k(x,\tau)u(x,t-\tau)\mathrm{d}\tau,\ x \in \Omega \subset \mathbb{R}^3,\ t > 0 \tag{9.3}$$

for a bounded domain Ω with a smooth boundary from a given Dirichlet–Neumann operator were studied (by Dirichlet-to-Neumann maps method).

It would be interesting to study the possibility of using this method for Eq. (9.3) with the operators $\tilde{L}$, L instead of Δ and to find sufficient conditions for the coefficients of these operators, under which the kernel $k(x,t)$ is uniquely determined.

Problem 3 An interesting problem arises for the inverse problem studied in Sect. 1.5 for the case when

$$f(x,z) \equiv 0,\ x \in \mathbb{R},\ 0 < z < H,\ \ g_i(t,x) = \delta(t,x),\ (t,x) \in \mathbb{R}^2,\ i = 1,2$$

with initial–boundary conditions (1.149), where $\delta(t,x)$ is the Dirac function, concentrated at the point $t = 0$, $x = 0$. In this case, the problem of finding the kernels of the system of two-dimensional Maxwell integro-differential equations (1.148) with conditions (1.149), (1.150) is an open problem and the study of which is important in applications.

Problem 4 In all problems of Chap. 2, it is assumed that the unknown kernel does not depend on one of the spatial variables, on the orthogonal to which the Neumann boundary condition is specified. This condition is a delta δ-function and its derivatives. One should also study such problems for distributed incoming data for different spatial domains. It is also interesting to consider the problem of Sect. 2.3 in the case of the three-dimensional Maxwell integro-differential system (2.69).

Problem 5 An analysis of dynamic equations describing processes with memory shows that to the right sides of equations and systems of hyperbolic equations are added operators of the convolution type of some function depending on the time variable and the elliptic part of the corresponding hyperbolic operators on the left side of equations. In applications, it is important to study the inverse problem of Sect. 3.4 with the equation

$$u_{tt} - \Delta u - \int_0^t k(\tau)\Delta u(x,y,t-\tau)\mathrm{d}\tau = 0,\ (x,t) \in \mathbb{R}^{n+1},\ y > 0,$$

in which Δ is n-dimensional Laplacian in the variable $x = (y, x_1, \ldots, x_n)$, instead of (3.49).

Problem 6 In Sect. 4.1, it is investigated the problem of determining the wave propagation velocity and the memory of the layered medium from equalities (4.1)–(4.3). The problem of this section is also important from the point of view of applications and is open if we consider instead of (4.1) the equation

$$u_{tt} - \left(a^2 u_x\right)_x - \left(a^2 u_y\right)_y$$
$$= \int_0^t k(\tau) \left(\left(a^2(y) u_x\right)_x - \left(a^2(y) u_y\right)_y \right) (x, y, t - \tau) \mathrm{d}\tau = 0,$$
$$(x, y, t) \in \mathbb{R}^2_+ \times \mathbb{R} \tag{9.4}$$

with conditions (4.2) and (4.3).

Problem 7 Section 4.2 discusses the problem of determining the regular part of the instantaneous source physically simulating a point explosion at the boundary of the domain and the memory of the medium, i.e., functions $f(t)$, $k(t)$.

7.1. The problem of this section is also important from the point of the view of applications and is open if we consider instead of (4.1) Eq. (9.4) with initial–boundary and overdetermination conditions (4.2) and (4.3), respectively.

7.2. It is also important to study the existence of a solution to the inverse problem of Sect. 4.2 in the global sense by the method of Chap. 3.

Problem 8 Chapter 5 is devoted to the inverse problems of recovering two multidimensional memory kernels in hyperbolic integro-differential equations, having a finite Fourier series with respect to one of the spatial variables. Such problems with equations of the types

$$\frac{\partial^2 u}{\partial t^2} - \Delta u - \int_0^t k(y, \tau) \Delta u(x, y, t - \tau) \mathrm{d}\tau$$

and

$$\frac{\partial^2 u}{\partial t^2} - \Delta u = \int_0^t k(x, y, \tau) \Delta u(x, y, t - \tau) \mathrm{d}\tau$$

instead of (5.1) and (5.37), respectively, are open.

Problem 9 Prove global unique solvability of the problem of determining the two-dimensional kernel of the thermoviscoelasticity equations system (7.126)–(7.131) for $x = (x_1, x_2, x_3)$, $(x_1, x_2) \in \mathbb{R}^2$, $x_3 > 0$ $t \in \mathbb{R}$, where

$$T_{ij}(x, t) = \sigma_{ij}[u] + \int_0^t h(x_2, t - \tau) \sigma_{ij}[u](x, \tau) \, \mathrm{d}\tau, \tag{9.5}$$

The inverse problem: find the kernel $h(x_2, t), t > 0$, appearing in (7.126) with (9.5) if the following additional information about the solution of problem (7.126)–(7.131) is known:

$$u_3(x_2, x_3, t)|_{x_3=+0} = g(x_2, t), \quad t > 0,$$

$g(x_2, t)$ is given function.

One can obtain conclusions from the results of Sects. 3.5 and 7.4.

Problem 10 Prove unique solvability of the three-dimensional linearized problem of determining the kernel in viscoelasticity equation for $x = (x_1, x_2, x_3), (x_1, x_2) \in \mathbb{R}^2, x_3 > 0, t \in \mathbb{R}$

$$\rho \frac{\partial^2 u}{\partial t^2} = \sum_{j=1}^{3} \frac{\partial^2 u}{\partial x_j^2} + \int_0^t k(x_1, x_2, t-\tau) \sum_{j=1}^{3} \frac{\partial^2 u}{\partial x_j^2}(x, \tau) d\tau, \tag{9.6}$$

under the following initial and boundary conditions:

$$u |_{t<0} \equiv 0, \tag{9.7}$$

$$\left[\frac{\partial u}{\partial x_3} + \int_0^t k(x_1, x_2, t-\tau) \frac{\partial u}{\partial x_3}(x, \tau) d\tau\right]\Bigg|_{x_3=+0} = -\delta'(t)/2. \tag{9.8}$$

The inverse problem: to determine the kernel $k(x_1, x_2, t), t > 0$, appearing in (9.6) if the following additional information about the solution of problem (9.6)–(9.8) is known:

$$u(x, t)|_{x_3=+0} = g(x_1, x_2, t), \quad t > 0,$$

$g(x_1, x_2, t)$ is given function.

If we suppose that $k(x_1, x_2, t) = k_0(x_1, x_2, t) + \varepsilon k_1(x_1, x_2, t)$, where $k_0(x_1, x_2, t) \equiv k_0(x_2, t)$ is known and ε is a small parameter, then one can solve this problem from results of Sect. 6.1.

Reference

1. Bukhgeim, A.L., G.V. Dyatlov, and V.M. Isakov. 1997. Stability of memory reconstruction from the Dirichlet–Neumann operator. *Siberian Mathematical Journal* 38 (4): 636–646.